INDUSTRIAL WASTEWATER TREATMENT, RECYCLING, AND REUSE

INDUSTRIAL WASTEWATER TREATMENT, RECYCLING, AND REUSE

VIVEK V. RANADE

VINAY M. BHANDARI

AMSTERDAM • BOSTON • HEIDELBERG • LONDON
NEW YORK • OXFORD • PARIS • SAN DIEGO
SAN FRANCISCO • SINGAPORE • SYDNEY • TOKYO

Butterworth-Heinemann is an imprint of Elsevier

Butterworth-Heinemann is an imprint of Elsevier
The Boulevard, Langford Lane, Kidlington, Oxford, OX5 1GB, UK
225 Wyman Street, Waltham, MA 02451, USA

First published 2014

Notices
Knowledge and best practice in this field are constantly changing. As new research and experience broaden our understanding, changes in research methods, professional practices, or
medical treatment may become necessary.

Practitioners and researchers must always rely on their own experience and knowledge in
evaluating and using any information, methods, compounds, or experiments described
herein. In using such information or methods they should be mindful of their own safety and
the safety of others, including parties for whom they have a professional responsibility.

To the fullest extent of the law, neither the Publisher nor the authors, contributors,
or editors, assume any liability for any injury and/or damage to persons or property as a
matter of products liability, negligence or otherwise, or from any use or operation of any
methods, products, instructions, or ideas contained in the material herein.

British Library Cataloguing in Publication Data
A catalogue record for this book is available from the British Library

Library of Congress Cataloging in Publication Data
A catalog record for this book is available from the Library of Congress

ISBN: 978-0-08-099968-5

For information on all Butterworth-Heinemann publications
visit our website at **store.elsevier.com**

Printed and bound in the UK
14 15 16 17   10 9 8 7 6 5 4 3 2 1

CONTENTS

10. Application of Anaerobic Membrane Bioreactor (AnMBR) for Low-Strength Wastewater Treatment and Energy Generation   399

Janardhan Bornare, V. Kalyanraman, R.R. Sonde

11. 3D TRASAR™ Technologies for Reliable Wastewater Recycling and Reuse   435

Manish Singh, Ling Liang, Atanu Basu, Michael A. Belsan, G. Anders Hallsby, William H. Tripp Morris

PREFACE

CSIR-National Chemical Laboratory (NCL) is a premier research laboratory in the area of chemical and allied sciences in India. CSIR-NCL interacts closely with the chemical industry in India and abroad and develops knowledge base and intellectual properties to address the relevant problems of this industry. We have recently started a large and ambitious program entitled Indus MAGIC (innovate, develop, and up-scale modular, agile, intensified, and continuous processes; see www.indusmagic.org for more information). As a part of this program, we worked closely with the fine and specialty chemicals sector to identify industry needs. "Industrial wastewater treatment, recycling, and reuse of treated water" was identified as one of the key needs and was incorporated as one of the major sub-programs of Indus MAGIC. As a part of this work, we organized a workshop entitled Indus Water (see http://induswater.ncl.res.in for more information) at CSIR-NCL. The workshop brought together several experts on industrial wastewater treatment, recycling, and reuse from research institutes, academia, and industry. This book essentially originated from the Indus Water workshop.

Wastewater treatment has remained an important subject for several decades not just for industrial operations, but also for the society in general. The subject will continue to maintain significant importance because the theme of wastewater treatment, recycling, and reuse is perceived to have a huge positive impact at the industrial and societal levels, particularly in developing countries. It is not an exaggeration to say that the future of the chemical industry is very strongly related to how effectively industrial wastewater is treated, recycled, and reused. This heightened need, therefore, requires coordinated efforts, especially through dissemination of knowledge on the fundamentals and practices of wastewater treatment, recycling, and reuse. This book attempts to do this by providing information on existing methods and technologies, up-gradation of existing technologies, advent of newer technologies, and focus areas for further developments. It also highlights opportunities in the existing technologies along with industrial practices and real-life case studies. An attempt is made to provide an appropriate blend of academic research and industrial information that is required for translating ideas into practice.

The book deals with specific aspects in developing wastewater treatment methodologies for effective separations using new/novel adsorbents, ion exchange processes, coagulation/formulations, membrane separations, cavitation, and biological methods, in isolation and in combination. Emphasis is given to industrial applications and the removal of pollutants in wastewater treatment. Advanced computational tools and modeling methodologies have tremendous potential for optimizing industrial wastewater treatment (see, for example, a recent Indo-EU collaborative project on advanced tools for water treatment [www.newindigo-atwat.com for more details]). Some aspects of these advanced tools are covered in this book. More recently, wastewater has also been viewed as a renewable source for production of various useful products and energy. A sustainable hybrid wastewater treatment methodology, capable of removing multiple and refractory pollutants along with renewable energy generation, has immense potential in future industrial operations. This book attempts to focus on these and other aspects to realize environmentally benign and globally competitive processes and technologies.

The material in this book has been arranged in four areas, encompassing physico-chemical methods of treatment including advanced oxidation processes (Chapters 1–5); biological treatment, adsorbents derived from biomass, and advanced membrane separations (Chapters 6–11); process simulations for water optimization (Chapter 12); and finally, some case studies on zero liquid discharge solutions (Chapter 13) and some comments on future directions in this area (Chapter 14). An attempt is made to evolve general guidelines that will be useful for solving practical issues in industrial wastewater treatment, recycling, and reuse.

Chapter 1 provides an overview and outlines the current scenario on water consumption and strategies for water treatment and reviews various methods of treatment to develop the theme of wastewater treatment, recycling, and reuse. Chapter 2 provides detailed information on various physico-chemical methods of treatment and highlights how effectively these methods can be applied to real-life situations with specific examples of industry sectors. Chapter 3 elaborates the theme and application of hydrodynamic cavitation and advanced oxidation process in wastewater treatment. Chapters 4 and 5 both discuss the application of advanced oxidation processes in industrial practice.

Chapter 6 describes reuse of wastewater for harnessing bioenergy—an approach that represents a paradigm shift in conventional wastewater treatment. Overview of the urban water scenario, treatment methods, and potential for reuse in the industrial setup are presented in Chapter 7. Chapter 8 presents the philosophy of new material developments (derived from biomass) that may be useful for industrial wastewater treatment, in general, and phenolic wastewater treatment, in particular, with elaborate discussion on synthesis, characterization, and application.

Chapters 9 through 11 describe the use of polymeric materials such as membranes for wastewater treatment in various advanced forms—both in materials as well as applications. In this regard, Chapters 9, 10, and 11 specifically highlight industrial wastewater treatment issues and applications of newer methodologies with specific industry examples.

Chapter 12 describes process simulation for water systems optimization, the subject that can complete the water management theme discussed by various contributors in this book. Case studies of possible zero liquid discharge solutions based on evaporation technologies are presented in Chapter 13. Chapter 14 finally summarizes the past, present, and future for industrial wastewater treatment, recycling, and reuse. We hope that this book will stimulate further work in this very important area from an industrial as well as from a scientific point of view.

The intended users of this book are chemical and environmental engineers working in chemical and petroleum industries and industrial R&D laboratories as well as chemical engineering scientists/research students working in the field of wastewater treatment. Some prior background in separation processes and industrial practices is assumed. The material included in this book may be used in several ways and at various stages of designing of wastewater treatment plants as well as in fundamental research projects on understanding and developing various technologies for water treatment. It may also be used as a basic resource of methodologies for wastewater treatment, recycling, and reuse and making decisions in practice. The content may also be useful as a study material for an in-house course, operation and optimization of wastewater treatment plants, or a companion book while solving practical problems.

There are many people to thank who made this book possible. First of all, we would like to thank all the contributors to this book. We are also grateful to many of our associates, colleagues, and collaborators with whom we worked on different research and industrial projects. We would like to acknowledge financial support from CSIR for the Indus MAGIC (CSC123) project that allowed us to undertake our work on wastewater treatment and develop this book. Dr. Bhandari would like to express his profound gratitude to Professor V. A. Juvekar for being a constant source of inspiration for him in research. Many of our colleagues and students have contributed to this book in different ways. In particular, Dr. (Ms.) Laxmi

Gayatri Sorokhaibam has been of great assistance throughout the progress of this book—collecting information, critically reading the manuscripts, and providing useful comments. Mr. Hiremath has also helped in generating some useful data, especially on cavitation. We also wish to thank the editorial team at Elsevier for their patience and understanding during the long process of developing this book.

Vivek V. Ranade, Vinay M. Bhandari
Pune, March 2014.

CONTRIBUTORS

Shrikant Ahirrao
Praj Industries Limited, Pune, India

M. Ahmaruzzaman
Department of Chemistry, National Institute of Technology, Silchar, India

G. Anders Hallsby
Nalco Company, Naperville, Illinois, USA

Atanu Basu
Nalco Water India Ltd., Pune, India

Michael A. Belsan
Nalco Company, Naperville, Illinois, USA

Vinay M. Bhandari
Chemical Engineering and Process Development Division, CSIR-National Chemical Laboratory, Pune, India

Haresh Bhuta
XH2O Solutions Pvt. Ltd., Ahmedabad, India

Udo Birkenbeul
Waste water treatment, Bayer Technology Services GmbH, Germany

Janardhan Bornare
R.D. Aga Research, Technology and Innovation Centre, Thermax Ltd., Pune, India

Nitin Chandan
AMP Technologies (A Division of Aquatech Systems Asia Pvt. Ltd.), Pune, India

Parag R. Gogate
Chemical Engineering Department, Institute of Chemical Technology, Mumbai, India

V. Kalyanraman
R.D. Aga Research, Technology and Innovation Centre, Thermax Ltd., Pune, India

Johannes Leonhauser
Filtration, Membrane Technology and Waste Water, Bayer Technology Services GmbH, Leverkusen, Germany

Ling Liang
Nalco (China) Environment Solutions Co. Ltd, Shanghai, China

Aniruddha B. Pandit
Chemical Engineering Department, Institute of Chemical Technology, Mumbai, India

Jijnasa Panigrahi
Helium Consulting Private Limited, Pune, India

Jyoti Pawar
Bayer Technology Services, Thane, India

Dipak V. Pinjari
Chemical Engineering Department, Institute of Chemical Technology, Mumbai, India

Pavan Raina
AMP Technologies (A Division of Aquatech Systems Asia Pvt. Ltd.), Pune, India

Vivek V. Ranade
Chemical Engineering and Process Development Division, CSIR-National Chemical Laboratory, Pune, India

Virendra K. Saharan
Chemical Engineering Department, Malaviya National Institute of Technology, Jaipur, India

R. Saravanane
Environmental Engineering Laboratory, Department of Civil Engineering, Pondicherry Engineering College, Pondicherry, India

A. Seshagiri Rao
Department of Chemical Engineering, National Institute of Technology, Warangal, Andhra Pradesh, India

Sharad C. Sharma
Helium Consulting Private Limited, Pune, India

Manish Singh
Nalco Water India Ltd., Pune, India

R.R. Sonde
R.D. Aga Research, Technology and Innovation Centre, Thermax Ltd., Pune, India

Laxmi Gayatri Sorokhaibam
Chemical Engineering & Process Development Division, CSIR-National Chemical Laboratory, Pune, India

Nilesh Tantak
Aquatech Systems (Asia) Pvt. Ltd., Pune, India

William H. Tripp Morris
Nalco Company, Naperville, Illinois, USA

S. Venkata Mohan
Bioengineering and Environmental Science (BEES), CSIR-Indian Institute of Chemical Technology (CSIR-IICT), Hyderabad, India

CHAPTER 1

Industrial Wastewater Treatment, Recycling, and Reuse: An Overview

Vivek V. Ranade, Vinay M. Bhandari
Chemical Engineering and Process Development Division, CSIR-National Chemical Laboratory, Pune, India

1.1 WATER USAGE IN INDUSTRY

Water is a precious commodity that was once available almost free of cost. However, times have changed, and today water is free for neither people in society nor for industry. In fact, for industry, the cost of water has risen to such a level that it is now considered the same as for any other raw material used in the industry. Water fulfills several roles and functions in all types of industries. Almost all the water used in industries ends up as industrial wastewater. Release of this industrial wastewater into the environment creates a significant footprint and may also create various other hazards. This is especially true for chemical and allied process industries. It is therefore imperative that every effort is made to reduce water usage and to treat wastewater to make it reusable or at least safer to discharge into the environment. This book focuses on providing comprehensive information about state-of-the-art industrial wastewater treatment, recycling, and reuse. This chapter provides an overview of key aspects and technologies for wastewater treatment, recycling, and reuse with a particular focus on chemical and allied process industries.

Chemical and allied process industries use water extensively, thus making them *water-intensive industries*. The main uses of water are listed in Table 1.1. These needs for water are satisfied using the following sources:
- Surface water/groundwater
- Seawater
- Recycled water (industrial wastewater/urban sewage).

The following section provides a brief overview of water availability and usage. Aspects of wastewater treatment and the need for recycling and reuse of used water are then briefly discussed.

Table 1.1 Typical water uses in chemical and allied industries

Usage	Volume	Extent of contamination
Reactant	Low	High
Solvent	Low	High
Cleaning/stripping agent	Medium	Medium
Cooling water	Large	Low
Boiler water	Large	Low

In the fertilizer industry water is used as a reactant. Production of nitric acid (used for nitrogen fertilizers) using oxidation of ammonia gas requires a reaction of NO_2 with water. Similarly, most commonly used wet process for production of phosphoric acid (required for phosphate fertilizers) also require water as a reactant. The quantity of wastewater generated in many industries varies substantially from process to process and is substantially higher in developing countries. This is also dependent on the nature of the contaminants and the level of their concentration in the wastewaters. For example, the steel industry in India consumes 25–60 m^3 water/ton of steel, which is 8–10 times higher than that in developed countries. Cooling waters from the steel and coke industry, therefore, constitute a high volume and can have contaminants in the form of toxic components such as cyanide, ammonia, phenols, and metals. In pharmaceutical industries, wastewater is generated mainly from cleaning equipment. Although the volumes are not very large, the generated wastewater is significantly polluted because of the presence of high organic contaminants found in medicinal compounds, solvents, and other materials.

1.1.1 Overall Water Availability

Of the total available water on earth, 97.5% is salty and is not usable as such; of the remaining 2.5% of fresh water, only a marginal part, ~1%, is available for human consumption. Since 1950, the world population has doubled, and water consumption has increased sixfold; industrial consumption has also grown rapidly. To a great extent, recently, parts of the world have already started feeling the "water crunch." It is believed that by 2025, India, China, and select countries of Europe and Africa will face water scarcity. A recent UN report indicates that by 2025, two-thirds of the population of the world could face water stress (UNEP, 2007; Water Scarcity, www.un.org/waterforlifedecade/scarcity.shtml, Watkins, 2006). The scarcity of water could be in the form of physical scarcity, where water availability is limited and demands are not met, or it could be in the form of economic scarcity, where although water is available, there are no means/infrastructure to provide water of the quantity and quality needed. As far as the global water scenario is concerned, as a whole, there may not be water scarcity. However, since the distribution of water across the globe is not uniform, parts of world are increasingly facing water scarcity. The origin of water scarcity can be natural in some regions because of reduced rainfall or climate changes. The human factor, however, is most critical in aggravating this problem by wasting water, polluting water resources, and/or inappropriately managing water. According to recent reports (Corcoran et al., 2010; UN Water, 2008), it is believed that the total wastewater—combining sewage, industrial, and agricultural—discharged globally is tens of millions of cubic meters per day. It is also believed that a significant portion of all wastewater in developing countries is discharged untreated, resulting in large pollution of rivers and other water bodies, consequently endangering living species including any surrounding population dependent on these water sources. Some recent reports, such as UN Water, 2008 and World Water Assessment Programme, 2009, also suggest that nearly 80–90% of all the wastewater in developing countries is discharged directly into surface water bodies. In India, nearly 6.2 million m^3 of untreated industrial wastewater is generated every day (of the total $\sim$44 million m^3/day wastewater). Only $\sim$26% of domestic and $\sim$60% of industrial wastewaters are believed to be treated in India (Bhardwaj, 2005; Grail, 2009; Kaur et al., 2012; Kamyotra and Bhardwaj, 2011; CPCB, 2005a,b; CPCB 2007b, www.ais. unwater.org).

India is projected to be severely water stressed by the year 2025 (Figure 1.1) and thus needs to carefully evaluate water management options.

Because India is a developing country, its industrial development is rapid. India is also an agriculture-based country. Both these aspects require increased water consumption, and with increasing population, India is poised to become water stressed by 2020, according to a recent FICCI report (2011). Figure 1.2 shows typical water consumption patterns for India (Amarasinghe et al., 2007), and Figure 1.3 shows a typical water demand pattern worldwide. While developed countries consume much less water for agriculture, of the order of just 30% of the total, developing countries such as India consume water for agriculture at a rate of 80%. Indian states such as Maharashtra, Uttar Pradesh, Andhra Pradesh, Tamil Nadu, and Karnataka use $\sim$90% water for irrigation. This is mainly because these states produce the most water-intensive crops, accounting for $\sim$70% of total crops in India.

Further, although the industrial sector in developing countries consumes close to 10% of the water, for India, according to a recent survey

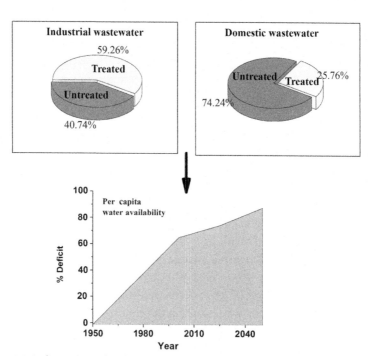

Figure 1.1 India projected to be water stressed by the year 2025.

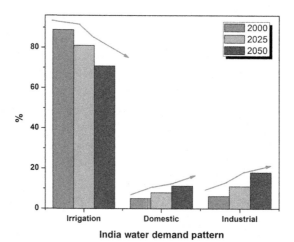

Figure 1.2 Typical water consumption pattern, India.

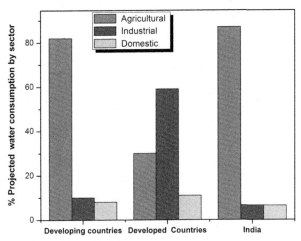

Figure 1.3 Water demand worldwide.

(Amarasinghe et al., 2007), the present industrial water consumption accounts for ~6%, which will increase to ~8.5% in 2025 and >10% (~18%) in 2050. Thus, in India, the major water consuming sector, as of today and in the near future, is the agricultural sector. It is believed that the production of most water-consuming crops, such as sugar cane, will increase by ~80% from 2000 to 2050 (water–intensive crops such as sugar cane, rice, and food grains constituted ~90% of India's crop output in 2008). In a way, this can be advantageous for water recycling and reuse because treated wastewater can be made suitable for agricultural use. This is in contrast to the general view, discussed as part of this water recycling and reuse theme, where the recycled water is intended to be used in the same industry, either as processed water or for other utilities. Where water recycling and reuse is practiced in Indian industry, the preference is for using treated wastewaters for horticulture and gardening, while some industries, such as the thermal power sector, use treated waters for ash handling. In the mining industry, treated wastewaters can be used for washing ore. Apart from pollution control, one of the main reasons for wastewater treatment is the scarcity of water. In view of the growing water requirements for the agricultural sector, a major water-consuming area, water for industry is going to become scarcer and scarcer in the future. Just as in the rest of the world, the major source of water for industry is surface water. In India, nearly 41% of industries depend on surface water. Other sources of water include groundwater (~35%) and municipal water (~24%) (see Figure 1.4).

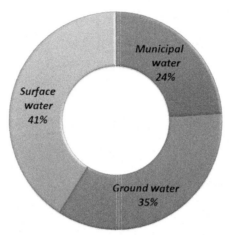

Figure 1.4 India's sources of water for industry.

It is quite evident that the availability of both surface water and groundwater is declining, and industries have to look for alternative sources of water supply if they are not self-sufficient through water recycling and reuse. A recent trend in Mumbai, the financial capital of India, indicates increased dependency of industry on treated municipal water, that is, sewage water. Among the different industry sectors, thermal power plants consume the most water and are termed water-intensive. In India, the thermal power plant sector is a major power-generating source and accounts for ~65% of the total installed power capacity as per 2008 records. Further, it is expected to grow at a rapid rate in view of a huge power demand and is likely to constitute ~75% as the total power sector expands. Not only is the thermal power sector most water intensive, but it also discharges huge amounts of wastewater and is one of the highest contributors to industrial wastewater discharge.

As mentioned earlier, India generates nearly 6.2 million m^3 of industrial wastewater every day, which requires treatment before discharge to meet environmental norms. Recent data, shown in Figure 1.5 (CPCB, 2009–10; Frost and Sullivan, 2011, www.frost.com), indicates a huge gap between wastewater generated and wastewater treated, highlighting the need for better water management. If the wastewater is not treated, agricultural output could be lowered if it is used for irrigation. Conversely, the runoff waters from agricultural land, which may contain many hazardous chemicals such as pesticides and fertilizers used in crop production, can also pose the threat of surface water pollution.

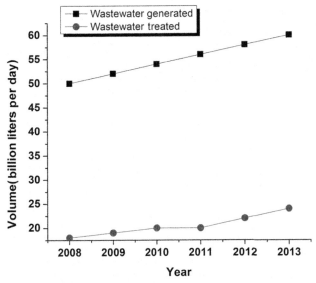

Figure 1.5 India's mega cities: domestic and industrial wastewater, generated vs. treated.

1.1.2 Industrial Water Usage

The most water-polluting industries, along with details of the pollutants they produce/discharge, have to be managed through environmental pollution control measures listed in Table 1.2 and shown in Figure 1.6. These industries also represent industry sectors where a strategy for wastewater treatment, recycling, and reuse is most crucial.

From Table 1.2 and Figure 1.6 it is evident that thermal power plants, steel plants, and the pulp and paper industries belong to the class of the most water-intensive industries (UNIDO, 2011; World Water Assessment Programme, 2009). In fact, most water-intensive industries are in reality the highest contributors to industrial wastewater generation. For example, the normal paper industry consumes $\sim$300 m^3 water per ton of product and also generates a similar amount of wastewater; the sugar industry requires up to 15 m^3 water per ton of sugar. Similarly, the mining industry requires $\sim$40 m^3 water per ton of ore, and the petroleum industry requires 10–300 m^3 water per ton of product. The chemical fertilizer industry also contributes greatly, requiring up to 270 m^3 water per ton of product, and the iron and steel industry requires water to the tune of 20–60 m^3 per ton of product, while the high-quality paper industry needs a maximum

Table 1.2 Major Polluting Industries and Nature of Pollutants

Major Polluting Industries	Nature of pollutants
Cement Mills (>200 t/day)	Dust particles, alkali, sulfur oxides, nitrogen oxides, heavy metals, waste soil, byproduct gypsum, coal ash
Sugar	Floor washing waste, sugar cane juice, molasses
Thermal power plants	Fly ash, heavy metals, coal, oil, suspended solids
Distilleries	Glucose, polysaccharides, ethanol, glycerol, amino acids, proteins, caramels, high concentration of salts, organic matter, sulfates
Fertilizers	Organics, ammonia, nitrate, phosphorus, fluoride, cadmium/other heavy metals, and suspended solids
Petroleum/ Petrochemicals	Oil, acid, soda sludge, hydrogen sulfide, lead sludge, hydrocarbons, spent filter clay, ethylene glycol, 1,4-dioxane
Mining industry	Heavy metals like copper, lead, zinc, mercury, cadmium oxide, calcium oxide, sodium oxide, barium oxide, cuprous oxide, zinc oxide, sulfates, chlorine, lithium oxide, manganese oxide, magnesium oxide, silica, gypsum, hydroxides, carbonates, cyanide, sulfur
Integrated iron and steel	Ammonia, cyanide, benzene, naphthalene, anthracene, phenol, cresol, heavy metals
Pulp and high-quality paper	High concentration of chemicals such as sodium hydroxide, sodium carbonate, sodium sulfide, bisulfide, elemental chlorine, chlorine dioxide, calcium oxide, HCl, organic halides, toxic pollutants, lime mud, wood processing residuals, traces of heavy metals, pathogens
Tanneries	Organics, heavy metals such as Cr, ammoniacal nitrogen, acids, salts, sulfides, suspended solids, dyes, fats, oil
Pharmaceuticals	Polycyclic aromatic hydrocarbons (PAHs), arsenic trioxide, chlorambucil, epinephrine, cyclophosphamide, nicotine, daunomycin, nitroglycerin, melphalan, physostigmine, mitomycin C, physostigmine salicylate, streptozotocin, warfarin over 0.3%, uracil mustard, halogenated/ nonhalogenated solvents, organic chemicals, sludge and tars, heavy metals, test animal remains
Dye and dye intermediates/ textiles	Complex mixture of salts, acids, heavy metals, organochlorine-based pesticides, pigments, dyes, PAHs
Pesticides	Volatile aromatics, halomethanes, cyanides, haloethers, heavy metals, chlorinated ethane, phthalates, PAHs

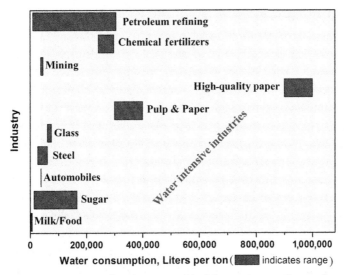

Figure 1.6 Water intensive industries most critical for water recycling and reuse.

amount of water up to 1000 m^3 per ton of product and is one of the most water-intensive in the industry sector (Chaphekar, 2013). Thus, among the list of waste-generating industries, some of the most hazardous wastewater comes from sectors such as mining, pulp mills, tanneries, refineries, sugar production/distillery, and pharmaceuticals. The food and agriculture industries produce high biological oxygen demand (BOD) wastewaters. Typically, food industries generate 0.6 to 20 m^3 wastewater/ton of product such as bread/butter/milk or fruit juice. This is an order of magnitude less than the chemical/petrochemical industries. The nature of pollutants and levels of BOD/chemical oxygen demand (COD) also vary drastically in these industry classes.

There have been process changes in the industry sector to minimize wastewater generation. However, many times, this alters the quality of wastewater so that it contains more refractive pollutants than the previous process; for example, in the distillery industry, the conventional process generates ~15 L of wastewater per liter of alcohol. With a process modification, the industry can produce much less wastewater, to the tune of 5 L per liter of alcohol; however, the BOD levels are significantly altered: ~90,000 mg/L compared to ~40,000 mg/L previously. It is necessary that the process changes be tuned not just to lower wastewater generation but also to reduce the concentration of refractory pollutants in the wastewater to make it more amenable to treatment.

1.1.3 Treatment, Recycling, and Reuse

The industrial development model in India has gone through major revision in recent years. Previously, the industrial model concentrated mainly on the development of small-scale industries, thereby providing developmental opportunities to rural sectors. It is known that the chemical industry is a scale-sensitive industry, and its profitability increases with the scale of operation. Thus, giving incentives to small-scale chemical industries drastically affected their ability to cater to their wastewater treatment needs. This has resulted in a very large number of industries that practically faced threats of closure for not meeting pollution control norms. Being a small-scale industry has its own limitations by having its own research and development for in-house operations and process developments, especially for wastewater treatment. Further, providing solutions for wastewater treatment has dented its profitability and sustainability substantially because of extra expenditures for investing in land, equipment, and operations. This is therefore an excellent case of economic development versus environmental sustainability. In India, nearly 40% of the wastewater generated by the most polluting industries comes from small-scale industries (Murty and Kumar, 2011). As discussed earlier, the size of these industries makes the installation of a standard effluent treatment plant unaffordable for many. In order to resolve the predicament of economic policy versus environmental pollution control, common effluent treatment plant schemes have been set up. This facilitates the sustainability of the small-scale industries by allowing them to treat their effluents collectively.

Environmental sustainability is also directly related to the industry and its profitability. Industry, in the basic definition of business, is required to make a profit in order to be sustainable. To continue to be sustainable, industry must progress through improved processes, research and development, decreasing energy costs, and improving salability of its products. Zero waste is a myth, by and large, and every industry generates waste in some form. The treatment of waste eats into the profitability of the industry, and for this reason, many times waste treatment takes a back seat in the overall operation or the industrial process. It is important to recover valuable chemicals and other materials from waste to improve profitability and to recycle important constituents of the waste stream where possible. In its theme of recycle and reuse, water is the last and very important significant component that can be recovered, recycled, and reused back in the same industry or elsewhere. However, this is not always possible. Further, where it is possible, the percentage of recovery of valuable chemicals/materials varies from industry to industry, and hence the dent in profitability (or increase in profitability in a positive sense) of the industry also varies from case to case. In some cases, the cost of effluent treatment could be so prohibitive that the sustainability of the industry itself becomes crucial.

In India, in view of its developing status, industry as well as the agricultural sector are growing and require water, demand for which is also growing nearly at the same rate. Thus, if industry's water share is to increase from the present 6–7% to ~18% in 2050, it will be largely at the expense of agricultural water share, which is ~80% at present, in general, and >90% in the Indian states producing water-intensive crops. In view of a huge growth in population, from the present 1.2 billion to ~1.7 billion by 2050, it is expected that the demand for food grains will be rising substantially, requiring a proportionate rise in agricultural production. This is exactly where the conflict of interest is likely to occur and is crucial from the point of view of sustainability. For example, a water-intensive crop such as wheat requires ~350–600 m^3 of water per ton of grain, and the sugar cane crop requires 1400–2500 m^3 of water per ton cane. Further, polluted waters are a threat to agriculture, and a decrease in the order of 40–50% in rice production has been reported due to using the polluted waters of the Musi river for irrigation (Grail, 2009). It is to be noted that a rice crop requires a ~12–18 mm water column in the field for almost half of crop life. The prolonged period of exposure to polluted waters is also an immense threat to the food and subsequently to human life. The use of untreated sewage wastewater, if used for agriculture, is similarly hazardous from a human health point of view. Conservation of water and wastewater treatment, recycling, and reuse in different formats is the only answer for realizing sustainability.

Thus, in this context, it becomes necessary to look at the sustainability of both the environment and industry (Figure 1.7) simultaneously and holistically. This is where the development and application of various treatment methodologies become important. In view of the growing concern for the environment and increasingly stricter government norms worldwide, it has become necessary for industries to adopt the most suitable technologies for treating their wastewaters.

For industries and the environment to become sustainable together, it is necessary that there is a strong commitment at all levels from industry to government. Further, the environment management system can be decentralized for effective operations. The solution to environmental problems can be tried from the lowest level in the chain, then taken further up the

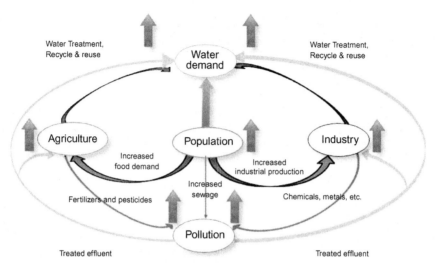

Figure 1.7 Sustainability—a proper perspective.

ladder to encompass the overall environmental scenario. Issues pertaining to industrial water pollution and effluent treatment are related to water safety and security as well and impact economic and social growth. Stricter and enforceable government regulations on pollution control and access to affordable and effective solutions to industrial wastewater treatment are essential. Further, for sustainable growth, there is a need to accept environmental challenges by way of developing green processes, minimizing wastewater generation, and creating awareness along with employment for sustainable development and safeguarding quality of life.

Although it is imperative that industry is needed for human survival, growth, and progress, it cannot be at the cost of human security. Thus, even though pollution may not be life threatening at present, the industry needs to carefully assess its adverse effects on the environment and make suitable long-term projections. If the overall effects point toward unsustainability of the environment, such industry then becomes unsustainable from a long-term point of view. This viewpoint is irrespective of the economic soundness of the industry or current sustainability criteria. To resolve such issues, there is a need to change the entire focus in such a way that water recycling and reuse is an integral part of our functioning so that there remains no scarcity of water and no damage to the environment, and the environment remains safe for living in.

1.2 CHARACTERIZATION OF INDUSTRIAL WASTEWATER

Characterization of wastewaters is the first step in the process of finding solutions to their treatment, recycling, and reuse. The next section here deals with the strategy of wastewater treatment and management. In this section, we identify the major issues pertaining to such characterization and its importance in the total methodology of water treatment, recycling, and reuse.

The first starting point in the wastewater characterization is the identification of the source of generation. Identifying the source is important so that corrective action, even if it is from a process point of view, such as process modification or increase in efficiency, can be taken to eliminate or reduce pollutant levels at that stage. If this is not possible, identifying the nature of the pollutant can help in the assessment of risk associated with mixing that stream with other plant effluents. If there are priority pollutants, it is recommended that those streams be separated from general plant effluents and treated separately using an appropriate method. It is absolutely essential that the nontoxic and reusable water not be allowed to mix with any other polluted stream. It is also advisable that the segregation of the streams be done in accordance with the nature of pollutants and priority goals such that recycling and reuse of the water is facilitated. For example, a stream containing high ammoniacal nitrogen can be separated from the streams not containing any appreciable ammoniacal nitrogen. Further, it may not be out of place to emphasize proper sample collection for the effluent. Usually, a chemical plant comprises many buildings performing specific roles for process reactions and utilities. The wastewater generated from each section is therefore different, both qualitatively and quantitatively, in terms of pollution levels. Proper sampling is therefore essential for accurate identification of the streams containing pollutants, their volumes, and the differences in nature of effluent streams. This will assist in devising a suitable strategy for management of waste streams, discussed later in this chapter. Table 1.3 provides a format for identifying the main elements of industrial or sewage wastewater that would be most useful with the desired limits, which are set differently depending on place of discharge, nature of the industry, and government regulations.

Toxic wastewater streams should never be allowed to mix with other wastewater streams; otherwise, the entire volume will have to be considered as toxic, which can increase the load on effluent treatment and pollution control tremendously. Toxicity is generally expressed as toxicity units (TUs), which is 100 divided by the toxicity measured:

$$TU = 100/LC_{50} \text{ or NOEL (no-observed-effect level)}$$

Table 1.3 Recommended Format for Effluent Characterization

Parameter	Normal reading	Maximum value	Desired value[a]	Unit
Wastewater flow rate			NA	Lph
pH of wastewater			6.5–8.5	
Color of wastewater			No noticeable color	
Odor			No offensive odor	
Density			~1 @20 °C	g/cc
Viscosity			1 @20 °C	cP
Chemical oxygen demand, COD			<250 for most industrial waters	mg/L
Biological oxygen demand, BOD			10–30 for most industrial waters	mg/L
Total suspended solids, TSS			100	mg/L
Total dissolved solids, TDS			2100	mg/L
Ammoniacal nitrogen/total nitrogen			30–50 for most industrial waters	mg/L
Total phosphorus			<2	mg/L
Metals present (e.g., Cu, Mn)			(Cu, Cr, Ni, Pb, etc.) usually well below 5	mg/L
Total inorganics content Specify nature of components			Sulfates—1000 Chlorides—1000	mg/L
Total organics content Specify nature of components			Phenolics <5	mg/L
Oil content			10–20	mg/L
Toxic/nontoxic			Nontoxic	
Any other information			Suitable for discharge/ suitable for recycling/ reuse	

[a]Varies from place to place/government standards/discharge point (http://law.epa.gov.tw/en/laws/480770486.html)

LC_{50} or the NOEL (no-observed-effect level) is expressed as the percent effluent in the receiving water. Effluent having an LC_{50} of 10% represents 10 TUs. The term *toxic organics in wastewaters* includes synthetic organic compounds such as pesticides, herbicides, and chlorinated hydrocarbons. This wastewater is generated where these chemicals are produced by the manufacturers and industries/people using these chemicals. These toxic pollutants persist over a longer period of time in a natural environment and have to be destroyed/eliminated from the streams preferably by incineration.

Odors associated with wastewater are difficult to quantify. The odor can be qualitatively differentiated on the basis of its origin: A wide variety of compounds give off foul odors apart from their possible harmful nature. It is often easy to detect very low concentrations of odorous substances in the air (sulfides/other sulfur compounds, ammonia, amines, etc.). Hydrogen sulfide toxicity is comparable to that of hydrogen cyanide; even a low level of exposure to the gas induces headaches and nausea, as well as possible eye damage. At higher levels, life-threatening conditions set in, and a number of fatal accidents, attributable to the buildup of sulfide in sewage systems, have been reported. Hydrogen sulfide gas is also soluble, and weak acids that form after it is absorbed can cause corrosion.

The discussion below applies to nontoxic wastewaters, in general.

There are separate and independent parameters for the characterization of wastewaters. These mainly include organic components, inorganics, and total dissolved and suspended solids. At the outset, one needs to resolve issues pertaining to contaminants requiring specific pretreatment such as pH, alkalinity, acidity, and suspended solids. The presence of organics in the wastewaters is conventionally indicated in terms of BOD or COD/TOD (total oxygen demand). In recent years, the total organic carbon (TOC) measurement, which represents the total organic fraction in terms of carbon, is also emphasized for better accuracy and evaluation of wastewater quality. The BOD is representative of the total oxygen requirement for oxidizing those chemicals that can be oxidized by bacterial/biological means. These are typically biodegradable substances such as food organic matter. Since not all chemicals are biodegradable, it is required that we have another comprehensive parameter that can represent the amount of pollution due to organics. This parameter is COD, which represents the oxygen requirement for oxidizing all the chemicals in the wastewaters. Obviously, COD is always a higher value than BOD. However, there are certain chemicals, such as benzene and its derivatives, which are difficult to oxidize completely. Therefore, even COD measurement is incomplete and inaccurate at times. Also, it should be noted that inorganics such as sulfides and ferrous iron also get chemically oxidized, and thus their oxygen demand also gets incorporated in the COD value. So, the COD value in such cases is not truly representative of the presence of organics in wastewaters. In this respect measurement of the TOC is important and is truly representative of organics in the wastewaters. Also, it should be noted that since COD values many times also include oxygen demand of other components, it is higher than TOC. It is also important to identify volatile organic carbons and the presence of specific priority pollutants, in addition to the total organic content, to facilitate effluent treatment strategy.

Although for any wastewater treatment, complete characterization is desirable for devising treatment methodology, many times, only a few of the parameters are measured and monitored. The description of all the parameters and analytical procedures is beyond the scope of this review, details of which are available elsewhere in the literature (Rice et al., 2012). Hence, only important parameters for industrial wastewaters are discussed here. The most common parameters to be monitored for nontoxic wastewaters are BOD, COD, and ammoniacal nitrogen. If there are any specific pollutants such as heavy metals or priority chemicals that are harmful, these have to be removed to well below the statutory limits prescribed by the government for those specific chemicals. For industrial wastewaters in India, BOD and COD limits of 10–30 mg/L and 250 mg/L respectively are usually considered as satisfactory for discharge. The limit for the ammoniacal nitrogen is generally less than 50 mg/L. Thus, measurement of these parameters and monitoring is essential for safe discharge of water. For water recycling and reuse, further stringent limits will be required depending on the nature of the process and plant. Alternatively, treated wastewaters can also be used for other agricultural purposes.

For industrial purposes, COD is considered to be a reliable parameter that can be monitored conveniently. The measurement of COD by ordinary laboratory procedures requires digestion of the sample for 2 h and titration/spectroscopic measurement. It does not require the use of expensive chemicals in most cases. The COD value is very important from the point of view of selection of treatment methodology. The chemical industry by virtue of its diverse nature generates wastewaters with CODs ranging from a few hundred to several thousands. As a general rule, the higher the COD of wastewater, the more difficult it can become to treat it. However, measurement of BOD can help in deciding if the wastewater can be treated biologically, chemically, or by employing both methods. This is indicated by the ratio of BOD to COD. The ratio keeps on changing depending on the biodegradation process, which has a definite rate even under normal conditions. As the biodegradation progresses, the BOD goes on decreasing; ideally, it should become zero when biodegradation is complete.

In contrast to COD measurement, acceptable standard BOD measurement practices require 5 days for incubation. The 5-day period is set as the standard for comparison, although the test can take longer than 5 days and is sometime continued for more than 5 days in specific cases. Usually, for industrial wastewaters, the 5-day BOD is up to 80% of the total BOD value. The BOD measurement involves oxygen demand by the bacteria for oxidizing the chemicals. There are certain drawbacks or limitations that might cause BOD measurement to fail. When industrial wastewaters contain toxic heavy metal ions, cyanides, or other substances that may be harmful to the living microorganisms oxidizing the chemicals, the BOD measurement value can be highly misleading and incorrect. Further, if the nature of a chemical species present in the wastewaters is such that it is difficult to degrade in the presence of bacteria, the 5-day BOD test can give a misleading result. This is the case with specialty molecules, including biodegradable polymers, that require a long period of degradation. Apart from this, there may be chemicals that can interfere with the BOD determination. Inconsistent laboratory practices with respect to the addition of bacteria, the presence of algae, and high ammonia levels in the sample can result in inconsistent and unreliable BOD values. Also, while measuring BOD or COD for oxidant-containing wastewaters, especially with the presence of hydrogen peroxide, it is essential to determine the residual concentration of hydrogen peroxide prior to measurement of BOD and COD because H_2O_2 will interfere in the analysis of both BOD and COD. The presence of hydrogen peroxide in water will liberate residual oxygen during the analysis of BOD, consequently creating a lower value of BOD than actual. Similarly, in the measurement of COD by the standard potassium dichromate method, it will react with it, thereby creating a higher value of COD than actual.

The inorganic pollutants in general include heavy metals such as iron, copper, manganese, chromium, zinc, and lead. Some of the metal pollutants, such as arsenic, chromium, and mercury, can be potentially toxic. The inorganic pollutants can also include the presence of ammonia. Total dissolved solids (TDS) and salts such as chloride also need to be included in inorganic pollutant characterization because some of these can act as potential inhibitors. A very important problem with respect to many chemical industries is reducing ammoniacal nitrogen from wastewaters. Ammoniacal nitrogen (NH_3-N) measures the amount of ammonia, a toxic pollutant. Ammonia can directly poison humans and upset the equilibrium of water systems. The nitrogen in sewage is assessed as ammoniacal nitrogen. This indicates the amount of nitrogenous organic matter that has been converted to ammonia. The average strength of crude domestic sewage will have a combined nitrogen content of 40–60 mg/L, while industrial wastewaters can have ammoniacal nitrogen up to a few thousand mg/L. Ammoniacal nitrogen removal can be carried out by biological or physico-chemical means, or a combination of these methods. Available technologies include adsorption, chemical precipitation, membrane filtration, reverse osmosis (RO), ion exchange, air stripping, breakpoint chlorination, and biological nitrification and denitrification (Metcalf and Eddy, Inc. 1991). Conventional methods such as chlorination, electrodialysis, RO, and distillation, however, are not efficient and are cost intensive. For wastewaters containing a large ammoniacal nitrogen content, if other pollutants are satisfactorily removed, the water can be used as a nitrogenous fertilizer. The fertilizer effect can be augmented by suitable addition of phosphate, which can substantially enhance its application directly as a fertilizer for agriculture. Conventionally, stripping/ion exchange or biological processes are used for removal of ammonia. A pH of 9–10 is required for air-stripping ammonia. Further, it may result in air pollution. In biological processing, the time requirement is very high and even sludge of 10 days may not appreciably oxidize ammonia, indicating lower performance in the biological processes.

Thus, the total volume of wastewater as well as the chemical analyses indicating the organic and inorganic components along with toxicity evaluation is essential for conceptualizing the scheme of overall wastewater treatment, recycling, and reuse. It would also help in plant design for treating wastewater once the appropriate methodology is identified based on the above.

1.3 STRATEGY FOR WASTEWATER MANAGEMENT

Key issues in industrial wastewater treatment, recycling, and reuse in today's context can be briefly outlined as follows (Figure 1.8):
1. Can we avoid liquid discharge?
   - (Aspire to zero liquid discharge)
2. Can we reuse treated water?
   - (At least as cooling water or boiler makeup if not as process water)
3. How do we manage process economics?
   - (In terms of money and space)
   - Where is the dedicated space for the process plant?
   - Treatment cost: $\sim$ Rs X/liter
4. Beyond 3Rs
   - (Reduce, recycle, and reuse)

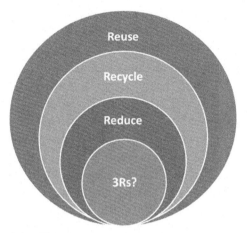

Figure 1.8 Key issues in effluent treatment.

- Can we recover useful chemicals and/or energy from industrial wastewater?
- Conservation of water?

It is imperative that we have answers to not just one or two issues listed above, but to all issues for proper preservation and utilization of our existing water resources and effective and efficient utilization of water for industrial applications in general, thereby eliminating or minimizing threats to mankind from the scarcity of water and the dangers of environmental pollution. Thus, there is a need to look beyond the popular approach of the 3Rs and find effective ways to eliminate pollution-related issues along with recovery of useful chemicals or energy to maximize process benefits and profitability.

The challenge in development of sustainable waste management technologies for chemical and allied industries is not simple for the very reason that most of the existing wastewater treatment methodologies differ significantly from each other, not just in their principles, but in their form of application and, most importantly, in process economics with huge differences in land, equipment, and material requirements. Today, the industrial-wastewater engineer must be familiar with the manufacturing process and the chemistry of the raw materials, products, and byproducts. In recent years growth in industrial activity has significantly altered the composition of wastewaters. Pollutants that are resistant to biological oxidation have become predominant (e.g., synthetic detergents, petrochemicals, synthetic rubber), requiring the development of new nonbiological processes and approaches to water-pollution control. Ready-to-use solutions are available only in a few cases. If biological treatment is possible, it should be preferably and effectively applied using modern methods. The current trend is for reducing plant space requirements along with cost by utilizing hybrid/new technologies. Process integration is required for fine and specialty chemicals with high COD and non-biodegradable chemicals. Wastewater as a source of energy is being increasingly considered. A new strategy is required involving novel materials, methods, and process integration options/technology for wastewater treatment encompassing an entire spectrum of chemical and allied industries. It may not be out of place at this stage to emphasize the need to concentrate more on water-intensive industries rather than scattered efforts in various directions to make major impacts in this regard.

1.3.1 Hierarchical Approach for Solving Pollution Problems

Strategies for wastewater management or, for that matter, any waste management problem, can be best addressed through a hierarchical approach, which involves identification and solution at every stage of operation (Figure 1.9). At the base is pollution prevention/waste minimization, which is and should be the most preferred approach. In wastewater treatment, this can be referred to as ZERO waste: disposal or generation. However, zero

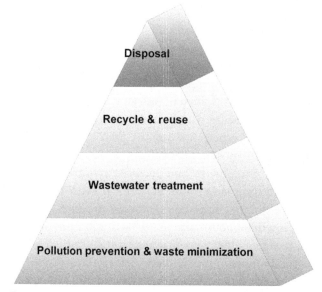

Figure 1.9 Environmental pollution control—hierarchical approach.

waste is largely a myth; hence the maximum one can achieve is to approach zero discharge. This base fundamental recommendation involves, primarily, modification in the process part such that waste generation is avoided at all the processing stages. This could imply changes in raw materials, changes in solvents used in the process, changes in processing conditions/catalysts such that no impurities are formed, and near total conversion to the desired product. Process examination may reveal changes or alternatives that can eliminate or reduce pollutants, or at least decrease the volumes to be treated or concentrations of the pollutants. To a great extent, substantial efforts have been continuously directed in this direction for several decades, and many times multiple options are available for the selection of any process for manufacturing. Thus, a careful selection of the process based on technical and societal issues along with economics can provide the most suitable option in many cases. Further, there should be enough scope to include further improvements as and when solutions to problems are available.

Next in the hierarchical approach comes wastewater treatment. As discussed earlier, it is imperative that some waste gets generated in the process. It is then essential to treat the wastewater and to determine whether this water can be recycled and reused. This would imply reuse in the same industrial setup—preferred—or it could also be reuse in other forms. Many times, this theme is applicable to waste that does not contain any harmful substances. For example, if the wastewater has useful amounts of only ammoniacal nitrogen and phosphates that can serve as nutrients, it can be directly used for gardening or agricultural purposes.

Third in the hierarchical pyramid comes recycling and reuse of treated (waste) waters. This is quite challenging at times and requires careful evaluation of all the possible options before implementation. The second and third levels in the pyramid are interlinked or should be interlinked, but this may not be possible.

The fourth and last option in the hierarchical triangle is disposal. Waste material that is absolutely untreatable and cannot be recycled in any way has to be disposed of in a safe manner. Incineration and landfill are the common examples of end disposals.

A wastewater flowchart is recommended that provides data for a complete wastewater management plan. It involves the following steps:

1. *Chemical reaction/reactant/catalyst modification for eliminating/reducing pollutants.* Effluent treatment is the last step in plant operations and therefore has a lower degree of freedom as far as pollutant quality and quantities are concerned, especially for hazardous effluents or for effluents

containing refractory pollutants. This problem can be completely circumvented through appropriate process modification that may involve raw materials/solvent substitution, or recovery of by-products. The goal of process modification is to qualitatively change wastewaters from toxic to nontoxic and use better and more efficient catalysts for optimization of reactions and raw material use.

2. *Reaction/separation equipment modification*: Along with process modifications, product yield can be improved by incorporating the most efficient and environment friendly equipment and control devices. This involves mainly installing equipment that produces little or no waste, suitable modification or newer designs for equipment to enhance recovery or recycling options, and improving the operating efficiency of equipment.

3. *Separate flushing of water stream*: Segregation of clean water is very important, especially from the recycling and reuse point of view. There are certain wastewater streams that result from washing or flushing and may not contain appreciable quantities of pollutants. Such water can be separately treated and reused for the cleaning/flushing of plant equipment or for any other suitable purpose. Further, segregation of clean water offers substantial reduction in the requirement of fresh water and subsequently reduction in cost of fresh water.

4. *Avoid leaks/spills/wastages*: This is quite easy and comes under efficient plant operation/maintenance. It can save appreciable quantities of water every day apart from ending the streams in the effluent that need treatment.

5. *Isolation and segregation of noncompatible waste streams*: This can also help in water recycling and reuse through the appropriate selection of methods for biologically treatable waste and chemically treatable wastewaters, consequently drastically reducing the load on any single effluent treatment option. In general, for ordinary wastewaters, primary treatment combined with coagulation/clarification followed by biological or chemical treatment methods is usually satisfactory to make water suitable for discharge. If priority pollutants are present, necessitating the final polishing step in the form of methods such as adsorption/ion exchange to meet statutory limits, these have to be considered before discharge to surface waters. Alternatively, polishing methods could also be used for water recycling and reuse in the same plant facility by appropriately reducing the pollutant levels. The toxic wastewaters have to be treated using special methodologies or ultimately destroyed by incinerating.

6. *Isolate concentrated and dilute streams*: It is desirable to isolate and segregate concentrated waste streams for recovery of valuable components. Such

streams also emerge from regeneration of adsorption/ion exchange processes. The recovery of valuable products from wastewaters can be attractive propositions in terms of value addition to the existing process and increasing economic feasibility of wastewater treatment operations.

7. *Plant-/process-specific measures*: Reduction and recycling of waste are inevitably site and plant specific. A number of recommendations have been made in the literature in this regard that include installing closed-loop systems, recycling on site for reuse, recycling off site for reuse, and exchanging wastes for further application.

1.4 SEPARATION PROCESSES AND CONVENTIONAL METHODS OF WASTEWATER TREATMENT

A number of separation processes have been well established in the area of chemical engineering separations and in wastewater treatment. The drive is to improve:

- Purity or lower impurities level by removing selected component(s)
- Energy efficiency by using most appropriate separation technology
- Environmental safety and compatibility for meeting regulations
- Economic viability by not putting strain on the process and profitability
- Sustainability (recycling and reuse) of the industry, in general.

In the area of wastewater treatment, typically the operations are classified as primary, secondary, and tertiary based on the nature of separation processes selected and outcome of the process. As a general rule, primary treatments are size-based separations using physical methods such as sedimentation/filtration for basic cleanup. The secondary treatment mainly involves physicochemical methods and/or biological methods and is capable of removing 85–95% of BOD/COD and TSS from the wastewaters. Tertiary treatment involves the final polishing of the effluent by removing toxic/harmful pollutants to desired levels; more than 99% removal can be achieved at the end of tertiary treatment.

The primary processes produce wastewaters that are not suitable for discharge or for recycling and reuse, the main objective being to produce water quality suitable for treatment involved in secondary and tertiary separations. The primary processes mainly have a mandate for protecting processes and materials that are to be used in secondary/tertiary treatments in order to avoid process failure. A prominent example of this is pH modifications/filtration/clarification before sending the stream for membrane separations/adsorption/ion exchange. The separation processes that are typically considered in primary treatment involve size based separation, by and large, involving physical driving force for effecting separation. They mainly comprise screening, sedimentation, thickening, precipitations, centrifugation, cyclone separations, and filtration.

In secondary and tertiary treatment stages, more advanced separation processes are used with huge variation in the nature of process and equipment. These processes include evaporation, distillation, absorption, extraction, adsorption, ion exchange, crystallization, cavitation, biological processes, and membrane separations. The separation processes employed here can be classified on the basis of driving force such as thermal driving force (distillation, evaporation, pervaporation) and pressure-driven processes such as membrane separation-microfiltration (MF), ultrafiltration (UF), nano-filtration (NF), and RO or electrical forces such as electrodialysis. Physico-chemical methods form the most important class of separation processes that play a crucial role in the area of wastewater treatment. These mainly include processes that exploit both physical and chemical reactions/interactions for effecting desired separation. This important class includes a wide variety of processes, such as coagulation/flocculation, extractions, reactive separations, oxidations, and cavitation. Separation processes such as adsorption and ion exchange also come into physico-chemical methods of treatment employing both surface forces and chemical/electrostatic attraction. Coagulation, adsorption, ion exchange, and some of membrane separations (ion exchange membranes) belong to the class of charge-based separations where separation is largely effected by neutralizing the charges and is specifically applicable for the removal of charged bodies/ions from the solution. Depending on the nature of the effluent, one or more separation processes are employed for meeting end objectives of discharge/water recycling/reuse.

Figure 1.10 shows an analysis of the trends in the application of different separation processes in wastewater treatment, depicting the number of publications made over last 40 years. It is evident that biological methods (aerobic and anaerobic-combined together) top the list, closely followed by individual physico-chemical methods such as adsorption, oxidation, membrane separation, coagulation, ion exchange, extraction, and last, cavitation. From the analysis of the plots, the spectrum can be easily divided into two zones.

Zone I: Corresponds to well-established processes in the area of wastewater treatment and where there has been continued interest both in terms of application and in process modification. This class incorporates processes such as biological processes, adsorption, oxidation, membrane separation processes, coagulation, and ion exchange.

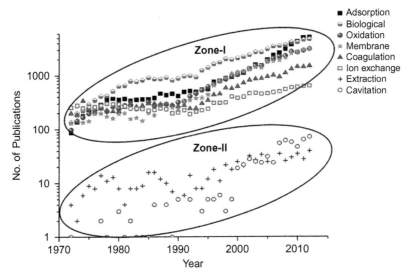

Figure 1.10 Trends in separation processes for wastewater treatment.

Zone II: Corresponds to processes such as extraction and cavitation, which are less commonly employed in the industrial wastewater treatment. For extraction, typical applications remove pollutants such as phenol and such acids/organics from the wastewaters. The important drawback of the extraction processes lies in the fact that the selection of an extractant is very crucial, and the major cost is in the recovery of extractant, apart from the cost of extractant itself. Further, loss of an extractant in the wastewater can pose further pollution problems. As far as the cavitation process is concerned, it is evident that up to 2000, there were a very few publications concerning wastewater treatment, and there has been a marked increase in interest after 2000. However, in terms of industrial applications in wastewater treatment, there is still a very long way to go.

There has been huge interest in membrane separations in wastewater treatment with a number of modifications in the form of membranes and technology (Figure 1.11). MF applications outnumber several other processes, mainly due to the fact that these are required for removal of suspended solids and for removal of larger molecules. After MF, there has been a lot of attention to RO, which is again crucial in water recycling and reuse. In comparison, research in the area of other membranes, especially NF, is far from being well developed.

Similarly, in the area of biological separations in wastewater treatment, there is marked interest in the area of anaerobic treatment as compared to

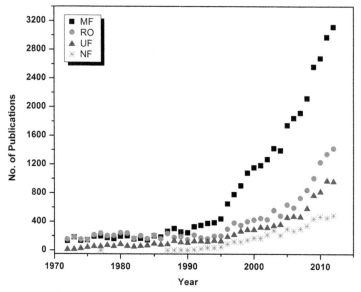

Figure 1.11 Membrane separation—research trends in wastewater treatment.

aerobic treatment (Figure 1.12). The main reason for this is increased attention to utilization of waste as a source of energy. This view is further strengthened by the fact that in recent years, the gap between the anaerobic processes and aerobic processes has widened further. The graph also shows increased interest in the area of membrane bioreactors (MBRs), especially since 2000. In fact, increased installation of MBRs has been seen in recent years for the treatment of wastewaters, especially in Asia.

It is pertinent to note that there is also increased focus on judicious combinations of aerobic and anaerobic processes for exploiting the advantages of both processes and maximizing benefits from wastewater treatment.

Overall, if one compares the contribution of physico-chemical methods of wastewater treatment with that of biological methods, Figure 1.13, on the basis of numbers of publications, shows the difference between the two methods. Since the advent of treatment methods and application in wastewater treatment, the physico-chemical methods, combined together, have an order of magnitude higher usage when compared to biological methods. Figure 1.13 also shows that in recent years, this difference between the two methodologies has nearly remained the same with physico-chemical methods enjoying the upper hand in wastewater treatment applications. In the earlier years, operation of biological wastewater treatment was not so common, and at the same time processes such as coagulation, adsorption, and ion exchange have established

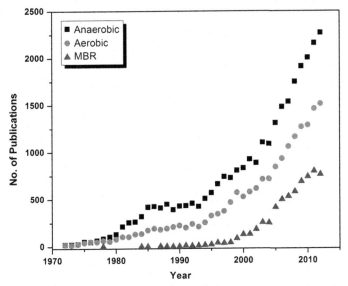

Figure 1.12 Biological processes—research trends in wastewater treatment.

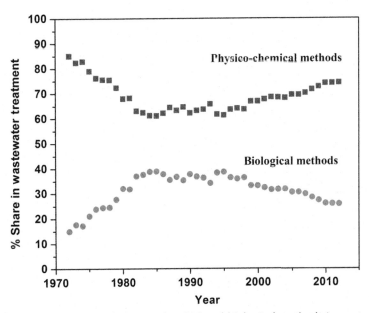

Figure 1.13 Contribution of physico-chemical and biological methods in wastewater treatment.

their place in the industry. Thus, the difference between the physico–chemical methods and biological methods in terms of usage and interest was significantly large. With improved biological processes and operations due to efficient microorganisms and better understanding, the use of biological processes has increased substantially in industrial wastewater treatment and can be considered stable at present. However, although application of physico-chemical methods decreased corresponding to the increase in biological processes in the 1980s, recent trends again show an upward rise in the use of physico-chemical methods. This can be attributed mainly to the rise of refractory pollutants in wastewaters that are difficult to degrade using biological methods.

With the addition of newer methodologies, such as cavitation, and also with the development of newer materials for membranes (composite membranes), adsorbents, ion exchange resins, and so on, and also in view of more refractory chemicals being found in wastewaters that are difficult to degrade biologically, the physico-chemical methods are likely to retain their high position in the area of industrial wastewater treatment, recycling, and reuse. The commonly used treatment methods are briefly discussed in the following section.

1.4.1 Coagulation/Flocculation

Coagulation is a process in which destabilization of colloidal particles present in the fluid is achieved by addition of salts, which reduce, neutralize, or invert the electrical repulsion between particles. Coagulants can be broadly classified as inorganic and organic. Coagulation is one of the most commonly employed methods in effluent treatment. However, use of coagulants in wastewater treatment containing refractory pollutants is a complex problem, and no general solutions are available yet.

1.4.1.1 Commonly Used Coagulants
Inorganic
   Aluminum salts (alum)
   ($Al_2(SO_4)_3 \cdot 14H_2O$ or $Al_2(SO_4)_3 \cdot 18H_2O$ (alum))
   Ferric and ferrous salts
   ($FeCl_3$, $Fe_2(SO_4)_3$, $FeSO_4 \cdot 7H_2O$)
   Lime ($Ca(OH)_2$)
Organic
   Cationic polymers
   Anionic and non-ionic polymers

In most cases coagulation has been effective in removing color, especially from wastewaters containing dissolved solids and charged matter. However, high chemical dosages are usually required, and large volumes of sludge must be disposed of, in general, for inorganic coagulants, resulting in high cost of sludge disposal. Alternatively, organic coagulants such as polydiallyldimethylammonium chloride can be used and are known to enhance coagulation efficiencies in some cases. A near-zero production of sludge in the case of organic coagulants almost eliminates sludge disposal problems and significantly reduces treatment costs. Thus, the formulation of inorganic and organic coagulants can provide a better techno-economically feasible operation in wastewater treatment. Typically, inorganic coagulants are aluminum sulfate, aluminum chloride, ferric chloride/sulfate, calcium/magnesium oxide, and PAC. In general, inorganic coagulants produce smaller and lighter flocs that require more time to settle. This is reflected in the sludge volume, which is always greater with inorganic coagulants. Another disadvantage with most inorganic coagulants is that they are pH sensitive and therefore work only in a narrow pH range. Some of the disadvantages of common inorganic coagulants can be eliminated with the use of organic coagulants or formulations of both inorganic and organic coagulants.

It is to be noted that coagulation alone is not generally a complete solution to wastewater treatment problems. It is again a charge-based separation method that works on the principle of charge neutralization. However, in the floc formation process, some of the uncharged particles and organics get physically trapped and removed, consequently improving the process performance. Recent trends in this area involve development of suitable coagulant formulations for specific cases, development of coagulant aid/flocculant aid, and newer polymeric coagulants with varying basicity for wider applicability.

1.4.2 Adsorption

Solid adsorbents have been used since ancient times for various separation/purification applications. Today, they find applications in drinking water treatment, the food industry, the pharmaceutical industry, the chemical and petrochemical industries, and wastewater treatment.

In the adsorption process, adsorption of organics/inorganics can take place on the surface of the adsorbent, thereby effecting removal of pollutants from the wastewaters. Adsorption in wastewater treatment is typically considered as a physico-chemical process involving selective attachment of specific molecules on the surface of the adsorbent. The attraction of a specific molecule is believed to be due to action of surface forces that are responsible for the interaction with it. Apart from purely physical forces such as van der Waals forces, both physisorption and chemisorption along with electrostatic attraction can play an important role in the overall adsorption process. The contribution of physisorption, chemisorption, and electrostatic attraction can vary depending on the nature of the adsorbent (synthetic/biomass derived); the nature of the substrate molecules and surface molecules; surface modification; presence of acidic or basic groups on surfaces; and doping of the metal ions of specific functionality on the surfaces (see, for example, Ahmaruzzaman and Laxmi Gayatri, 2011; Bhandari et al., 2006; Mane et al., 2003, 2006). The most important current applications of adsorption are odor and color removal, removal of acids, removal of metals, and removal of refractory pollutants in the chemical industry.

It is very important that specific interactions of pollutants with the sites available on the surface of the adsorbent be understood in order to achieve maximum removal and therefore the most efficient process performance for COD/ammoniacal nitrogen removal. To accomplish this, it is necessary to characterize process effluents in terms of the nature of contaminants along with their concentrations and adsorbent material. Apart from these aspects, the process parameters such as pH of the solution and temperature also affect the performance and therefore need to be properly addressed while designing a wastewater treatment strategy. Adsorbents can be classified in two major sections:

Inorganic Adsorbents
   Zeolites
   (A, X, Y, ZSM-5, silicalite, ALPO)
   Oxides
   (Silica, alumina)
Organic Adsorbents
   Activated carbon
   (powder, granules, molecular sieves, carbon fiber)
   Polymeric adsorbents
   Ion exchange resins
   Biomass-derived adsorbents

Selection of the most suitable adsorbent is very important from the point of view of techno-economic feasibility. A huge number of commercial adsorbents are available in the market, and proper selection is usually made based on laboratory studies and prior experience. Although the type of adsorbents is relatively small, hundreds of these products are being produced commercially with differences in performance, surface/pore characteristics, functionalities, activation procedures, and other features. There is a wide scope for the development and characterization of new materials and modification of materials and process development with the ultimate aim of recovering valuable components or mitigating environmental concerns. Specialty materials pose difficulty in evaluating them in terms of actual surface interactions they offer in effluent treatment and for the removal of pollutants simply on the basis of fundamental parameters such as surface area, pore size, and size distribution. However, laboratory data in terms of capacity, extent of removal, and regeneration of the adsorbent is usually sufficient for suitable design. The selection and design procedure is similar to that for ion exchange resins, which are discussed in the following section.

1.4.3 Ion Exchange

Ion exchange is commercially carried out for water softening/treatment and various ion removal applications for dilute solutions.

The conventional ion exchange process depends upon the metathetical exchange of ionic species and the capacity of resin to exchange; this is an important parameter in the process design. However, for wastewater treatment, it is quite attractive to employ weak base ion exchange resins that, although they do not work in accordance with the conventional ion exchange mechanism, remove both organics and acidic ionic species, thereby resulting in COD reduction to a great extent. The mechanism in the case of acid removal proceeds through protonation of ionogenic species on the weak base resins by the proton of the acid and subsequent addition of anion to the exchange site through electrostatic interaction (Bhandari et al., 1992a,b, 1993, 2000).

$$R + H^+ \rightarrow RH^+$$

$$RH^+ + A^- \rightarrow RH^+A^-$$

Where A^- is the anion of the acid, HA. R is a resin group. RH^+ and RH^+A^- are protonated species and acid salt of the resin respectively. The treatment with a base, such as sodium hydroxide, regenerates the resin in original free base form. The removal of non-ionics/organics, however, takes place mainly through surface interaction with the polymer matrix and does not involve any ionic interaction as such. The design of the ion exchange process thus involves the efficient combination of cation and anion exchange resins for effective salt removal, removal of ionics, and removal of organics, which finally results in a large reduction in COD levels and also removal of ammoniacal nitrogen.

In the cation exchange process, ammoniacal nitrogen typically gets removed in the following way:

$$R - H^+ + NH_4NO_3 \rightarrow R - NH_4^+ + HNO_3$$

$$R - H^+ + (NH_4)_2SO_4 \rightarrow R - NH_4^+ + H_2SO_4$$

The regeneration of the resin is carried out using acid to reverse the above reaction:

$$R - NH_4^+ + HNO_3 \rightarrow R - H^+ + NH_4NO_3$$

Selection of suitable resins for obtaining high operating capacity with ease of regeneration is very important in the design. High operating capacity results in lowering the quantity of resin required for operating the treatment process, and easy regeneration results in lowering regenerant consumption and also minimizing the secondary waste stream.

The design of the ion exchange process involves selection of an appropriate resin as a first step. Not all resins are suitable for any specific application. It is necessary to make proper selection of the resin once objectives of wastewater treatment are clearly defined (e.g., salt removal, removal of ionic species, or removal of organics in the present context). The second step involves characterization of the resins through capacity determination, size analysis (important from the pressure drop point of view in the final design), and pore size/size distribution (important from the kinetics/rate point of view). The important parameter is the maximum or theoretical capacity of the resin and the operating capacity that is required for the design of the column. The third step requires ion exchange breakthrough studies in wastewater treatment, an important parameter in plant operation. The nature of the breakthrough curve gives a lot of information required for the design of ion exchange systems in terms of length of used/unused bed. The sharpness of the breakthrough profile indicates near total capacity utilization for the resin while a more dispersed/spread nature of the breakthrough curve indicates poor capacity utilization (which in turn indicates that the resin may not be suitable for actual application). Other aspects, such as fouling of the resin and extent of color removal, also need careful evaluation. The final step requires regeneration studies for the selected resin.

Typically, for adsorption/ion exchange, the useful number of experiments can be seen as:

Adsorbent-1/Resin-1	×	Effluent × 3	×	Min. 2 Parameters	=	Total Expt.
3		3		2 × 3	=	54
Regeneration 3	×	–	×	2	=	06
				Total No. of Expt.	=	60

In light of the above typical experimental analysis that may be required for evaluation of any application, design of experiments is a critical step. Apart from the conventional evaluation, the solution to environmental problems will continue to require inexpensive ion exchange materials that can be used once only and discarded. Developing a highly selective material at low cost is a daunting challenge in today's context.

1.4.4 Membrane Separation

Membrane separation is based on permselectivity, which is determined by differences in the transport rate of various components through the membrane. This permeation rate in turn is determined by the structure of the membrane, the size of the permeating component, the chemical nature and the electrical charge of the membrane material and permeating component, and the driving force due to the chemical or electrochemical potential gradients (that is, concentration pressure and electrical potential differences). The use of different membrane structures and driving forces has resulted in a number of different membrane processes:

Conventional
- Reverse osmosis
- Nano/ultra/micro-filtration

Relatively Recent Developments
- Pervaporation
- Membrane distillation
- Dialysis/electrodialysis
- Emulsion liquid membranes
- MBR
- Hybrid membrane system.

Today, membranes are used on a large scale in three distinct areas:

1. Applications in which the use of membranes is technically feasible, but where they must compete with conventional separation processes on the basis of overall economics. This, for instance, is the case in seawater desalination and the treatment of certain wastewater streams (UNEP, 2008). Here, membranes must compete with processes such as distillation and biological treatment.

2. Applications for which alternative techniques are available, but membranes offer a clear technical and commercial advantage. This is the case in the production of ultrapure water and in the separation of certain food products.

3. Applications where there is no reasonable alternative to membrane processes. This is the case in certain drug delivery systems and artificial organs.

For membrane separations, established process applications include gaseous separations such as O_2/N_2, H_2/CH_4, Olefin/N_2, and liquid separations such as desalination and various applications in the beverage industry. In wastewater treatment, MF, UF, and RO have established niche areas for themselves. Apart from these, numerous applications in combined processes such as MBRs are being increasingly used.

Compared to conventional procedures, membrane processes are often more energy efficient, simpler to operate, and yield higher-quality products. The environmental impact of membrane processes is relatively low. No hazardous chemicals are used in the processes that require a discharge step, and there is no heat generation. However, membrane separations have their own limitations that need to be carefully weighed before considering the application. These include the following:

1. A major disadvantage of membranes, especially in water- and wastewater-treatment processes, is that the long-term reliability has not been completely proven.

2. Membrane processes sometimes require excessive pretreatment due to their sensitivity to concentration polarization, chemical interaction with water constituents, and fouling.

3. Membranes are mechanically not very robust and can easily be destroyed by a malfunction in the operating procedure.

4. Another critical issue is the process cost. In general, membrane processes are quite energy efficient. However, the energy consumption is only part of the total process costs. Other factors determining the overall economics of a process include the investment-related cost, which is determined by the cost of the membranes and other process equipment and their useful life under operating conditions, and various pre- and post-treatment procedures of the feed solutions and the products.

In wastewater treatment, mostly MF membranes are used for the removal of large size species and RO membranes to obtain pure water. There are many other variations possible in wastewater treatment such as membrane distillation, which is commercially applied for removal of hydrochloric acid from pickling wastewater streams. A large number of newer installations are in the area of MBRs. The MBR process essentially has an in-line combination of a biological wastewater treatment-activated sludge process with a membrane process, the two stages working in tandem to treat wastewater biologically and to separate biomass physically from the wastewater in a single step. In the polluted effluents from industrial sources, such as petroleum refineries, plastic/coke oven industries, and phenolic resin manufacturing processes, organic chemicals (up to several grams per liter) can be removed by cells immobilized on microporous hollow fibers forming a membrane-attached biofilm-microporous membrane bioreactor (MMBR) for biodegradation applications. Further, different engineering designs are available, such as suspended basket or crossflow tubular MBRs for specific applications, which may enhance overall process performance. Compared to other membrane separations, the engineering principles underlying MBRs are mature enough to ensure reliability. MBRs have been used to treat a wide range of municipal and industrial wastewaters and currently are believed to be installed at more than 1000 sites in Asia. Adequately reliable equipment and technological support are commercially available to meet existing and developing demand.

1.4.5 Cavitation

Although various physico-chemical and biological methods are available to treat wastewater, many of the conventional treatment techniques employ large quantities of chemicals for treatment. Subsequent disposal of these chemicals poses problems in the conventional treatment methods. The cavitation method is a relatively recent physico-chemical method for treating wastewater. There are certain organic pollutants, especially in dye/pigment/textile wastewaters that are considered refractory compounds—difficult to remove/degrade by using conventional methods of chemical/biological treatment. For such pollutants, newer techniques have to be explored, such as cavitation, where extreme conditions generated by collapsing cavities can break down pollutants and organic molecules. Typically, hydrodynamic and sonochemical or acoustic cavitation are found useful in destruction of organics. Cavitation generates strong oxidizing conditions due to production of hydroxyl radicals. The impact of cavitation processes can be dramatically increased by combining them with other oxidation processes employing catalysts or additives. This process can work very well, especially in treating wastewaters containing refractory pollutants and/or having unusually high COD. Newer, specially designed devices such as the vortex diode can provide an effective water treatment method (Ranade et al., 2006, 2013). Further, cavitation can offer considerable economic benefits compared to other conventional physico-chemical methods of treatment.

Cavitation is the formation, growth, and collapse of cavities, releasing a large amount of energy and generating oxidizing agents in wastewater. Cavities generated undergo a series of radical reactions with complex organic matter present in wastewater leading to the destruction of contaminants and decolorization of wastewater. Physico-chemical changes in the fluid take place due to transient temperature and pressure conditions with strong oxidizing conditions as active chemical radicals and hydrogen peroxide are formed. The process is somewhat analogous to advanced oxidation processes (AOPs), with local high magnitude pressure pulse, 100–5000 atm., and extremely high temperatures of 1000–15,000 °K. However, the overall liquid medium can be maintained close to ambient conditions. The cavitation technology can be combined with an array of other technologies such as Fenton, ozone, wet air oxidation, and others for providing a very effective platform for solving a variety of wastewater treatment problems. Thus, many hybrid technologies are possible, such as cavitation + oxidation; cavitation + coagulation; cavitation + adsorption/ion-exchange; cavitation + membrane, and cavitation + biological treatment.

The cavitation technology not only offers promising methodology on its own for industrial wastewater treatment, but it can also be integrated effectively with other conventional methods for achieving a complete, techno-economically feasible solution for water recycling and reuse.

1.4.6 Advanced Oxidation Processes

AOPs have been in commercial practice, especially for refractory pollutants that are difficult to remove using conventional physico-chemical methods. These are mainly useful for highly toxic and non-biodegradable wastes. Although there is significant variation in the form of catalyst and reactor configurations used for OPs, Fenton oxidation and photo-Fenton have been more successfully applied in industrial wastewater treatment. AOPs operate through the generation of hydroxyl radicals and other oxidant species to degrade organic compounds in wastewater. There are different methods for the generation of hydroxyl radicals, which include:

- H_2O_2 (Fenton)/O_3
- Hydrodynamic and acoustic cavitation
- Homogeneous ultraviolet irradiation
- Heterogeneous photocatalysis using semiconductors
- Radiolysis/electric and electrochemical methods.

The hydroxyl radical attacks the organic molecules by abstracting a hydrogen atom or by addition to the double bond, finally reducing them to carbon dioxide and water.

The salient features of most AOPs are summarized below:

Fenton oxidation

- The oxidative decomposition and transformation of organic substrates by H_2O_2/Fe^{2+} (known as Fenton's reagent)
    - Produces OH radicals at acidic pH
    - Ambient temperatures
    - Optimal Fe/H_2O_2 ratio is 1:10
    - The OH radical attacks organic molecule and they are mineralized to CO_2 and H_2O
- Emerging technology for wastewater treatment
- Useful for highly toxic and non-biodegradable wastes
- Employed successfully for black olive and mineral oil wastewaters

Oxidation with Ozone (O_3)

- Removes organics in wastewater treatment
- Elimination of color, smell, and taste; disinfection
- An ozone molecule can react with many organic compounds, particularly those unsaturated or containing aromatic rings or heteroatoms
- Ozonation has been found to be very efficient for the decolorization of textile wastewaters

Wet air oxidation

- For non-degradable or slowly degradable or toxic substances
    - Low pressure wet oxidation
        - $T < 200\,°C$
        - $P = 5–20$ bar
        - At COD > 6 mg/L: heat recovery from enthalpy of oxidation is comparable to the total energy requirement of the process
        - SO_3^{2-}/HSO_3^-, phenol, amino and hydroxyl substituted phenols, wastewater from dye manufacturing
        - AOX decreases, COD decreases, BOD increases
    - High-pressure wet oxidation
        - $T > 200\,°C$
        - $P > 20$ bar
        - At COD > 50 mg/L: heat recovery from enthalpy of oxidation is comparable to the total energy requirement of the process
- Good for high concentrations of contaminants/may have corrosion issues.

Some organic compounds, for example, acetic/maleic/oxalic acids, acetone, or chloride derivatives such as chloroform are not attacked by OH· radicals. In such cases, application of AOP may not be suitable. AOPs are generally cost intensive, mainly because of high operating cost. However, they require much less space and many times less capital cost as compared to many of the physico-chemical processes and biological processes. The oxidation products are generally less complex and can be treated by conventional biological methods.

1.4.7 Incineration

Toxic organic materials such as pesticides, herbicides, and chlorinated hydrocarbons are difficult to remove by the conventional processes. Incineration, by default, can be considered as an option only when the pollutants are most difficult to degrade biologically or cannot be economically removed by any of the physico-chemical methods (adsorption/ion exchange/extraction/membrane separation, etc.) of separation. Because there are strict restrictions on the levels of toxic compounds in wastewaters, destructive methods such as incineration (thermal/catalytic) are then found as the only techno-economically feasible technology. The organics present in the wastewaters can be oxidized in this process along with some of the inorganics using extremely high temperatures in the presence of oxygen. The recommended temperatures for incineration of chlorinated hydrocarbons and pesticides range from 980 to 1500 °C. A sustained high temperature prevents the emission of degradation products. The incinerator stack gases generally contain HCl vapors that require the installation of scrubbers. In general, flue gas treatment along with proper disposal or reuse for the slag/ash produced during incineration is always required.

Incineration technology for wastewater treatment should be considered only if there is a sufficient load of organics and inorganics for burning. Evaporation of water is sometimes considered for improving process performance. For spent wash treatment, concentration followed by incineration is one option. However, in any case, incineration is a highly energy intensive process, and the requirement for proper technology to treat such a difficult stream is also a challenging task, especially if the pollutant concentrations and economics of the process cannot justify the size of the incinerator. For these reasons, many times, design of such an incinerator facility for wastewater treatment is highly plant specific requiring tailor-made designs. Incineration design is usually complex and involves altogether different chemistry and engineering when compared to the conventional methods of treatment.

1.4.8 Biological Method of Treatment

1.4.8.1 Aerobic Treatment

A commonly practiced biological treatment methodology involves aerobic treatment, which is simple in its concept and operation. A high degree of substrate conversion is the primary aim in aerobic wastewater treatment (e.g., the activated-sludge process, Environmental Protection Agency, 1997). A major disadvantage of aerobic processes is large amounts of biomass (clarification sludge) that is formed simultaneously because of the aerobic nature of the phenomenon. The fundamental advantage of an aerobic treatment is that oxidative degradation of the carbon substrates provides the energy required for propagation of the microorganisms that act as the biocatalyst. The organisms involved in biological wastewater treatment are invisible to the naked eye. Depending upon their structures and cellular components, they can be subdivided into bacteria, fungi, plants, with viruses representing a special group of their own. The bacterial cells can be said to represent "biochemical reactors," wherein heterotrophic organisms oxidize organic compounds with the aid of oxygen. The microorganisms also perform the role of sorbents that bind both organic substances and heavy metals, subsequently aiding their decomposition and/or removal.

For efficient operation of aerobic processes, the following aspects are to be carefully evaluated:

- The concentration of nutrients in the wastewater should not be too low.
- As many suitable organisms as possible should be present within the bulk (bacterial retention) to ensure stable biocenosis.
- The bacteria must be provided with an adequate supply of oxygen to support aerobic metabolism (aeration).
- The three reaction partners—bacteria, substrate, and dissolved oxygen—must be brought together using appropriate design.
- The growth rates for the different bacteria must be well matched (sludge age).
- The optimal environmental conditions (pH, temperature) for biological degradation must be established.
- The C:N:P ratio required for the growth of cell substance must be assured, and all required trace elements must be present.

The commonest form of aerobic process is the activated sludge process, which is now approximately 100 years old. It involves the following steps:

1. Wastewater aeration in the presence of a microbial suspension.
2. Solid-liquid separation following aeration.
3. Discharge of clarified effluent.

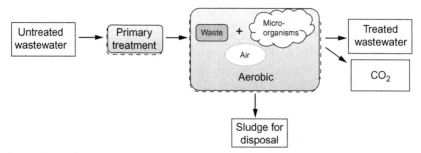

Figure 1.14 Schematic of aerobic biological wastewater treatment process.

4. Disposal of excess biomass and return of remaining biomass to the aeration tank.

Figure 1.14 represents a conventional process flow diagram of the activated sludge process. A number of process variations are possible and are being practiced in different industries; for example, a re-aeration tank and contact tank can be replaced by an aeration tank, physical surfaces can be made available so that the microorganisms can attach and grow, and different reactors can be configured. Among the important variables of the activated sludge process are the mixing regime, the loading rate, and the flow scheme. The mixing regime, plug flow, or complete mixing, is important from the point of view of efficient oxygen transfer and for better kinetics of the process. The design needs to consider solid retention time (SRT), organic loading rate, and food-to-microorganism ratio (S/X). A longer SRT generally corresponds to better biodegradation. There are various designs of activated sludge processes; some commercial variations are listed below:

1. Step aeration
   - Influent addition at intermediate points provides more uniform organic removal throughout the tank.
2. Tapered aeration
   - Air is added in proportion to BOD exerted and flow is tapered along the length.
3. Contact stabilization
   - Biomass adsorbs organics in contact basin and settles out in the secondary clarifier; the thickened sludge is aerated before returning to the contact basin.
4. Pure-oxygen activated sludge
   - Oxygen added under pressure keeps the dissolved oxygen level high.

5. Oxidation ditch
   – Similar to extended aeration, except the aeration is done by brush-type aerators, thus reducing electricity usage.
6. High rate
   – Short detention time and high F/M ratio in aerator to maintain culture in the log-growth phase.
7. Extended aeration
   – Long detention time and low F/M ratio in aerator to maintain culture in the endogenous phase.
8. Sequencing batch reactor
   – Only one reactor for both aeration and sludge settlement—lower capital cost. A special form of ASP that operates in batch mode with sequencing.

The activated sludge process generates sludge that requires further treatment and processing before it can be finally disposed of. A common method of disposal requires thickening of the waste sludge, further processing by anaerobic or aerobic digestion, and dewatering before land application or land disposal. Typical maximum discharge limits for BOD, non-filterable residue, total phosphorus, and total nitrogen for activated sludge plants are shown in Figure 1.15.

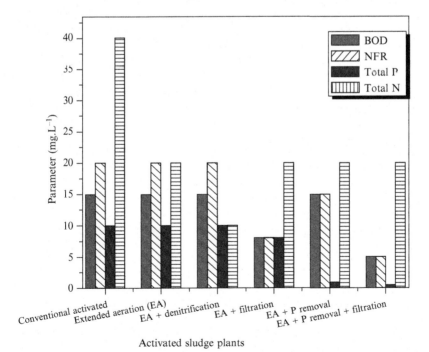

Activated sludge plants

Figure 1.15 Maximum discharge limits for activated sludge plant.

1.4.8.2 Anaerobic Treatment

The anaerobic treatment of industrial wastewater, as shown schematically in Figure 1.16, has become increasingly important in recent years as a result of environmental protection legislation, rising energy costs, and problems with the disposal of excess sludge formed in aerobic treatment processes. In anaerobic fermentation, the formation of methane, alcohols, ketones, and organic acids is important. Biogas as a byproduct is utilized to meet the energy requirements. This gas contains methane and carbon dioxide in a 1:1 to 3:1 ratio, together with hydrogen sulfide to an extent that depends on the sulfate content of the substrate, as well as traces of nitrogen and hydrogen. The growth of anaerobic microorganisms is a function of numerous factors, including residence time, temperature, redox potential, pH, and nutrient composition. In the process of anaerobic methane generation in the wastewater treatment, at least three groups of microorganisms are involved in the degradation of complex organic molecules—acidogenic bacteria, acetogenic bacteria, and methanogenic bacteria. First, biopolymers are hydrolytically degraded by the process of acidogenesis to give soluble monomers. Acidogenesis is followed by acid formation, with the simultaneous generation of hydrogen and carbon dioxide, a step referred to as acetogenesis. It is at this point that methane formation can commence, and the reaction of hydrogen with carbon dioxide occurs to produce methane. The complete degradation chain is in reality much more complicated and requires balanced cooperation among various microorganisms. The kinetics of the individual steps is inadequately understood even today because substrate mixtures are complex in most cases and collecting reliable kinetic data is time consuming and difficult.

Given the same organic load, excess sludge formation is lowered by a factor of ~10 in an anaerobic process. The fundamental advantages of anaerobic wastewater treatment are the following:

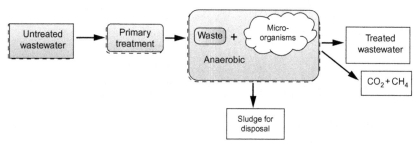

Figure 1.16 Schematic of anaerobic biological wastewater treatment process.

1. Cost and energy-intensive oxygen transfer is avoided. No large aerators are required.
2. Space–time yields (bioreactor performances) are far better than in the activated-sludge process because of the limitations imposed by oxygen transfer.
3. Since oxygen is not used, aerosol formation is avoided apart from stripping out the volatile components.
4. There is effective removal of heavy metals through reductive precipitation (as heavy-metal sulfides) rather than oxidative precipitation.
5. Energy is utilized in the form of biogas.
6. The anaerobic process represents true waste disposal with very little sludge generated. Nearly 95% of the organic contamination is converted into a combustible gas.

A popular mode involves a two-stage operation: stage I for acidogenic microorganisms and stage II for mainly methanogenic microorganisms. The methanogenic operation is usually implemented in the form of an upflow anaerobic sludge blanket (UASB) or fixed-bed reactor. Although a single-stage operation requires lower initial investment than a two-stage system, today, the trend in the anaerobic treatment of highly contaminated wastewater is toward a two-stage process design. There have been a number of approaches for retention and recycling of the biomass that serves as catalyst for different steps involved in the anaerobic process. The most important methods for decoupling the residence times for liquid substrate and biomass in anaerobic wastewater treatment include either internal retention or external separation and recycling.

Biomass retention (internal)

1. Sedimentation by pellet formation
   - UASB
2. Filtration
   - Membrane anaerobic reaction system
   - Rotor-fermenter
3. Immobilization by adsorption
   - Fixed-bed reactor
   - Fixed-bed loop reactor
   - Anaerobic film reactor
   - Fluidized-bed reactor
   - Hybrid concepts (UASB/fixed bed)
   - Stirred-tank reactor with suspended carriers
4. Immobilization by inclusion or covalent bonding

Biomass recycling (external separation)
5. Sedimentation by chemical or physical separation
    - Anaerobic contact process (flocculation/lamella separator)
    - Centrifugation
6. Flotation

The scale-up of anaerobic wastewater treatment has not been studied in a systematic way. Adequate information on scale-up and designing of anaerobic reactors with integrated provision for biomass retention (e.g., UASB reactors, fixed-bed and fluidized-bed reactors) is not available.

1.4.8.3 Biological Treatment: Combination of Aerobic and Anaerobic Operations

Aerobic treatment is more suitable for low-strength wastewaters, while anaerobic treatment is more suitable for high-strength wastewaters. There are a number of advantages and disadvantages associated with each of these processes that are well documented. In order to maximize the impact, many variations in the form of combined operations involving both aerobic and anaerobic treatments exist. One variation includes sequencing of the two methods, recycling of sludge, and optimizing the benefits of both the operations, resulting in better-treated water quality along with reduced sludge generation. One such scheme is schematically shown in Figure 1.17. It is to be noted that such sequencing can be highly useful in cases where one fraction of the wastewater can only be aerobically degraded and the other only anaerobically degraded. A sequential operation can also help in substantial nitrogen removal in the process. It can also result in a reduced odor problem. Some designs suggest combination of anaerobic + aerobic or anaerobic + aerobic + anaerobic for further improved efficiency. A typical commercial example of such a combination of processes is distillery wastewater treatment, which is conventionally done using an anaerobic process for generation of biogas followed by aerobic treatment for meeting wastewater standards, utilizing high-strength wastewaters for anaerobic treatment followed by low-strength wastewaters for aerobic processes (Pant and Adholeya, 2007).

1.4.9 Hybrid Separations

The chemical industry can be conveniently defined as a combination of reaction and separation processes. Technologies for process improvement should involve improvement in any or both of these operations, either in isolation or in the form of process integration. Thus, new research is needed

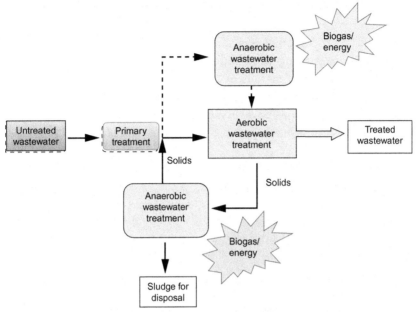

Figure 1.17 Combined aerobic/anaerobic process for wastewater treatment.

to develop better separation technologies in physico-chemical treatments with an emphasis on adsorption, ion exchange, oxidation, cavitation, and membrane separation along with the development of hybrid systems. It should be noted that in the area of adsorption, the focus had been largely on process development and not on materials, while in the membranes area, the focus had been mostly on the material science while the process application area lacked attention. As mentioned earlier, the energy costs in advanced effluent treatment methods are tremendous, and thus emphasis is required today on new adsorbent and ion exchange materials with increased capacity, selectivity, and kinetics to improve separations at substantially reduced cost. Effective regeneration of materials is also critical to efficient operation and increased life. Demonstration of performance in real-world systems is crucial since most of the work reported deals predominantly with clean/ideal systems. This is especially true for membrane separations since membrane fouling problems, temperature stability, chemical/mechanical stability, and separation characteristics are tested in real-world systems, where most failures occur. All these techniques of adsorption/ion exchange/membrane separation along with newer processes such as cavitation are most suitable for development of hybrid systems where reaction-separation occurs (i) in one step or (ii) in combination with two or more separation techniques.

Hybrid separations and technologies are required mainly for the following:

- To increase process efficiency along with cost effectiveness
- To comply with stringent pollution control norms
- To take advantage of various new developments in the separations area including new devices/processes.

In view of the discussion as above on various separation processes, the following hybrid processes can be envisaged in combination with individual processes:

- Coagulation +
  - Newer formulations combining inorganic and organic coagulants for improved performance
  - Process integration with other technologies such as coagulation + cavitation = cavigulation
  - Hybrid technologies here can substantially improve the secondary treatment step
- Adsorption/Ion Exchange +
  - Newer materials/nanomaterials coupled with other separation processes for final polishing to meet statutory requirements
  - Hybrid technologies here can substantially improve the tertiary/polishing treatment step
- Membrane +
  - Advanced materials/membranes including high temperature ceramic membranes
  - Can be combined effectively with most other physico-chemical methods of separation
  - Hybrid technologies here are being increasingly used in biological effluent treatment, for example, MBR, MMBR, EMBR
- Cavitation +
  - New devices for improved performance
  - Can be combined with many processes such as coagulation, oxidation, and others
  - Hybrid technologies here can be applied in both secondary and tertiary treatment steps
- Biological +
  - The efficiency of the biological oxidation techniques is often hampered by the presence of bio-refractory materials, whereas AOPs promise degradation of almost all the contaminants
  - A hybrid method of AOPs to reduce the toxicity of the effluent up to a desired level followed by biological oxidation is needed for the future!

1.5 INDUSTRY SECTORS WHERE WASTEWATER TREATMENT, RECYCLING, AND REUSE CAN HAVE A HIGH IMPACT

The theme of industrial wastewater treatment, recycling, and reuse is relevant for all chemical and allied industries. It is even more crucial for some of the sectors where large volumes of wastewater are generated. The key issues here could be small concentrations but significant health hazards (metals removal—all industries associated with metal processing), toxic pollutants/organics and color issues (dyes and textiles), or industries processing foods where low to high volumes of wastewaters are generated that can be easily treated, recycled, and reused. In this regard, a brief discussion on the specific aspects in these representative sectors will be highly useful.

1.5.1 Removal of Metals

The toxic and recalcitrant nature or difficulty in biodegradability of metals necessitates the utmost importance for their removal from industrial effluents. High concentrations of metals are undesirable, due to the chronic toxicity associated with many metals like Cu, Ni, Cr, Pb, Hg, to name a few. The main toxic effect of heavy metals can adversely affect mental and learning abilities in humans, apart from creating other health problems. This important area therefore attracted attention of a number of researchers and a good amount of literature exists in this regard. Application of different physico-chemical methods such as ion exchange (Maturana et al., 2011), adsorption (Chen et al., 2012), membrane filtration (Mungray et al., 2012), solvent extraction (Regel-Rosocka et al., 2006), precipitation, and others have been reported for the removal of metallic species from wastewater streams. The cost of operation plays major role in deciding the separation methodology.

Often, heavy metals are removed by the precipitation method (Lewis, 2010) by using caustic soda or lime as the precipitating reagent. The metals are removed in the form of hydroxides where the pH of the medium plays an important role. Metals that are of interest in wastewater treatment include Zn, Cr, Pb, Ni, Cd, Cu, Fe, and Al (Akbal and Camci, 2011). The removal of Cr and Zn is more effective at a lower pH, below neutral. Cr is a common pollutant in the plating industry, and it is desirable to reduce hexavalent Cr to trivalent Cr for effective removal.

Some metals that are relevant to wastewater treatment are briefly discussed in Table 1.4. (source for discharge limits: (www.cpcb.nic.in/generalstandards.pdf)

Table 1.4 Examples of important metal removal in wastewater treatment

Metal	Occurs in Effluents of	Discharge Limits	Treatment Methods
Mn	Ceramics, pickling, ore processing, glass making, alloy, electrical coil manufacturing industry, and dry cell batteries	1–5 mg/L as Mn; 0.05 mg/L as minimum soluble manganese allowable in fresh water	Oxidizing the water soluble Mn(II) to insoluble Mn(IV) in the form of MnO_2 using $KMnO_4$, can be coupled with coagulation. Key parameters: coagulant dose, initial concentration of soluble Mn, and $KMnO_4$. Low pH is unfavorable. Removal of over 90% could be obtained at elevated pH > 9. Further Mn removal: using chelating polymer such as polyacrylic acid or using hydrogen peroxide + NF. Adsorbents like zeolites, activated carbons, manganese oxide coated sand (Lee et al., 2009), iron oxide coated sand, $KMnO_4$ activated carbon, polymeric adsorbents such as amberlite XAD-7 and XAD-8 may be used for Mn removal. Biological methods involving microbes that are able to oxidize soluble Mn(II) to insoluble MnO_2 can remove up to 95%.
Pb	Storage battery industry, paints, lead additives, oils, and mining industry	0.05–0.1 mg/L for surface water, EPA limit is set to 0.015 mg/L for drinking water (Gurel et al., 2005)	Currently, the widely applied method of wastewater treatment for lead removal consists of the electrochemical method, biosorption, coagulation, and ion exchange. The major difficulty is regeneration of the resin for the ion exchange technique and pH adjustment in case of coagulation because lead hydroxide is only moderately soluble. On an

Table 1.4 Examples of important metal removal in wastewater treatment—cont'd

Metal	Occurs in Effluents of	Discharge Limits	Treatment Methods
			average, storage battery lead cells have initial concentration ~4 mg/L and pH 1–2. Recently, magnetic metal oxide nanoparticles such as γ-Fe_3O_4 have been devised for removal of heavy metals like Pb with maximum removal efficiency at pH 5.5 (Cheng et al., 2012). Adsorption with activated carbon and hybrid processes such as magnetic nanoadsorbent plus adsorption are gaining importance in light of ever-increasing stringent regulations for discharge of this particular heavy metal.
Ni	Electroplating, silver refineries, storage batteries	0.1–3 mg/L for surface and sewer waters	Adsorption methodology is considered to be the most economical for nickel removal. Precipitation using chemical coagulants used to be a widely practiced method for Ni removal as well. Electrocoagulation is preferred over chemical coagulation (Meunier et al., 2006). The electrical consumption is the main parameter in cost evaluation of the electrocoagulation method. Studies have shown that electrocoagulation using an Fe–Al electrode has high Ni removal efficiency of almost 98–99% at pH 3 and 7 respectively. The removal of nickel ions can be achieved using membrane technology by

Continued

Table 1.4 Examples of important metal removal in wastewater treatment—cont'd

Metal	Occurs in Effluents of	Discharge Limits	Treatment Methods
Fe	Electroplating industries, and many other industries	1–3 mg/L for most discharge forms	Pretreatment with filtration or carbon adsorption is necessary because Fe ions are highly prone to scale forming or fouling of the membranes. When specific recovery of metal ions from electroplating wastes is needed, application of adsorption on activated carbon or ion exchange resins becomes a difficult task. The key method is to oxidize ferrous to ferric form because the different oxidation states are found to precipitate at different pH with Fe^{3+} precipitating at pH 4 and Fe^{2+} at pH 7.
Zn	Brass and alloy industry, run-off from batteries, pigment, fungicides	0.1–5.0 mg/L for discharge in fresh waters	Chemical precipitation (hydroxide or sulfide precipitation) followed by agglomeration of the colloidal or insoluble particles by sedimentation; filtration is used. Reverse osmosis, ion exchange, and carbon adsorption methods are also used. The secondary

(continued at top) using $Mg(OH)_2/Al_2O_3$ composite nanostructure membranes or UF with polyethyleneimine additives (Sun et al., 2009). Electroflocculation without the use of chemical coagulants is also gaining importance. However, the longer settling time of the flocs and turbidity is a major drawback.

Table 1.4 Examples of important metal removal in wastewater treatment—cont'd

Metal	Occurs in Effluents of	Discharge Limits	Treatment Methods
			chemical precipitation or ion exchange of Zn ions may be used as fertilizer. Biological treatment, although considered as cost effective, has high sensitivity to pH, temperature, and seasonal fluctuations, which poses limitations in its wide-scale application. Chemical precipitation into metal hydroxide also has its own limitation because an excessive acid or alkaline condition may dissolve the hydroxides. On the other hand, sulfide precipitation using calcium sulfide as the precipitating agent can operate over a broad range of pH because sulfides are insoluble over a wide range of pH. Recently, NF using RO and UF has also been considered a promising alternative for removal of Zn from industrial effluents.
Cr	Leather, dye, electroplating, photographic industries	Total Cr 0.1–2 mg/L for fresh water discharge, hexavalent Cr 0.1 mg/L	Adsorption is a widely used method for the removal of toxic hexavalent Cr from aqueous phase. Cr(VI) is believed to be carcinogenic and 10–100-fold more toxic than Cr(III). It is also more difficult to remove from aqueous streams. For effective removal of chromium, a combination of various physico–chemical methods such as chemical precipitation,

Continued

Table 1.4 Examples of important metal removal in wastewater treatment—cont'd

Metal	Occurs in Effluents of	Discharge Limits	Treatment Methods
			adsorption/ion exchange, coagulation, and RO is required. For Cr(VI), the biosorption process is often recommended, although process integration using adsorption and electrocoagulation is suggested for near total removal (Ouaissa et al., 2013). A number of adsorbents such as sawdust, bagasse fly ash, various agricultural sorbents, synthesized zeolites, low-cost activated carbons, and biomass have also been studied by various researchers.
Cu	Hydrometallurgical processes, pharmaceuticals, dyes, and petrochemicals	0.05–3 mg/L	Separation processes such as liquid–liquid extraction (electrowinning) are used to recover Cu metals from their leach solution in wastewater. Application of physico–chemical methods, in the conventional form as well as process integration, can be the preferred option for copper removal as well as its recovery. Often biological methods are preferred when metals are present in trace amounts because higher metal concentrations may prove toxic for microbes. Although the electrolytic method is considered a costly method in terms of energy consumption, the additional effect of ultrasound is believed to

Table 1.4 Examples of important metal removal in wastewater treatment—cont'd

Metal	Occurs in Effluents of	Discharge Limits	Treatment Methods
Cd	Batteries, electrochemical cells, semiconductor devices, incinerators, leachates from landfill sites, plating, pigment, gypsum manufacturing	0.01–2 mg/L	reduce the cost substantially (Farooq et al., 2002) for low concentrations. Limited successes have been obtained with the ion exchange, RO, solvent extraction, chemical precipitation, and electrolytic methods. Emerging technology includes application of a new hybrid form of adsorbents, ceramic polymers having functional ligands on solid supports such as silica and activated carbons with oxidized surface groups. The cationic form of cadmium ions, Cd^{2+}, is believed to be linked with the oxygen-containing functional groups such as the carboxylic acid group. More detailed studies on the mechanism of diffusion of metallic ions interacting with adsorbent surfaces need to be undertaken. Some reports suggest polymer-modified magnetic nanoadsorbents for removal of metals, including cadmium.

Overall, industrial wastewater effluents usually contain mixtures of various toxic metals. It is indeed a huge task to eliminate the pollution effects of these toxic heavy metals by bringing down concentrations to desired levels. A process integration approach by combining various conventional as well as newly developed technologies, especially nanotechnology, can help substantially in pollution control and recovery of metals, quite apart from bringing down the cost of operation.

1.5.2 Dye Wastewater Treatment

1.5.2.1 Indian Scenario

The dye sector in India has major contributions from the dyes/dye intermediates-producing industry and the pigment industry as well as from end user industries that are closely associated with this entire sector, such as textiles, paint/ink, paper, and leather, and polymers and plastics. It is reported that ~70% of dyestuff gets consumed in textiles alone. In India, the dyestuff industry is fragmented, where most of the units belong to the small-scale industry sector, which is a major concern from the environmental pollution control point of view because small industries have a limited capacity for effluent treatment. Most of the dye industries are concentrated in western India, and over 80% of the total capacity comes from the state of Gujarat, which houses about 750 units. The capacity and production volumes of dyes and pigment, put together, are well above several thousand tons per annum, contributing to a ~12% share in the world market. In the last few years, although the growth of dye industry has been ~6% on an average, production of some of the dyes such as acid and direct dyes and reactive and vat dyes has remained nearly constant. Dyes such as sulfur dyes have shown over a 100% increase in the 5 years from 2002 to 2007. The organic pigment sector has also shown an increase of ~100% in production in this period (DCPC, Annual Report, 2011–12; CPCB, 2004; 2007a, 2009–10a; Department of Chemicals and Petrochemicals, Annual Report 2011–12; Performance of Chemical and Petrochemical Industry at a glance- 2001–2007).

1.5.2.2 Global Scenario

The dyestuff industry is one of the larger polluting industries, which has led to the closing down or shifting of units to emerging economies. For example, a majority of the international producers have shifted technology to developing nations such as China, India, Indonesia, Korea, Taiwan, and

Thailand. DyStar, Ciba, Specialty Chemicals, and Clariant are some of the world's leading manufacturers of dyes. Regulatory barriers have nearly stopped progress in opening fundamentally new dyestuff manufacturing facilities. The manufacturers are, however, continuing to add to their production of reactive and specialty dyes. Among the developing nations, China's production of dyes and organic pigments is increasing at a fast rate. In 2004, the production share of the United States and Europe was reported to be 24% and 22% respectively.

1.5.2.3 Dye Wastewater Treatment: Overview and Recommendations

Typically, the dye industry handles different types of dyes, most prominent of which are acid and direct dyes, disperse dyes, sulfur dyes, reactive dyes, and vat dyes, while the dye intermediate industry produces H-acid, vinyl sulfone, and gamma acid as the main products. A conventional water pattern in small- to medium-scale dyestuff manufacturers is shown in Figure 1.18.

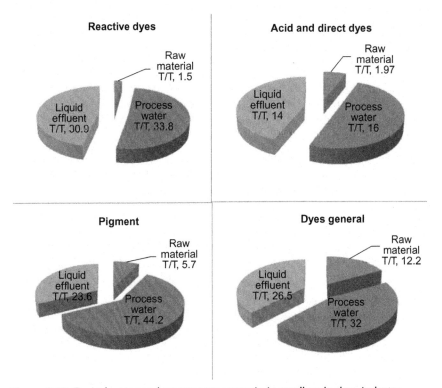

Figure 1.18 Typical water and wastewater scenario in small scale dyes industry.

The dye industry generates wastewaters to the tune of 8–75 m^3 wastewater per ton of product, and dyes such as reactive dyes are prominent along with dyes and pigment in producing large quantities of liquid effluent. Typically, the raw materials used in the dye industry include organic compounds such as benzene, xylene, toluene, aniline, anthtraquinone, and naphthalene, along with other chemicals such as acids (sulfuric, nitric, HCl, acetic acid), ammonia, sodium hydroxide, and sodium/potassium salts (carbonates and sulfates). Some solvents, such as alcohols, are also used in the process and have their presence in the liquid effluent even after their recovery. There is significant use of components, mainly as catalysts or in complexing, containing heavy metals such as copper, nickel, zinc, and iron, traces of which eventually find their way into liquid effluents. Many times metal content as high as 3000–4000 mg/L is found in the wastewaters of such industries. The wastewaters are also characterized by high COD, ranging from a low of 3000 to as high as 32,000 mg/L or even more. The wastewaters are many times characterized by high ammoniacal nitrogen content with values as high as 6000–11,000 mg/L. Ammoniacal nitrogen can create complexities in the selection of suitable wastewater treatment methodologies.

The treatment methodology for dye industry wastewaters is complex and diverse depending on the specific characteristics of the effluent; by and large, no general solutions are available. A number of physico-chemical methods—filtration, coagulation, adsorption, and biological methods— are generally employed in combination. It is recommended that components that can be recovered, such as solvents, should be recovered prior to treatment. By-products such as gypsum salt, iron sludge, spent acid/dilute acid solutions of HCl or other acids, and ammonia solutions can be recovered and can be value-addition to the wastewater treatment process. The lowering of salts/recovery of salts also reduces the load on wastewater treatment. An effluent having high ammoniacal nitrogen content is usually difficult to treat using conventional methods and needs to be separated from common effluent for separate processing. Process changes in the operations, such as replacing precipitation of reactive dyes by salting out with spray drying, can drastically reduce the wastewater generation, which otherwise is huge and has high COD and TDS. Many times, the reactive dyes are considered refractory pollutants that are difficult to degrade using conventional physico-chemical and biological methods. Newer methods of treatment such as AOPs and cavitation can be attractive alternatives in such cases.

The overall Indian production of dyes is expected to grow significantly in the future. However, strict effluent treatment and environmental norms form an impending bottleneck for the growth of this industry sector.

In recent years, the sustainability of the industry has been threatened because of not meeting pollution standards. It is imperative that the industry sector implement proper effluent treatment methods, not just to reduce pollution, but also to contribute to reducing water stress through water recycling and reuse.

1.5.3 Food Industry

One important industry sector that conventionally uses biological methods of treatment, mostly activated sludge processes, is the food industry. This is typically a seasonal industry, and therefore there is no regular generation of wastewater throughout the year similarly to the chemical industry. Further, wastewaters generated differ significantly in the type of pollutant, concentration, and volume, depending on the nature of food processing and manufacturing. The wastewater from this industry emanates a foul odor and has somewhat high BOD, of the order of 400 mg/L or higher. For the starch processing industry, BOD and COD are significantly higher, of the order of 10,000 and 20,000 mg/L respectively. Important food industries include brewery and beverages; vegetable oil–producing industries; the dairy industry, producing milk and milk derivatives such as cheese and butter; the starch industry; and the confectionary industry. Oil is also the main pollutant in the wastewaters of the vegetable oil and starch industries; this needs to be removed prior to biological treatment. Slaughterhouses and the meat-processing industry belong to a different class that produces wastewater with biological material such as blood containing pathogens, hormones, and antibiotics. The problem of finding antibiotics in sewage wastewaters because of improper disposal in cities is also approaching significant proportions and requires careful attention because the desired concentrations for many of these pollutants are below picogram levels. The wastewater volume generated in the beverage industry is believed to be $\sim$10 m^3 per ton product, for milk and the dairy industry $\sim$10-20 m^3 wastewater per ton of product, while the wastewater volume in brewery industry such as beer production is of the order of 10–30 m^3 water per ton of product (Examples of food processing wastewater treatment, www.env.go.jp/earth/coop/coop/document/male2_e/007.pdf). The BOD level in wastewater is also an order of magnitude higher in the brewery industry as compared to that in the beverage industry. In general, the treatment methodology follows the following steps:

1. Preliminary treatment for the removal of oils/solids
2. Aerobic/anaerobic treatment

3. Treatment of excess sludge
4. Release/recycling of treated wastewater
5. Sludge disposal.

The preliminary treatment generally requires a combination of various methods such as coagulation, sedimentation, and filtration. The choice of biological method depends on various factors. The conventional activated process was highly popular, especially before 2000. However, increased awareness of the cost of operations has made use of the anaerobic process more attractive in recent years. The organic load in food industry wastewaters is usually very high, thereby making wastewater more suitable for anaerobic processes for generation of energy. The high volume of sludge generated during conventional ASP is also driving industries to adopt new changes such as the MBR and specific modifications for near zero-sludge generation. A polishing treatment in the form of adsorption is typically used for meeting stringent standards for effluent discharge.

1.6 INDUSTRIAL WASTEWATER TREATMENT PROCESS ENGINEERING

Industrial and sewage wastewater treatment essentially falls in the category of dilute separations wherein ppm concentrations of pollutants are involved. The wastewaters can be further divided into subclasses:

1. Wastewaters requiring removal of ionic species from aqueous streams. This would include salts and inorganics such as metal ions in the wastewaters.
2. Wastewater streams requiring removal of organic pollutants from aqueous streams.

Identification of wastewaters containing components of high value for removal and recovery is an important aspect that needs attention prior to the treatment of the streams. The separation operations for removal and recovery of high-value materials mainly involve adsorption, ion exchange, and membrane separations (UF/NF/RO). Although these operations are used currently in the industry, there are a number of limitations, and applications in most cases can be said to be underdeveloped in terms of availability of suitable material, capacity, selectivity, chemical and thermal stability, effective regeneration, and proper understanding of the phenomena.

In general, for the removal of suspended solids (TSS), sedimentation, MF, filtration, and coagulation processes are most suitable and commonly employed. Multimedia filtration, coagulation, and UF are also useful for the removal of colloids to a great extent. The effluent contains dissolved solids (TDS) mostly in the form of salts/inorganics/heavy metals that can be removed by processes such as adsorption, ion exchange, chemical precipitation, and the more expensive membrane-based methods such as NF, RO, and electrodialysis. A major part of the pollution load is in the form of organics (BOD/COD/TOC) that can be removed using oxidations, adsorption/ion exchange, chemical precipitation, and higher-end membrane separations (e.g., UF/NF and RO), apart from biological processes (such as aerobic and anaerobic processes). Most of the methods generate secondary waste streams in the form of sludge or a secondary stream, though on a much smaller scale compared to the wastewater treated. These secondary streams have to be disposed of using appropriate means. For industrial wastewater treatment process engineering, typically the following formats are most popular.

Primary treatment (Basic clean-up/physical methods)
- Filtration (dual media, carbon, sand filters)
- Screening, grit removal, and sedimentation for sludge/solid removal

Secondary (removes 85–95% of BOD/COD and TSS; 20–40% P; 0–50% N)
- Neutralization/stabilization
- Coagulation/clarification
- Biological treatment (aerobic/anaerobic)
- Physico-chemical methods
- Hybrid separations

Tertiary/Polishing treatment (removes >99% of pollutants)
- Adsorption/ion exchange/membranes
- Hybrid separations.

Industrial wastewater process engineering involves identification of process stages in accordance with the nature and concentration of pollutants and desired goals for wastewater treatment. Typically, goals for transforming wastewater to make it suitable for discharge into surface water bodies differ significantly from those for water recycling and reuse. The cost of operation and maintenance is of prime importance in process engineering and design. A brief review of industrial wastewater treatment for some specific industries is given in Table 1.5, mainly representing textile and dye industry wastewater treatment (Amaravati ETP- Das, www.fibre2fashion.com/industry-article/3/215/textile-effluent-treatment-a-case-study-in-home-textile-zone1.asp; Egypt Textile ETP- Seif and Malak, 2001; Cotton Textile, Mumbai- Babu, 2008; Dye wastewater for recovery of salt- CPCB India, 2007; Arulpuram

Table 1.5 Examples of Industrial Wastewater Treatment

No	Industry or type of industrial wastewater	Process	Wastewater content	Quantity of effluent to be treated	COD			BOD			TDS		
					Inlet	Outlet	% Removal	Inlet	Outlet	% Removal	Inlet	Outlet	% Removal
1	Amaravati Effluent Treatment Plant	Primary + secondary	Different kinds of solid and liquid waste from different textile plants	200 m³/day	300–400	140–250	38–54%	100–150	20–30	80%	2500–3000	1800–2100	30–40%
2	Egypt Textile Treatment (Seif and Malak, 2001)	Primary + secondary + tertiary	–	–	475–1835	35–45	95%	295–1280	25–30	91%	650–1940	550–1250	40%
3	Cotton Textile Industry, Mumbai (Babu, 2008)	Primary + secondary	Reactive dye + alkali	2500 m³/day	4992	476	90%	Not reported			Not reported		
4	Dye Wastewater Treatment to recover water and Glauber salt (CPCB, 2007a)	Primary + tertiary	–	500 m³/day	9366	81	99%	1823	30	98%	17,000	1426	92%
5	Zero Discharge in Arulpuram Plant (Hussain, 2012)	Primary + secondary + tertiary	Dye + dissolved salts + metals + alkali	Design capacity; 5500 m³/day	1000–1200	Below detectable level	~100%	400–500	Below detectable Level	~100%	6000–7000	282	95.30%

Table 1.5 Examples of Industrial Wastewater Treatment—cont'd

No	Industry or type of industrial wastewater	Process		Wastewater content	Quantity of effluent to be treated	COD	BOD	TDS
6	Raymond Zambaiti Ltd. Textile Industry, Kolhapur (Desai and Kore, 2011)	Primary + secondary	Screening + equalizing + mixing + flocculation + aeration + sludge thickener + FAB I + FAB II + settling + chlorination + sludge thickener	Starch, waxes, dyes, urea, NaOH, H_2O_2, reducing and oxidizing agents, binders, oils, gums, PVA, fats	1214	Inlet 1422 Outlet 224 % Removal 84.0%	Inlet 298 Outlet 36 % Removal 88%	Inlet 2000 Outlet 1900 % Removal 5%
7	Silk and dyeing Mill in Shaoxing Zhejiang (China) (Wang et al., 2011)	Primary + secondary	Regulating tank + reaction tank + coagulation + tube sedimentation tank + effluent tank	Reactive dye, direct dye, acid dye, sulfur dye	500 t/d	Inlet 1114–1153 Outlet 613–680 % Removal 55–59	Not reported	Not reported
8	Printing and Dyeing Mill in Beijing (Wang et al., 2011)	Primary + secondary	Regulation + coagulation + sedimentation + floatation + filtration	Acid dye, dispersed dye, reactive dye, sulfur dye	120 t/d	Inlet 228–353 Outlet 171–264 % Removal 75%	Not reported	Not reported

plant ETP- Hussain, 2012; Raymond Textile Kolhapur ETP- Desai and Kore, 2011; Silk and Dyeing Mill, Shaoxing- Wang et al., 2011; Printing and dyeing mill, Baijing- Wang et al., 2011). It is evident that for discharge of wastewaters in surface water bodies such as rivers, often an appropriate combination of primary and secondary methods is quite adequate. However, in the case of priority pollutants and for satisfactory water recycling and reuse, usually tertiary treatments are essential.

It is important to design proper sequencing of different operations to meet the desired levels of pollutants in the treated water, keeping cost considerations in mind. For example, bioprocesses for ethanol production generate wastewaters with high BOD amenable for treatment using biological methods. Here, anaerobic treatment is done first in view of the high organic load, and then the aerobic process is applied to meet the desired effluent standards. In the first step, nearly 80–90% of COD is removed, and then the partially treated wastewater with reduced organic load is sent for aerobic treatment. This scheme of treatment is more efficient: High-strength wastewaters are used for gathering energy in the form of biogas, while an aerobic process is employed to perform the remaining task of getting clear water without appreciable odor problems (which is otherwise difficult with an anaerobic process alone). If the treated waters still continue to have odor problems, a tertiary treatment/polishing is essential.

It is also important that a comparison of two or more methods is required for techno-economic feasibility analysis of the wastewater treatment. It is quite possible that both physico-chemical methods and biological methods may be able to achieve desired effluent goals. In that event, it is necessary to closely evaluate cost performance of the methods for the selection made. An example of this is shown in Figure 1.19 for effluent from a steel plant (Sirajuddin et al., 2010). Two methods have been compared, carbon adsorption from the class of physico-chemical methods and a biological method. It can be seen that efficiency of carbon treatment is better in all respects for BOD and COD removal. Integrated adsorption and biological treatment methods can provide excellent solutions in removing BOD, COD, organics, and pharmaceutical and pesticide pollutants apart from inorganics such as some heavy metals, ammonia, and nitrates (Ferhan, 2011).

1.6.1 Newer Modifications in the Existing Methods

Separation technology is an essential component in most chemical process operations and is considered to be the most critical component

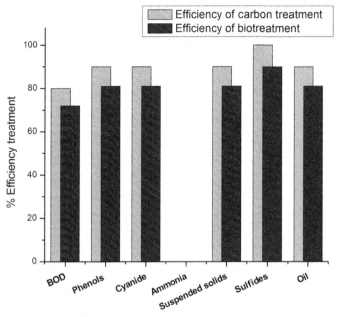

Figure 1.19 Comparison of carbon adsorption and bio-treatment methods for steel plant.

in situations involving dilute separations such as wastewater treatment, bio-separations, and gas separation technologies. It is a widely accepted fact that significant improvements can be obtained for most of the separation processes with proper scientific modifications, either through material modifications or through process modification. Zeolite is an excellent example where huge benefits have been obtained through material modifications, whereas a process such as pressure-swing adsorption is an excellent example of a huge transformation in separation science through process modification—in both cases, a number of commercial processes have been in operation in chemical and allied industries in today's world. Apart from engineering key modifications in the existing separations, process integration is an interesting and fairly wide open area for research and development wherein there is a huge possibility for the intelligent combination of two or more operations. The advent of nanomaterials has also increased the scope of investigating advanced separations with altogether new perspectives. Thus, there are ample possibilities for development of new/novel materials such as adsorbents, resins, membranes, hybrid membranes, cavitation devices, and nanomaterials quite apart from process integration options for encompassing the entire spectrum of separation science and technology in chemical and allied industries.

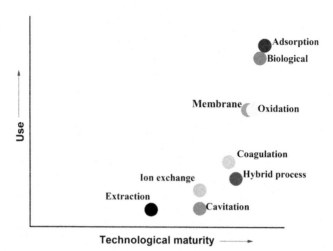

Figure 1.20 Wastewater treatment processes: technology maturity analysis.

Although there has not been much progress in developing new separation processes, a lot of new process schemes are being developed through modification of existing methods and intelligent combinations of different separation methods or reaction and separation methods. These newer separation processes combine or redefine finer aspects of existing methods, for example, distillation, extraction, ion exchange, adsorption, membrane separation, cavitation, and biological processes. Figure 1.20 qualitatively presents the status of different technologies where the use of the technology is plotted against its technological maturity. It can be seen that processes such as adsorption and biological processes have securely established themselves in terms of commercial use and are also technologically mature, indicating their wide acceptability in the industry. Processes such as membrane separation, oxidation, and coagulation are considered technologically mature to a lesser extent and require further development in terms of materials for increasing their use in the industry. Ion exchange and extraction have less use in wastewater treatment, mainly because of their fundamental limitations. Interestingly, cavitation and hybrid separations, although sufficiently technologically mature, find less applicability in the industry. The plausible reason for this is the lack of awareness and adequate knowledge on design and application of such systems. Broadly, one needs to evolve the following options for accomplishing enhanced water treatment technologies:

1. Development of new adsorbent materials having high selectivity, improved stability, and with more favorable geometries and better kinetics with substantially reduced cost.

2. Development of designer solvents for improved extractions along with better understanding of the process and enhanced physical property database.

3. Development of new robust membranes with increased surface area per unit volume and membranes suitable for effluent treatment that can withstand dirty liquids. Newer configurations are needed for reducing fouling problems and for improved performance, along with confidence-building measures in the application of newer methodologies such as MBR.

4. Development of new ion exchange materials/separations with better selectivity for refractory pollutants and improved regeneration characteristics for increased profitability, especially in the removal of heavy metals and valuable chemicals.

5. Development of newer cavitation devices and processes for effluent treatment.

6. Development of an efficient strategy using judicious selection of separation method and an apposite combination of methods ranging from precipitation to extraction/adsorption/ion exchange/cavitation/biological.

7. Development of hybrid technologies where controlling parameters common to two or more processes in combination are still not clear in most cases.

8. Development of a strategy for continuous monitoring of the cost of operation and measures for cost reduction. In principle, combining successive process steps can certainly cut the cost and improve treated water quality. In the wastewater treatment area, there is huge opportunity for process intensification/integration along with individual process/material modifications.

9. Development of reactive separation for effluent treatment, which is not seriously considered, probably for the reason that very low concentrations are involved. Here, newer devices of hydrodynamic cavitation that allow oxidation reaction on a miniaturized scale can be most useful. Also, combinations of reaction and separation processes—reactive distillation, reactive absorption, reactive extraction or reactive membrane separation—are possible. This area, however, still requires a lot of developmental efforts for practical industrial wastewater treatment applications.

10. Development of wastewater treatment applications in the areas of electrochemical and nano technology.

Electrochemical technology, an area still not so common in wastewater treatment, can treat a vast range of contaminants of high concentrations such as salts, cyanides, nitrites, PCBs, phenols, heavy metals, hydrocarbons, and fats. It uses only electricity and can operate at room temperature and atmospheric pressure. The energy consumption may, however, depend on the COD level of the wastewater. This method becomes relevant when removal of heavy metals from industrial effluents from processes relating to, for example, metallurgical, metal coatings, dyes, and batteries are concerned. This technology can be used to recover precious metals such as Ag from photographic materials, heavy toxic metals like Pb from battery industries, and others. The recovery of metal is achieved by means of metallic deposition on the cathode in the electrochemical reactor. The innovative aspect of this technology is that it can basically treat toxic waste of high organic concentration as well as recover metals with greater purity in an environmentally friendly manner because it avoids emission of gases, sulfur, and metal particles, unlike pyrometallurgy. The current state of the technology implies that it is being tested at the pilot and pre-industry level with some installations working in some parts of world. Siemens Corporate Technology in Erlangen uses an electrochemical method of wastewater treatment, where water molecules are converted to hydroxyl radicals that act as cleaning agents by attacking almost every carbon-containing structure of organic substances (Waidhas,   www.siemens.com/innovation/en/publikationen/publications_ pof/pof_fall_2008/rohstoffe/abwasser.htm). The resultant pollutant fragments may be digested by bacteria. The reactors are generally steel shells where wastewater is pumped between oppositely charged electrodes whose potential difference creates hydroxyl radicals at the positively charged electrode and liberates hydrogen gas at the negative electrode. The process setup is expected for application on an industrial scale because the pilot plant setup can effectively treat 200 L/h of wastewater. The electrochemical method is suitable for the highly concentrated wastewater from the textile, paper, and pharmaceutical industries. The greater the concentration of pollutants in the water, the more effective is the process because at higher concentrations, more particles adhere to the electrode surface, increasing the decomposition quota per kilowatt-hour. The electrochemical method did not gain much importance initially because of the extensive capital investment and expensive electricity requirement. However, in view of stringent environmental regulations regarding wastewater discharge and drinking water standards, the electrochemical method for wastewater has gained importance over the past two decades.

The application of nanomaterials and nanotechnology in industrial wastewater treatment also needs careful evaluation. Although this area appears to hold much potential for future applications, only a few studies have been undertaken so far for industrial wastewater treatment. Nanode-position of materials such as titanium dioxide and zinc sulfide has potential for use as a photocatalyst for removal of harmful pollutants from the waste-waters. Two of the important applications in this area include removal of dyes from the wastewaters and treatment of pharmaceutical wastewater containing specialty molecules. Newer materials in the form of carbon nano-tubes/nanofibers are also commercially available as adsorbents and can find increasing use in this area. Further development in the area of nanocompo-sites also holds a lot of potential for future applications. An increasing number of studies have been reported in recent years on the application of nanoma-terials, such as silver nanoparticles for the removal of mercury (Sumesh et al., 2011); cellulose-manganese nanocomposite for removal of lead (Maliyekkal et al., 2010); and iron oxide nanoparticles for removal of metal ions (Shen et al., 2009). Adsorption technology using nanoparticles for the treatment of wastewaters containing inorganic pollutants such as metals, Cd, Co, Cu, and others, apart from the removal of organic pollutants in dyes and pes-ticides, has been reviewed by Imran (2012), Ali et al. (2012). However, com-mercial applications of this technology in wastewater treatment is still largely lacking, mainly due to the fact that not much information is available on materials handling and application, and potential health and environmental effects, apart from life cycle analysis of the processes and materials on a commercial level.

1.7 ADVANCED MODELING FOR WATER TREATMENT

Apart from developing newer methods for treating industrial wastewater, there is an increasing trend of using advanced computational modeling for developing better water treatment solutions. Advanced computational models play a crucial role in achieving these objectives. Typically, advanced models are applied to:

- Get more out of existing assets
- Reduce spatial footprint
- Reduce capital costs for new assets
- Conform to new efficiency and environmental norms
- Plan for "off design" operating points/sudden changes
- Reduce the cost of treatment.

Various advanced models are used to achieve these objectives. Broadly, these models may be classified in two categories:

- Process models aim to simulate the overall process including the wastewater treatment plant and try to minimize water requirements as well as optimize water treatment plant operation.
- Equipment models aim to use computational fluid dynamics (CFD) models to optimize and intensify individual equipment used in the treatment plant.

Both these types of model are widely discussed in the published literature. The use of process models for optimizing water usage and treatment is discussed in detail in Chapter 12 of this book. These are therefore not discussed here. CFD models in the last two decades have evolved into an important tool to optimize water treatment, recycling, and reuse applications. CFD models essentially use mass, momentum, and energy conservation equations to represent real-life processes (see Ranade, 2002, for a comprehensive discussion on computational flow models and their applications). Recent advances in understanding the physics of flows, numerical methods to solve complex partial differential equations and computing power, and the CFD simulations (numerical solutions of the model equations using computers) allow users to make *a priori* predictions of the flow field and therefore have opened up several new opportunities to optimize a variety of processes. There are several possible applications in the area of water treatment, recycling, and reuse. Some of the possible applications are the following:

1. Designing of water handling systems and pipe networks; flow distribution
2. Flocculation/coagulation/clarifiers/settling tanks
3. Cavitation devices for water treatment
4. Chlorine, oxygen, ozone, or hydrogen peroxide contacting/reactors including surface aerators
5. Anaerobic and aerobic digesters.

In each of the applications mentioned above, there are several opportunities to use CFD models for performance enhancement. For example, in settling tanks or clarifiers, it is essential to design the configuration in such a way that settled particles are not re-entrained. In a conventional clarifier, the incoming wastewater stream at the top creates turbulence in the clarification zone. The hydraulic energy of the incoming flow needs to be dissipated in order to avoid turbulence in the clarifier (which may cause re-entrainment). CFD models can be used to devise optimal configuration to avoid re-entrainment as well as to realize additional advantages. For example, Pophali (2007) has used CFD models to develop an improved clarifier. A sample of his simulated results is shown in Figure 1.21.

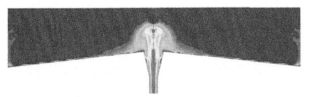

Figure 1.21 Improved clarifier designed using CFD simulations (Pophali, 2007).

The CFD simulations were used to evolve several new features such as a low-level gradually enlarged inlet that ensures hydraulic energy dissipation through plume formation, improved flocculation because of the plume, uniquely sloped bottom of the clarifier, and specially devised sludge removal. The design has also resulted in a patent (Pophali et al., 2009).

In recent years, there have been large numbers of publications on using CFD models for variety of applications. Applications of CFD models for flocculation in water treatment are reviewed by Bridgeman et al. (2009). Fayolle et al. (2007) have demonstrated use of CFD models for predicting oxygen transfer in aeration tanks. Ranade et al. (2006, 2013) have used CFD models to devise a patented design of vortex diodes for water treatment and disinfection. Dedicated vertical applications based on CFD simulations for water treatment are being developed (see for example, www.tridiagonal.com/products/mixit.html). There is also a dedicated Linkedin group on applications of CFD for wastewater applications (www.linkedin.com/groups/CFD-Wastewater-4664268/about).

It is not possible or necessary to summarize all of the publications/reviews here. Please refer to Ranade (2002) to gain an understanding of the general philosophy of using CFD models for performance enhancement. The application of such advanced computational tools with the creative use of hybrid treatment technologies has the potential to significantly bring down treatment cost and spatial footprint.

1.8 COST OF WASTEWATER TREATMENT AND POSSIBLE VALUE ADDITION

The challenge of removal or recovery of materials from dilute streams is unique in several respects and is expected to become a much larger problem in the future. Government regulations, as well as the growth of biotechnology and the pharmaceutical sector, will be forcing increased attention on this area. Separation processes generally represent 40–70% of both capital and operating costs and account for approximately 45% of energy cost in the chemical and petrochemical industries. The cost of treating wastewaters is also quite significant, requiring complex design and operations apart from maintenance. Thus, separation science and technology in wastewater treatment is no longer just an operation to purify/recover products but also an important tool for improving cost competitiveness, boosting energy efficiency, and increasing productivity with the ultimate goal of addressing environmental concerns. Separations from dilute solutions have been receiving wide attention recently because of the need for recovery of small quantities of high-value materials (e.g., in the pharmaceutical industry) and removal of contaminants from dilute aqueous and organic streams. In the case of high-value materials, it becomes value addition to the existing process, while for low-value chemicals, it is challenging to find economically feasible separation processes to recover materials or to address environmental concerns.

Cost calculations in most wastewater treatments are not straightforward. They require not just an understanding of the process operations but also of upstream fallouts of the process modifications and further effects down the line after effluent discharge/recycling/reuse. It is easy to calculate the cost of processing for a specific pollution problem involving a particular stream containing defined pollutants with known concentrations, volumes for treatment, and method of treatment. However, often there are process changes that alter the composition of the wastewaters. Further, a number of separation processes compete with each other, and outlet water characteristics of each are very different. It can be said that employing a much cleaner process using environmentally friendly reactions and separations would provide a much simpler and cost-effective option in environmental pollution control rather that allowing pollutants to get discharged into the water pools/rivers and affect people and the environment, the cost of which is incalculable. The situation can be disastrous with toxic pollutants.

A common belief in wastewater treatment relates to only considering treating wastewater to make it suitable for discharge in surface waters/rivers/water ponds. The fact that wastewater can be considered as a useful and potential resource is grossly ignored. Some important recommendations in this regard are:

1. Wastewater treatment, recycling, and reuse for the purpose of agriculture can be quite straightforward for many industrial and sewage wastewaters. The cost benefits include reduced freshwater requirements and thus lowered costs of obtaining water, conserving fresh water resources, improving soil conditioning through nitrogen/phosphates in the recycled water, and preventing discharge of pollutants to surface and groundwater. The

overall economic benefits can be very high for industry as well as for society. Industries that have wastewater suitable for such purposes can ally with local corporation bodies for the treatment, recycling, and reuse necessary and come up with designs for appropriate distribution of fresh and recycled water. Typical concentrations of elements in treated wastewater effluent from conventional sewage treatment processes are 50 mg/L of nitrogen, 10 mg/L of phosphorus, and 30 mg/L of potassium. Assuming an application rate of 5000 m^3/ha/year, the fertilizer contribution of this effluent would be 250 kg/ha/year of nitrogen, 50 kg/ha/year of phosphorus, and 150 kg/ha/year of potassium. Thus, the effluent would supply all of the nitrogen and much of the phosphorus and potassium normally required for agricultural crop production. Other valuable micronutrients and the organic matter contained in the effluent would also provide benefits (Corcoran et al., 2010). Recently, CPCB India attempted to evaluate the value associated with municipal wastewaters with the assumption that both water and nutrients have value and that there is no loss of nutrients during treatment. It found that the fertilizer equivalent along with wastewater worth Rs 1091.20 million are discharged in to the coastal waters from coastal cities and towns annually (CPCB, 2009–10b; http://nptel. iitm.ac.in/courses/105105048/M11_L14.pdf). This one example should be an eye opener for both environmental bodies and industries and clearly highlights the importance of water recycling and reuse in an appropriate manner. A recent report by IWMI (Amerasinghe et al. 2013) also highlights value additions due to domestic wastewater treatment, while cautioning about adverse effects of wastewater usage in irrigation. It is to be emphasized that the adverse effects can be practically eliminated if the wastewaters are accurately characterized as discussed in the earlier sections and suitably treated (for augmentation of N, P, and K as well as for removal of heavy metals and other harmful pollutants) before considering its use for agricultural purposes.

2. Process modifications can substantially reduce the cost of wastewater treatment apart from a change of methodology for effluent treatment, land/space requirements, equipment costs, and legal binding legislation concerning pollution levels. Such changes are far better than end-of-pipe solutions in the form of stringent methods of treatment, such as oxidation, that are highly energy intensive. Use of easily biodegradable raw materials, fewer refractory pollutants generated for simpler treatments, and easy recycling and reuse can be potentially attractive options from a cost-consideration point of view.

In most cases, chemical industries are located far from the urban areas, and often the effluent standards are less strict in these locales. If the land is easily available, conventional biological processes can prove to be more economical, especially with the use of a lagoon, which offers significant cost reduction due to the simplicity of the process (less capital/land cost), less manpower/processing costs, and ease of operation. Temperature/climate can be a major consideration however, especially in cold regions and can adversely affect the microbial activity and process performance.

In summary, value addition from wastewater treatment, recycling, and reuse comes mainly from:

1. Recovery of chemicals, such as acids
2. Recovery of metals
3. Recovery of energy in the form of biogas, power
4. Recovery of water as fertilizer
5. Recovery of water as such for reuse.

Some reports have presented the cost of wastewater treatment for sewage and for some specific industrial wastewaters such as dye wastewaters. These include capital costs, operating costs for conventional operations such as primary treatment, UF systems, and RO. In our opinion, these calculations are highly case and process specific and can be misleading at times, with severe limitations from a generalization point of view. It is therefore advisable to address the cost calculation issue from a specific separation process point of view as a preliminary criterion for the selection of a particular method. In this regard, some of the conventional separation processes have been compared qualitatively in terms of energy consumption in Figure 1.22.

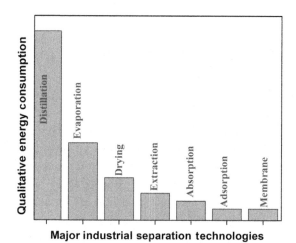

Figure 1.22 Energy consumption pattern in separation processes.

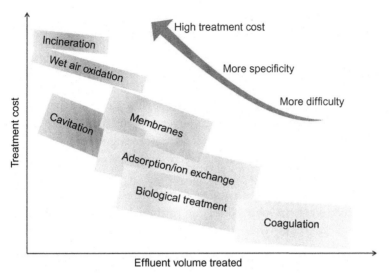

Figure 1.23 Industrial wastewater treatment-cost analysis.

A data analysis of this type would help practicing environmentalists to make decisions in terms of process selection from the energy point of view.

A more useful comparison of different processes in wastewater treatment has been reported by Casper et al. (1999; Figure 1.23). According to this analysis, incineration cost is the highest followed by energy-intensive processes such as evaporation and wet air oxidation. The treatment costs vary to some extent with respect to the effluent load, and increased load is expected to dictate the cost of operation. Biological treatment of anaerobic type, because there is payback in the form of energy, substantially reduces the overall cost of operation. A more generalized comparison for different individual processes and combined processes is, however, required for better understanding, which is lacking at present.

A complete evaluation of cost also requires accounting of different pricing patterns for domestic, industrial, and agricultural water. Many countries other than India have different pricing mechanisms for these three water-consuming sectors. Generally, the agricultural sector receives subsidized water while industry water prices vary depending on location and availability (and are usually highly priced). The present overview clearly indicates the availability of suitable technology for industrial wastewater treatment in almost all cases and at times ways and means for water-saving. However, it is also felt that there is increased need for greater awareness in this regard, especially from a cost point of view; more information on cost comparisons and analyses of different methodologies is lacking at present. This will facilitate transmission and assimilation of the most suitable technologies in the industries, which will impact positively and reduce environmental pollution apart from helping water conservation.

In view of the above analysis on various methods of wastewater treatment, their characteristics, advantages and disadvantages, and costs of operations, and on the basis of commercial acceptance, a qualitative analysis of major commercial physico-chemical and biological methods is presented in Table 1.6. The analysis represents a 5-point rating for each of these processes on different aspects pertaining to economic, environmental, technical, and sociocultural issues. The status indicated by 4- and 5-open circle symbols represent the developed nature of the process, while the 3-star black status indicates considerable room for further development. Filled circle symbols indicate status where the application requires significant efforts to make it commercially attractive. It appears that hybrid processes combining two or more processes would have significant potential from a commercial viewpoint. While processes such as coagulation and adsorption are well placed in the overall scheme of wastewater treatment, oxidation and cavitation need efforts to increase their commercialization and acceptance. Similarly, biological processes need improvements in final effluent quality, quite apart from the land requirement factor. Membrane separations, though, look excellent from a separation point of view, but they need a lot of effort in treatment of wastewaters in real-life situations.

1.9 SUMMARY

Industrial wastewater treatment, recycling, and reuse is an important theme in today's context, not just to protect the environment from pollution, but also to conserve water resources so that water stress is reduced. A number of technologies are available to treat industrial wastewaters, and judicious decision making is required for selecting appropriate technologies. The selection depends on the desired goals of wastewater treatment: recovery of valuable chemicals from wastewater, possible water recycling and reuse, complying with the statuary norms for discharge into water bodies, and economics of the treatment process. The number and diverse nature of pollutants in wastewater makes the task of selection rather difficult. The overview presented here should be useful in identifying key issues and facilitating selection of appropriate processes for treating wastewater. Various technologies are discussed in detail in the chapters following.

Table 1.6 Sustainability Criteria: Current Status

Indicators	Coagulation	Adsorption	Membrane	Cavitation	Oxidation	Biological		
						Aerobic	Anaerobic	Hybrid
On 5-point rating scale								
Economic								
Cost/ Affordability	○○○○	★★★	••	○○○○	★★★	○○○○	○○○○	★★★
Environmental								
Effluent Quality	★★★	○○○○	○○○○	★★★	○○○○	★★★	★★★	○○○○○○
Organics	○○○○	○○○○	★★★	○○○○	○○○	○○○	○○○	○○○○
Inorganics/ Metals	•	○○○○	○○○○	••	•	•	•	○○○
Chemical Usage	★★★	○○○○	○○○○	○○○○	•	○○○○	○○○○	○○○○
Energy	○○○○	○○○○	★★★	★★★	○○○○	○○○	○○○	○○○
Land Usage	••	○○○○	○○○○	○○○○	○○○○	•	••	○○○
Technical								
Reliability	○○○○	○○○○	★★★	○○○○	○○○○	○○○○	○○○○	○○○○
Ease of Operation	○○○○	○○○○	○○○	★★★★	★★★	○○○	○○○	○○○
Adaptability	○○○○	★★★	★★★	○○○○	★★★	★★★	★★★	○○○○
Scale-up	○○○○	○○○○	○○○○	○○○	★★★	○○○○	○○○○	○○○○
Sociocultural								
Acceptance	○○○○	○○○○	★★★	•	★★★	○○○○	○○○○	★★★
Expertise	○○○○	○○○○	○○○	○○○○	★★★	○○○○	○○○○	○○
Overall sustainability	○○○○	○○○○	★★★	○○○	○○○○	○○○○	○○○○	○○○

ACRONYMS

Bn	billion
BOD	biological oxygen demand
COD	chemical oxygen demand
CPCB	Central Pollution Control Board (India)
IPCC	Intergovernmental Panel on Climate Change
NOEL	No-observed-effect level
TDS	total dissolved solids
TOC	total organic carbon
TSS	total suspended solids
UN	United Nations
UNDP	United National Development Program
UNEP	United Nations Environment Program
UNESCO	United Nations Educational, Scientific and Cultural Organization
UN-HABITAT	United Nations Human Settlements Program
WHO	World Health Organization
WWAP	World Water Assessment Program

REFERENCES

Ahmaruzzaman, M., Laxmi Gayatri, S., 2011. Activated neem leaf: a novel adsorbent for the removal of phenol, 4-nitrophenol and 4-chlorophenol from aqueous solutions. J. Chem. Eng. Data 56, 3004–3016.

Akbal, F., Camci, S., 2011. Copper, chromium and nickel removal from metal plating wastewater by electrocoagulation. Desalination 269, 214–222.

Ali, I., Asim, M., Khan, T.A., 2012. Low cost adsorbents for the removal of organic pollutants from wastewater. J. Environ. Manage. 113, 170–183.

Amarasinghe, U A, Shah, T, Turral H, Anand, B. K, 2007, India's Water Future to 2025-2050: Business-as-Usual Scenario and deviations, Colombo, Sri Lanka: International Water Management Institute. 47p. (IWMI Research Report 123).

Amerasinghe, P., Bhardwaj, R.M., Scott, C., Jella, K., Marshall, F., 2013. Urban Wastewater and Agricultural Reuse Challenges in India. International Water Management Institute (IWMI), Colombo, Sri Lanka, 36 pp. (IWMI Research Report 147), viewed 28 May 2013, http://www.iwmi.cgiar.org/Publications/IWMI_Research_Reports/PDF/PUB147/RR147.aspx.

Babu, B.V., 2008. Effluent Treatment: Basics and a Case Study, http://discovery.bits-pilani.ac.in/~bvbabu/BVBabu_Water_Digest_Jan_2008.pdf.

Bhandari, V.M., Juvekar, V.A., Patwardhan, S.R., 1992a. Sorption studies on ion exchange resins. Ind. & Eng. Chem. Res. 31, 1060–1073.

Bhandari, V.M., Juvekar, V.A., Patwardhan, S.R., 1992b. Sorption studies on ion exchange resins. II. Sorption of weak acids on weak base resins. Ind. & Eng. Chem. Res. 31, 1073–1080.

Bhandari, V.M., Juvekar, V.A., Patwardhan, S.R., 1993. Sorption of dibasic acids on weak base resins. Ind. & Eng. Chem. Res. 32, 200–206.

Bhandari, V.M., Yonemoto, T., Juvekar, V.A., 2000. Investigating the differences in acid separation behaviour on weak base ion exchange resins. Chem. Eng. Sci. 55 (24), 6197–6208.

Bhandari, V.M., Ko, C.H., Park, J.G., Han, S.S., Cho, S.H., Kim, J.N., 2006. Desulfuriza-tion of diesel using ion exchanged zeolites. Chem. Eng. Sci. 61 (8), 2599–2608.

Bhardwaj, R.M., 2005. Status of wastewater generation and treatment in India 2005. IWG-Env Joint Work Session on Water Statistics, Vienna, 20–22 June 2005, viewed 28 May 2013, http://www.umweltbundesamt.at/fileadmin/site/umweltthemen/wasser/IWG_ENV/3b-Status_of_Wastewater_India.pdf

Bridgeman, J., Jefferson, B., Parsons, S.A., 2009. Computational fluid dynamics modeling of flocculation in water treatment: a review. Eng. Appl. Comp. Fluid Mech. 3 (2), 220–241.

Casper C, Deibele, L, Fathmann, ATV-Handbuch Industrieabwasser, Grundlagen, 4. Auflage, Ernst & Sohn, A Wiley Company, Berlin 1999, 144-146.

Central Pollution Control Board, 2005a. Performance status of common effluent treatment plants in India, viewed 28 May 2013, http://cpcb.nic.in/upload/Publications/Publication_24_PerformanceStatusOfCETPsIinIndia.pdf.

Central Pollution Control Board, 2005b. Status of sewage treatment in India, viewed 08 May 2013, http://www.cpcb.nic.in/newitems/12.pdf.

Central Pollution Control Board, Ministry of Environment & Forests, Annual Report 2009–10a, viewed 28 May 2013, http://www.cpcb.nic.in/upload/AnnualReports/AnnualReport_40_Annual_Report_9-10.pdf.

Central Pollution Control Board, Ministry of Environment & Forests 2007a, Advance methods for treatment of textile industry effluents, Resource Recycling Series: RERES/&/2007, viewed 28 May 2013, http://www.cpcb.nic.in/upload/NewItems/NewItem_89_27.pdf.

Central Pollution Control Board, Ministry of Environment & Forests, 2007b. Evaluation of Operation and Maintenance of Sewage Treatment Plants in India 2007, Control of Urban Pollution Series: CUPS/68/2007, viewed 28 May 2013, http://www.cpcb.nic.in/upload/NewItems/NewItem_99_NewItem_99_5.pdf.

Central Pollution Control Board, Ministry of Environment and Forests, Govt. of India, 2009–10b. Status of water supply, wastewater generation and treatment in Class I cities and Class II towns of India. Series: CUPS/70/2009-10, viewed 28 May 2013, http://www.cpcb.nic.in/upload/NewItems/NewItem_153_Foreword.pdf.

Central Pollution Control Board, 2004. Dyes and dye Intermediate Sector, Identification of Hazardous Waste Streams, their characterization and waste reduction options in Dyes & Dye Intermediate Sector, viewed 28 May 2013, http://www.ecacwb.org/editor_upload/files/IoHWS%20their%20Characterization%20and%20WRO%20in%20D&DI%20Sector.pdf.

Chaphekar, S.B., 2013. Education for water conservation. In: National Conference on Water Sanitation and Recycling. Maharashtra Chamber of Commerce, Industry & Agriculture, Mumbai.

Characteristics of sewage and treatment required, viewed 09 May 2013, http://nptel.iitm.ac.in/courses/105105048/M11_L14.pdf.

Chen, Y.X., Zhong, B.H., Fang, W.M., 2012. Adsorption characterization of lead(II) and cadmium(II) on crosslinked carboxymethyl starch. J. Appl. Polym. Sci. 124, 5010–5020.

Cheng, Z., Tan, A.L.K., Tao, Y., Shan, D., Ting, K., Yin, X.J., 2012. Synthesis and char-acterization of iron oxide nanoparticles and applications in the removal of heavy metals from industrial wastewater. Int. J. Photoenergy 2012, 1–5.

Corcoran, E., Nellemann, C., Baker, E., Bos, R., Osborn, D., Savelli, H. (Eds.), 2010. Sick water? The Central Role of Wastewater Management in Sustainable Development, UNEP UN-HABITAT, GRID-Arendal, viewed 28 May 2013, http://www.unep.org/pdf/SickWater_screen.pdf.

Das, S. 2008. Textile Effluent Treatment: A case study in home textile zone, viewed 29 May 2013, http://www.fibre2fashion.com/industry-article/3/215/textile-effluent-treatment-a-case-study-in-home-textile-zone1.asp.

Department of Chemicals and Petrochemicals, Annual Report 2011–12, viewed 23 May 2013, http://chemicals.nic.in/Annual%20Report%202011-2012.pdf.

Desai, P.A., Kore, V.S., 2011. Performance evaluation of effluent treatment plant for textile industry in Kolhapur of Maharashtra. Univ. J. Environ. Res. Tech. 1 (4), 560–565.

Examples of food processing wastewater treatment, viewed 10 May 2013, http://www.env.go.jp/earth/coop/coop/document/male2_e/007.pdf.

Environmental Protection Agency, 1997. Wastewater Treatment Manual – 'Primary, Secondary and Tertiary Treatment', Ireland, viewed 28 May 2013, http://www.epa.ie/pubs/advice/water/wastewater/EPA_water_%20treatment_manual_primary_secondary_tertiary1.pdf.

Farooq, R., Wanga, Y., Lina, F., Shaukatb, S.F., Donaldsonc, J., Chouhdary, A.J., 2002. Effect of ultrasound on the removal of copper from the model solutions for copper electrolysis process. Water Res. 36, 3165–3169.

Fayolle, Y., Cockx, A., Gillot, S., Roustan, M., Héduit, A., 2007. Oxygen transfer prediction in aeration tanks using CFD. Chem. Eng. Sci. 62, 7163–7171.

Ferhan, C., 2011. Activated Carbon for Water and Wastewater Treatment: Integration of Adsorption and Biological Treatment, first ed. Wiley, viewed 28 May 2013, http://onlinelibrary.wiley.com/book/10.1002/9783527639441.

FICCI Water Mission, 2011. Water use in Indian industry survey, New Delhi, viewed 29 May 2013, http://www.ficci.com/Sedocument/20188/Water-Use-Indian-Industry-Survey_results.pdf.

Frost & Sullivan 2011, www.frost.com.

Grail Research, Water—The India Story, 2009, viewed 09 May 2013, http://www.grailresearch.com/pdf/ContenPodsPdf/Water-The_India_Story.pdf.

Gurel, L., Altas, L., Buyukgung, H., 2005. Removal of lead from wastewater using emulsion liquid membrane technique. Environ. Eng. Sci. 22, 411–420.

Hussain, I.S., 2012. Case Study of a zero liquid discharge facility in textile dyeing effluents at Tirupur. In: One day National Workshop at CETP-Hyderabad, viewed 28 May 2013, http://www.igep.in/live/hrdpmp/hrdpmaster/igep/content/e48745/e49028/e51431/e51468/SajidHussain.pdf.

Imran, A., 2012. New generation adsorbents for water treatment. Chem. Rev. 112, 5073–5091.

Kamyotra, J.S., Bhardwaj, R.M., 2011. India Infrastructure Report 2011, Municipal Wastewater Management in India, viewed 9 May, 2013, http://www.idfc.com/pdf/report/2011/Chp-20-Municipal-Wastewater-Management-In-India.pdf.

Kaur, R., Wani, S.P., Singh, A.K., Lal, K., 2012. Wastewater production, treatment and use in India, viewed 28 May 2013, www.ais.unwater.org/ais/pluginfile.php/./CountryReport_India.pdf.

Lee, S.M., Tiwari, D., Choi, K.M., Yang, J.K., Chang, Y.Y., Lee, H.D., 2009. Removal of Mn(II) from aqueous solutions using manganese-coated sand samples. J. Chem. Eng. Data 54, 1823–1828.

Lewis, A.E., 2010. Review of metal sulphide precipitation. Hydrometallurgy 104, 222–234.

Maliyekkal, S.M., Lisha, K.P., Pradeep, T.A., 2010. Novel cellulose-manganese oxide hybrid material by in situ soft chemical synthesis and its application for the removal of Pb(II) from water. J. Hazard. Mater. 181, 986.

Mane, J.D., Kumbhar, D.L., Barge, S.C., Phadnis, S.P., Bhandari, V.M., 2003. Studies in effective utilization of biomass. Chemical modification of bagasse and mechanism of color removal. Int. Sugar J. 105 (1257), 412.

Mane, J.D., Modi, S., Nagawade, S., Phadnis, S.P., Bhandari, V.M., 2006. Treatment of spentwash using chemically modified bagasse and color removal studies. Bioresour. Technol. 97 (14), 1752–1755.

Maturana, A.H., Peric, I.M., Rivas, B.L., Pooley, S.A., 2011. Interaction of heavy metal ions with an ion exchange resin obtained from a natural polyelectrolyte. Polym. Bull. 67, 669–676.

Metcalf, Eddy, 1991. Wastewater Engineering: Treatment, Disposal and Reuse, third ed. McGraw-Hill Co, New York.

Meunier, N., Drogui, P., Montane, C., Hausler, R., Mercier, G., Blais, J.F., 2006. Comparison between electrocoagulation and chemical precipitation for metals removal from acidic soil leachate. J. Hazard Mater. B137, 581–590.

Monitoring and Evaluation Division Department of Chemicals & Petrochemicals Ministry of Chemicals & Fertilizers Government of India New Delhi, Performance of Chemical and Petrochemical Industry at a glance-2001–2007, viewed 23 May, 2013, http://chemicals.nic.in/stat0107.pdf.

Mungray, A.A., Kulkarni, S.V., Mungray, A.K., 2012. Removal of heavy metals from wastewater using micellar enhanced ultrafiltration technique: a review, Central European. J. Chem. 10, 27–46.

Murty, M.N., Kumar, S., 2011. Water Pollution in India – An economic appraisal, viewed 09 May, 2013, http://www.idfc.com/pdf/report/2011/Chp-19-Water-Pollution-in-India-An-Economic-Appraisal.pdf.

Ouaissa, Y.A., Chabani, M., Amrane, A., Bensmaili, A., 2013. Removal of Cr (VI) from model solutions by a combined electrocoagulation sorption process. Chem. Eng. Technol. 36, 147–155.

Pant, D., Adholeya, A., 2007. Biological approaches for treatment of distillery wastewater: a review. Bioresour. Technol. 98, 2321–2334.

Pophali, G.R., 2007. Operation Dynamics and Process Control in Secondary Clarifiers; A Novel Design. Ph.D. Thesis, Nagpur University.

Pophali, G.R., Kaul, S.N., Nandy, T., Devotta, S., 2009. Australian Patent No. 2007330354. Circular Secondary Clarifier for Wastewater Treatment and an Improved Solids-Liquid Separation Process Thereof.

R.O.C. (Taiwan) Environmental Law Library, Effluent Standards, viewed 26 April, 2013, http://law.epa.gov.tw/en/laws/480770486.html.

Ranade, V.V., 2002. Computational Flow Modeling for Chemical Reactor Engineering. Academic Press, San Diego, CA, USA.

Ranade, V.V., Pandit, A.B., Anil, A.C., Sawant, S.S., Ilangovan, D., Madhan, R., Pilarisetty V.K., 2006. Apparatus for filtration and disinfection of seawater/ship's ballast water and a method of same. U.S. Pat. Appl. Publ. US 20080017591 A1 24 Jan 2008, 19 pp. (English). (United States of America.) CODEN: USXXCO. INCL: 210767000; 210416100. APPLICATION: US 2007-726399 20 Mar 2007. PRIORITY: IN 2006-DE734 20 Mar 2006.

Ranade, V.V., Kulkarni, A.A., Bhandari, V.M., 2013. Vortex diodes as effluent treatment devices. PCT Int. Appl. WO 2013054362 A2 20130418.

Regel-Rosocka, M., Cieszynska, K., Wisniewski, M., 2006. Extraction of zinc (II) with selected phosphonium ionic liquids. Przem. Chem. 85, 651–654.

Rice, E.W., Baird, R.B., Eaton, A.D., Clesceri, L.S., 2012. Standard Methods for Examination of Water and Wastewater, 22nd ed. APHA®, AWWA® & WEF®, Washington DC.

Seif, H., Malak, M., 2001. Textile wastewater treatment. In: Sixth International Water Technology Conference (IWTC), Egypt, pp. 608–614.

Shen, Y.F., Tang, J., Nie, Z.H., Wang, Y.D., Renc, Y., Zuo, L., 2009. Preparation and application of magnetic Fe_3O_4 nanoparticles for wastewater purification. Sep. Purif. Technol. 68, 312–319.

Sirajuddin, A., Rathi, R.K., Chandra, U., 2010. Wastewater treatment technologies commonly practiced in major steel industries in India, viewed 28 May 2013, http://www.kadinst.hku.hk/sdconf10/Papers_PDF/p537.pdf.

Sumesh, E., Bootharaju, M.S., Anshup, Pradeep, T., 2011. A practical silver nanoparticle-based adsorbent for the removal of Hg^{2+} from water. J. Hazard. Mater. 189 (1–2), 450–457.

Sun, L., Miznikov, E., Wang, L., Adin, A., 2009. Nickel removal from wastewater by electroflocculation-filtration hybridization. Desalination 249, 832–836.

UN Water 2008, Tackling a global crisis: International Year of Sanitation 2008, viewed February 2010, http://www.wsscc.org/fileadmin/files/pdf/publication/IYS_2008__ tackling_a_global_crisis.pdf.

UNEP 2007, Global Environment Outlook Geo4 environment for development, viewed 28 May 2013, http://www.unep.org/geo/geo4/media/.

UNEP, 2008. Desalination Resource and Guidance Manual for Environmental Impact Assessments. United Nations Environment Programme, Regional Office for West Asia, Manama, and World Health Organization, Regional Office for the Eastern Mediterranean, Cairo http://www.unep.org.bh/Newsroom/pdf/EIA-guidance-final.pdf.

United Nations Industrial Development Organization, 2011. Introduction to treatment of tannery effluents—What every tanner should know about effluent treatment, Vienna, 2011 viewed 28 May 2013, http://www.unido.org/fileadmin/user_media/ Publications/Pub_free/Introduction_to_treatment_of_tannery_effluents.pdf.

Waidhas, M., (Viewed on 2 January, 2014). http://www.siemens.com/innovation/en/ publikationen/publications_pof/pof_fall_2008/rohstoffe/abwasser.htm.

Wang, Z., Xue, M., Huang, K., Liu, Z., 2011. Textile dyeing wastewater treatment. In: Hauser, P.J. (Ed.), Advances in Treating Textile Effluent. InTech, viewed on 28 May 2013, http://cdn.intechopen.com/pdfs/22395/InTech,Textile_dyeing_wastewa-ter_treatment.pdf.

Water Scarcity http://www.un.org/waterforlifedecade/scarcity.shtml.

Watkins, K., 2006. UNDP Human Development Report, 2006, Beyond scarcity: Power, poverty and the global water crisis, 440 pp, viewed 30 May 2013, http://hdr.undp. org/en/media/HDR06-complete.pdf.

World Water Assessment Programme, 2009. The United Nations World Water Develop-ment Report 3: Water in a Changing World. UNESCO/Earthscan, Paris/London, viewed 10 May, 2013, http://unesdoc.unesco.org/images/0018/001819/181993e.pdf.

CHAPTER 2

Advanced Physico-chemical Methods of Treatment for Industrial Wastewaters

Vinay M. Bhandari, Vivek V. Ranade
Chemical Engineering and Process Development Division, CSIR-National Chemical Laboratory, Pune, India

2.1 INTRODUCTION

Industrial wastewater treatment is a vast area encompassing a number of physical, physico-chemical, and biological methods of treatment. The selection and practice of any particular method depends largely on a variety of factors, the most important being quality of wastewater in terms of number and nature of pollutants and desired level of reduction of these pollutants. In this context, it is usually presumed that the only goal of industrial wastewater treatment is removal of pollutants from the effluent to such levels that it meets the prescribed norms of discharge into surface waters. However, where the nature of the pollutant is such that the recovery becomes attractive and contributes to value addition to the existing process, the objectives need to be modified to removal and recovery of the pollutants to the desired levels. Another goal that is becoming increasingly important and relevant is not just removal/recovery of pollutants but that of water as well. Water recycling and reuse should also be considered in addition to elimination of pollution. Some of these aspects have already been discussed in Chapter 1.

Wastewater processing often requires a combination of different methods because a single method is seldom satisfactory. The overall treatment methodology generally consists of primary methods, secondary methods, and tertiary methods, each having a specific task to perform on the effluent. Primary methods usually modify effluent quality to suit a specific requirement of further processing and include pH modification, clarification, removal of suspended particles, and to a certain extent, removal of part of the Chemical Oxygen Demand (COD) from the effluent. The secondary treatment includes both physico-chemical and biological methods, either alone or in combination. For example, coagulation, oxidation, membrane separations, cavitation, and/or biological methods mainly perform the major task of removing pollutants from the effluent. These processes are sometimes able to accomplish the desired reduction in the pollutant levels. Occasionally, adsorption, ion exchange, or other processes also form part of the secondary treatment, depending on the nature and concentration of the pollutants. The tertiary treatment mainly involves polishing to remove the final traces of the pollutants; this typically uses only physico–chemical methods such as adsorption, ion exchange, and membranes as processing options. The tertiary treatment methods are important from the point of view of water recycling and reuse because they can achieve the final reusable water quality.

2.1.1 Selection of Method

If the problem of wastewater treatment is well defined, the selection of physico-chemical or biological methods can be straightforward. For example, the nature of the pollutants, their concentrations, and the ratio of biological oxygen demand (BOD) to COD primarily dictate the selection strategy. However, many times, the strategy can involve changes in the manufacturing process itself, consequently redefining the wastewater treatment problem because the nature and concentration of the pollutants are drastically altered in such cases because of changes in solvent, catalyst, or raw material. It is therefore instructive to evaluate the overall processing of the chemical plant (shown schematically in Figure 2.1) in the context of wastewater treatment.

As seen in Figure 2.1, the central theme of any chemical plant is a reaction process that has inputs in the form of reactants, solvent/catalyst (if any), and utilities such as cooling water, steam, and inert gases. The central part is surrounded by separation from all the sides. The reaction process may require separation in the form of removal of impurities from the raw materials (pretreatment). After the reaction is over, the product mass again requires separation to get desired product with required specification apart from separation required for treatment of gaseous/liquid effluent (post-treatment). The reaction component of the process therefore deals with selection of suitable raw materials, reaction process, and reactors; the final objective is to have a most efficient and environmentally friendly reaction process. Similarly, the separation component of the process involves selection of suitable separation methods and equipment with the final objective of achieving the most efficient and environmentally friendly process. The separation and reaction parts here are not independent; any change in the reaction zone automatically alters the separation zone and vice versa.

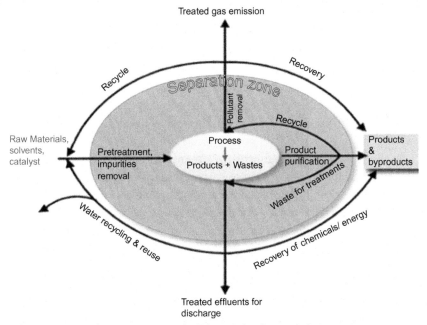

Figure 2.1 Schematic representation of chemical industry and processing.

This interrelationship is extremely important in defining the industrial wastewater treatment problem. Overall, the problem in definition originates at the reaction step and is most critical: a core issue. Often, there are changes in the reaction zone from time to time (especially in manufacturing of fine and specialty chemicals), subsequently requiring modifications in the separation zone. As a consequence, the industrial wastewater treatment methodology inherently requires flexibility to absorb these changes in the processing without affecting the efficiency and final goals of pollution control. Thus, the following points highlight the need for and the extent of environmental pollution control for industrial wastewater treatment:

1. Replace/substitute for hazardous and toxic chemicals in reaction, if possible.
2. Separation often requires the most effort and cost.
3. Effluent treatment can get complicated because of the presence of toxic pollutants and/or number of pollutants such as metals and organics existing at the same time, requiring multiple solutions, sequential solutions, or parallel solutions.
4. Failure to control pollution can even threaten the existence of the industry at times.

2.1.2 Devising a Solution for Industrial Wastewater Treatment

A systematic approach to wastewater treatment with a specific emphasis on water recycling and reuse along with the recovery of valuable chemicals is very important, especially for fine and specialty chemical and water-intensive industries. This involves the basic characterization of the wastewater and understanding volumes and variations in the pollutant levels or concentrations. A detailed approach in this regard has already been discussed in Chapter 1.

A schematic flowchart of the main elements in the decision process for the selection of an appropriate technology is presented in Figure 2.2. It is important to define the problem and goals in light of the available data on the process and effluent. The next step is to list all the methods that are technically feasible to treat the effluent. If recovery is important, the specific methods for recovery must be identified and prioritized. Once this selection is done, the next step is to identify suitable methods, materials, and processes, followed by distributing the pollutant removal load among primary, secondary, and tertiary methods of treatment. It is essential that the interrelationships between these methods be understood in order to maximize the efficiency of the overall effluent treatment process. With this evaluation, the technologies for primary, secondary, and tertiary (if required) treatment are identified; then a techno-economic feasibility analysis needs to be carried out. At this stage, process integration options should also be considered. Because these alter the entire treatment strategy, the process of selecting appropriate methods/materials again has to be restarted and needs iterations until the most effective and techno-economically feasible solution is reached.

The physico-chemical methods of treatment that can be considered are the following:
1. Coagulation
2. Adsorption
3. Ion exchange
4. Extraction
5. Membrane separations
6. Oxidation
7. Cavitation.

Usually a combination of one or more processes is required to attain the goals of the effluent treatment process. Further, many of the physico-chemical methods can be aptly integrated with biological methods of treatment; this aspect also requires careful consideration.

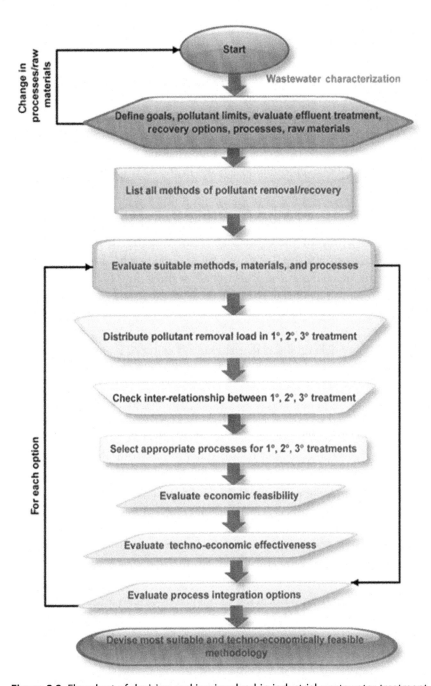

Figure 2.2 Flowchart of decision making involved in industrial wastewater treatment.

In general, the majority of the processes that have specific interactions playing a role in the removal of pollutants (such as coagulation, adsorption, ion exchange, electro-oxidation, membrane separations with ion exchange membranes) belong to a class of charge-based separations that manipulate electrostatic interactions with the pollutant species to effect removal. A recent survey of patents has placed emphasis on the development of sorption and ion exchange processes (Berrin, 2008) as well as water management methods that can make use of the above information. The physico-chemical methods listed above are briefly discussed in the next section.

2.2 ADVANCED COAGULATION PROCESSES

Coagulation is one of the most widely used treatment methods and is considered a mature technology in today's context. It has numerous applications in the area of wastewater treatment for the removal of pollutants. It is typically employed as a primary or secondary treatment step, depending on the effluent's nature, either in the form of chemically induced coagulation (conventional) or electric charge induced coagulation (electro-coagulation). Coagulation, by default, refers to a chemically induced phenomenon and is most commonly practiced in industrial wastewater treatment in the chemical, pulp and paper, dye and textile, petrochemical, pharmaceutical, food, mineral processing, and metal industries. In recent years, electro-coagulation has also been used to treat wastewater in industries such as food, pulp and paper, and dye, as well as wastewaters containing metals and oil. In commercial practice, coagulation and flocculation are two common terminologies that are, many times, interchanged and used to suggest a single process. However, it should be understood that the principal process is coagulation, and flocculation is an inherent part of the coagulation process as a whole.

Coagulation refers to a separation process based on charge modification where part or full reduction of repulsive forces on colloids takes place through the addition of specific chemicals referred to as coagulants. It can also be termed *destabilization of colloids* that otherwise are in stable suspension form. As opposed to this, flocculation refers to a process, purely physical in nature, where partly or fully neutralized particles come in contact with each other to form flocs, and agglomeration of the flocs takes place. The agglomerates are heavier than water and therefore settle and separate from water easily. Thus, coagulation as a whole represents a physico-chemical process where removal of colloidal charged matter takes place through charge modification/charge neutralization. The chemicals that promote the process of coagulation are referred to as coagulants, while those specifically promoting the flocculation process are known as flocculants. Flocculants increase the efficiency of the separation process through improved floc quality, which increases the settling of flocs and thereby better removes the pollutants.

A number of pollutants that are present in industrial wastewaters are colloidal in form and are charged species. They are:

- Organic materials
- Metals and metal oxides
- Insoluble toxic compounds
- Stable emulsions (emulsified oil)
- Material producing color, turbidity
- Other substances such as drug residues, bacteria and viruses, algae.

Some of the pollutants, although not in colloidal form, can also behave as colloids. A significant portion of the pollutants in wastewaters, therefore, constitute colloidal matter (up to 70% or sometimes even more). The colloids in wastewater are stable because of the electrical charge that they carry that prevents them from settling or getting separated from the wastewaters. The charges can be positive or negative. As far as wastewater treatment is concerned, most colloidal particles in wastewater are negatively charged, implying the necessity of cationic coagulants for neutralization. The charge typically comes from lattice imperfections, ionizable groups that become ionic species in water (e.g., amino or hydroxyl groups), or from ionic species that can get adsorbed on the surface of the colloid. The size of the colloidal particles is approximately $0.001-1$ µm, representing the range between a molecule and a bacterium. Because, for colloids, surface properties dominate and prevent particles from coming together, the primary objective of the coagulation process is to alter the surface properties through modification of charges in such a way that they agglomerate to a larger size and are separated from solution. This eventually effects reduction in color, turbidity, and COD through removal of pollutants. An example of this can be seen from reported results on polymer waste from latex manufacture that was coagulated with 500 mg/L ferric chloride and 200 mg/L lime at pH 9.6 resulting in COD and BOD reductions of 75% and 94%, respectively, from initial values of 1000 and 120 mg/L (Munter, 2000).

The surface charge and charge density are critical parameters in the removal of pollutants in colloidal form. This charge neutralization in coagulation is mathematically quantified using the value of the zeta potential parameter, which can be used for predicting stability and is pH dependent. However, in wastewater treatment, it is to be noted that it is practically impossible to know the precise value of the zeta potential to get efficient coagulation, although lower zeta potential values indicate effective coagulation. As a result, carrying out coagulation in real-life operations is, by and large, an empirical approach on the basis of laboratory data on various coagulants and coagulation behavior under different conditions of pH, concentration, and other parameters.

2.2.1 Types of Coagulant

There are two distinct types of coagulant: inorganic and organic (Table 2.1). The most common inorganic coagulants used in wastewater treatment are simple coagulants such as alum or ferric chloride or more complex ones such as polyaluminum chloride (PAC), while among organic coagulants the most commonly used are polymers such as poly-DADMAC (poly diallyldimethylammonium chloride) or polyacrylamide.

Organic coagulants or polyelectrolytes (also often referred to as "polymers" in practice, although even in inorganic coagulants there are polymers such as PAC, polyaluminum sulfate (PAS) with varying molecular weight) are synthetic or organic polymeric molecules having ionizable groups or charged groups along a polymer chain. Polyelectrolytes of natural origin are typically derived from starch products or of biological origin such as extracts of plants/fruits or alginate from algae, while synthetic organic coagulants are synthetically polymerized from monomers (e.g., polyamine, sulfonate).

2.2.2 How Coagulants Work and How to Select Coagulant

Coagulants function broadly through dissociation, neutralization, and an association mechanism. For example, with an inorganic coagulant, when added to a solution the salts dissociate into ions. The cationic species (such as Al^{3+}, Fe^{3+}), resulting from dissociation exist in their hydrated forms (such as $[Al(H_2O)_6]^{3+}$ or $[Fe(H_2O)_6]^{3+}$). A critical quantity of coagulant is needed for the ions from these salts to react with the OH^- or bicarbonate and carbonate ions in solution to produce the corresponding insoluble hydroxides $(Al(OH)_3$ or $Fe(OH)_3)$, which then precipitate. The solubility of $Al(OH)_3$ or $Fe(OH)_3$ is a function of the pH. The cationic species neutralize negative charges on the pollutant species, and the process requires a just-sufficient dose corresponding to quantity of colloids and surface charge, especially in the adsorption/charge neutralization mechanism. Any dose below or above this optimum concentration will again stabilize the solution adversely, affecting removal of pollutants from wastewaters. There have been various mechanisms proposed for the coagulation process that mainly incorporate

Table 2.1 Examples of Coagulants

Type	Functionality/ Characteristics	Example
Inorganic		
Aluminum salts (alum)	Al^{3+}	$Al_2(SO_4)_3 \cdot 14H_2O$ or $Al_2(SO_4)_3 \cdot 18H_2O$
Ferric and ferrous salts	Fe^{2+}/Fe^{3+}	$FeCl_3$, $Fe_2(SO_4)_3$, $FeSO_4 \cdot 7H_2O$
Lime	Ca^{2+}	$Ca(OH)_2$
Polymeric	Al/OH ratio; MW Al^{3+}; MW Al/OH ratio; MW Al/Si ratio	Polyaluminum chloride Polyaluminum sulfate Polyaluminium silicate chloride
Organic (Polyelectrolytes)		
Cationic	Amine	Polyethylene amine hydrochloride Functionalized polyacrylamide
	Quaternary	Poly diallyldimethylammonium chloride
Non-ionic	Polyalcohol	Polyvinylalcohol
	Amide	Polyacrylamide
Anionic	Carboxylic	Polymethacrylic acid
	Sulfonic	Polyvinylsulfonate
Natural Polymers		
Starch	Carbonyl groups/polysaccharide skeleton, cationic/ anionic	
Guar gum	High mol. wt., non-ionic	
Chitin derivatives/ Chitosan	Cationic polyelectrolyte	
Hybrid Coagulants		
Inorganic-Inorganic	Mol. wt., mixed functionality, functionality ratio, etc.	
Inorganic-Organic	Mol. wt., mixed functionality, functionality ratio, etc.	
Inorganic-Natural polymers	Mol. wt., mixed functionality, functionality ratio, etc.	
Organic-Organic	Mol. wt., mixed functionality, functionality ratio, etc.	
Organic-Natural polymers	Mol. wt., mixed functionality, functionality ratio, etc.	

double-layer suppression, adsorption and charge neutralization, adsorption and intra-particle bridging, and enmeshment or sweep flocculation (Tripathy and De, 2006). However, in real practice any one mechanism may not be construed as the only/single mechanism of coagulation. Usually in electrical double-layer suppression, there is little or no dependence on coagulant dosage after the minimum/optimum dosage. In adsorption/ charge neutralization, it is found that excessive dosage results in charge reversal and a negative effect on process performance. For organic coagulants, the coagulation mainly takes place due to the adsorption/charge neutralization mechanism through interactions of charged/ionogenic sites of polymers that have capacity to adsorb or cover colloidal pollutant species. For effective coagulation, coagulant dosage, pH of solution, and mixing are critical process parameters.

Since *a priori* selection of coagulant for any particular wastewater is not possible on the basis of theoretical analysis, it is imperative that selection be made using an empirical approach through a set of experiments performed on various coagulants and under different processing conditions. This is conventionally carried out using "jar tests" (Figure 2.3). The jar test simulates the coagulation/flocculation process in a batch mode, and a series of batch tests are carried out using different process parameters such as pH, coagulant type, and dosage (with/without coagulant aid/flocculant) to get the most suitable performance in terms of coagulant dosage, pH, best settling, sludge properties (minimum sludge volume), and so on. Finally, an economic analysis needs to be performed to select suitable coagulant and optimum processing parameters. Figure 2.3 demonstrates performance of coagulation (visually) in dye wastewater treatment for different coagulant doses showing clearly stepwise reduction in color and sludge.

Figure 2.3 Jar test apparatus for coagulation experiments.

2.2.3 Advances in Coagulation Process and Practice

The use of a single coagulant, usually inorganic, is seldom satisfactory to get the desired coagulation impact in wastewater treatment. However, since the performance of coagulation process directly affects the overall performance of the other effluent treatment steps, improving coagulation is a crucial factor from the point of view of overall effluent treatment strategy. The use of inorganic coagulants and the practice of coagulation is quite straightforward in drinking and municipal sewage water treatment. However, this is not the case with most industrial wastewaters where the number and nature of pollutants is varied and complex; usually, advanced methodologies are required for treating these high-strength wastewaters. The advances in these mainly incorporate newer coagulants with specific properties (such as inorganic and organic polymers, with specific molecular weight or Al/OH ratio), coagulant formulations (intelligent mix of inorganic and organic coagulants to get benefits of both), coagulant aid, flocculants and flocculation, mixing, and newer equipment designs and solid-liquid separation. Some other modifications, such as electro-coagulation, are also being increasingly considered for wastewater treatment.

Thus, in the context of the conventional coagulants listed above, many newer coagulant types and hybrid forms are being continuously researched and enter the market (Tzoupanos and Zouboulis, 2008; Lee et al., 2012). A review of various hybrid materials comprising inorganic-inorganic, inorganic-organic, organic-organic, and inorganic-natural polymers has been presented recently by Lee et al. (2012). This has been very instructive for researchers working in this area. Some natural coagulants, such as seed extracts, are also being investigated, although far from being considered for commercial application at this point of time (Bhuptawat et al., 2007). Polymer form is increasingly considered to be key factor in improving coagulation because the large molecular weight and controlled basicity/charged groups/charge density help to achieve efficient coagulation along with better flocculation. The most prominent advances among inorganic coagulants include polymers such as PAC, PAS, and polyaluminumchloro-sulfates, with substantial variation in their properties due to molecular weight and basicity. Typically, for PAC, three commercial variations are available: low basicity, medium basicity, and high basicity based on Al/OH ratio (Gao et al., 2005). Recently, Yuan et al. (2006) compared performance of PAC with polyferric chloride and poly-silicate-aluminum-ferric chloride (PSAFC) in the treatment of dye wastewater. Although generalization on the best coagulant in this regard could not be concluded in the study, it was suggested that

PSAFC would improve performance over PAC under certain conditions. All these variations have profound impact on the performance behavior in the coagulation process and manufacturers usually specify which product suits which wastewater treatment application. These improvisations directly impact the process efficiency, and with improved performance, reduced dosage, and wider pH applicability along with better economics, these products are finding increased acceptability over conventional coagulants.

Further advancement in the coagulation process can be achieved through the use of additives that are responsible, by and large, for improving flocculation and thereby easy, speedy separation with compact sludge volume. These additives can be inorganic or organic. In fact, most organic polymers perform the role of flocculant more than as coagulant.

Apart from the development of newer coagulants, a more common approach in recent years is to integrate advantages of inorganic and organic coagulants in the form of coagulant formulation (through physical blending or chemical/functional modifications). This substantially improves cost effectiveness apart from process performance. In general, inorganic coagulants produce smaller and lighter flocs that require more time to settle. Sludge volume is always greater with inorganic coagulants. Most inorganic coagulants are pH sensitive and therefore work only in a narrow pH range. Some of the disadvantages of common inorganic coagulants can be eliminated with the use of organic coagulants or formulations of both inorganic and organic coagulants. Any two or more coagulants can be combined to enhance the effect of coagulation. It makes sense to combine inorganic and organic coagulants to exploit the advantages of both types.

Although many new coagulants and formulations have been recently developed for a variety of industrial wastewater treatment applications, there is still a huge potential for further modifying the performance of existing materials/methods and developing newer hybrid materials in the form of coagulant formulations.

2.2.3.1 Electro-coagulation and Cavigulation

There are some new developments in the area of conventional chemical coagulation processes. Electro-coagulation, although conceptually not new, gained attention in recent years due to lower costs (mainly through reduced power consumption), better efficiencies, and compact reactor configurations. It belongs to a class of electrochemical processes wherein the electrical source is used to generate coagulating species. The principle is similar to chemical coagulation, except that in place of chemicals, here the aluminum or iron anode in the electrochemical reactor releases requisite Al^{3+} or Fe^{2+} cations in the solution upon application of an electric field. At the cathode, water cleavage takes place, generating hydroxyl ions that help in precipitation through formation of hydroxides in solution. The process is more complex when compared to conventional chemical process and is still not widely accepted commercially. Newer developments in this area include better electrodes, which dictate the coagulation process; improvements in the kinetics; and newer electrochemical reactor configurations. There have been some studies proving the efficiency of electro-coagulation for heavy metal removal from wastewater and process integration using adsorption and electro-coagulation for the removal of Cr(VI) of the order of 97% in the pH range of 3–6 (Ait Ouaissa et al., 2013).

Cavigulation is a process that combines two physico-chemical operations, namely coagulation and cavitation. The first principles of cavigulation are still not clearly understood. At present, the approach to cavigulation appears to be rather empirical, where coagulation and cavitation in combination are believed to offer better conditions for degradation of pollutant species through physical and chemical destruction.

2.2.4 Case Study: Dye Wastewater Treatment

The use of coagulants in dye wastewater treatment is a complex problem, and no general solutions are yet available. Textile and dyeing industries use many kinds of dyes, both reactive and non-reactive, and discharge large amounts of highly colored wastewater. Such effluent must be treated prior to discharge in order to comply with the environmental protection laws for the receiving waters. The biological treatment processes that are frequently used to treat textile effluents are generally efficient for meeting limits of BOD and suspended solids removal. However, these methods are largely ineffective for removing color from the wastewaters. Process design aspects of dye wastewater treatment involving various physico-chemical methods have been discussed in recent literature (Joo et al., 2007; Fung et al., 2012).

Application of the coagulation process can be accomplished in such cases through the use of coagulant formulations that yield superior performance. Coagulation has been effective in removing color especially from wastewaters containing dissolved solids and charged matters. However, high chemical dosages are usually required, and large volumes of sludge must be disposed of, in general, for inorganic coagulants, resulting in a high cost for sludge disposal. A near-zero production of sludge in the case of organic coagulants significantly reduces treatment costs with significant enhancement in coagulation efficiency. Thus, formulation of inorganic and organic coagulants can provide a better techno-economically feasible operation in wastewater treatment. This is especially true for reactive dyes, which pose the greatest color removal problems (even a concentration as low as 0.005 ppm is visibly detectable). Further, reactive dyes are not easily biodegradable; even after extensive wastewater treatment, color is likely to remain in the effluent. In the absence of any effective conventional treatment methodology for reactive dyes, it is imperative to develop an effective formulation in the form of combined coagulants. The objective in developing such formulation would be to have an efficient coagulation process operating over a wide pH range with better settling properties and minimum sludge volume.

The efficacy of development of coagulant formulations in altering the coagulation process performance is shown in Figures 2.4, 2.5, and 2.6 using data from our own studies (Ashtekar, 2007; Ashtekar et al., 2010). The coagulant formulations here were made by physical blending at ambient temperature, one of the most popular techniques that does not involve any new chemical bond formation. Figure 2.4 shows coagulation efficiency using an inorganic coagulant, PAC, in the removal of different dyes as a function of coagulant concentration. As stated earlier, in an electrical double-layer suppression-type mechanism common to inorganic coagulants, there is little or no dependence on coagulant dosage after minimum/optimum dosage of PAC, which is 100 ppm for most dyes reported. This is, however, not always the case, and for all the dyes and contributions of other mechanism types can be seen at substantially high concentrations in the form of reduced efficiency for some dyes. In comparison, it is to be noted that ferrous sulfate was largely

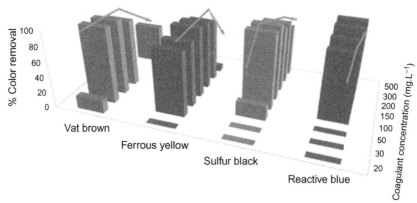

Figure 2.4 Dye removal behavior using inorganic coagulant poly-aluminum chloride.

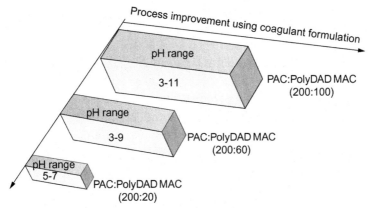

Figure 2.5 Process improvement using binary coagulant formulations.

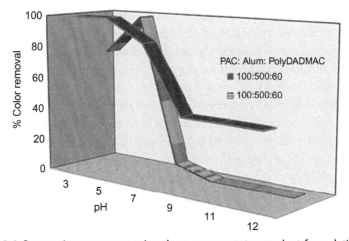

Figure 2.6 Process improvement using three-component coagulant formulations.

ineffective, while for alum, the required dosage was >500 ppm. The behavior here highlights three key issues:

1. For dye removal, the use of inorganic coagulants is quite effective, and close to 100% removal is possible.
2. Inorganic polymer PAC performs much better as compared to conventional inorganic coagulants such as ferrous sulfate or alum.
3. Even with PAC, the minimum required dose varies with nature of dye.

The use of the coagulant PAC alone has many limitations, such as applicability in a very limited range of pH with best results at pH ~5. The issues pertaining to such limitations can be best resolved using suitable coagulant formulation strategy. Figure 2.5 highlights changes and range modifications that can occur using a formulation of inorganic and organic coagulants viz. PAC and poly DADMAC for reactive blue dye removal with high efficiency. Analysis of data using a PAC concentration of 200 ppm and varying the concentration of poly DADMAC from 20 to 100 ppm shows significant modification in the operating pH range. This may be attributed to the contribution of an organic coagulant in the formulation. As mentioned earlier, the combined formulation works better with wider pH range than that with individual coagulants. This aspect is further supported by the fact that when PAC or poly DADMAC alone is not effective, the formulation of PAC and poly DADMAC works effectively.

The inorganic + organic coagulant formulation strategy can be further developed in the form of two or more inorganic and organic coagulants. This is demonstrated in Figure 2.6. A formulation of PAC, alum, and poly DADMAC (100 ppm inorganic and 60 ppm of poly DADMAC) only slightly modifies the color removal behavior. However, a little higher concentration of alum (500 ppm) changes the color removal behavior drastically, and the new formulation works over a wider pH range of 3–12 with very high removal efficiency in pH 3–7.

Inorganic-organic coagulant formulations are therefore considered to be ideal advanced materials in coagulation processes that allow expanding pH range coupled with lowered dosage to bring out overall improvement in process efficiency at reduced cost as compared to individual coagulants (Lee et al., 2012).

2.3 ADVANCED ADSORPTION AND ION EXCHANGE PROCESSES

Adsorption can be defined in simple terms as selective concentration of one or more components (adsorbate(s)) of either a gas or a liquid mixture on the surface of a material (adsorbent). By definition, adsorption is clearly a surface phenomenon incorporating wide variation in the surface interactions that range from very-low-strength van der Waals forces to strong bond formation. The number and density of adsorption sites, the strength of adsorption, the nature of the adsorbing species, and the nature of the adsorbent material are important parameters that dictate not just the overall adsorption process but also the regeneration of the adsorbent by desorption of the adsorbed species. In industrial wastewater treatment applications, adsorption is commonly performed as a column chromatography operation with adsorption and regeneration in a cyclic manner. Adsorption is considered today as one of the most important and crucial operations in secondary and tertiary wastewater treatment options, especially for the removal of refractory pollutants.

2.3.1 Adsorbent: Screening and Selection

The selection of a suitable adsorbent is the most important step in devising adsorption separation for effluent treatment, and it is largely dictated by the nature of pollutants such as acids, organics, and metals. A number of commercial adsorbents, both hydrophilic and hydrophobic types, are available. These include activated carbon (powder, granules, activated carbon fiber), zeolites (A, X, Y, ZSM-5, silicalite, ALPO), inorganic oxides (silica, alumina), and polymers. Adsorbent surface area typically ranges from 100 to 1500 m^2/g. In conventional separation applications, usually adsorbent with high surface is considered better. However, for effluent treatment applications, it is the affinity of the surface to the pollutant species that is most important; thus, the capacity to remove specific pollutants is the criterion for selection. Adsorbents come in different shapes, such as granules, spheres, cylindrical pellets, flakes, and/or powders, with size variations from 50 μm to 1.2 cm, porosity 30–85%v, and average pore diameters in the range 10–200 Å. For selection of an adsorbent, adsorption capacity (e.g., gram of adsorbate adsorbed per gram of adsorbent at equilibrium), and adsorbent life (number of adsorption–desorption cycles for which the adsorbent can be used without significant loss in its capacity) are important practical considerations. The choice of a suitable adsorbent is highly case specific, and careful identification/definition of the problem is essential for devising the proper solution. Ideally, the adsorbent is required to have the following properties, which form the basis of selection:

1. High selectivity for pollutants to be removed.
2. High capacity so that quantity of adsorbent required is less.
3. Favorable kinetics for rapid sorption, requiring suitable structural characteristics.
4. Chemical and thermal stability, low solubility to minimize operational losses.
5. Hardness and mechanical strength to prevent crushing and erosion to lower physical losses.
6. Ease of handling.
7. High resistance to fouling for long life, especially important in effluent treatment.

8. Ease of regeneration, important for techno-economic feasibility.
9. Low cost is most crucial aspect in wastewater treatment, dictating constant search for newer materials.

Activated carbons are probably the most used adsorbent materials, derived from a range of raw materials, available in variety of forms ranging from powdered to nanofiber type, and sold as inexpensive materials to highly expensive adsorbents depending on the type. They are most widely used for adsorption of organic pollutants, metal removal, and for color removal applications. It is reported that the adsorption capacity of carbon per cycle usually ranges from 0.25 to 0.87 kg COD removed per kg of carbon applied (Guyer, 2011). Regeneration is usually carried out using thermal, steam, or solvent extraction, acid or base treatment, or chemical oxidation. For adsorptive wastewater treatment, thermal regeneration is the preferred route. Sometimes, reactivation using high temperatures is required, increasing cost of treatment (usually to reopen pores that are blocked due to adsorption of pollutant species). Activated carbons can be a very good adsorbent, especially for adsorption of pollutants from pesticides, herbicides, aromatics, chlorinated aromatics, phenolics, chlorinated solvents, high-molecular-weight organics, amines, esters, ethers, alcohols, surfactants, and soluble organic dyes, practically covering majority of industrial wastewater treatment requirements, although there is some limitation in removal for low molecular weight or high polarity compounds (Cecen, 2011). Thus, adsorption finds widespread use in the treatment of wastewaters of most industries, for example, petrochemicals, chemicals, pesticides, dyes and textiles, pharmaceuticals, food, and inorganic mineral processing. In view of the wide range of pollutants that can be removed by activated carbons, adsorption using these materials is very frequently used as a process in secondary treatment to remove pollutants that are toxic to microorganisms of biological treatment—a process integration strategy—or as an independent tertiary treatment process for the removal of pollutants to achieve desired pollution control limits. The other adsorbents, such as inorganic adsorbents and polymeric adsorbents, are also gaining importance. A recent review on inorganic adsorbents has suggested detailed studies to explore the feasibility of replacing activated carbon with low-cost inorganic adsorbents, such as modified clays and zeolites for water and wastewater treatment that can be used for removal of metals and trace organics (Jiang and Ashekuzzaman, 2012). Recent developments in the area of nanoadsorbents and applications in the removal of inorganic and organic pollutants have been discussed by Ali (2012), indicating advantages in terms of low dose and high rate for nanomaterials as adsorbents. However, both these recent reviews have highlighted further research in the area of inorganic adsorbents and nanomaterial adsorbents in view of limited information on applications in industry wastewater treatment.

2.3.2 Equilibria and Kinetics of Adsorption

The fundamental principles of adsorption are well known and have been well reported in standard textbooks (Ruthven, 1984). Therefore, only information pertaining to wastewater treatment is discussed here. The design of the adsorption process in wastewater treatment involves understanding of sorption equilibria and kinetics apart from other aspects for commercial viability. The equilibrium capacity gives the maximum theoretical capacity that can be obtained, while in real-world operations, more useful terminologies such as operating capacity and breakthrough capacity are commonly used. The mathematical models for equilibrium and kinetics are most useful for designing wastewater systems, and it is convenient to have the maximum possible information through appropriate models for accurate design, scale-up, and physical understanding of the phenomenon so that it can be extended to similar systems.

2.3.2.1 Adsorption Isotherm

Adsorption isotherm is in simple terms a mathematical expression of equilibrium adsorbate loading on adsorbent as a function of concentration at constant temperature. In wastewater treatment, a highly favorable isotherm is generally not preferred because it adversely affects regeneration. Isotherm considerations in single use adsorbent not requiring regeneration are however, totally different requiring irreversible isotherm.

The adsorption isotherms can be classified mainly from Type I to Type V, depending on the nature of sorption curve (IUPAC recommendation). Different equations are available to describe adsorption isotherms, and some common forms are listed in Table 2.2.

The Langmuir model is one of the best known, theoretically understood, and widely applied models. Developed by Langmuir in 1916 (Langmuir,

Table 2.2 Commonly used forms of adsorption isotherm

Langmuir $\frac{C_e}{q_e} = \frac{1}{a_L \cdot b_L} + \frac{C_e}{a_L}$	Freundlich $q_e = K\, C_e^{1/n}$
Dubinin-Radushkevich $\ln q_e = \ln q_D - B_D \left[RT \ln\left(1 + \frac{1}{C_e}\right) \right]^2$	Redlich-Peterson $\ln[K_R(C_e/q_e) - 1] = \ln a_R - \beta \ln C_e$

1916), originally, it was a theoretical equilibrium isotherm relating to the amount of gas adsorbed per unit mass of the adsorbent. It is based on a uniform, monolayer, and finite adsorption site. The Langmuir adsorption isotherm assumes that there is no interaction among molecules adsorbed on the neighboring adsorption sites. The Langmuir constant, a_L (mg g^{-1}) gives the theoretical adsorption capacity, while C_e and q_e have their usual physical meanings. b_L (L g^{-1}) is another Langmuir constant whose value can be determined from the plot of C_e vs. C_e/q_e.

The Freundlich equation (Freundlich, 1906) is an empirical equation based on adsorption on heterogeneous sites. It is also one of the most common two-parameter isotherms applied for wastewater treatment. It assumes that stronger binding sites are occupied first, and the strength of adsorption decreases with the degree of occupation. The constant, K (mg g^{-1})· (L mg^{-1})$^{1/n}$ in the Freundlich expression, is related to the capacity of the adsorbent for the adsorbate; $1/n$ is a function of the strength of the adsorption, and it indicates the affinity between the adsorbent and adsorbate. A value of $1/n$ below unity implies that the adsorption process is chemical; if the value is above unity, adsorption is more of physical process; the more heterogeneous the surface, the more the $1/n$ value approaches zero. The values of k for several priority pollutants (e.g., nitrobenzene, styrene, chlorobenzene, bromoform) are in the range 60–360 mg/g and the values of $1/n = 0.12$–0.98 (Munter, 2000).

The Dubunin-Radushkevich (Dubinin, 1960) is an empirical adsorption isotherm that is now mostly applied for metal ion adsorption from an aqueous system. It is used to predict whether the adsorption is physisorption or chemisorption from its mean free energy per molecule of the adsorbate, E (kJ mol^{-1}). The energy of adsorption is computed by the relationship

$$E = \frac{1}{\sqrt{2B_D}}$$

The D-R equation takes into account the temperature effect as the factor. In the isotherm, q_D (mol g^{-1}) is the theoretical saturation capacity, B_D (mol^2 J^{-2}) is a constant related to the mean free energy per mole of the adsorbate, and R is the molar gas constant (8.314 J K^{-1} mol^{-1}).

The Redlich-Peterson model is a three-parameter isotherm model (Redlich and Peterson, 1959) compromising the features of both Langmuir and Freundlich. It can be applied over a wide range of concentrations for several solid-liquid adsorption equilibrium data.

The sorption isotherms give important information about the process. Typically, the Langmuir isotherm belongs to a Type I isotherm that indicates a uniform homogeneous surface with monolayer adsorption without interaction between the species getting adsorbed. Type II and Type IV are typically multilayer adsorptions that are most common in effluent treatment, while Type III and Type V indicate strong attraction/bonding between the molecules.

2.3.2.2 Adsorption Kinetics

While equilibrium data is useful in knowing maximum capacity and thereby aiding preliminary selection of the material, for practical application and for final selection of the material, it is essential to know the rate of the adsorption, that is, change of concentration with respect to time. This requires accounting for various mass transfer steps and evaluating mass transfer coefficients for the same. For adsorption kinetics, understanding of the following three contributing steps is required.

A. Film transport to surface of particle
B. Pore diffusion
C. Adsorption on the surface.

For desorption, the above steps occur in reverse order. Usually, pore diffusion is a controlling step in most adsorption processes. For commercial adsorbents that have a large surface area, the maximum contribution is of micropores (<2 nm), consequently high capacity in micropores. However, this results in poor kinetics. The kinetics can be improved by mesoporosity (2–50 nm), where adsorbents typically have moderate capacity with specific kinetics (pore size specificity), or by macroporosity in materials (>50 nm) that have less capacity/surface area but better kinetics. A detailed discussion on surface removal of pollutant species in porous materials can be found in the section on the ion exchange process.

2.3.3 Recent Advances in Adsorption Processes

The most predominant and extensively researched adsorption process scheme includes material/surface modification.

A. New materials/material modifications
  - New adsorbents from biomass, polymers, nanomaterials
  - Immobilization of enzymes
  - Chemical Vapor Deposition (CVD)/nano-deposition
  - Ion exchanged/impregnated zeolites, carbons, and others

B. Surface modification
- Acids/base/specialty materials
  (Often undisclosed)
  The adsorption process and its application to wastewater treatment is also a relatively mature technology. At present, there is reasonably good understanding of the following aspects:
- Fundamentals of adsorption and adsorption processes (forces responsible for adsorption, factors favoring adsorption, adsorption equilibria and kinetics, temperature/pressure dependence, heat of adsorption, molecular simulation pressure drop, cost)
- Synthesis and characterization: materials and methods
- Regeneration: chemicals and cost
- Process scheme: fixed bed/fluidized bed operation
- Advantages and disadvantages of adsorption separation
- Selection of adsorbents (listing of guidelines)
- Economic evaluations and comparison with other competing processes.

Significant improvements can be obtained for most of the adsorption processes with proper scientific modifications, either through material modifications or through process modifications. Surface modification of inorganic/organic materials has a fundamental role in industrial and environmental processes. The surface chemical modification of carbon is of great interest in order to produce materials with specificity and essentially control the type of pollutants that are adsorbed. This modification has been mainly carried out by oxidative methods, producing a more hydrophilic structure with a large number of oxygen-containing groups (Rios et al., 2003). The surface modification of adsorbent materials is believed to improve quantity and quality of the adsorbing sites and also at times alter functionality of the surface or impart ion exchange properties. Zeolite is an excellent example where huge benefits can be obtained through material modifications. A process like pressure-swing adsorption is an excellent example of a huge transformation in separation science through process modification wherein adsorption and desorption can take place by swinging pressure alone. While pressure-swing adsorption is established in gas separations, it finds little or no applications in wastewater treatment, although thermal swing operations have been investigated and tried. Zeolites typically have useful properties through their cage-like defined pore structures and ionic framework. The functional properties and advancement in the zeolites can be obtained in the production process through change in parameters such as Si/Al ratio, structure and porosity of the material, ionic species, and reaction conditions apart from further modifications, for example by exchanging different ions. In addition to engineering key modifications in the existing process, process integration is an interesting and fairly wide open area for research and development wherein there is huge opportunity for studying intelligent combinations of two or more operations. The advent of nanomaterials has also increased the scope of investigating advanced adsorptive separations with altogether new dimensions. In the future, new adsorbent materials are required to have high capacity, high selectivity, improved stability, and more favorable geometries (Adler et al., 2000). The technical challenges for wastewater treatment applications include:

1. Developing newer, efficient, and cost-effective adsorbents for better performance.
2. Developing efficient ways to regenerate and reuse adsorbents.
3. Developing tailor-made adsorbents for complex systems, especially for the removal of refractory pollutants.
4. Predicting adsorption behavior through modeling and simulation.

As far as adsorptive process advances are concerned, it is believed that cost and efficacy of pollutant removal will drive further growth in this area.

2.3.4 Ion Exchange

The ion exchange process is mainly used to separate ionized molecules (organic as well as inorganic) from aqueous solutions as well as contaminants in organic streams. Ion exchange has been used industrially for many years in the form of cationic or anionic resins with strong or weak acidic or basic groups to remove ions from dilute solutions (Helfferich, 1962). In wastewater treatment, this process is commonly employed as a tertiary or polishing method for the removal of specific pollutants to desired levels.

For industrial wastewater treatment, the ion exchange process mostly employs synthetic ion exchange resins in bead form (Figure 2.7). The resin beads are spherical, by and large, with size typically ranging from 0.1 to 1 mm in radius. The backbone can be made of polymers such as polystyrene, epoxy, phenol-formaldehyde, or polyacrylate-polystyrene; polyacrylates are most widely used in practice. The bead contains a large network of pores, similar to that discussed in Section 2.3.1. On the basis of pore size, the resins are classified as microporous, mesoporous, or macroporous. The polymer backbone in resin is essentially hydrophobic, while the pore phase, due to the presence of ionic/ionogenic functional groups, is essentially hydrophilic.

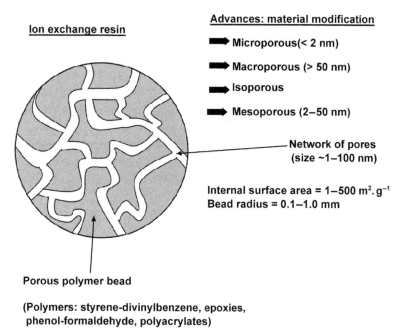

Figure 2.7 Schematic representation of ion exchange resin bead.

In ion exchange resins, functionality of the matrix is more important than the surface area. Although the polymer backbone is also believed to have some role in the removal of contaminants, for all practical purposes, surface functionality on the polymers is crucial in deciding the separation of pollutant species from the solution.

The functional groups of the resin are located on the walls of the polymer; this is schematically shown in Figure 2.8, along with a listing of common functional groups. On the basis of functionality, the ion exchange resins are classified as strong acid resins, weak acid resins, strong base resins, and weak base resins. Selection of the resin is generally made on the basis of functionality and the capacity of the resin along with porosity considerations. For example, for established water treatment using ion exchange application, cations in the water are removed by cationic resins, and anions are removed by anion exchange resins. The capacity of the resin from the specified class (usually operating capacity) dictates the quantity of resin required and cost, while the rate of exchange also dictates selection of suitable resin among the classes, for example selecting microporous resin versus macroporous resin from a kinetics point of view. It is to be noted that there are hundreds of resins in each class, and selection of any resin for specific application is

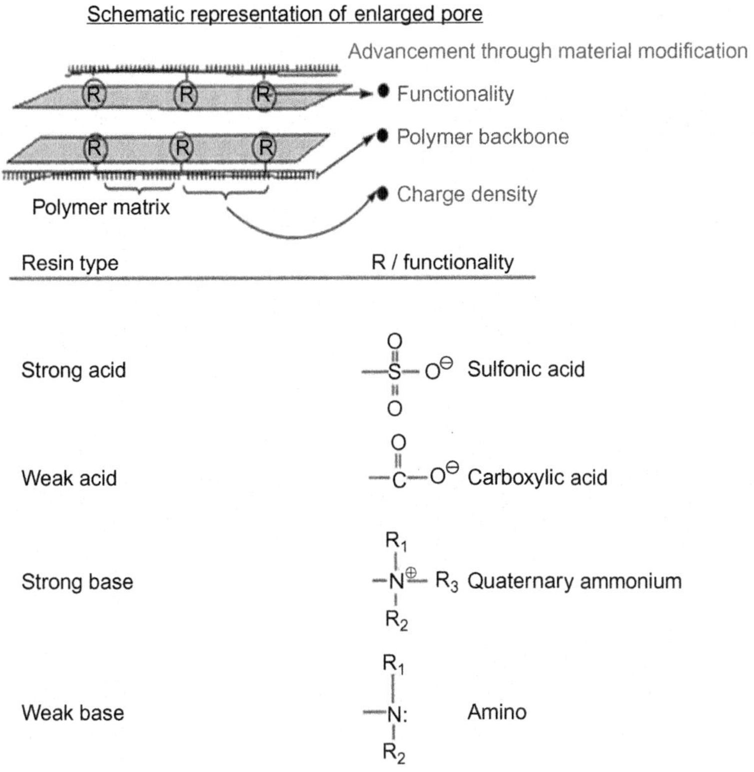

Figure 2.8 Ion exchange resin: enlarged pore and functionality.

not straightforward, many times requiring preliminary experimental screening and prior experience.

Ion exchange, in its conventional form, is a metathetical process where stoichiometric exchange of ions takes place. The exchange of a hydroxyl ion with a chloride ion of the acid on a strong base resin (polymer is indicated with the notation P, while alkyl groups are denoted by R_1, R_2, and R_3) can be shown as:

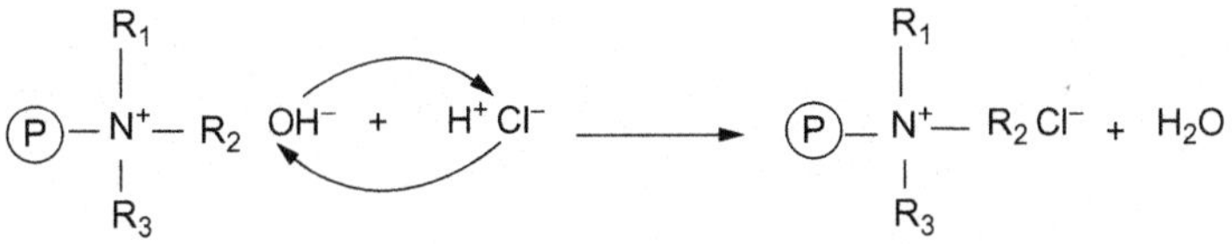

However, weak base resins are the exception to this format; here the resins offer ionogenic sites on the polymer matrix that can get ionized with the proton of the acid subsequently attaching the anion to it without any exchange of ions as shown below:

$$\text{\textcircled{P}}-\underset{\underset{R_2}{|}}{\overset{\overset{R_1}{|}}{N}}: \; + \; HCl \; \longrightarrow \; \text{\textcircled{P}}-\underset{\underset{R_2}{|}}{\overset{\overset{R_1}{|}}{N}}H^+ \; Cl^-$$

In this respect, it can be said that weak base resin is not a true ion exchange resin and differs significantly in its form and in action. The difference mainly arises from the fact that it contains nitrogen with lone pair of electrons (free ionogenic base group) that can accept a hydrogen ion to get ionized. The electrostatic attraction forces then allow the binding of anions from the solution. This process is termed *sorption*, and the mechanism of the same in the case of acid removal is well discussed in the literature for various acids such as strong inorganic acids, weak organic/carboxylic acids, and polybasic acids (Bhandari et al., 1992a,b, 1993, 1997, 2000). As compared to strong base ion exchange resins, the weak base resins have larger sorption capacity and are easy to regenerate. This aspect makes them highly useful in wastewater treatment, especially for the removal of acids from the wastewater streams.

The theory and practice of ion exchange process is well known and quite well developed. As mentioned in Section 2.3.1, knowing the kinetics of the process is essential for final selection of the resin. For example, when capacity may be high, but rates of exchange are very low, the resin may not be suitable for the application, and a comparatively low capacity resin with better rates of exchange will have to be chosen. Ion exchange kinetics bears no resemblance to chemical reaction kinetics in the usual sense. For both adsorption and ion exchange, it is essential to know:
- The mechanism of the process.
- The rate-determining step.
- What rate laws are obeyed.
- How the rate can be predicted.

Similarly to adsorption, ion exchange is also often a diffusion-controlled process. For simple undissociated species, Fick's law of diffusion can be conveniently used, whereas for ionic species, usually the Nernst-Planck equation or other equations that allow accounting for ionic interactions such as Stefan-Maxwell equation are more appropriate. Ion exchange reactions are instantaneous. If the different rate steps involved in adsorption and ion exchange are to be compared using order-of-magnitude analysis, the diffusion step is the slowest step in the overall mechanism (Figure 2.9). Thus, in

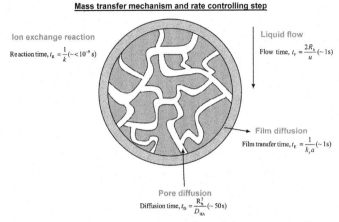

Figure 2.9 Ion exchange: depicting mass transfer and the rate controlling step.

these processes, it can be safely assumed that diffusion is the rate–controlling step. A number of diffusion models, both pseudo–homogeneous and hetero-geneous, have been developed for various processes. The main disadvantage in mathematical modeling of sorption processes is that many times, gener-alized models are not available and a system specific mathematical treatment has to be developed.

Ion exchange treatment, similarly to adsorption, is commonly practiced in plants using a fixed-bed column chromatographic operation (Figure 2.10). The flow of liquid can be upflow or downflow, and well-established principles of column design for fixed-bed operation are applicable. It should be pointed out that in the ion exchange column, the resin bed is normally only filled up to 50% to have free space for the expansion of the resin bed due to swelling. Free space is also required in the backwashing step, which results in fluidization of the bed due to upflow backwash. The kinetics and pressure drop consider-ations dictate the selection of suitable resin bead size.

In the column chromatographic operation, the resin bed is progressively consumed in terms of available capacity through the exchange of pollutant species. The unreacted zone moves in the direction of flow, as shown in Figure 2.10. When the reaction front reaches the end of column, the pol-lutant starts appearing in the exit solution, and the concentration profile in the effluent takes the shape of a breakthrough curve. The lower limit of the permissible concentration of the pollutant is usually set as the breakthrough point, and the column is switched for the regeneration cycle, once the breakthrough point is reached. Thus, the nature of the breakthrough curve

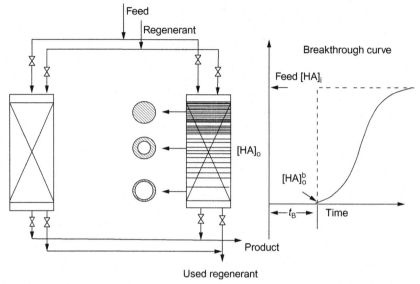

Figure 2.10 Ion exchange operation: Column and breakthrough curve analysis.

dictates the column utilization, and the sharp/rectangular form of the curve is most desirable. However, in reality, this is rarely possible and typically dispersive fronts are common. The larger the dispersive front, the more unutilized the resin bed is in the column, implying significant underutilization of the resin capacity, even if the resin has a theoretical high capacity of sorption. The nature of the breakthrough curve can be modified using different changes in the process parameters; breakthrough curves also will depend on the material itself. Therefore, in real applications, breakthrough capacity or operating capacity are more important parameters than theoretical equilibrium capacity (which is rarely achieved). Further, it is to be understood that 100% regeneration is practically never achieved due to excessive chemical requirements. Many times, capacity up to 80% is regenerated as a compromise between regenerant usage and capacity.

2.3.5 Ion Exchange: Advances and Applications in Wastewater Treatment

Recently, newer synthetic polymers without functional groups have been developed and used commercially. These have a very large surface area and defined pore structure for better performance and regenerability. For selective removal of heavy metals, newly developed chelating resins that help in removal of these toxic metals from wastewaters are widely used

(Neumann and Fatula, 2009). In the plating industry, ion exchange processes are widely employed for removal of chromium and for water recycling and reuse. Ammonia removal from wastewaters using ion exchange has also been explored and has been found effective (Jorgensen and Weatherley, 2003).

In recent years, new resins that are more selective for specific ions have been developed. They have shorter diffusion paths and better exchange rates. Ion exchange provides a good separation technology for dilute systems requiring less frequent regeneration.

The need to recover or remove contaminants from dilute solutions will increase in the future due to stringent regulations applying to waste streams (Adler et al., 2000). Today, the requirement of clean water for the growing population has already reached a demanding stage. Many problems associated with the application of ion exchange in wastewater treatment need to be addressed. The most prominent among the issues include the cost of ion exchange that linearly scales with column size and the cost and difficulties in the regeneration of materials. Highly selective ion exchange materials pose serious difficulty in regeneration: The more favorable the loading, the more difficult the unloading is the operating rule. The capacity, selectivity, and specificity of existing ion exchange material issues are major impeding factors to improved efficiencies in treating dilute solutions of wastewaters, be they from the chemical or biochemical industry. A new approach for devising more robust, high-capacity, high-selectivity exchangers with ease of operation needs to be found. It is a recognized fact that the solution to environmental problems will require very inexpensive ion exchange materials, some that can even be used once and discarded (Adler et al., 2000). This aspect is driving research in the area of ion exchange materials derived from biomass, such as chemical modification of bagasse. Developing a highly selective material at low cost is therefore a serious challenge for the future.

2.3.6 Case Study: Adsorption/Ion Exchange for Acid Removal

Organic acids are major commercial chemicals, and their recovery/separation from reaction mixtures, fermentation broths, and wastewaters has been a challenging problem for over four decades. Wastewater streams containing low concentrations of acids are inevitably encountered in acid manufacturing plants, industries where acids are used as raw material or as catalyst, fermentation processes, the metal plating industry, and others (Table 2.3). Wastewater usually contains a single acid or mixture of acids from 0.5%

Table 2.3 Some examples of industrial wastewater streams

Industry	Acids	Composition	Reference
Pulping process	Formic, acetic	4–10%	Othmer (1958)
Synthetic fatty acid	Formic, acetic, propionic, butyric	5, 5, 2.5, & 2% respectively	Matsarenko et al. (1969)
Industrial effluents	Formic, acetic, propionic	0.5–5%	Helsel (1977)
Manufacture of formic, acetic acids Oxidation of organics e.g., p-xylene, cyclohexane	Formic, acetic, oxalic	<10%	Parulekar et al. (1982); Ricker et al. (1980); Wadekar and Sharma (1981b); Kawabata et al. (1981)
Dyes and pigment industry Acid leached wastewater	Sulfuric, phthalic, oleic	<1%	–
Electroplating industry	Hydrochloric, sulfuric	High, up to 10%	Bonev and Nenov (2006)

to 4%, depending upon the source of generation. Removal of acids from aqueous streams containing low to medium concentrations of acids—organic, inorganic, or both—is an important problem in chemical processing industries, necessitating selection of suitable separation methodologies on the basis of the composition of the stream. Ion exchange, adsorption, solvent extraction, membrane separation, reactive distillation, reactive extraction, membrane-based solvent extraction, and emulsion liquid membrane (ELM) separation are considered as promising separation methods for devising suitable strategies. Ion exchange resins, in general, and weak base resins, in particular, are most commonly employed for lower concentrations. Because large volumes that contain low concentrations of acids need to be treated, ion exchange is the most convenient method of treatment in many cases. The primary goal in such cases is largely removal of acids from the solution, not separation. However, acid separation is important in acid manufacturing units where another acid is usually obtained as a by-product, for example, formic acid from the manufacture of acetic acid. Thus, the selection of an ion exchange resin for any particular application, whether removal or separation, is an important consideration. A number of experimental and theoretical studies have been reported on the sorption of various acids (Helfferich, 1962; Adams et al., 1969; Höll and Sontheimer, 1977; Hübner et al., 1978; Rao and Gupta, 1982a,b; Helfferich and Hwang, 1985; Bhandari et al., 1992a,b, 1993, 1996, 1997, 2000; Cloete and Marais, 1995; Juang and Chou, 1996; Bhandari, 1998; Husson and King, 1999; Sonawane et al., 2009). The anomalous nature of sorption equilibria has also been indicated with polybasic acids (Bhandari et al., 1997). The experimental and theoretical information on acid sorption is rather incomplete due to the following:

1. The majority of the studies have been reported on a single component acid, not on acid mixtures, and there is little information on wastewater treatments containing acids.
2. The selection of a resin for any acid removal/separation in wastewater treatment has not been discussed. Not many new materials have been investigated, developed, reported, or discussed.
3. Less information is available on experimental and theoretical aspects of sorption equilibria, especially with reference to the differences in the sorption behavior and behavior modifications.
4. Not much information is available in open literature on adsorption/ion exchange in industrial wastewater treatment and techno-economic feasibility analysis.

For the removal of acids from dilute solutions, weak base resins have been recommended due to their high capacity of removal and ease of regeneration. It has also been suggested that the resins with high basicity are more suitable in this regard because a near rectangular type of sorption isotherm can be expected, indicating total sorption of acids corresponding to theoretical resin capacity, especially for stronger acids (Bhandari et al., 2000). Commercial weak base resins are classified into weak, medium, and high basicity on the basis of ion exchange equilibria.

The importance of basicity depends on type of acid and is found to be less pronounced for very weak acids, implying that for removal of weak acids such as phenols and cresols, the capacity and the basicity of resin are equally important considerations in selection.

It is necessary to know how material selection and material modification can affect the removal of acids. This is briefly explained here for both adsorption and ion exchange using the example of lactic acid removal (Bhandari et al., 2006), where it was found that resins like Dowex MWA-1 can provide an excellent option for acid removal, especially at low concentrations typical of wastewater treatment. The separation of lactic acid/lactate salt can be accomplished mainly using solvent extraction, ion exchange, adsorption, distillation, and membrane separation (Scholler et al., 1993; Aljundi et al., 2005; Schlosser et al., 2005; Lee and Kim, 2011). Joglekar et al. (2006) have recently reviewed these different methods for the removal of lactic acid and have confirmed that the uptake of lactic acid for sorption by Dowex MWA-1 is much higher than that for extraction by Alamine 336. The main limitation of the method of solvent extraction is in the form of the toxicity of the solvent, especially with respect to microorganisms apart from solvent losses due to miscibility of the two phases and operating problems. Adsorption studies using silicalite molecular sieves were reported recently by Aljundi et al. (2005). However, their capacities were very low—of the order of 55 g/kg adsorbent as compared to the typical capacity of 210 g/kg in the ion exchange resin Amberlite IRA-35. Sorption capacity with commercial polymeric adsorbents such as XAD-4, XAD-7, and XAD-16 are very much lower—with/without modification, although sorption capacity can be significantly altered with surface modification (Figure 2.11). A high sorption capacity of ∼360 g/kg can be obtained using carbon-based/modified carbon adsorbents. With ion exchange, the sorption capacity for weak base ion exchange resins is also high— ∼360 g/kg on Dowex MWA-1 (Figure 2.12). Thus, from many materials in the adsorbent and ion exchange resin class, the ion exchange resin appears to be most suitable for acid

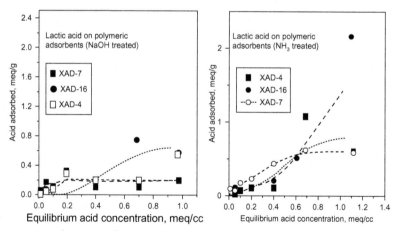

Figure 2.11 Adsorption of lactic acid on polymeric adsorbent with surface modification (Bhandari et al., 2006).

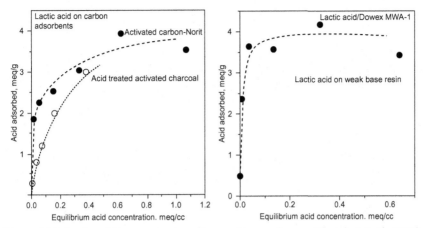

Figure 2.12 Lactic acid removal: adsorption and ion exchange (Bhandari et al., 2006).

removal from dilute streams. The results clearly highlight the importance of selection of proper material and that surface properties/modification can play important roles in enhancing removal capacity.

Extraction is not considered favorably in wastewater treatment due to solvent contamination that can be a "treatment problem" in itself. However, if environmentally friendly solvents (biodegradable) can be tailor made, solvent extraction can also provide a suitable alternative to the existing processes. This is important since many organic acids with varying acid strength are commonly encountered as organic pollutants. In contrast to wastewater streams, especially in process separations, mixtures of organic acids such as cresols and xylenols are obtained in significant quantities and recovery/separation of these can be difficult by conventional means. Here, separation methods such as dissociation extraction and ion exchange that exploit differences in the pKa values of the components in a mixture could be commercially important operations in the recovery/separation of acids from aqueous or organic streams. Dissociation extraction is a technique for the separation of organic acid/base mixtures on the basis of differences in the dissociation constants of the components. Dissociation extraction is particularly useful in the separation of organic compounds that closely resemble each other in their physical properties but differ in ionizing character. A number of studies have been reported on the separation of organic acids using dissociation extraction. Jagirdar and Sharma (1981) and Gaikar and Sharma (1985) reported the separation of 2,6-xylenol and o-methoxy phenol using a strong extracting agent with benzene, anisole, and n-octanol as solvents. The separation of acid, phenol, chlorophenol, and dichlorophenol was reported by various researchers (Anwar et al., 1971a,b, 1979, 1995; Laddha and Sharma, 1978; Wadekar and Sharma, 1981a,b; Gaikar and Sharma, 1985; Gaikar et al., 1989; Guy et al., 1994; Malmary et al., 1995; Modak et al., 1999), and the separation of xylenols was reported by Coleby (1971). In most studies, solvent selection was found to be crucial. Jagirdar and Sharma (1980) explored the possibility of the recovery/separation of various carboxylic acids from dilute aqueous solutions using an extracting agent such as tri-n-octylamine in various solvents. Extraction studies in the presence of inorganic acids and inorganic salts for improving the extraction efficiency have also been reported for the recovery of carboxylic acids (Ingale and Mahajani, 1996). However, in most cases, separation factors were not large. Solvent extraction, although it appears to be a potentially attractive process for the recovery of acids, still has many unresolved problems with respect to proper selection of a solvent and solvent recovery due to high affinity of acids for water.

While methods such as adsorption/ion exchange and solvent extraction can be used for removal and recovery of acids, destructive methods such as oxidation and cavitation can be used to eliminate acids from dilute wastewater solutions. Extreme conditions of cavitation can break down pollutants and organic molecules to bring COD down to desired levels.

2.4 OTHER ADVANCED PHYSICO-CHEMICAL METHODS OF TREATMENT

2.4.1 Membrane Separations

Membranes are extensively used in industrial wastewater treatment. Compared to the adsorption and ion exchange processes, membrane separation is commonly employed as a primary treatment as well as a fine polishing step in tertiary treatment. In fact, it is one of the processes that has the explicit capacity to yield pure water that may be recycled and reused.

Conventional membrane processes are pressure-driven processes that require specific pressures to effect separation, depending on the type of membrane. However, many a times, it is the combination of two or more driving forces such as pressure, concentration and electrical potential that may also be interdependent. Conventional membrane separations are essentially physical processes that employ a sieving mechanism based on the size of the pores. These physical separation membrane processes are classified as microfiltration (<2 atm pressure, can separate suspended solids), ultrafiltration (pressure up to 10 atm, can remove organics of high molecular weight), nanofiltration (pressures up to 20 atm, can remove many divalent/polyvalent ions and organics of medium molecular weight), and finally reverse osmosis (pressures as high as 50–70 atm, can remove practically all pollutant species, yielding pure water). Thus, depending on the separation requirement, a desirable membrane process is selected. Because reverse osmosis can give practically pure water, it finds maximum use in the area of desalination and drinking water and is less preferred in wastewater treatment due to its high cost. For all practical purposes, industrial wastewater treatment mostly uses microfiltration and ultrafiltration, while nanofiltration is an emerging area in this field.

Although the conventional membrane separation principle indicates a purely physical separation mechanism, many new developments in this area have membranes suitably modified to enhance the performance in terms of flux, reduced fouling, and higher selectivity. In this sense, membrane separations are transformed as physico-chemical separation processes. Ion exchange membranes and modified membrane forms such as facilitated transport membranes represent this transformation from physical separation to physico-chemical separation. The charge effects, apart from physical separation, have a direct impact on equilibria and transport in these membranes and can be attributed to the ionization of functional groups present in the membrane and/or to the adsorption of ions from solution when the membrane is immersed (Mafé et al., 1993). The research on modification of membranes has been an important theme in the last few decades, and applications of ion exchange membranes for acid recovery and metal removal from various wastewaters (e.g., pickling wastewater) have been highlighted for a long time (Sata, 1991; Muthukrishnan and Guha, 2006; Navarro et al., 2008; Wang et al., 2008). A modification of commercial composite membrane by deposition of acrylic acid was shown to yield COD removal of 84% as compared to 67% before the modification (Cho and Ekengren, 1993). However, whether physical or physico-chemical, membranes have occupied a niche area in the world of wastewater treatment, and their role will be increasingly important in future wastewater treatment technologies, especially through process integration and intensification such as membrane bioreactors (MBRs), membrane distillation, dialysis/electrodialysis, ELM separation systems, and so on. A number of polymeric membranes have been studied for the treatment of alcohol industry wastewaters, and nanofiltration polyterphthalate NF45 membrane has shown high ability to remove organics (98%) (Madaeni and Mansourpanah, 2006). A study on multiple membrane separation in plating wastewater treatment indicated a feasible industrial application for heavy metal removal in the future (Zuo et al., 2008).

The application of membrane technology involves two main components, membrane element and membrane housing (membrane module). The basic membrane element comprises a single membrane or hybrid membrane system and is characterized in terms of membrane material, membrane porosity, and physical and chemical characteristics of the membrane. The membrane is the heart of the membrane separation system and dictates the separation behavior. The membrane material can be organic or inorganic. The membrane module is important from the point of view of membrane application in wastewater treatment; it can come in different formats such as a flat sheet module or spiral wound or hollow fiber. Usually, in wastewater treatment, organic membranes such as cellulose acetate, polypropylene, polyamides, acrylonitrile, and polytetrafluoroethylene are most commonly employed. The versatility and functionality of membranes make precise application of membrane technology most challenging, in spite of the theoretical possibility for achieving pollution control goals. In many applications, membrane processes have to compete directly with one or more conventional physico-chemical processes. As compared to conventional processes, membrane processes are often believed to be more energy efficient and simpler to operate, and to yield higher quality products with low impact on the environment, because no hazardous chemicals are used that have to be discharged. However, this may not be true in all cases, and there are many limitations even today, especially in industrial wastewater treatment areas that severely restrict the application of membranes. The important drawbacks of this technology are:

- In wastewater treatment processes, the long-term reliability has not completely been proven.
- It may require excessive pretreatment due to sensitivity to concentration polarization, chemical interaction with water constituents, and fouling.
- Mechanical robustness is an issue due to the possibility of easy damage by a malfunction in the operation.
- Process cost is a critical issue, although membrane separations, in general, are considered quite energy efficient. The need to account for the contribution of all direct and indirect costs such as energy consumption, investment in membranes and modules, the cost of other process equipment, the useful life under operating conditions, and various pre- and post-treatment process costs in the calculation of overall cost is important and can lead to higher costs.

The process calculations for membrane separations are discussed widely in the literature and in standard textbooks (see, e.g., Noble and Stern, 1995). From a practical application point of view, the most important parameters include membrane selection and type, trans-membrane pressure drop, operating flux, and fouling characteristics for getting desired water quality at the required rate. Fouling of membranes is a serious problem in wastewater treatment and dictates pretreatment needs, separation behavior, and the cost of operation apart from impacting the engineering design of the system.

2.4.1.1 Membrane Variants in Wastewater Treatment

There are some membrane separation types that differ significantly from the conventional membrane separations. These mainly include:

- *Membrane distillation*: This is similar to conventional distillation. The main difference is that the feed and the product are separated by a porous hydrophobic membrane, either at different temperatures or at different compositions. Because of the vapor pressure difference, water evaporates on one surface of the membrane and passes through the membrane in the form of vapor and gets condensed on the other side of the membrane. Membrane distillation is commercially used for recovery of hydrochloric acid from waste pickling liquors containing sulfuric acid and other ionic species such as chromium, nickel, cobalt, and zinc (Lawson and Lloyd, 1997).

- *Pervaporation*: This is an energy-efficient combination of membrane permeation and evaporation. It is commonly used for removal of organics from aqueous streams.

- *ELM separation systems*: These comprise an emerging separation technology with potential applications in wastewater treatment. Chemicals that can be removed/recovered from industrial streams using ELM separation systems are organic acids, phenols, cresols, and amines, as well as metallic ions such as lead, copper, cadmium, and mercury (Lee and Hyun, 2010).

- *Membrane contactors*: These are used in solvent extraction. Here, the membrane acts as a barrier between the feed and the stripping solution.

- *MBRs*: These may be considered to involve a physico-biological process and not a physico-chemical process in the conventional sense (combination of a bioreactor + inline membrane separation step). MBRs combine two familiar technologies: activated sludge and membrane filtration. Significant engineering expertise can be applied to MBR design and operation and principles underlying MBRs are familiar enough to ensure reliability. MBRs have been used to treat a wide range of municipal and industrial wastewaters, and currently they are believed to be installed at more than 1000 sites in Asia, Europe, and North America (Kumar and Roy, 2008). Enough reliable equipment and technological support are commercially available to meet existing and developing demand.

2.4.1.2 Membranes in Wastewater Treatment: Future Needs

The application of membrane separations in wastewater treatment today requires development of low-cost membranes with high surface area to volume; and development of high-temperature membranes (ceramic, metal), nanocomposites, chemically inert membrane material, material suitable for hydrophilic compounds in dilute streams, and mixed organic/inorganic membranes. Designing or selecting the most appropriate membrane material for a particular application is a crucial step since the choice of material can change the efficiency of a process by several orders of magnitude. Not many theoretical tools are available yet in this regard, and computer-aided membrane design is still an emerging field. It is recommended that a database search along with mathematical modeling and simulation can help in the selection process. Mechanical robustness of the membranes is important since any physical damage to the membrane element can create serious problems in the operation, requiring replacement of the membrane. It is also required that commercialization of processes using membranes be demonstrated for different industrial effluents to address scale-up issues. Further, it is felt that the integration of membrane separation with other physicochemical methods can provide a satisfactory solution in many wastewater treatment applications.

2.4.2 Advanced Oxidation Processes

Advanced oxidation processes (AOPs) have been commercially used in wastewater treatment, especially for the removal of refractory pollutants. They are preferably considered in the category of tertiary treatment processes. When used as a secondary treatment step, the role of an AOP is to degrade toxic pollutants and convert them into forms that can be treated by other methods, such as biological treatment. The goals here differ completely from conventional goals where complete degradation of organic compounds is sought rather than partial degradation of compounds amenable for treatment. AOPs are less preferred as technology if other methods are applicable due to technology and cost reasons.

AOPs have been shown to be effective for the destruction of many refractory pollutants (Hoffmann et al., 1995). Using an oxidation process in wastewater treatment is based on the attack of a highly reactive and oxidizing agent such as a hydroxyl radical on organic/inorganic pollutant species, resulting in the destruction of the pollutants through oxidation reactions. The process has two inherent components, the first being the generation of an oxidizing agent, for example, hydroxyl radicals, and the second being the oxidation reaction. A number of oxidizing agents can be employed: the most conventional are oxygen, chlorine, hydrogen peroxide, and hypochlorite. The hydroxyl radical is a comparatively strong oxidant that can destroy many pollutants otherwise difficult to destroy. Different methodologies have been developed for the generation of suitable oxidizing species which are then allowed to react (Gogate and Pandit, 2004; Kim et al., 2004; Metcalf & Eddy, Inc. 2003). In wastewater treatment, a number of AOPs that are commonly used include Fenton processes, photo-Fenton processes, ozone oxidation processes, oxidation using UV or ozone and UV, oxidations using a peroxide such as hydrogen peroxide + UV or ozone + hydrogen peroxide + UV, and photo-oxidations using TiO_2. These and other such varied combinations are being tried for mineralization of pollutants and for wastewaters containing phenolics. AOPs have been found to be highly effective using such combinations. Ozone is hazardous, needs to be generated *in situ*, and the off gases need strict monitoring. H_2O_2, on the other hand, is convenient to use and is available as a solution with good shelf life. Wet air oxidation is suitable for wastewaters containing organics in the concentration range that is rather low for incinerating but high or toxic for biological treatment. Wet air oxidation is conventionally carried out in the liquid phase. The process requires severe temperature and pressure conditions ($125-300°C$ and $0.5-20$ MPa) and oxidizes organics and oxidizable inorganic pollutants using a gaseous source of oxygen (e.g., air). A number of catalysts, usually metal based, have also been reported for wet air oxidation processes to reduce the severity of process conditions such as temperature or pressures and to enhance reaction. However, compared to most other processes, the wet air oxidation process requires high capital and operating costs and may have difficulty in degrading refractory pollutants. The process finds useful application in the pulp and paper industry. Heterogeneous photocatalysis is an emerging field that is believed to overcome many of the drawbacks of the traditional water treatment methods and provides an alternative for mineralization of pollutant species with reduced cost and without use of hazardous chemicals.

2.4.2.1 Electro-oxidation

Electro-oxidation is a relatively recent development along similar lines as electro-coagulation where anodic oxidation on the anode surface is employed in place of direct chemical/photochemical oxidation. An alternate variation of this process involves only the generation of an oxidizing species using the electrochemical route followed by direct oxidation by the generated oxidizing agents. Thus, the basic mechanism is the same while the location and the form of the oxidation reaction are slightly different. Similarly to the conventional oxidation process, common oxidizing agents are the hydroxyl radical, hydrogen peroxide, chlorine, and ozone. The design of the electrochemical cell, selection of electrodes, operational parameters, and cost of power are critical parameters for the application of electro-oxidation in wastewater treatment. Typically, the available electro-oxidation processes are believed to be effective in degrading refractory pollutants that are difficult to degrade otherwise. These mainly include toxic compounds and ammoniacal nitrogen. However, information on practical application in this area is scarce and requires substantial research and developmental work for industrial wastewater treatment, especially with respect to the nature of effluents/pollutants and their concentrations.

2.5 CAVITATION

Cavitation is a phenomenon of formation, growth, and collapse of micro bubbles (cavities) within a liquid. Cavities are generated by decreasing local pressure (either because of desorption of dissolved gases or evaporation of liquid). The cavities, when traveling in a region of higher pressure, implode (collapse) under certain conditions, resulting in very high pressure and temperature near the location of collapse. Realization of such very high pressure (>1000 atm) and very high temperature (>5000 K) is usually harnessed for a variety of applications ranging from disinfection of water to the carrying out of reactions (Pandit and Joshi, 1993; Suslick et al., 1997; Didenko et al., 1999; Gogate and Pandit, 2004, 2005; Ranade et al., 2008, 2013).

Cavitation is also used to treat industrial effluents. Collapse of cavities generates hydroxyl ions, which are useful in realizing many benefits in effluent treatment including the following:
- Reduction in COD
- Reduction in color
- Reduction in ammoniacal nitrogen
- Increase in BOD to COD ratio (and therefore enhancing the effectiveness of bio-digesters).

Although cavitation can be realized in variety of ways, acoustic cavitation and hydrodynamic cavitation are widely used in practice. Acoustic cavitation, based on piezo-electric ultrasound horns, is more suitable for specialty and small-scale applications. Higher energy requirements and difficulties in scale-up associated with acoustic cavitation do not make it attractive for effluent treatment applications. Hydrodynamic cavitation, however, is inherently suitable for effluent treatment applications.

Hydrodynamic cavitation is realized by creating low pressure regions in the flow domain. There are mainly the following two ways with which such low pressure regions can be generated:

- Increasing the linear velocity of liquid by forcing it to flow through constrictions such as orifices or venturi. The constrictions are designed in such a way that velocity at the throat (smallest flow area zone) is large enough to generate cavities. These cavities will collapse further downstream of the constriction, resulting in the desired effect.
- Increasing the tangential velocity of a liquid by forcing it to flow through a device like a vortex diode, which has an outlet port from the center of the vortex (tangential flow). The dimensions of the diode and outlet port are designed in such a way that a low pressure region is generated because the highly swirling flow in the diode chamber is adequate to generate cavities. The cavities escape the diode chamber via the outlet port and then collapse as they enter the high-pressure region.

The efficacy of cavitation essentially depends on the number density of generated cavities and the intensity of the collapse of generated cavities. The cavitation process and its various applications are discussed in detail in Chapter 3. In order to avoid duplication, here we restrict the scope of the discussion to the type of cavitation (cavitation realized by tangential flow) not covered in Chapter 3.

2.5.1 Cavitation Using Tangential Flow/Vortex Diodes

Recently, Ranade et al. (2008, 2013) have developed a cavitation device in the form of a vortex diode for effluent treatment and other applications. A vortex diode is a disk-shaped chamber with a tangential port and a cylindrical axial port. As compared to conventional cavitating devices that depend on constriction for pressure changes, the vortex diode relies on fluid vortex phenomena for its operation. The chamber is characterized by its diameter and height along with curved surface, with or without internals, which decide the chamber volume. The flow entering the device through the tangential port sets up a vortex and establishes a large pressure drop across the device. The tangential velocity of flow increases from the periphery toward the center, creating a swirling flow. The flow in vortex diodes can be classified as confined vortex flow. The principle of working in the vortex diode is shown schematically in Figure 2.13. One form of vortex diode, along with the cross sectional views, is shown in Figure 2.14.

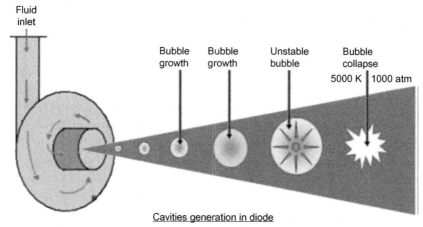

Figure 2.13 Vortex diode: working principle.

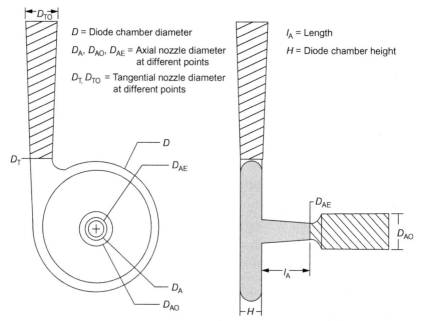

Figure 2.14 Construction details of a vortex diode (Ranade et al., 2008).

In the vortex diode, D represents the diode chamber, while H is the chamber height. The diameters at different points at the inlet and outlet are represented in the figure by the shaded regions and can be parameters for specific design. As can be seen, the design can have many variations in the size and shape of chamber and modifications that can alter the flow pattern within the device. The flow through the device is complex, and some studies using computational fluid dynamics have been carried out to obtain insight into the flow pattern and its impact on the cavitation process (Bashir et al., 2011; Kulkarni et al., 2008; Pandare and Ranade, 2013). However, as far as application of cavitation technology to wastewater is concerned, the approach is largely empirical with preliminary experimental studies required for obtaining useful data on the degradation of pollutants and the optimization of process parameters, similarly to coagulation processes.

A schematic diagram of an experimental unit comprising different cavitation reactors is shown in Figure 2.15. The setup includes a wastewater storage tank, a high-pressure pump, and a single cavitation reactor/set of cavitation reactors with isolation valves along with tools for measurement of pressure and flow. The reactors can be any hydrodynamic cavitating

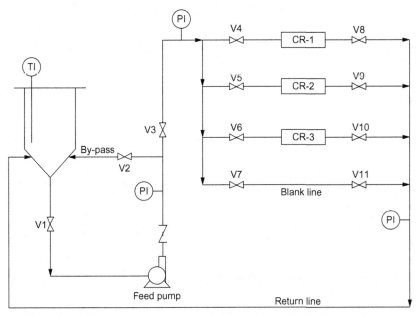

PI - Pressure indicator
TI - Temperature indicator

CR-1 : Cavitation reactor (vortex diode)
CR-2 : Cavitation reactor (orifice)
CR-3 : Cavitation reactor (venturi)

Figure 2.15 Experimental setup for cavitation studies on wastewater treatment.

device such as orifice type, venturi type, or vortex diode. Further, they can be used in isolation, in sequence, or in parallel. The process parameters that need to be optimized include flow and pressure drop apart from the number of passes that are required for effecting the desired reduction in the effluent COD. Some other process parameters such as addition of an oxidizing agent, a coagulant, in situ, or the form of a combination of processes can also be evaluated. For practical applications, the cost of the technology is dictated by the pressure drop across the device and the number of passes.

The working of the vortex diode in cavitation is significantly different and is expected to yield cavities that are different from the other devices, both in quality and quantity. This would be reflected in the extent and nature of degradation of pollutants, not just for types of cavitation such as acoustic and hydrodynamic but also among different cavitating devices.

2.5.2 Application of Cavitation in Dye Wastewater Treatment

There are certain organic pollutants, especially in dye/pigment/textile waste-waters, that are considered refractory compounds—difficult to remove/degrade by using conventional methods of chemical/biological treatment. For such pollutants, newer techniques have to be explored such as cavitation, where extreme conditions of cavitation can break down pollutants and organic molecules. Typically, hydrodynamic and sonochemical or acoustic cavitation are found useful in the destruction of organics. Cavitation generates strong oxidizing conditions due to the production of hydroxyl radicals and also hydrogen peroxide. Although significant work has been reported in the area of sonochemical reactors and their application in wastewater treatment, the implementation for actual industrial practice is still negligible because of the high cost of treatment and operational difficulties, especially in power dissipation. The impact of cavitation processes can be dramatically increased by combining them with other oxidation processes employing catalysts or additives. It has been reported that cavitation coupled with coagulation, called cavigulation, can be effective in water treatment and pollutant removal (http://bestotc.com/493/uncategorized/cavigulation-cavitation-technologies-solution-to-the-water-crisis-refining-the-desalination-process; Chakinala et al., 2009; Mishra and Gogate, 2010; Ranade et al., 2013; Saharan et al., 2011, 2013; Sawant et al., 2008; Wang et al., 2011; Xu et al., 2010). Thus, process intensification can work wonders if cavitation and other suitable methods are integrated, especially in treating wastewaters containing refractory pollutants and/or having unusually high COD.

A comparison of synthetic dye wastewater treatment for different types of dyes with hydrodynamic cavitation (vortex diode as cavitating device) and acoustic cavitation using our experimental investigations is shown in Figures 2.16–2.18 (Hiremath et al., 2012). The performance of the process was evaluated using the results on the extent of reduction in the chemical oxygen demand and in color. Three different dyes, i.e., methyl red, congo red, and reactive red were used to see if the nature of the dye has any effect.

It is evident that, in all the cases, hydrodynamic cavitation using the vortex diode performs far better compared to acoustic cavitation, both in terms of COD removal and color removal. In the case of methyl red dye (initial COD of 100 mg/L), the COD removal is above 61% as against 19% with acoustic, and color removal is close to 40% as against 16%, more than double. Similarly, for congo red dye (initial COD of 286 mg/L), these figures are 62% and 52% for COD and color using hydrodynamic cavitation, again more than double that observed with acoustic cavitation. Reactive dyes are considered to be refractory pollutants that are difficult to treat, and here, too, the results with hydrodynamic cavitation are far superior with more than 70% reduction in the COD of the wastewaters (initial COD of 456 mg/L). The consistency in obtaining better results using hydrodynamic cavitation with the vortex diode clearly highlights the utility of such new devices in treating industrial wastewater.

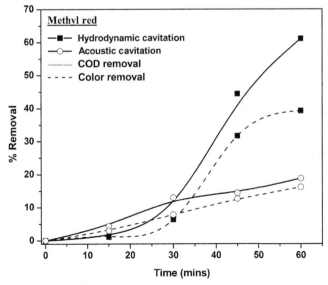

Figure 2.16 Comparison of hydrodynamic and acoustic cavitation: Methyl red dye.

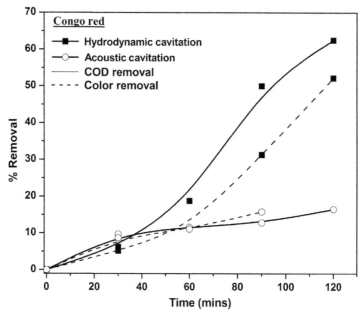

Figure 2.17 Comparison of hydrodynamic and acoustic cavitation: Congo red dye.

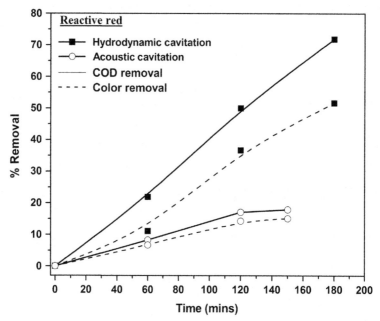

Figure 2.18 Comparison of hydrodynamic and acoustic cavitation: Reactive red dye.

2.5.3 Application of Cavitation in Reducing Ammoniacal Nitrogen

A very important problem with respect to many chemical industries is reducing ammoniacal nitrogen in wastewaters. Ammoniacal nitrogen (NH_3-N) is a measure for nitrogen as ammonia, a toxic pollutant. Ammonia can directly poison humans and upset the equilibrium of water systems. The nitrogen in sewage/industrial wastewaters is assessed as ammoniacal nitrogen. This indicates the amount of nitrogenous organic matter that has been converted to ammonia. The average strength of crude domestic sewage will have a combined nitrogen content of 40-60 mg/L. Ammoniacal nitrogen removal can be carried out by biological, physical, or chemical methods or a combination thereof. Available technologies include adsorption, chemical precipitation, membrane filtration, reverse osmosis, ion exchange, air stripping, breakpoint chlorination, and biological nitrification and denitrification (Metcalf & Eddy, Inc. 1991). Conventional methods, however, are not efficient and are cost intensive. Physico-chemical treatment or ion exchange/adsorption is preferred over other methods because it is stable, easy to maintain, and reliable. Aguilar et al. (2002) investigated physico-chemical removal of ammoniacal nitrogen by coagulation-flocculation using activated silica, powdered activated carbon, and precipitated calcium carbonate. They found very low ammonia removal of around 3–17%, but albuminoid nitrogen (nitrogen in the form of proteins) removal was appreciable (74–89%), and the addition of coagulant aids reduced the sludge volume to 42%. Ion exchange resins and some cheaper alternative natural and waste materials can be used to replace high-cost materials. Various researchers have studied the effectiveness of a variety of low-cost materials for ammonia removal such as clay and zeolite (Aziz et al., 2004; Çelik et al., 2001; Demir et al., 2002; Rožić et al., 2000; Sarioglu, 2005), limestone (Aziz et al., 2004), natural and waste materials such as waste paper, refuse cement, and concrete (Ahsan et al., 2001). Thus there is huge opportunity for development of new materials, processes, and process integration options for cost-effective industrial separations along with water recycling strategy.

2.5.4 Case Study: Hydrodynamic Cavitation Using a Vortex Diode in Real Industrial Effluent Treatment

It is instructive to evaluate the performance of newer devices such as the vortex diode in treating real industrial wastewaters. A detailed study was carried out using industrial wastewater from the dye and pigment industry (Hiremath et al., 2012). The important characteristics of this wastewater are given below:

- Dark brown color with unpleasant odor
- Presence of organics, salt, and other unknown contaminants
- High COD (~36,000 mg/L).

For conducting an effluent trial using the vortex diode, the following process parameters were set.

- Flow rate = 3800 LPH
- Initial temperature = 28 °C
- Pressure drop = 155 kPa.

It was found that COD removal is highly dependent on the nature of the industrial effluent. This is not unexpected since there is lot of variation in the quality of wastewater, including dissolved and suspended solids apart from unknown pollutants present in the wastewater such as organics and metals. This, therefore, requires in-depth analysis of the wastewater and causes uncertainty in the removal behavior since most of the time data is not made available in this regard for every pollutant present in the wastewater. The presence of solids in the wastewater is believed to cause damage to the cavitation device through mechanical erosion in prolonged operations and needs to be accounted for. Nontoxic wastewater is traditionally characterized in terms of major parameters such as BOD, COD, and ammoniacal nitrogen. The present case study, therefore, is an attempt to verify the usefulness of cavitation technology in treating industrial wastewater mainly through reduction in COD and ammoniacal nitrogen. An attempt was also made to compare different wastewater treatment technologies in treating the same industrial wastewater in order to get at least some idea about the implementation of these technologies alone or used in a process integration strategy. The general observations made in this regard can be compiled as below:

Reduction in COD

- Coagulation: ~65%
- Ion exchange/adsorption: up to 90% or more
- Cavitation: 20–95%
- Process integration (Ion Exchange/Adsorption + Cavitation): >97%.

It may be noted here that although adsorption or ion exchange can be useful in removing a large percentage of COD in some cases, the capacity for removal is limited, requiring huge quantities of material and making it impractical for direct application. Thus these processes can be best used in integration with other processes such as cavitation. The progress of color removal with time using cavitation alone is shown in Figure 2.19.

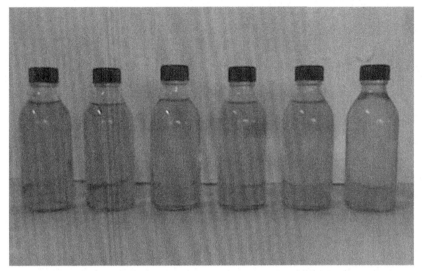

Figure 2.19 Examples of the extent of color removal using hydrodynamic cavitation.

Overall, with the use of an appropriate technology or a combination, more than 95% reduction in color and more than 97% COD reduction in some cases can be accomplished for real industrial wastewaters. The final result in the above example of industrial wastewater treatment is shown in Figure 2.20, where practically clear water can be obtained using process integration with cavitation and ion exchange.

As far as the removal of ammoniacal nitrogen is concerned, the hydrodynamic cavitation with a vortex diode appears to be far more effective and produced near total removal of ammoniacal nitrogen from the real industrial wastewaters in some cases (Table 2.4, Figure 2.21). The performance was again substantially dependent on the type of effluent. However, the utility of the vortex diode in hydrodynamic cavitation appears to be beyond doubt (Ranade et al., 2013).

The hydrodynamic cavitation device is simple in construction, without any moving parts, and easy to operate. More importantly, hydrodynamic cavitation devices are amenable to scale-up and can be designed for treating large volumes of industrial effluents. The vortex diodes have already been tested using up to 40 m^3/h flow rates (Ranade and Bhandari, 2013). Hydrodynamic cavitation is therefore perceived to be an energy efficient and quite effective technology for wastewater treatment, especially in view of its capability to degrade refractory pollutants. It can also be conveniently combined

| Effluent before treatment | Effluent after treatment (Ion exchange + Cavitation) |

Figure 2.20 Industrial effluent treatment using process integration.

Table 2.4 Industrial effluent treatment using cavitation

Sample	Effluent	Initial Ammoniacal Nitrogen Content, ppm	% Reduction After Cavitation
S1	Dyes and pigment unit	32.3	99
S2	Dyes industry	23.4	98
S3	Specialty chemical industry	57.5	94
S4	Specialty chemical industry	44	57
S5	Distillery industry	2050	36.5
S6	Dyes and pigment industry	1100	29
S7	Specialty chemical industry	33	24

with processes such as coagulation, adsorption, ion exchange, and oxidation/reaction for best overall results on effluent treatment.

2.6 COST CONSIDERATIONS

Coagulation processes have immense potential for removal of both pollutants and color, and, to a large extent, can be effectively employed for removal of

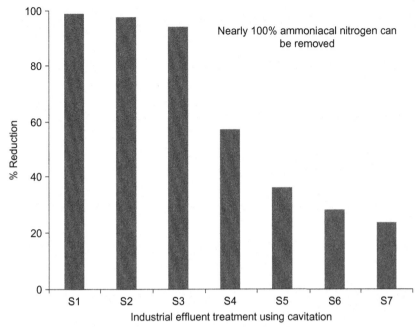

Figure 2.21 Reduction of ammoniacal nitrogen in different industrial effluents from the specialty chemicals sector using hydrodynamic cavitation.

color. From a cost point of view, common inorganic coagulants are less expensive, while organic coagulants are relatively expensive. Alum and ferric coagulants are among the least expensive materials, although dosage requirements would be high compared to polymeric coagulants, and ferric coagulants can pose corrosion problems. Specific formulations/hybrid coagulants are available in the market that have high efficiency in specific cases; however, these can be expensive. Thus, the process requires careful selection of material for its techno-economic feasibility.

Adsorption and ion exchange are economically feasible processes for color removal in dye industry effluents and/or decolorization of textile effluents. The technical aspects in the selection of specific material involve dye/adsorbent interaction, adsorbent surface area, particle size (relevant from a pressure drop point of view in plant operation), temperature (if thermal regeneration is involved), pH, and contact time (to fix cycle time for process/regeneration). Adsorbents that contain amino nitrogen tend to have a significantly larger adsorption capacity in acid dyes (Zaharia and Suteu, 2012). Although activated carbon is widely used in such applications for its techno-economical feasibility, other low-cost adsorbents mainly derived from biomass, for example, modified bagasse, fly ash, clay, rice husk, and neem leaf powder, are also used for such applications (Ahmaruzzaman and Gayatri, 2011; Mane et al., 2003, 2006). However, factoring cost considerations in these applications should essentially consider the costs associated with the entire removal process rather than the cost of adsorbents alone. Many times, the biomass-derived adsorbent need not be regenerated and can be destroyed/burnt as fuel. Typically, the most critical techno-economic elements for consideration in figuring the cost of adsorbents are the nature of adsorbents, size of adsorbent (pressure drop vs. rates of removal), cost of regeneration, life of adsorbent, operating range of pH, and temperature for most satisfactory selection.

The costs of cavitation operations can be substantially reduced when compared to other methods of effluent treatment and drastically so when compared to intensive processes such as oxidation. A typical cost calculation using hydrodynamic cavitation using a vortex diode as a cavitating device is as follows:

$$\text{Cost of treatment/m}^3 \text{ of effluent} = \frac{N_c \Delta P P_E}{36\eta}$$

where N_c is number of circulations needed for the treatment, ΔP is pressure drop in atm, P_E is price of electricity per kWh, and η is efficiency of the pump. For a typical situation in India, taking the price of electricity as Rs. 10 per kWh, pump efficiency as 0.66, and pressure drop across the diode as 2.4 atm, the cost of treatment comes out at N_c Rs./m^3 of effluent. Typically, 5–10 recycles are adequate; therefore, the cost is about Rs. 5–10/m^3 of effluent.

Overall, coagulation should be used as an essential step in effluent treatment, not just to get maximum COD removal, but also to reduce the overall cost of industrial wastewater treatment, because it has limited goals and affordable cost. Adsorption and ion exchange are defined processes for specific pollutants and are required to achieve the desired final objectives of a wastewater treatment process. Here, the cost of material is significant and needs to be optimized, mainly through the selection of appropriate materials and process integration. Oxidation processes are again specific processes for the removal of refractory pollutants and are generally cost intensive due to the cost of catalysts and severe process conditions. Membrane processes, as a general sieving process, are not cost intensive but physicochemical processes, for example facilitated transport, hybrid membranes, and ion exchange membranes, can substantially add to the process costs, not just because of the cost of membrane material, but also because of associated costs of operation, and in such cases, these processes no longer remain a competitive option. Cavitation, in general, is a low-cost operation both in terms of device cost and cost of operation. Further, cavitation can be satisfactory as a single technology for effluent treatment, or it can be employed efficiently in hybrid combinations with most other conventional separations, thereby proving to be potentially cost-effective and thus techno-economically feasible option for industrial wastewater treatment.

In most cases, two or more processes can provide techno-economically feasible options. An appropriate combination of treatment methods combined with process integration often leads to effective water treatment solutions (not just meeting the regulatory norms, but even adding to the profitability of the overall process).

2.7 SUMMARY

Physico-chemical methods have important role to play in today's industrial wastewater treatment scenario and will continue to have increased use because of established practice and continued improvisation. Although some methods, such as adsorption, ion exchange, and coagulation, have been considered relatively matured over the years, they will still continue to dominate the wastewater treatment mainly through improved materials, devices, and practices. Further, new emerging technologies such as membrane separations and cavitation will have to compete with the existing established processes, especially through improved process efficiencies and economics. However, the most challenging aspect for industrial wastewater treatment in the future will be process integration, which can bring out intelligent combinations of different physico-chemical processes as well as physico-chemical processes coupled with biological processes. The application of physico-chemical methods is not just relevant from the point of view of industrial wastewater treatment for meeting pollution control norms but also relevant from the point of view of water recycling and reuse. This is especially important in view of the scarcity of water resources in several parts of the world that would demand water conservation, quite apart from environmental pollution control.

Nomenclature

a_L Constant in Langmuir equation (mg g^{-1})
a_R Constant in Redlich-Peterson equation

B_D	Constant in Dubunin-Radushkevich equation $(mol^2 J^{-1})$
b_L	Constant in Langmuir equation $(L\ g^{-1})$
C_e	Equilibrium concentration in solution phase $(mol\ m^{-3})$
D_{HA}	Diffusion coefficient $(m^2\ s^{-1})$
d	Diameter (m)
d_T, d_{TO}	Tangential nozzle diameter at diode inlet and pipe inlet of vortex diode
d_c	Diameter of diode chamber (m)
E	Energy of adsorption $(kJ\ mol^{-1})$
HA	Acid species
$[HA]$	Concentration of acid $(mol\ m^{-3})$
$[HA]_i$	Initial concentration of acid $(mol\ m^{-3})$
$[HA]_o$	Outlet concentration of acid $(mol\ m^{-3})$
$[HA]_o^b$	Breakthrough concentration of acid $(mol\ m^{-3})$
H	Height of diode chamber (m)
K	Constant in Freundlich equation
k	Reaction rate constant (s^{-1})
K_R	Constant in Redlich-Peterson equation
$k_s a$	Film mass transfer coefficient (s^{-1})
N_p	Number of passes
$1/n$	Constant in Freundlich equation
P_s, P_i	Power $(J\ s^{-1})$
Q	Flow rate $(m^3\ s^{-1})$
Q_e	Equilibrium concentration in adsorbent phase $(mg\ g^{-1})$
q_D	Theoretical saturation capacity in Dubunin-Radushkevich equation $(mol\ g^{-1})$
R	Molar gas constant
R_b	Radius of bead (m)
r_c	Radius of curvature (m)
T	Temperature
t	Time (s)
t_B	Breakthrough time (s)
t_D	Diffusion time (s)
t_R	Reaction time (s)
t_F	Flow time (s)
t_E	Film transfer time (s)
u	Flow velocity $(m\ s^{-1})$
V	Volume (m^3)
β	Constant in Redlich-Peterson equation
η	Pump efficiency
ΔP	Pressure drop

REFERENCES

Adams, G., Jones, P.M., Millar, J.R., 1969. Kinetics of acid uptake by weak-base anion exchangers. J. Chem. Soc. A: Inorg., Phys., Theor. 2543–2551.

Adler, S., Beaver, E., Bryan, P., Robinson, S., Watson, J., 2000. Vision 2020: 2000 Separations Roadmap. Center for Waste Reduction Technologies of the American Institute of Chemical Engineers, New York, pp. 10016–5991.

Aguilar, M.I., Saez, J., Llorens, M., Soler, A., Ortuno, J.F., 2002. Nutrient removal and sludge production in the coagulation-flocculation process. Water Res. 36, 2910–2919.

Ahmaruzzaman, M., Gayatri, S.L., 2011. Activated neem leaf: a novel adsorbent for the removal of phenol, 4-nitrophenol, and 4-chlorophenol from aqueous solutions. J. Chem. Eng. Data 56, 3004–3016.

Ahsan, S., Kaneco, S., Ohta, K., Mizuno, T., Kani, K., 2001. Use of some natural and waste materials for waste water treatment. Water Res. 35, 3738–3742.

Ait Ouaissa, Y., Chabani, M., Amrane, A., Bensmaili, A., 2013. Removal of Cr(VI) from model solutions by a combined electrocoagulation sorption process. Chem. Eng. Technol. 36, 147–155.

Ali, I., 2012. New generation adsorbents for water treatment. Chem. Rev. 112, 5073–5091.

Aljundi, I.H., Belovich, J.M., Talu, O., 2005. Adsorption of lactic acid from fermentation broth and aqueous solutions on zeolite molecular sieves. Chem. Eng. Sci. 60 (18), 5004.

Anwar, M.M., Hanson, C., Pratt, M.W.T., 1971a. An improved dissociation extraction process for the separation of acidic and basic isomers. In: Process Intensification Solvent Extraction Conference, vol. 2, 911 pp.

Anwar, M.M., Hanson, C., Pratt, M.W.T., 1971b. Dissociation extraction. I. General theory. Trans. Ins. Chem. Eng. 49, 95.

Anwar, M.M., Cook, S.T.M., Hanson, C., Pratt, M.W.T., 1979. Separation of 2,3- and 2,6-dichlorophenols by dissociation extraction. In: Proceedings International Solvent Extraction Conference, 1977, vol. 2, 671 pp.

Anwar, M.M., Arif, A.S., Pritchard, D.W., 1995. Separation of closely related organic acids and bases dissociation extraction. Solvent Extr. Ion Exch. 13, 127–142.

Ashtekar, V.S., 2007. Studies in coagulation for removal of colour from dye wastewater. M. Tech. Thesis, University of Pune.

Ashtekar, V.S., Bhandari, V.M., Shirsath, S.R., 2010. How to develop effective coagulant formulations for dye wastewater treatment? – Key parameters and effect analysis. In: IIChE Conference CHEMCON-2010.

Aziz, H.A., Adlan, M.N., Zahari, M.S., Alias, S., 2004. Removal of ammoniacal nitrogen (N-NH3) from municipal solid waste leachate by using activated carbon and limestone. Waste Manage. Res. 22, 371–375.

http://www.balticuniv.uu.se/index.php/boll-online-library/831-swm-2-water-use-and-management[Accessed: 12 August 2013].

Bashir, T.A., Soni, A.G., Mahulkar, A.V., Pandit, A.B., 2011. The CFD driven optimisation of a modified venturi for cavitational activity. Can. J. Chem. Eng. 89, 1366–1375.

Berrin, T., 2008. New technologies for water and wastewater treatment: a survey of recent patents. Recent Patents Chem. Eng. 1, 17–26.

http://bestotc.com/493/uncategorized/cavigulation-cavitation-technologies-solution-to-the-water-crisis-refining-the-desalination-process/ [Viewed 21 January, 2014].

Bhandari, V.M., 1998. Implications of weak donnan potential in ion-exchange reactions. An alternate strategy for modeling sorption processes. Sep. Sci. Technol. 33, 2009–2024.

Bhandari, V.M., Juvekar, V.A., Patwardhan, S.R., 1992a. Sorption studies on ion exchange resins. 1. Sorption of strong acids on weak base resins. Ind. Eng. Chem. Res. 31, 1060–1073.

Bhandari, V.M., Juvekar, V.A., Patwardhan, S.R., 1992b. Sorption studies on ion exchange resins. 2. Sorption of weak acids on weak base resins. Ind. Eng. Chem. Res. 31, 1073–1080.

Bhandari, V.M., Juvekar, V.A., Patwardhan, S.R., 1993. Sorption of dibasic acids on weak base resins. Ind. Eng. Chem. Res. 32, 200–206.

Bhandari, V.M., Sawarkar, C.S., Juvekar, V.A., 1996. Ion exchange separations – selectivity in acid sorption. In: Varma, et al., (Eds.), Advances in Chemical Engineering. Proceedings of International Conference. Allied Publishers, 81 pp.

Bhandari, V.M., Juvekar, V.A., Patwardhan, S.R., 1997. Ion-exchange studies in the removal of polybasic acids. Anomalous sorption behavior of phosphoric acid on weak base resins. Sep. Sci. Technol. 32, 2481–2496.

Bhandari, V.M., Yonemoto, T., Juvekar, V.A., 2000. Investigating the differences in acid separation behaviour on weak base ion exchange resins. Chem. Eng. Sci. 55, 6197–6208.

Bhandari, V.M., Vaidya, S.H., Bobade, G.V., Ahmad, T., 2006. Studies in adsorptive separation of lactic acid using conventional and modified adsorbents. In: IIChE Conference, Chemcon-2006, Bharuch.

Bhuptawat, H., Folkard, G.K., Chaudhari, S., 2007. Innovative physico-chemical treatment of wastewater incorporating Moringa oleifera seed coagulant. J. Hazard. Mater. 142, 477–482.

Bonev, B., Nenov, V., 2006. Copper recovery from acidic wastewater by ED. Acta Metall. 12, 39–44.

Cecen, F., 2011. Water and wastewater treatment: historical perspective of activated carbon adsorption and its integration with biological processes. In: Cecen, F., Aktasr, O. (Eds.), Activated Carbon for Water and Wastewater Treatment: Integration of Adsorption and Biological Treatment. first ed WILEY-VCH Verlag GmbH & Co. KGaA, Weinheim.

Çelik, M.S., Özdermir, B., Turan, M., Koyuncu, I., Atesok, G., Sarikaya, H.Z., 2001. Removal of ammonia by natural clay minerals using fixed and fluidized bed column reactors. Water Sci. Technol. Water Supply 1 (1), 81–99.

Chakinala, A.G., Gogate, P.R., Burgess, A.E., Bremner, D.H., 2009. Industrial wastewater treatment using hydrodynamic cavitation and heterogeneous advanced Fenton processing. Chem. Eng. J. 152, 498–502.

Cho, D.L., Ekengren, Ö., 1993. Composite membranes formed by plasma-polymerized acrylic acid for ultrafiltration of bleach effluent. J. Appl. Polym. Sci. 47, 2125–2133.

Cloete, F.L.D., Marais, A.P., 1995. Recovery of very dilute acetic acid using ion exchange. Ind. Eng. Chem. Res. 34, 2464–2467.

Coleby, J., 1971. Industrial Organic Processes. In: Hanson, C. (Ed.), Recent Advances in Liquid–Liquid Extraction. Pergamon Press, Oxford.

Demir, A., Gunay, A., Debik, E., 2002. Ammonium removal from aqueous solution by ion-exchange using packed bed natural zeolite. Water SA 28, 329–336.

Didenko, Y.T., McNamara, W.B., Suslick, K.S., 1999. Hot spot conditions during cavitation in water. J. Am. Chem. Soc. 121, 5817–5818.

Dubinin, M.M., 1960. The potential theory of adsorption of gases and vapors for adsorbents with energetically non-uniform surface. Chem. Rev. 60, 235–266.

Freundlich, H.M.F., 1906. Uber die adsorption in losungen. Z. Phys. Chem. 57, 385–471.

Fung, K.Y., Lee, C.M., Ng, K.M., Wibowo, C., Deng, Z., Wei, C., 2012. Process development of treatment plants for dyeing wastewater. AIChE J. 58, 2726–2742.

Gaikar, V.G., Sharma, M.M., 1985. Dissociation extraction: prediction of separation factor and selection of solvent. Solvent Extr. Ion Exch. 3, 679–696.

Gaikar, V.G., Mahapatra, A., Sharma, M.M., 1989. Separation of close boiling point mixtures (p-cresol/m-cresol, guaiacol/alkylphenols, 3-picoline/4-picoline, substituted anilines) through dissociation extractive crystallization. Ind. Eng. Chem. Res. 28, 199–204.

Gao, B.-Y., Chu, Y.-B., Yue, Q.-Y., Wang, B.-J., Wang, S.-G., 2005. Characterization and coagulation of a polyaluminum chloride (PAC) coagulant with high Al13 content. J. Environ. Manag. 76, 143–147.

Gogate, P.R., Pandit, A.B., 2004. A review of imperative technologies for wastewater treatment I: oxidation technologies at ambient conditions. Adv. Environ. Res. 8, 501–551.

Gogate, P.R., Pandit, A.B., 2005. A review and assessment of hydrodynamic cavitation as a technology for the future. Ultrason. Sonochem. 12, 21–27.

Guy, M., Annick, V., Arlette, R., 1994. Recovery of tartaric and malic acids from dilute aqueous effluents by solvent extraction technique. J. Chem. Technol. Biotechnol. 60, 67.

Guyer, J.P., 2011. Introduction to Advanced Wastewater Treatment. Continuing Education and Development Inc., New York.

Helfferich, F.G., 1962. Ion Exchange. McGraw Hill, New York.

Helfferich, F.G., Hwang, Y.L., 1985. Kinetics of acid uptake by weak base anion exchangers. Mechanism of proton transfer. In: American Institute of Chemical Engineers Symposium, Ser. 81, No. 242, pp. 17–27.

Helsel, R.W., 1977. Waste recovery: Removing carboxylic acids from aqueous wastes. Chem. Eng. Prog. 73, 55–59.

Hiremath, R.S., Bhandari, V.M., Ranade, V.V., 2012. New methodologies for industrial wastewater treatment. In: 2-day Workshop on Industrial Wastewater Treatment, Recycle & Reuse, June 15–16, 2012, National Chemical Laboratory, Pune.

Hoffmann, M.R., Martin, S.T., Choi, W., Bahnemann, D.W., 1995. Environmental applications of semiconductor photocatalysis. Chem. Rev. 95, 69–96.

Höll, W., Sontheimer, H., 1977. Ion exchange kinetics of the protonation of weak acid ion exchange resins. Chem. Eng. Sci. 32, 755–762.

Hübner, P., Kadlec, V., Dukla, C., 1978. Kinetic behavior of weak base anion exchangers. AIChE J. 24, 149–154.

Husson, S.M., King, C.J., 1999. Multiple-acid equilibria in adsorption of carboxylic acids from dilute aqueous solution. Ind. Eng. Chem. Res. 38, 502–511.

Ingale, M.N., Mahajani, V.V., 1996. Recovery of carboxylic acids, C2-C6, from an aqueous waste stream using tributylphosphate (TBP): Effect of presence of inorganic acids and their sodium salts. Sep. Technol. 6, 1–7.

Jagirdar, G.C., Sharma, M.M., 1980. Recovery and separation of mixtures of organic acids from dilute aqueous solutions. J. Sep. Process Technol. 1, 40–43.

Jagirdar, G.C., Sharma, M.M., 1981. Separation of close boiling mixtures of heterocyclic amines and LTC tar acids by dissociation extraction. J. Sep. Process Technol. 2, 37.

Jiang, J.-Q., Ashekuzzaman, S.M., 2012. Development of novel inorganic adsorbent for water treatment. Curr. Opin. Chem. Eng 1, 191–199.

Joglekar, H.G., Rahman, I., Babu, S., Kulkarni, B.D., Joshi, A., 2006. Comparative assessment of downstream processing options for lactic acid. Sep. Purif. Technol. 52, 1–17.

Joo, D.J., Shin, W.S., Choi, J.-H., Choi, S.J., Kim, M.-C., Han, M.H., et al., 2007. Decolorization of reactive dyes using inorganic coagulants and synthetic polymer. Dyes Pigments 73, 59–64.

Jorgensen, T.C., Weatherley, L.R., 2003. Ammonia removal from wastewater by ion exchange in the presence of organic contaminants. Water Res. 37, 1723–1728.

Juang, R.-S., Chou, T.-C., 1996. Sorption of citric acid from aqueous solutions by macroporous resins containing a tertiary amine equilibria. Sep. Sci. Technol. 31, 1409–1425.

Kawabata, N., Yoshida, J.-I., Tanigawa, Y., 1981. Removal and recovery of organic pollutants from aquatic environment. 4. Separation of carboxylic acids from aqueous solution using crosslinked poly(4-vinylpyridine). Ind. Eng. Chem. Prod. Res. Dev. 20, 386–390.

Kim, T.H., Park, C., Yang, J., Kim, S., 2004. Comparison of disperse and reactive dye removals by chemical coagulation and Fenton oxidation. J. Hazard. Mater. 112, 95–103.

Kulkarni, A.A., Ranade, V.V., Rajeev, R., Koganti, S.B., 2008. CFD simulation of flow in vortex diodes. AIChE J. 54, 1139–1152.

Kumar, A., Roy, P.K., 2008. Environmental Issues and Solutions. Daya Publishing House, Delhi.

Laddha, S.S., Sharma, M.M., 1978. Separation of close boiling organic acids and bases by dissociation extraction: chlorophenols, N-alkylanilines and chlorobenzoic acids. J. Appl. Chem. Biotechnol. 28, 69–78.

Langmuir, I., 1916. The constitution and fundamental properties of solids and liquids. J. Am. Chem. Soc. 38, 2221–2295.

Lawson, K.W., Lloyd, D.R., 1997. Membrane distillation. J. Membr. Sci. 124, 1–25.

Lee, S.C., Hyun, K.-S., 2010. Development of an emulsion liquid membrane system for separation of acetic acid from succinic acid. J. Membr. Sci. 350, 333–339.

Lee, S.C., Kim, H.C., 2011. Batch and continuous separation of acetic acid from succinic acid in a feed solution with high concentrations of carboxylic acids by emulsion liquid membranes. J. Membr. Sci. 367, 190–196.

Lee, K.E., Morad, N., Teng, T.T., Poh, B.T., 2012. Development, characterization and the application of hybrid materials in coagulation/flocculation of wastewater: a review. Chem. Eng. J. 203, 370–386.

Madaeni, S.S., Mansourpanah, Y., 2006. Screening membranes for COD removal from dilute wastewater. Desalination 197, 23–32.

Mafé, S., Manzanares, J.A., Reiss, H., 1993. Donnan phenomena in membranes with charge due to ion adsorption. Effects of the interaction between adsorbed charged groups. J. Chem. Phys. 98, 2325–2331.

Malmary, G.H., Monteil, F., Molinier, J.R., Hanine, H., Conte, T., Mourgues, J., 1995. Recovery of aconitic acid from simulated aqueous effluents of the sugar-cane industry through liquid–liquid extraction. Bioresour. Technol. 52, 33–36.

Mane, J.D., Kumbhar, D.L., Barge, S.C., Phadnis, S.P., Bhandari, V.M., 2003. Studies in effective utilization of biomass. Chemical modification of bagasse and mechanism of colour removal. Int. Sugar J. 105, 412, No. 1257.

Mane, J.D., Modi, S., Nagawade, S., Phadnis, S.P., Bhandari, V.M., 2006. Treatment of spentwash using chemically modified bagasse and colour removal studies. Bioresour. Technol. 97, 1752–1755.

Matsarenko, V.A., Lebedinskaya, N.A., Dudkova, L.I., Gushchina, L.I., 1969. Recovery of water soluble C1–C4 acids from acid wastewaters of the production of synthetic fats. Khim. Prom. 45, 12.

Metcalf & Eddy, Inc., 1991. Wastewater Engineering: Treatment, Disposal and Reuse, third ed. McGraw-Hill Co., New York.

Metcalf & Eddy, Inc., 2003. Wastewater Engineering: Treatment, Disposal and Reuse, fourth ed. McGraw-Hill Co., New York.

Mishra, K.P., Gogate, P.R., 2010. Intensification of degradation of Rhodamine B using hydrodynamic cavitation in the presence of additives. Sep. Purif. Technol. 75, 385–391.

Modak, S.Y., Rane, V.C., Juvekar, V.A., Bhandari, V.M., Yonemoto, T., 1999. Separation of cresols using dissociation extraction. Kagaku Kogakkai Shuki Taikai Kenkyu Happyo Koen Yoshishu 32, 947.

Munter, R., 2000. Industrial wastewater treatment. In: Lundin, L.C. (Ed.), Water Use and Management. Uppsala University, p. 240, ISBN: 91-973579-4-4.

Muthukrishnan, M., Guha, B.K., 2006. Heavy metal separation by using surface modified nanofiltration membrane. Desalination 200, 351–353.

Navarro, R., González, M.P., Saucedo, I., Avila, M., Prádanos, P., Martínez, F., et al., 2008. Effect of an acidic treatment on the chemical and charge properties of a nanofiltration membrane. J. Membr. Sci. 307, 136–148.

Neumann, S., Fatula, P., 2009. Principles of Ion Exchange in Wastewater Treatment. Asian Water, March 2009; Pages 14–19.

Noble, R.D., Stern, S.A., 1995. Membrane Separation Technology: Principles and Applications. Elsevier, Amsterdam.

Othmer, D.F., 1958. Acetic acid recovery methods. Chem. Eng. Prog. 54, 48.

Pandare, A., Ranade, V.V., 2013. Flow in Vortex Diode. CSIR-NCL Internal Report.

Pandit, A.B., Joshi, J.B., 1993. Hydrolysis of fatty acids: effect of cavitation. Chem. Eng. Sci. 48, 3440.

Parulekar, S.J., Sharma, M.M., Joshi, J.B., Shah, Y.T., 1982. Separation processes for recovery of valuable chemicals from aqueous effluents. J. Sep. Process Technol. 3, 2–29.

Ranade, V.V., Bhandari, V.M., 2013. Application of Vortex Diode for Enhancing Gas Yield of Anaerobic Digester: Field Trials with 40 m^3/h Flow Rate. NCL Internal Report.

Ranade, V.V., Pandit, A.B., Anil, A.C., Sawant, S.S., Ilangovan, D., Madhan, R., Pilarisetty, V.K., 2008. Apparatus for filtration and disinfection of seawater/ship's ballast water and a method of same. U.S. Pat. Appl. Publ. US 20080017591 A1 24 Jan 2008, 19 pp. (English). (United States of America.) CODEN: USXXCO. INCL: 210767000; 210416100. APPLICATION: US 2007-726399 20 Mar 2007. PRIORITY: IN 2006-DE734 20 Mar 2006.

Ranade, V.V., Kulkarni, A.A., Bhandari, V.M. 2013. Vortex diodes as effluent treatment devices, PCT Int. Appl. (2013) WO 2013054362 A2 20130418.http://www.google.com/patents/WO2013054362A3?cl=en [Accessed: 20th August 2013].

Rao, M.G., Gupta, A.K., 1982a. Ion exchange processes accompanied by ionic reactions. Chem. Eng. J. 24, 181–190.

Rao, M.G., Gupta, A.K., 1982b. Kinetics of ion exchange in weak base anion exchange resins. In: AIChE Symposium Series, No. 219, vol. 78, 96 pp.

Redlich, O., Peterson, D.L., 1959. A useful adsorption isotherm. J. Phys. Chem. 63, 1024–1026.

Ricker, N.L., Pittman, E.F., King, C.J., 1980. Solvent extraction with amines for recovery of acetic acid from dilute aqueous industrial streams. J. Sep. Process Technol. 1, 23.

Rios, R.R.A., Alves, D.E., Dalmázio, I., Bento, S.F.V., Donnici, C.L., Lago, R.M., 2003. Tailoring activated carbon by surface chemical modification with O, S, and N containing molecules. Mater. Res. 6, 129–135.

Rožić, M., Cerjan-Stefanović, Š., Kurajica, S., Vančina, V., Hodžić, E., 2000. Ammoniacal nitrogen removal from water by treatment with clays and zeolites. Water Res. 34, 3675–3681.

Ruthven, D.M., 1984. Principles of Adsorption & Adsorption Process. John Wiley and Sons, New York.

Saharan, V.K., Badve, M.P., Pandit, A.B., 2011. Degradation of reactive red 120 dye using hydrodynamic cavitation. Chem. Eng. J. 178, 100–107.

Saharan, V.K., Rizwani, M.A., Malani, A.A., Pandit, A.B., 2013. Effect of geometry of hydrodynamically cavitating device on degradation of orange-G. Ultrason. Sonochem. 20, 345–353.

Sarioglu, M., 2005. Removal of ammonium from municipal wastewater using natural Turkish (Dogantepe) zeolite. Sep. Purif. Technol. 41, 1–11.

Sata, T., 1991. Ion exchange membranes and separation processes with chemical reactions. J. Appl. Electrochem. 21, 283–294.

Sawant, S.S., Anil, A.C., Krishnamurthy, V., Gaonkar, C.A., Kolwalkar, J., Khandeparker, L., et al., 2008. Effect of hydrodynamic cavitation on zooplankton: a tool for disinfection. Biochem. Eng. J. 42, 320–328.

Schlosser, S., Kertesz, R., Martak, J., 2005. Recovery and separation of organic acids by membrane based solvent extraction and pertraction. Sep. Purif. Technol. 41, 233–266.

Scholler, C., Chaudhuri, J.B., Pyle, D.L., 1993. Emulsion liquid membrane extraction of lactic acid from aqueous solutions and fermentation broth. Biotechnol. Bioeng. 42, 50–58.

Sonawane, S.H., Chaudhari, P.L., Ghodke, S.A., Parande, M.G., Bhandari, V.M., Mishra, S., Kulkarni, R.D., 2009. Ultrasound assisted synthesis of polyacrylic acid-nanoclay nanocomposite and its application in sonosorption studies of malachite green dye. Ultrason. Sonochem. 16, 351–355.

Suslick, K.S., Midleleni, M.M., Reis, J.T., 1997. Chemistry induced by hydrodynamic cavitation. J. Am. Chem. Soc. 119, 9303–9304.

Tripathy, T., De, B.R., 2006. Flocculation: a new way to treat the waste water. J. Phy. Sci. 10, 93–127.

Tzoupanos, N.D., Zouboulis, A.I., 2008. Coagulation-flocculation processes in water/wastewater treatment: the application of new generation of chemical reagents. In: Proceedings of the 6th IASME/WSEAS International Conference on Heat Transfer, Thermal Engineering and Environment (HTE'08), Rhodes, Greece, August 20–22.

Wadekar, V.V., Sharma, M.M., 1981a. Separation of close boiling substituted phenols by dissociation extraction. J. Chem. Technol. Biotechnol. 31, 279–284.

Wadekar, V.V., Sharma, M.M., 1981b. Separation of close boiling organic acids/bases; binary and ternary systems; substituted anilines; binary systems with thermally regenerative extractants; chlorophenols. J. Sep. Process Technol. 2, 28–32.

Wang, J., Yue, Z., Economy, J., 2008. Novel method to make a continuous micro-mesopore membrane with tailored surface chemistry for use in nanofiltration. J. Membr. Sci. 308, 191–197.

Wang, X., Jia, J., Wang, Y., 2011. Degradation of C.I. Reactive Red 2 through photocatalysis coupled with water jet cavitation. J. Hazard. Mater. 185, 315–321.

Xu, R., Jiang, R., Wang, J., Liu, B., Gao, J., Wang, B., Han, G., Zhang, X., 2010. A novel method treating organic wastewater: air-bubble cavitation passing small glass balls. Chem. Eng. J. 164, 23–28.

Yuan, Y.-L., Wen, Y.-Z., Li, X.-Y., Luo, S.-Z., 2006. Treatment of wastewater from dye manufacturing industry by coagulation. J. Zhejiang Univ. Sci. A 7, 340–344.

Zaharia, C., Suteu, D., 2012. Textile organic dyes – characteristics, polluting effects and separation/elimination procedures from industrial effluents – a critical overview. In: Puzyn, T., Mostragszlichtyng, A. (Eds.), Organic Pollutants Ten Years After the Stockholm Convention – Environmental and Analytical Update. InTech, Rijeka, Croatia.

Zuo, W., Zhang, G., Meng, Q., Zhang, H., 2008. Characteristics and application of multiple membrane process in plating wastewater reutilization. Desalination 222, 187–196.

CHAPTER 3

Advanced Oxidation Technologies for Wastewater Treatment: An Overview

Virendra K. Saharan[1], Dipak V. Pinjari[2], Parag R. Gogate[2], Aniruddha B. Pandit[2]
[1]Chemical Engineering Department, Malaviya National Institute of Technology, Jaipur, India
[2]Chemical Engineering Department, Institute of Chemical Technology, Mumbai, India

3.1 INTRODUCTION

In today's industrial arena, new molecules are developed and manufactured to meet the ever-increasing human demand; thus, a lot of new hazardous molecules are being added continually to our water bodies through effluents coming from different manufacturing units. The waste coming from various chemical-producing industries, including pesticides, pharmaceutical, petro-chemical, and other process units, contains complex molecules that are bio-refractory in nature and thus cannot be completely degraded by conventional biological processes. With more stringent rules and regulations, these industries do not have any choice but to treat these chemicals up to a safe dischargeable limit. Various methods are available to treat waste effluent; the operation of a typical wastewater treatment unit utilizes a combination of various physical, chemical, and biological processes.

The processes involved in the operation of a wastewater treatment unit consist of primary treatment (e.g., screening, mixing, flocculation, sedimentation, flotation, and filtration), secondary treatment (e.g., aerobic, anaerobic, anoxic, and facultative processes), and tertiary/advanced treatment (e.g., adsorption, ion exchange, membrane filtration, disinfection, and oxidation using chemicals). Most of these processes, apart from biological treatment and chemical oxidation, do not involve chemical transformations and therefore generally transfer waste components from one phase to another, thus causing secondary loading of the environment. On the other hand, the high fabrication and maintenance costs of advanced tertiary treatment make these processes uneconomical. Such technoeconomic limitations make these processes unviable; therefore, new techniques that can overcome these techno-economic limitations need to be developed.

In the last two decades a lot of research work has been carried out for the development of new technologies, especially in the area of advanced oxidation processes (AOPs) for the degradation of complex biorefractory pollutants for complete mineralization or as a pretreatment (Babuponnusami and Muthukumar, 2012; Gogate and Pandit, 2004a,b; Kusvuran and Erbatur, 2004; Pang et al., 2011; Thiruvenkatachari et al., 2007). AOPs are defined as processes that involve the generation and use of the hydroxyl radical ($^{•}$OH) as a strong oxidant to destroy (oxidize) compounds that cannot be oxidized by conventional oxidants such as gaseous oxygen, ozone, and chlorine. The hydroxyl radical reacts with the dissolved constituents, initiating a series of oxidation reactions until the constituents are completely mineralized to CO_2 and H_2O.

Different AOPs have been developed and tested for the degradation of different pollutants (inorganic and organic compounds) present in the wastewater. These processes include cavitation (generated either by means of ultrasonic irradiation or using constrictions such as valves, orifices, and venturi in hydraulic devices) (Adewuyi, 2001; Hua and Hoffmann, 1997; Joshi and Gogate, 2012; Saharan et al., 2011; Sivakumar and Pandit, 2002; Wang et al., 2008; Weavers et al., 1998), photocatalytic oxidation (using ultraviolet radiation/near UV light/sunlight in the presence of a semiconductor catalyst) (Adewuyi, 2005; Cao et al., 2006; Konstantinou and Albanis, 2004; Lin et al., 2012), and Fenton chemistry (using the reaction between Ferrous sulphate and hydrogen peroxide (H_2O_2), i.e., Fenton's reagent) (Karci et al., 2012; Kusic et al., 2006; Xue et al., 2009). These AOPs can also be used in combinations termed as *hybrid methods* such as Ultrasound assisted Fenton, sono-photocatalytic, Photo-Fenton, and ozone/hydrogen peroxide to get the enhanced oxidation efficiency and overcome the limitations and difficulties of individual AOPs toward some specific pollutants (Adewuyi, 2005; Gogate and Pandit, 2004b; Pang et al., 2011). In this chapter, we review these AOPs and their combinations. The total or partial success of these AOPs in relation to some specific biorefractory pollutants is discussed and recommendations are made to arrive at an optimized treatment methodology/flow sheet.

3.2 CAVITATION

Cavitation is defined as the phenomenon of the formation, growth, and subsequent collapse of microbubbles or cavities occurring in an extremely small interval of time (microseconds) and at multiple locations in the reactor, releasing large magnitudes of energy. As a result of cavity collapse, hot spots are created, where the temperature and pressure can reach up to 10,000 K and 2000 atm; high velocity water jets (100–300 m/s) are also created (Adewuyi, 2001; Didenko et al., 1999; Pang et al., 2011). Some of the secondary effects resulting from the cavity collapse are chemical transformation (chemical bond breakage) that releases highly reactive free radicals, physical cleaning of solid surfaces, and enhancement in diffusive mass transfer rates. The destruction/oxidation of organic pollutants using cavitation takes place through two mechanisms: (1) thermal decomposition/pyrolysis of the volatile pollutant molecule in and around the collapsing bubbles and (2) oxidation of pollutant molecules by reactive free radicals (such as $HO^{\bullet}$, $O^{\bullet}$, and $HOO^{\bullet}$ radicals) generated during the cavity collapse (Hua and Hoffmann, 1997). The cavity contains gases and water vapors and other volatile contents that dissociate under cavitating conditions according to the reactions such as cleavage of water molecules (into $H^{\bullet}$ atoms and $^{\bullet}OH$ radicals) and dissolution of oxygen molecules. From the reactions of these entities ($O^{\bullet}$, $H^{\bullet}$, $^{\bullet}OH$) with each other and with H_2O and O_2 during the rapid quenching phase, $HO_2^{\bullet}$ radicals and H_2O_2 are formed. These radicals ($^{\bullet}OH$, $O^{\bullet}$, and $HOO^{\bullet}$) then diffuse into the bulk liquid medium where they react with oxidizable pollutants and oxidize them. The following are the possible reactions occurring as the result of cavity collapse (Hamadaoui and Naffrechoux, 2008; Pang et al., 2011):

$$H_2O +))) \rightarrow HO^{\bullet} + H^{\bullet} \qquad (3.1)$$

$$O_2 +))) \rightarrow 2O^{\bullet} \qquad (3.2)$$

$$O^{\bullet} + H_2O \rightarrow 2HO^{\bullet} \qquad (3.3)$$

$$HO^{\bullet} + H^{\bullet} \rightarrow H_2O \qquad (3.4)$$

$$2HO^{\bullet} \rightarrow O^{\bullet} + H_2O \qquad (3.5)$$

$$H^{\bullet} + O_2 \rightarrow HOO^{\bullet} \qquad (3.6)$$

$$2HO^{\bullet} \rightarrow H_2O_2 \qquad (3.7)$$

$$2HOO^{\bullet} \rightarrow H_2O_2 + O_2 \qquad (3.8)$$

Cavitation is classified into four types based on the method of generation: acoustic, hydrodynamic, optic, and particle. Among these, only acoustic and hydrodynamic cavitation (HC) have been found to be economically efficient and feasible in bringing about the desired chemical and physical changes, whereas optic and particle cavitation are typically used for single bubble cavitation, which fails to induce chemical change in a bulk solution.

3.2.1 Acoustic Cavitation

In the case of acoustic cavitation, also termed sonication (US), cavitation is produced using high-frequency sound waves, usually ultrasound, with frequencies in the range of 16 kHz to 2 MHz (Lorimer and Mason, 1987; Suslick, 1990). Alternate compression and rarefaction cycles of the sound waves result in various phases of cavitation such as generation of the bubble/cavity, growth phase, and finally the collapse. During the compression cycle, the average distance between the molecules decreases, while during rarefaction the distances increase. If a sufficiently large negative pressure is applied to the liquid, such that the average distance between the molecules exceeds the critical molecular distance necessary to hold the liquid intact, the liquid will break down and voids or cavities will be created. These cavitation bubbles may grow in size until the maximum negative pressure has been reached. In the following compression cycle of the sound wave, these cavities will compress, i.e., decrease in volume, and some of them may implosively collapse. Because these events occur over extremely small time intervals (micro- to nano-seconds), other transport processes are absent, and the final collapse stage is adiabatic in nature, thus producing very high local temperatures (up to 10,000 K) and pressures (up to 2000 atm).

Since 1990, there has been an increasing interest in the use of ultrasound to destroy organic contaminants present in wastewater (Fındık and Gündüz, 2007; Francony and Petrier, 1996; Hamadaoui and Naffrechoux, 2008; Merouani et al., 2010; Nagata et al., 2000; Shrestha et al., 2009; Wang et al., 2006; Weavers et al., 1998). Many researchers have reported that ultrasonic irradiation processes were capable of degrading various recalcitrant organic compounds, such as phenolic compounds, chloroaromatic compounds, aqueous carbon tetrachloride, pesticides, herbicides, benzene based compounds, polycyclic aromatic hydrocarbons, and organic dyes. The efficiency of the acoustic cavitation reactor depends on the intensity of the cavity collapse, which in turn depends on the total number of cavitational events occurring inside the reactor and the final cavity collapse pressure. This again depends on several operational parameters such as frequency of ultrasound, irradiating surface, intensity of sound waves, calorimetric efficiency of ultrasonic equipment (power dissipated into the system per unit power supplied), physicochemical properties of the liquid medium, and the presence of air and solid particles. For the best results, these parameters need to be optimized.

3.2.1.1 Reactors Used for Acoustic Cavitation

Different types of acoustic cavitational reactors are being used including ultrasonic horn, ultrasonic bath, and multiple frequency flow cells (Gogate et al., 2011; Hua and Hoffmann, 1997; Koda et al., 2003; Sutkar and Gogate, 2009; Weavers et al., 1998). Figure 3.1 gives the schematic representation of the most commonly used equipment for acoustic cavitation.

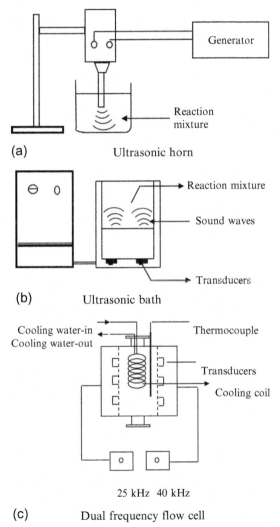

Figure 3.1 Schematic representation of the equipment used for acoustic cavitation.

(Continued)

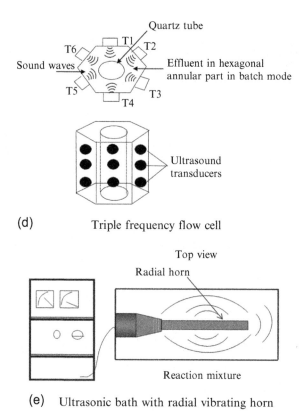

(d) Triple frequency flow cell

(e) Ultrasonic bath with radial vibrating horn

Figure 3.1, cont'd

In the acoustic cavitation reactor, the cavitational activity is found to be maximum only near to the irradiating surface of the horn/transducer; hence, the major challenge in the design of such a reactor is to dissipate supplied energy more uniformly. For higher transformational cavitational yield of an acoustic cavitation reactor, where the cavitational yield is defined as moles of compound oxidized/degraded per unit energy used, a larger irradiating surface area is recommended such that power supplied can be dissipated into a larger volume. Also, the use of equipment based on multiple frequencies/ multiple transducers (device used for converting the supplied electrical energy into sound energy and generating ultrasound with frequency in the range of 16 kHz to 2 MHz) has been reported to be more beneficial as compared to the equipment based on a single frequency (Gogate and Pandit, 2004a). This is due to the creation of efficient cavity dynamics using multiple frequencies. Ultrasonic horns vibrating in radial directions, which also gives the advantage of better energy dissipation due to a larger irradiating area, is another new development with a promising future for medium- to large-scale applications, but more work is required in terms of testing this equipment for operation at high frequency and high power dissipation.

3.2.1.2 *Optimization of Operating Parameters for Acoustic Cavitation*
3.2.1.2.1 Effect of Frequency

Higher frequencies of sound waves are suited for effective destruction of pollutants using acoustic cavitation up to an optimum value (Francony and Petrier, 1996; Hua and Hoffmann, 1997; Hung and Hoffmann, 1999; Petrier et al., 1996). A higher frequency leads to higher cavitational collapse intensity; hence, the total quantum of collapse pressure increases, leading to higher cavitational activity. But at a very high frequency, the cavity generation may reduce due to the short time of the rarefaction phase (higher frequency) and may require higher driving pressures. This can result in the immediate collapse of cavities without attaining maximum size (smaller life time), thereby causing reduced degradation efficiency (pollutant molecules experience cavitational conditions over a shorter time due to the shorter life span of the cavitational bubble). The optimum frequency found in the acoustic cavitation reactor depends on the type of pollutant to be treated and the concentration along with the reactor configuration (Weavers et al., 1998). Also, continuous operation with high frequencies at a larger scale of operation (and hence at higher power dissipation levels) leads to an erosion of the transducer surface due to the sudden and immediate implosion of the cavities present inside the reactor. Moreover, the power required for the onset of cavitation increases with an increase in the frequency of irradiation; hence the process may become uneconomical at much higher frequencies of irradiation. Instead of using a single high frequency transducer, reactors having multiple low frequency transducers located at different and strategic locations are found to be more energy efficient due to their ability to generate a similar cavitational intensity as in the high-frequency acoustic reactor. Sivakumar et al. (2002) have described a design where six transducers in total have been attached on the opposite faces of a rectangular cross-section irradiating 25 and 40 kHz either individually or simultaneously. They have reported that significantly higher collapse pressure pulse is generated at the end of the cavitational event for the multiple-frequency operation as compared to the single-frequency operation. This results in higher transformational yields. Thus, dual- or triple-frequency reactors are recommended for use that will also give similar results as a single very high frequency transducer, but with minimal erosion problems.

3.2.1.2.2 Effect of Irradiating Surface

In the acoustic cavitation reactor, most of the cavitational activity is present near the vicinity of the irradiating surface, and the liquid present in that area is only exposed to the cavitating conditions. The lower energy efficiency of the acoustic cavitational reactor is mainly due to poor energy dissipation in the surrounding liquid. Greater energy efficiency has been observed for ultrasonic probes with higher irradiating surfaces (lower operating intensity of irradiation), which results in uniform dissipation of energy (Gogate et al., 2001). Thus, for the same power density (power input into the system per unit volume of the effluent to be treated), power input to the system should be through larger areas of irradiating surface. Flow cells having transducers at multiple locations fitted at different heights and irradiating planes can be used to increase the irradiation surface and hence the energy transfer efficiency.

3.2.1.2.3 Intensity of Irradiation

The intensity of ultrasonic equipment is defined as the ratio of the system's power input to the transmitting area. Hence, intensity of irradiation can be varied either by changing the power input to the system or by changing the irradiating area of the transducers in the equipment. There exists a critical intensity at which the beneficial effect of cavitation starts to occur. This is due to the fact that the chemical reactions due to cavitation occur as a result of the formation of a certain minimum number of free radicals. This number of free radicals depends on the intensity of cavity collapse and the number of cavitating events, which in turn depends on the operating intensity of irradiation. Thus, the intensity of irradiation should not be decreased below a certain minimum cavitation intensity. Moreover, as said earlier, the two methods by which intensity can be changed also play an important role in the actual results. If the intensity is increased by increasing the power input to the system (P/V of the system), there will be an increase in the number of cavitation events; hence the cumulative pressure pulse (number of cavities generated multiplied by the collapse pressure due to a single cavity) will increase. In such a case, the degradation rates will be higher due to higher overall magnitude of the pressure energy released. However, very high power dissipation or larger intensity could result in lower cavitational intensity because of the formation of more cavities over a small area (number density of cavities), which start coalescing with each other and result in larger bubbles collapsing in the gas-liquid compressible dispersion, leading to a lower pressure pulse at the time of collapse (Fındık and Gündüz, 2007; Gogate et al., 2003; Hamadaoui and Naffrechoux, 2008). On the other hand, if the intensity is changed by changing the transmittance area of ultrasonic equipment, at lower intensities, the same power dissipation is taking place over a larger area, resulting in uniform energy dissipation and a larger active area of cavitation and higher cavitational yields.

3.2.1.2.4 Effect of Physico-chemical Properties of Liquid

The physico-chemical properties of the liquid medium, such as vapor pressure, surface tension, solvent viscosity, and presence of impurities/gases, also crucially affect the performance of the sonochemical reactors by altering the cavity dynamics. The cavity formation (inception) and the number of cavities being generated, the initial size of the nuclei/cavity, and the maximum size reached by the cavities before collapse depend mainly on these liquid phase physico-chemical properties. Threshold power requirement for cavitation inception can be defined as the minimum power required for the onset of the cavitation process, i.e., the formation of the cavities, and this should be as low as possible so that the energy effectively available for the growth of the cavities is larger (the total supplied energy is utilized for generation of cavitational nuclei and for the growth of cavities followed by subsequent collapse), leading to a higher collapse pressure pulse (Gogate and Pandit, 2000a). The total quantum of pressure/temperature pulse generated as a result of cavitation is the product of the pulse generated by the collapse of a single cavity (lower initial cavity size results in higher collapse pressure/temperature pulse) multiplied by the number of cavities generated in the reactor deciding the overall transformational energy delivery. It is hence advisable to have a large number of cavitational events occurring in the reactor with lower initial size of the cavitating nuclei. Higher surface tension, lower viscosity, and lower vapor pressure favor cavitation. Therefore, the liquid phase physico-chemical properties should be adjusted in such a way so as to lower the cavitation inception threshold, resulting in easy generation of cavities and at the same time increasing the number of cavities generated with lower initial size, allowing them to grow and collapse more violently. This will result in higher energy delivery to the system and a higher extent of transformational efficiency.

3.2.2 Hydrodynamic Cavitation

One of the alternative techniques for the generation of cavitation is the use of hydraulic devices where cavitation is generated by the passage of the liquid through a constriction such as a valve, orifice plate, or venturi (Gogate and Pandit, 2005; Moholkar et al., 1999).

Hydrodynamic cavitation (HC) can simply be generated by the passage of the liquid through a constriction such as an orifice plate. When the liquid passes through the reduced cross-sectional flow area of the constriction, the kinetic energy/velocity of the liquid increases at the expense of the pressure. If the throttling is sufficient to cause the pressure around the point of vena contracta to fall below the threshold pressure for cavitation (usually vapor pressure of the medium at the operating temperature), cavities are generated locally. Subsequently, as the liquid jet expands downstream of the constriction, the pressure recovers, and this results in the collapse of the cavities. During the passage of the liquid through the constriction, boundary layer separation occurs and a substantial amount of energy is lost in the form of a permanent pressure drop. Very high intensity turbulence occurs on the downstream side of the constriction; its intensity depends on the magnitude of the pressure drop, which, in turn, depends on the geometry of the constriction and the flow conditions of the liquid. The intensity of turbulence has a profound effect on the cavitation intensity (Moholkar and Pandit, 1997). Thus, by controlling the geometric and operating conditions of the reactor, one can produce the required intensity of the cavitation so as to bring about the desired chemical and/or physical change with maximum efficiency. Also, the collapse temperatures and pressures generated during the cavitation phenomena are a strong function of the operating and geometric parameters (Gogate and Pandit, 2000b).

Figure 3.2 shows the hydrodynamic generation of cavities schematically. The pressure-velocity relationship of the flowing fluid as explained by Bernoulli's equation can be exploited to achieve this effect. When flowing liquid passes through a mechanical constriction (either orifice or venturi; Figure 3.2a), its velocity increases, accompanied by an increase in the kinetic energy and a corresponding decrease in the local pressure (Figure 3.2b). If the throttling is sufficient to reduce the absolute local pressure below the vapor pressure (at the operating temperature), spontaneous vaporization of the medium in the form of microbubbles (nucleation) occurs. With continued lowering of the pressure, the cavity continues to grow by further vaporization or desorption of gases (usually some gases are dissolved

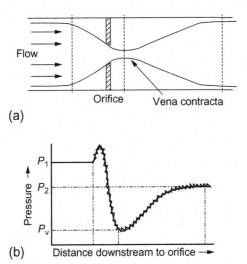

Figure 3.2 Fluid flow and pressure variation in a hydrodynamic cavitation (HC) setup.

in the medium) which reaches its maximum size at the lowest pressure. Subsequent increase (pressure recovery) in the pressure compresses this fully grown cavity and subsequently to collapse in near adiabatic fashion, thus generating extreme pressure and temperature conditions.

A dimensionless number known as the cavitation number (C_V) is used to relate the flow conditions to the cavitational intensity. The cavitation number is given by the following equation:

$$C_V = \left(\frac{p_2 - p_v}{\frac{1}{2} p v_0^2} \right) \tag{3.9}$$

where p_2 is the fully recovered downstream pressure, p_v is the vapor pressure of the liquid, and v_0 is the velocity at the throat of the cavitating constriction.

The number at which the inception of cavitation occurs is known as the cavitation inception number C_{vi}. Ideally speaking, cavitation inception occurs at C_{vi} equal to 1, and there are significant cavitational effects at C_v values of less than 1. In the earlier work by Gogate and Pandit (2000b), it has been shown that cavities oscillate under the influence of a fluctuating pressure field, and the magnitudes of pressure pulses generated are much less, insignificant or too small to bring about a desired chemical change for the case where C_v values are greater than 1. However, cavitation has been found to occur at higher cavitation numbers also, possibly due to the presence of dissolved gases or some impurities in the liquid medium acting as cavitational nuclei. Yan and Thorpe (1990) have studied the effect of the geometry of cavitating devices (orifice plates) on the inception of cavitation. They observed that for a given size orifice, the cavitation inception number remains constant within an experimental variation narrow bound for a specified liquid. The cavitation inception number does not change with the liquid velocity, is a constant for a given orifice size, and is found to increase with an increase in the orifice size and dimension. Moholkar and Pandit (1997) have discussed these observations in terms of the variation in the turbulent fluctuating velocity magnitude with the orifice dimensions specifically.

The major advantages of HC are the following:

- It is one of the cheapest and most energy-efficient methods of generating cavitation.
- The equipment used for generating cavitation is simple.
- Maintenance of such reactors is negligible.
- The scale-up of these types of reactor is relatively easy.
- Independent of the wastewater composition, wastewater having a high chemical oxygen demand (COD) can be treated more effectively for COD reduction or pretreatment.
- It can be used at multiple locations in the existing treatment process, i.e., before and after the biological treatment process. It can also serve multiple applications such as complete oxidation of refractory pollutants; breakdown of complex molecules into smaller biodigestible molecules that can be further degraded by conventional processes, hence increasing the efficiency of conventional processes; and also for disinfection, thus reducing the quantum of chemicals used for disinfection.

These advantages make HC a useful technique that can be utilized successfully at an industrial scale. Researchers have now started to look at HC as a future technology for wastewater treatment. In last few years focus has been shifted from ultrasound based reactors toward the study of HC for the degradation of organic pollutants as a potential technique to be used on a larger scale. There are few reports available on the applications of HC to degrade pollutants (Braeutigam et al., 2009; Bremner et al., 2008; Chakinala et al., 2009; Padoley et al., 2012; Patil and Gogate, 2012; Pradhan and Gogate, 2010; Saharan et al., 2012; Sivakumar and Pandit, 2002; Wang and Zhang, 2009; Wang et al., 2011a,b).

In HC, the cavitational yield (e.g., the amount of pollutant degraded/mineralized per unit energy dissipated) depends on the intensity of cavity collapse, which in turn depends on several parameters, such as number of cavitational events, the maximum size of the cavity reached before its collapse, and the surrounding pressure field. In HC, all these parameters depend on the geometry of the cavitational device and the operating parameters. Important parameters that decide degradation efficiency and the overall cavitational yield are the following:

- Inlet pressure and the cavitation number
- Physicochemical properties of the liquid and the initial radius of the nuclei
- Size and shape of the throat and divergent section (in the case of venturi)
- Percentage cross-sectional free area offered for the flow.

3.2.2.1 HC Reactor

Different HC setups with different types of cavitating device have been used such as the single-hole orifice plate, the multiple-hole orifice plate, the circular venturi, and the slit venturi. Figure 3.3 shows the schematic of the laboratory-scale HC setup. Different designs of hydrodynamic cavitating devices are shown in Figure 3.4. The design of a hydrodynamic cavitating device depends on the application for which it is required and needs to be

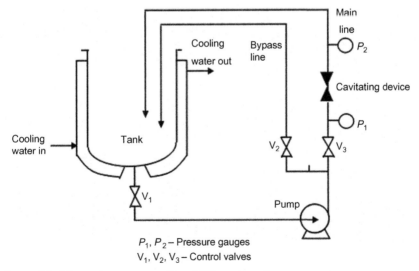

P_1, P_2 – Pressure gauges
V_1, V_2, V_3 – Control valves

Figure 3.3 Schematic representation of HC reactor setup.

first optimized in terms of the cavitational intensity required for the desired application. A typical HC setup consists of a closed loop circuit, including a feed tank, pump, pressure gauges, and valves. The bottom side of the feed tank is connected to the suction side of the pump. The discharge line from the pump is branched into two lines. The first line is for bypass and the second line is described as main line which contains a cavitating device. A bypass line is provided to control the flow through the main line containing the cavitation device. Control valves (V_1, V_2, and V_3) are provided at appropriate places in the bypass line and before cavitating devices in main line to control the flow through the cavitating device. Pressure gauges are provided to measure the inlet pressure (P_1) and fully recovered downstream pressure (P_2).

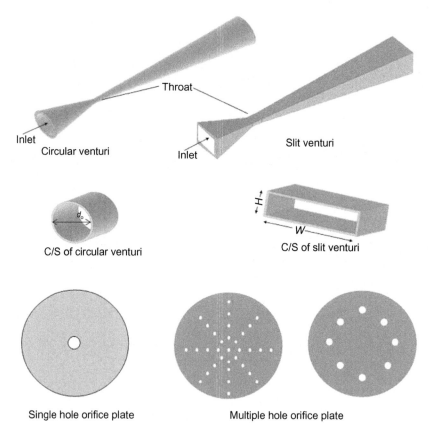

Figure 3.4 Schematic diagram of different cavitating devices.

3.2.2.2 Optimum Operating Conditions
3.2.2.2.1 Effect of Operating Pressure and Cavitation Number

The operating inlet pressures to the cavitating device and cavitation number are the two important parameters that affect the cavitational intensity generated in the reactor. The number of cavities being generated and the pressure/temperature pulse generated due to cavity collapse depend very much on the inlet pressure and the cavitation number. The operating inlet pressure and cavitation number also depend on the type of effluent to be treated because the physico-chemical properties (surface tension, density, etc.) of the effluent affect the cavity generation rate and its subsequent dynamic. A lower cavitation number or higher operating inlet pressure to the cavitating device is more useful for oxidizing organic pollutants because the number of cavitational events, and thus the final collapse pressure (which is equal to the number of cavities generated multiplied by the collapse pressure of single cavity, as explained before) also increases, resulting in more HO• radical generation. Decreasing cavitation number to a very low value may also not be effective due to conditions of super-cavitation and condition of optimum cavitation number exists. The optimum cavitation number was found to be in the range of 0.15–0.4 depending on the type of effluent to be treated and the geometry of the cavitating device used (Saharan et al., 2013; Senthilkumar et al., 2000; Sivakumar and Pandit, 2002). A cavitation number below the optimum number results in choked cavitation, with an outcome of reduced cavitational intensity. This should be avoided to obtain the maximum effect (Saharan et al., 2011).

3.2.2.2.2 Effect of Geometry of a Cavitating Device

The optimum cavitational yield of HC is dependent on several operating parameters: number of cavitational events occurring inside a cavitating reactor, residence time of cavity in the low-pressure zone (maximum size reached by the cavity before its collapse), and the rate of pressure recovery downstream of the throat (Bashir et al., 2011). These parameters depend on the geometry of the cavitating device and the flow conditions of the liquid, i.e., the scale of turbulence and the rate of pressure recovery. All these parameters need to be optimized considering interactive effect to get the enhanced cavitational yield from the HC because considering only one parameter in the design of a cavitating device would not result in the possible optimization of all cavitational conditions for the desired effects. This is because none of these parameters is independent. Thus, the cavitational condition can be altered by changing the ratio of the perimeter of cavitating holes to the cross-sectional area (α), and the ratio of the throat/orifice diameter to the pipe diameter and the divergent angle (in the case of a venturi). Different designs of cavitating devices can be constructed using the variations in these parameters which can aid in controlling the cavitational intensity.

The cavity inception and the number of cavities being generated depend on the size and shape of the constriction. To quantify this dependency, two parameters need to be considered in the analysis: (a) parameter α, which is defined as the ratio of the throat perimeter to the throat area and (b) β, which is the ratio of the throat area to the cross-sectional area of the pipe.

(a) Throat size and shape

Changing the shape of the throat, for example, the orifice plate and venturi, can be a efficient approach with the aim of maximizing α. It has been observed that for plates having the same flow area, it is advisable to use a plate with a smaller hole size opening, thereby increasing the number of holes in order to achieve a larger extent of the shear layer (Sivakumar and Pandit, 2002). The higher value of α results in the generation of a higher number of cavities. The value of turbulence pressure fluctuating frequency (f_T) also increases, leading to a more efficient collapse. On the contrary, for larger hole sizes, the frequency of turbulence (f_T) is likely to be much lower than the natural oscillation frequency of the generated cavity, resulting in a lower cavity collapse intensity. Also, parameter α can be increased by constructing different throat shapes such as rectangular and elliptical. Also, if there is a choice for the magnitude of the flow area, a lower percentage area should be chosen because with a decrease in flow area, the intensity of cavitation increases. The size of the required throat depends on the cavitation number. The area of the throat can be finalized based on the required cavitation number for the desired physicochemical transformation and volume of the effluent to be treated.

(b) Size and shape of the divergent section

Once the cavities are formed, they need to grow to a certain size before collapse, or in other words, the life of cavity should increase in order to get maximum cavitational yields. In the case of the orifice plate, because of the sudden pressure drop, cavities are forced to collapse immediately without attaining maximum size, which results in a reduced cavitational effect. In the case of the venturi, the pressure recovers smoothly because of the divergent angle, and cavities get enough time to grow to a maximum size before collapsing, thereby allowing cavities to undergo various cycles of expansion and collapse.

This exposes pollutant molecules to the cavitating conditions for a longer time. The divergent angle should be optimized for the desired cavitational intensity because a venturi with a larger divergent angle will behave as an orifice plate and will not enhance cavity growth. An optimum divergent angle should be between $10°$ and $15°$ (Bashir et al., 2011).

3.2.2.2.3 Effect of Physicochemical Properties of Liquid and Operating pH

Liquid phase properties is one of the very important aspect that affect cavitational processes, although the magnitude of the effect of all the liquid properties may not be the same. Most of the liquid properties affect cavitation in more than one way. For example, while an increase in the surface tension of the liquid increases the threshold pressure for cavitation, making the generation of cavities more difficult, the collapse of cavities is more violent. The opposing effects of liquid properties give ample scope for optimization. It should also be noted that the physicochemical properties of the liquid also decide the initial size of the nuclei, and the effect of the initial radius must also be considered when choosing a particular liquid medium and the process conditions. The state of the molecules, i.e., hydrophobic or hydrophilic, also plays an important role in the degradation of pollutant using HC. It has been stated that hydrophobic and more volatile compounds are more easily degraded through HC when compared to hydrophilic compounds. This is due to the fact that hydrophobic and volatile compounds can easily enter the gas-water interface region of cavities because of their hydrophobic nature. Thus, these compounds are more readily subjected to the $^{\bullet}OH$ radical attack and also to thermal decomposition. Thus, the overall decomposition of pollutant molecules is attributed to the pyrolysis and free radical attack occurring at both the cavity-water interface and in the bulk liquid medium. The hydrophilic compounds remain in the bulk liquid and thus can only degrade through the $^{\bullet}OH$ radicals reaching the bulk solution, and only about 10% of total generated OH radicals diffuse into the bulk liquid medium and rest are recombined to form H_2O_2. Thus, the concentration of $^{\bullet}OH$ radicals remains low in the bulk liquid medium, thereby giving a lower degradation rate for pollutant molecules. The solution pH and the presence of ionic species can alter the state of the molecules and their effect depends on the type of pollutant present in the solution. Therefore, experimental studies need to be conducted at the laboratory level to establish the effect of solution pH and ionic species on the degradation of pollutants to be treated using HC.

3.3 FENTON CHEMISTRY

The Fenton process involves the application of iron salts and H_2O_2 to produce hydroxyl radicals. Ferrous ion is oxidized by H_2O_2 to ferric ion, a hydroxyl radical, and a hydroxyl anion. Ferric ion is then reduced back (typically in the presence of irradiations) to ferrous ion, a peroxide radical, and a proton by the same H_2O_2. The Fenton reaction generally occurs in an acidic medium between pH 2 and 4 and involves the following possible steps (Masomboon et al., 2009; Rodriguez et al., 2003).

$$Fe^{2+} + H_2O_2 \rightarrow {}^{\bullet}OH + OH^- + Fe^{3+} \tag{3.10}$$

$$Fe^{3+} + H_2O_2 \rightarrow Fe^{2+} + H^+ + HOO^{\bullet} \tag{3.11}$$

$$Fe^{3+} + HOO^{\bullet} \rightarrow Fe^{2+} + H^+ + O_2 \tag{3.12}$$

$$Fe^{2+} + {}^{\bullet}OH \rightarrow Fe^{3+} + OH^- \tag{3.13}$$

$${}^{\bullet}OH + H_2O_2 \rightarrow H_2O + HOO^{\bullet} \tag{3.14}$$

$$Fe^{2+} + HOO^{\bullet} \rightarrow HOO^- + Fe^{3+} \tag{3.15}$$

$${}^{\bullet}OH + {}^{\bullet}OH \rightarrow H_2O_2 \tag{3.16}$$

$${}^{\bullet}OH + organics \rightarrow products + CO_2 + H_2O \tag{3.17}$$

The rate of reaction (3.10) is around 63 $M^{-1}\,s^{-1}$, while the rate of reaction (3.11) is only 0.01–0.02 $M^{-1}\,s^{-1}$ (Kang et al., 2002; Martinez et al., 2003). This indicates that ferrous ions are consumed more rapidly than they are produced. The hydroxyl radicals will degrade organic compounds through reaction (3.17), and H_2O_2 can also react with Fe^{3+} via reaction (3.11).

Many researchers have studied Fenton chemistry for the oxidation of different organic pollutants, including aromatic and phenolic compounds, pesticides, herbicides, and organic dyes (Bigda, 1996; Karci et al., 2012; Kusic et al., 2006; Ma et al., 2005; Martinez et al., 2003; Sun et al., 2007; Xue et al., 2009). In the Fenton-reagent-driven oxidation of organic pollutants, the important parameters that need to be considered to get the optimized results include ratio of H_2O_2 to ferrous ion concentration, operating pH, and concentration of pollutant. Although successful on a laboratory scale, this process finds lesser application on an industrial scale because of its ineffectiveness in reducing certain refractory pollutants such as acetic acid, acetone, carbon tetrachloride, methylene chloride, n-paraffins, maleic acid, malonic acid, oxalic acid, and trichloro-ethane, and due to the high cost of chemical reagents used in this process. Also the problem of higher total dissolved solids (TDSs; because of the added iron salt) need to be handled. With some adjustments to eliminate these drawbacks, Fenton chemistry can be successful in treating some organic pollutants on an industrial scale.

3.3.1 Reactor Used for Fenton Oxidation

A batch Fenton reactor essentially consists of a non-pressurized stirred reactor with metering pumps for the addition of acid, base, a ferrous sulfate catalyst solution, and industrial strength (35–50%) H_2O_2. It is recommended that the reactor vessel be coated with an acid-resistant material, because the Fenton reagent is very aggressive, and corrosion can be a serious problem. The pH of the solution must be adjusted for maintaining the stability of the catalyst. For many chemicals, the ideal pH for the Fenton reaction is found to be between 3 and 4, and the optimum catalyst to peroxide ratio is usually 1:5 wt/wt. Reactants are added in the following sequence: wastewater followed by dilute sulfuric acid (for maintaining acidic conditions) and catalyst in acidic solutions, followed by base or acid for the adjustment of pH at a constant value, and lastly, H_2O_2 (must be added slowly with proper maintenance of temperature). Because the wastewater compositions are highly changeable, the Fenton reactor needs some design considerations to give flexibility in terms of the operating parameters. The discharge from the Fenton reactor is fed into a neutralizing tank for adjusting the pH of the stream followed by a flocculation tank and a solid-liquid separation tank for adjusting the TDS content of the effluent stream. A schematic representation of the Fenton oxidation treatment is shown in Figure 3.5.

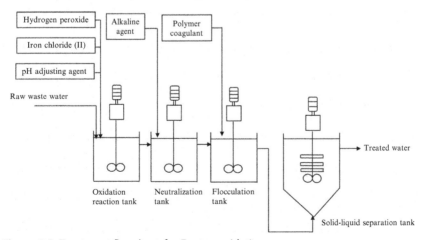

Figure 3.5 Treatment flowsheet for Fenton oxidation.

3.3.2 Optimum Operating Conditions

3.3.2.1 Operating pH

The pH of the system has been observed to significantly affect the degradation of pollutants and the optimum pH has been observed to be about 3 in the majority of cases (Tang and Huang, 1996; Venkatadri and Peters, 1993; Zhonga et al., 2009); thus it is recommended as the operating pH. At lower pH (pH < 2.5), the formation of $(Fe(II) (H_2O))^{2+}$ occurs, which reacts more slowly with H_2O_2 and, therefore, produces lesser quantum of hydroxyl radicals, thereby reducing the degradation efficiency. In addition, the scavenging of the hydroxyl radicals by hydrogen ions becomes important at a very low pH (Tang and Huang, 1996). The reaction of Fe^{3+} with H_2O_2 is also inhibited. At an operating pH of >4, the decomposition rate decreases because of the reduction in the free iron species in the solution, probably due to the formation of Fe (II) complexes with the buffer inhibiting the formation of free radicals. It can also be due to the precipitation of ferric oxyhydroxides (Lin and Lo, 1997), which inhibit the regeneration of ferrous ions. Also, the oxidation potential of $HO^\bullet$ radicals is known to reduce with an increase in the pH.

3.3.2.2 Number of Ferrous Ions

Usually, the rate of degradation increases with an increase in the concentration of ferrous ions, although the extent of increase is sometimes observed to be marginal above a certain concentration as reported by Lin et al. (1999) and Kang and Hwang (2000). Also, an enormous increase in ferrous ions will lead to an increase in the unutilized quantity of iron salts, which will contribute to an increase in the TDS content of the effluent stream; this is not permitted (Masomboon et al., 2009). Thus, laboratory-scale studies are required to establish the optimum loading of ferrous ions under similar conditions unless data is available in the open literature.

3.3.2.3 Concentration of H_2O_2

The concentration of H_2O_2 plays a more crucial role in deciding the overall efficacy of the degradation process. Usually, it has been observed that the percentage degradation of the pollutant increases with an increase in the dosage of H_2O_2 until it reaches an optimum value (Masomboon et al., 2009; Zhonga et al., 2009). However, care should be taken when selecting the operating oxidant dosage. The residual H_2O_2 contributes to COD, and hence excess H_2O_2 is not recommended as it would add to excess COD and also due to its $^\bullet OH$ scavenging effect at higher concentrations. Also, the presence of H_2O_2 is harmful to many of the microorganisms and will affect the overall degradation efficiency significantly where Fenton oxidation is used as a pretreatment to biological oxidation. One more negative effect of H_2O_2, if present in large quantities, is that it acts as a scavenger for the generated hydroxyl radicals. Thus, the loading of H_2O_2 should be adjusted in such a way that the entire amount is utilized; this can be decided based on laboratory-scale studies with the effluent in question.

3.4 PHOTOCATALYTIC OXIDATION

Photocatalytic or photochemical degradation processes are gaining importance in the area of wastewater treatment, since these processes result in complete mineralization with an operation at milder temperature and pressure conditions. The photo-activated chemical reactions are characterized by a free radical mechanism initiated by the interaction of photons of sufficient energy levels with the molecules of chemical species present in the solution, with or without the presence of the catalyst. The radicals can be easily produced using UV radiation by the homogeneous photochemical degradation of oxidizing compounds such as H_2O_2 and ozone. An alternative way to obtain free radicals is using the photocatalytic mechanism occurring at the surface of semiconductors (such as titanium dioxide), which substantially enhances the rate of free-radical generation and hence the rates of degradation (Bhatkhande et al., 2002; Mazzarino and Piccinini, 1999). A major advantage of the photocatalytic oxidation–based processes is the possibility of effectively using sunlight or near UV light for irradiation, which should result in considerable cost savings, especially for large-scale operations (Han et al., 2012; Konstantinou and Albanis, 2004). Various chalcogenides (oxides such as TiO_2, ZnO, ZrO_2, and CeO_2, or sulfides such as CdS and ZnS) have been used as photocatalysts so far in different studies reported in the literature (Bhatkhande et al., 2002). The surface area and the number of active sites offered by the catalyst (thus nature of the catalyst, i.e., crystalline or amorphous) is important because the adsorption of pollutants plays an important role in deciding the overall rates of degradation (Xu et al., 1999).

Photocatalytic reactions occur when charge separations are induced in a large bandgap semiconductor by excitation with ultra bandgap radiations (Daneshvar et al., 2003). In this way, the absorption of light by the photocatalyst greater than its bandgap energy excites movement of an electron from the valence band of the irradiated particle to its conduction band, producing a positively charged hole in the valence band and an electron in the conduction band. Because of the generation of positive holes and electrons, oxidation-reduction reactions take place at the surface of semiconductors. The photo-generated electrons could reduce the organic molecule or react with electron acceptors such as O_2 adsorbed on the catalyst surface or dissolved in water, reducing it to super oxide radical anion $O_2^{\cdot-}$. The photo-generated holes can oxidize the organic molecule to form R^+, or react with OH^- or H_2O, oxidizing them into $^\cdot OH$ radicals, which subsequently act as oxidizing species.

The scheme of various reactions due to the photocatalytic effect can be given as follows:

$$UV + MO \rightarrow MO(h^+ + e^-) \tag{3.18}$$

Here MO stands for metal oxide

$$h^+ + H_2O \rightarrow H^+ + {}^\cdot OH \tag{3.19}$$

$$2h^+ + 2H_2O \rightarrow 2H^+ + H_2O_2 \tag{3.20}$$

$$H_2O_2 \rightarrow HO^\cdot + {}^\cdot OH \tag{3.21}$$

Reductive reaction due to the photocatalytic effect have been given as follows:

$$e^- + O_2 \rightarrow {}^\cdot O_2^- \tag{3.22}$$

$${}^\cdot O_2^- + H^+ \rightarrow HO_2^\cdot \tag{3.23}$$

$${}^\cdot O_2^- + HO_2^\cdot + H^+ \rightarrow H_2O_2 + O_2 \tag{3.24}$$

$$HOOH \rightarrow HO^\cdot + {}^\cdot OH \tag{3.25}$$

Hydroxyl radicals are thus generated, which have the highest oxidation potential, and they react with organic pollutants to oxidize/mineralize them.

3.4.1 Reactor Used for Photocatalytic Oxidation

The photocatalytic process can be carried out by simply using slurry of the fine particles of the solid semiconductor material dispersed in the liquid phase in a reactor irradiated with UV light either directly or indirectly. The proper dispersion of the catalyst in the liquid phase can be achieved using either mechanical or magnetic stirrers. Aeration is usually maintained for scavenging the electrons ($HO_2^\cdot$ radicals are formed in the valence band) and prevent electron/hole charge recombination also helps in achieving the dispersion of the catalyst. The extent of dispersion can also be increased by ultrasonic irradiation of the slurry at a low frequency (e.g., 20 kHz) using an ultrasonic bath for approximately 10–15 min (Mazzarino and Piccinini, 1999). However, in the case of slurry reactors, the performance of the reactor might be severely affected by the low irradiation efficiency and low penetration depth of the incident irradiations because of the opacity of the slurry. Also, after the oxidation treatment, the solid catalyst needs to be separated from the liquid, which is not easy considering the small size of the catalyst particles (usually in the range of 100–500 μm). A further problem is the fouling of the catalyst due to the irreversible adsorption of the product on the catalyst surface. Thus, the application of slurry reactors for photocatalytic treatment on a large scale seems to be quite problematic.

An alternative to the use of a catalyst in suspended form is the use of a supported photocatalyst. The key advantages are the possibility of obtaining an active crystalline structure and the stability of the catalyst layer in the reacting media. Some of the laboratory equipment used for carrying out photocatalytic oxidation is shown in Figure 3.6.

3.4.2 Optimum Operating Conditions

3.4.2.1 Amount of Catalyst

An optimum catalyst concentration should be used because using excess catalyst reduces the amount of photo-energy being transferred in the medium due to the opacity offered by the catalyst particles. It should also be noted that the optimum value will be strongly dependent on the type and concentration of the pollutant, as well as the rate of generation of free radicals (determined by the operating conditions of the reactor), and laboratory-scale experiments are required to compute the optimum value.

3.4.2.2 Reactor Designs

Usually reactor designs should be such that uniform irradiation of the entire catalyst surface is achieved at the incident light intensity. This is a major problem associated with the large-scale designs. Moreover, nearly complete elimination of mass transfer resistances is another point that needs to be considered while designing large-scale reactors. Efficient reactor design must allow exposure of the greatest quantity of the activated catalyst to the illuminated surface and must permit a high density of active catalyst to come into contact with the liquid to be treated inside the reactor.

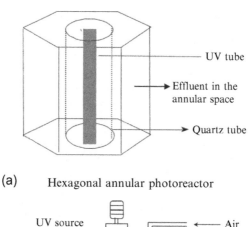

(a)   Hexagonal annular photoreactor

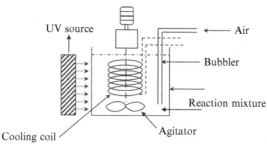

(b)   Simple suspended type reactors

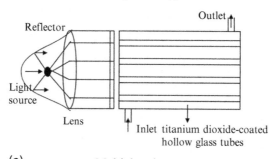

(c)   Multiple tube reactor

Figure 3.6 Schematic representation of equipment used in photocatalytic oxidation.

3.4.2.3 Wavelength of Irradiation

The threshold wavelength corresponds to the band gap energy for the semi-conductor catalyst, e.g., for the TiO_2 catalyst having a band gap energy of 3.02 eV, the ideal wavelength is 400 nm (Herrmann et al., 1999). Sunlight may also be used for excitation of the catalyst in some cases, as discussed earlier; this results in considerable cost savings (Yawalkar et al., 2001).

3.4.2.4 Radiant Flux
The reaction rate is directly proportional to the intensity of radiation; usually, linear variation is observed at low intensities and beyond a certain magnitude of optimum intensity (depending on the reactor conditions as well as the nature of the effluent), the rate of the reaction shows a square root dependence on the intensity (Ollis et al., 1991). Lower dependence on the intensity of irradiation is usually attributed to the increased contribution of the recombination reaction between the generated holes and electrons when their density is high. Usually, this is handled by choosing different polymorphs of TiO_2 (anatase, which is photocatalytically active, and rutile, which is photocatalytically inactive) in appropriate proportions. Another important factor regarding the incident light is the wavelength of irradiation. Shorter wavelengths are recommended for better results. Moreover, the angle of incidence of the UV light should always be 90° because maximum rates are observed at this angle of incidence (Ray and Beenackers, 1997).

3.4.2.5 Medium pH
The pH may affect the surface charge on the photocatalyst and also the state of ionization of the substrate and hence its adsorption. Medium pH has a complex effect on the rates of photocatalytic oxidation, and the observed effect is generally dependent on the type of the pollutant as well as the zero point charge (zPc) of the semiconductor used in the oxidation process, and, more specifically, on the electrostatic interaction between the catalyst surface and the pollutant. The adsorption of the pollutant, and hence the rates of degradation, will be maximum near the zPc of the catalyst (Subramanian et al., 2000). For some of the pollutants, which are weakly acidic, the rate of photocatalytic oxidation increases at lower pH because of an increase in the extent of adsorption of the pollutant on the catalyst surface under acidic conditions (Bhatkhande et al., 2002). Some of the pollutants, which undergo hydrolysis under alkaline conditions or undergo decomposition over a certain pH range, may show an increase in the rate of photocatalytic oxidation with an increase in the pH (Choi and Hoffmann, 1997). Fox and Duley (1993) and Davis and Huang (1991) reported that pH had a marginal effect on the extent of degradation over the range of pH used in their work. Since the effect of pH cannot be generalized, it is recommended that laboratory-scale studies should be performed for establishing the optimum conditions for the operating pH unless data are available in the literature with closely matched operating conditions, i.e., type of equipment as well as the range of operating parameters, including the composition of the effluent stream.

3.4.2.6 Effect of Ionic Species

The presence of ionic species may affect the degradation process via the effect on the process of adsorption of the contaminants on catalyst and its subsequent reaction with hydroxyl radical ions, and/or absorption of UV light. This is a very important point that needs to be considered because real-life industrial effluents will contain different types of salts at different concentrations. Generally, these salts are in ionized or dissociated forms. Many examples have been given in the literature (e.g., Kormann et al., 1991; Wei et al., 1990; Yawalkar et al., 2001) regarding the effects of various anions and cations. In general, it can be said that CO_3^-, HCO_3^- (they act as radical scavengers and also affect the adsorption process), and Cl^- (affects the adsorption step strongly and also partly absorbs UV light) ions have strong detrimental effects on the degradation process, whereas other anions such as sulfate, phosphate, and nitrate affect the degradation efficiency only marginally. Yawalkar et al. (2001) have studied the effect of SO_4^-, CO_3^-, Cl^-, and HCO_3^- ions on the overall degradation rates and reported that the detrimental effects are observed in the order $SO_4^- < CO_3^- < Cl^- < HCO_3^-$. For cations the results are generally contradictory, more on the negative side except for some reports of marginal enhancement due to the presence of Fe^{3+}/Cu^{2+} ions at very low concentrations (Wei et al., 1990). More work is required in this area before the generalized effect of cations on the efficacy of photocatalytic degradation can be well established.

3.5 HYBRID METHODS

The different AOPs discussed above are capable of degrading organic pollutants to a certain extent depending on the pollutant to be treated. All have to overcome some drawbacks for their efficient operation. The efficacy of the process depends strongly on the rate of free radical generation along with the extent of radical contact with the contaminant molecules. Thus, efficient design should aim at maximizing both these quantities. The similarity between the mechanism of destruction in the case of different advanced oxidation techniques and some of the common optimum operating conditions point toward the synergism between these methods and the fact that combinations of these AOPs should give better results as compared to individual techniques. Moreover, some of the drawbacks of the individual techniques can be overcome by some features of other techniques. For example, the efficiency of photocatalytic oxidation is severely hampered by two main factors, mass transfer limitations and fouling of the solid catalyst; however, if the photocatalytic oxidation technique is used in combination with ultrasonic irradiation, not only the rate of generation of hydroxyl radicals is increased (due to increased energy dissipation and generation of extreme conditions of temperature and pressure due to the cavitation phenomena), but, also, because of the acoustic streaming and turbulence created by ultrasonic irradiation, mass transfer resistance will be eliminated. Turbulence also helps in cleaning the catalyst, which increases the efficiency of the photocatalytic oxidation process. Many other combination techniques have been reported extensively in the literature for a variety of contaminants.

3.5.1 Cavitation Coupled with H_2O_2

In the case of cavitation (US/HC), the main mechanism for the destruction of organic pollutant is the reaction of ${}^{\bullet}OH$ radicals with the pollutant molecules. Thermal pyrolysis of molecules present near or inside the collapsing cavities is also possible. Hence, additional supplements of ${}^{\bullet}OH$ radicals should enhance the rate of degradation of pollutants. H_2O_2 is a commonly available oxidizing agent that can be used for the treatment of wastewater due to its high oxidation potential (1.78 V). The efficiency of H_2O_2 in oxidizing organic pollutants depends very much on the generation of ${}^{\bullet}OH$ radicals through the dissociation of H_2O_2. It is a well-known phenomenon that H_2O_2 molecules can readily dissociate into ${}^{\bullet}OH$ radicals under the extreme conditions of high temperature and pressure developed as a result of cavity collapse (Abbasi and Asl, 2008; Pang et al., 2011). If used individually, the efficiency of H_2O_2 in oxidizing organic pollutants is low because of poor dissociation of H_2O_2 into ${}^{\bullet}OH$ radicals under the conventional stirred conditions. But the efficiency of H_2O_2 in degrading organic pollutant can be enhanced significantly if used in combination with the cavitation process because of the formation of significant quantum of ${}^{\bullet}OH$ radicals as a result of dissociation of H_2O_2 under cavitational conditions and its enhanced microdiffusion into the bulk solution due to cavitation. The dissociation energy for the O-O bond in H_2O_2 is only 213 kJ/mol, which is significantly less than that of the O-H bond in H_2O, which is 418 kJ/mol (Pang et al., 2011). Thus, more ${}^{\bullet}OH$ radicals will be generated if H_2O_2 is used in combination with a cavitational process because the energy required for the dissociation of H_2O_2 will be available through the cavitation. The following reactions take place during the degradation of pollutant molecules using the combined cavitation and H_2O_2 process.

$$H_2O_2 +))) \rightarrow 2\,{}^{\bullet}OH \tag{3.26}$$

$$H_2O +))) \rightarrow H^{\bullet} + {}^{\bullet}OH \tag{3.27}$$

$${}^{\bullet}OH + {}^{\bullet}OH \rightarrow H_2O_2 \tag{3.28}$$

$${}^{\bullet}OH + H_2O_2 \rightarrow HO_2^{\bullet} + H_2O \tag{3.29}$$

$${}^{\bullet}OH + HO_2^{\bullet} \rightarrow H_2O + O_2 \tag{3.30}$$

$$HO_2^{\bullet} + H_2O_2 \rightarrow {}^{\bullet}OH + H_2O + O_2 \tag{3.31}$$

$$\text{Pollutant molecules} + {}^{\bullet}OH \rightarrow CO_2 + H_2O + \text{other intermediates} \tag{3.32}$$

$$\text{Pollutant molecules} + H_2O_2 \rightarrow CO_2 + H_2O + \text{other intermediates} \tag{3.33}$$

The efficiency of the combined process is very much dependent on the rate at which ${}^{\bullet}OH$ radicals are consumed by the pollutant molecules. It has been observed that the synergistic effects of the combined process of cavitation and H_2O_2 are obtained up to an optimum concentration of H_2O_2. Afterwards, the effects are reduced because of scavenging of the ${}^{\bullet}OH$ radicals by H_2O_2 itself and the recombination of ${}^{\bullet}OH$ radicals with other ${}^{\bullet}OH$ radicals present in the solution. Operating pH, the intensity of turbulence existing in the reactor, state (whether molecular or ionic) and nature (hydrophobic or hydrophilic) of the pollutant, and sometimes concentration of the pollutant and composition of the effluent stream, are the crucial factors to be analyzed before selecting the combination of ultrasound with H_2O_2 as the oxidation treatment scheme. It can be said that the use of H_2O_2 in conjunction with ultrasound is beneficial only till an optimum loading of H_2O_2 and where the free radical attack is the controlling mechanism of destruction. The optimum loading of H_2O_2 will be dependent on the nature of the effluent stream and the operating conditions, as mentioned above. Laboratory-scale studies are essential to establish this optimum for the effluent stream in question. HC coupled with H_2O_2 can be used on a larger scale due to its potential in generating cavitational conditions similar to those generated in acoustic cavitation and that too on a larger scale. HC also gives higher cavitational yield as compared to acoustic cavitation and is more energy efficient than acoustic cavitation.

Merouani et al. (2010) have reported that the degradation rate of rhodamine B increased by the addition of H_2O_2 to the ultrasound. They have observed that an optimum concentration of H_2O_2 existed to maximize the degradation rate of rhodamine B, using a combination of ultrasound and H_2O_2; excessive amounts of H_2O_2 could reduce its degradation rate.

Similar trends have been reported by Saharan et al. (2011) for the degradation of reactive red 120 while using the combined processes of HC and H_2O_2. They have reported the optimal molar ratio of dye to H_2O_2 as 1:60. Wang et al. (2011a) have also demonstrated the existence of the optimal loading of H_2O_2 for the degradation of reactive brilliant red K-2BP using a combination of HC and H_2O_2. The observed results for the combined process of cavitation and H_2O_2 illustrate that the scavenging of free radicals by H_2O_2 becomes a dominant process at a high concentration of H_2O_2, thereby lowering the extent of degradation. Also, operating at such high ratios of H_2O_2 may not be economically feasible and could pose safety hazards. Moreover, the retention of H_2O_2 in the effluent stream needs to be tested because it also presents as a possible contaminant and cannot be discharged as such. Hence optimum loading of H_2O_2 should be used for better efficiency in the combined process of cavitation and H_2O_2.

3.5.2 Cavitation Coupled with Ozone

Ozone (O_3) has an excellent potential for degrading organic pollutants due to its high oxidation potential (2.08 V). The degradation of organic pollutants through ozonation takes place via two routes: (1) At basic pH, ozone rapidly decomposes to yield hydroxyl and other radical species in solution which oxidize the pollutants and (2) under acidic conditions, ozone is stable and can react directly with organic substrates as an electrophile. Although ozone is a strong oxidant, there are certain limitations of using ozonation as an effective tool for the degradation of organic pollutants on an industrial scale. These limitations are (a) a high energy/intensity requirement to generate O_3, (b) pH sensitivity, and (c) selectivity for organic substrates; for example, O_3 preferentially reacts with alkenes and sites with high electronic density (Pang et al., 2011; Weavers et al., 1998). Therefore, many researchers have tried ozonation in combination with other process to overcome these drawbacks, including ozone combined with H_2O_2 and ozone combined with ultraviolet irradiation and ultrasound (Ince and Tezcanli, 2001; Kang and Hoffmann, 1998; Song et al., 2007; Vecitis et al., 2010; Wu et al., 2008). These combined processes generate $^{\bullet}OH$ radicals that have higher oxidation potential than ozone and are nonselective toward organic substances. Cavitation can be used in combination with ozone for efficiency improvement because of its ability to create hot spots at ambient conditions. It has been reported that ozone can readily be decomposed under the cavitational conditions of high temperature and pressure generated as a result of cavity collapse (Weavers et al., 1998). By the use of a cavitation process (either US or HC), ozone decomposes thermolytically in a cavitation bubble and yields molecular O_2 and atomic oxygen $O(^3P)$, which reacts with water molecule to form ˙OH radicals. Another advantage of the combined process of cavitation and ozone is the increased mass transfer of ozone from the gas phase to the bulk solution to react with a substrate because of the turbulence created as a result of high velocity water jet formation during cavity collapse and/or its oscillation. Hence, in this process, organic pollutants can be possibly degraded by the direct attack of ozone in aqueous solution and by the attack of ˙OH radicals generated as a result of the decomposition of ozone under the cavitation. The following reactions are known to occur during the combined process of cavitation and ozone.

$$O_3 +))) \rightarrow O_2 + O(^3P) \tag{3.34}$$

$$H_2O +))) \rightarrow \text{˙OH} + \text{H}^{\bullet} \tag{3.35}$$

$$O(^3P) + H_2O \rightarrow 2\,\text{˙OH} \tag{3.36}$$

$$\text{˙OH} + \text{˙OH} \rightarrow H_2O_2 \tag{3.37}$$

$$\text{˙OH} + \text{pollutant molecule} \rightarrow CO_2, H_2O, \text{etc.} \tag{3.38}$$

$$O_3 + \text{pollutant molecule} \rightarrow CO_2, H_2O, \text{etc.} \tag{3.39}$$

The combined process of cavitation and ozone produces two ˙OH radicals for every O_3 molecule and hence can significantly enhance the degradation efficiency of the combined process. In this combined process, ozone loading depends on the type of pollutant molecules to be treated and the initial concentration of the pollutant molecules. It has been reported that a dosage of O_3 exceeding the optimum value could result in the release of unreacted O_3 from the system (He et al., 2007). Hence, optimization of the O_3 dosage is necessary to minimize the energy consumption and the amount of O_3 in the exhaust gas. Figure 3.7 shows the schematic of a laboratory-scale setup used for the combined process of cavitation and ozonation. In the case of the combined process of US and O_3, a gas sparger is needed for the uniform distribution of ozone gas inside the US reactor. However, the poor energy efficiency of the US reactor makes these processes uneconomical on a larger scale. On the other hand, HC can be used more effectively on a larger scale in combination with O_3. The ozone gas can be injected directly at the throat or vena contracta of the cavitating device, thereby exposing O_3 directly to the cavitating conditions. This also

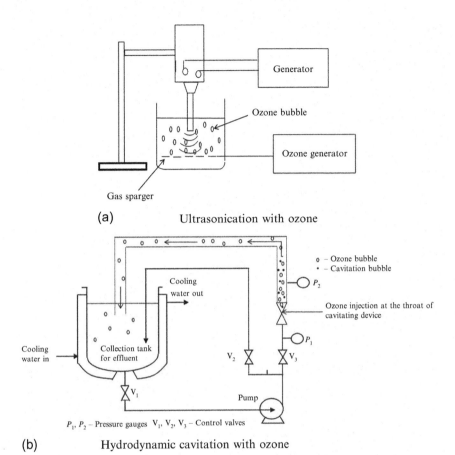

Figure 3.7 Schematic of the combined process of cavitation and ozone.

ensures minimal escape of unreacted O_3 due to rapid dissociation into $^\bullet OH$ radicals and improved mass transfer, ensuring maximum utilization of ozone for the oxidation of pollutant molecules.

3.5.3 Cavitation Coupled with Photocatalysis

In the case of photocatalytic oxidation, the most common problem is the reduced efficiency of a photocatalyst with continuous operation, possibly due to the adsorption of contaminants at the surface and the blocking of the UV activated sites, which makes them unavailable for the process of destruction. Process development for continuous cleaning of the catalyst surface during photocatalytic operation is necessary for the successful implementation of a photocatalyst on an industrial scale. Cavitation through US or HC is one such technique that can be used simultaneously with UV/ solar irradiation. Moreover, the photocatalytic oxidation technique is also affected by severe mass transfer limitations, especially in the immobilized catalyst type of reactors, which are generally preferred over slurry reactors to avoid solid catalyst separation problems. One further factor suggesting that the two techniques will give better results when operated in combination is the fact that for both the techniques (UV and cavitation), the basic reaction mechanism is the generation of free radicals and subsequent attack by these on the pollutant species. If the two modes of irradiations are operated in conjunction, more free radicals will be available for the reaction, thereby increasing the rates of degradation. Thus, the expected synergism between these two modes of irradiation can be possibly attributed to:

1. Cavitational effects leading to an increase in the temperatures and pressure at the localized cavity implosion sites.
2. Cleaning and sweeping of the photocatalyst surface due to micro turbulence and high velocity liquid jets created as a result of cavity oscillation and collapse.
3. Mass transport of the reactants and products is increased at the catalyst surface and in the solution, due to the facilitated transport by shockwave propagation.
4. Surface area is increased by the fragmentation or pitting of the catalyst and hence more actives sites will be available for the adsorption of the pollutant molecule at the photocatalyst surface.
5. More OH radicals are generated through the dissociation of water molecules under cavitational conditions, in addition to those generated by photocatalytic process, thereby increasing the concentration of OH radicals in the solution for the oxidation of pollutant molecules.
6. The organic substrate reacts directly with the photo-generated surface holes and electrons under cavitating conditions.

There have been many studies depicting the observed synergism and the enhanced rates of degradation for the combinatorial operation of sonochemical reactors and photocatalytic oxidation (Adewuyi, 2005; Cheng et al., 2012; Madhavan et al., 2010; Ragaini et al., 2001; Saien et al., 2010). For better efficiency or higher synergistic effect, these processes should operate simultaneously rather than having sequential irradiation of ultrasound followed by photocatalytic oxidation. The different reactor configurations used so far for evaluating the synergism between ultrasound and ultraviolet irradiation are depicted in Figure 3.8 (batch reactors) and Figure 3.9

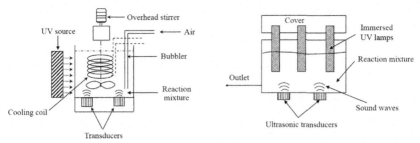

Figure 3.8 Typical schemes used for the combinatorial effects of UV and ultrasonic irradiations: Batch reactors.

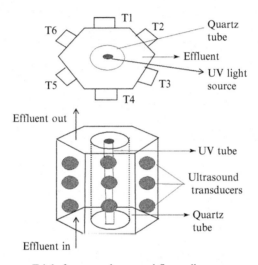

Triple frequency hexagonal flow cell

Figure 3.9 Typical schemes used for the combinatorial effects of UV and ultrasonic irradiations: Continuous reactors.

(continuous reactors). For wastewater applications, it is important to design continuous reactors rather than batch reactors or at least reactors operating in recirculating mode, because large quantities of effluents will have to be treated. The sonochemical element should be designed with power dissipation over a wider area (more energy efficiency and also larger cavitational yield); hence, ultrasonic bath-type reactors or parallel plate reactors with multiple transducers are recommended (Gogate et al., 2001, 2003). Multiple-frequency reactors have also been found to generate more intense and spatially uniform cavitation as compared to reactors with single frequency and/or single transducer operations; thus, these give better destruction efficiency (Sivakumar et al., 2002). For the photocatalytic element, usually immobilized or supported catalyst reactors are preferred, but slurry reactors can also be used if better separation characteristics or lower loading of photocatalysts is achieved. Most of the relevant studies have been done using ultrasound as a means of generating cavitation coupled with photocatalytic oxidation, whereas only few studies are available where HC has been used in combination with the photocatalysis (Wang et al., 2011b). In the case of HC, simultaneous processing is difficult to achieve because of complexity in designing and fabricating a hydrodynamically cavitating device mounted with a UV lamp; however, a sequential process of HC and photocatalyst is easy to design and construct. Figure 3.10 shows the schematic of an experimental setup for the combined processes of HC and photocatalyst. Another design can be where the lag phase between the HC and photocatalysis can be eliminated by making the flow loop for the photocatalytic rector, such that as soon as the effluent comes out of the HC reactor, it is exposed to UV light. This can be done by using transparent line (such as a glass line) that can be covered by a UV lamp, thus making it almost a simultaneous process rather than sequential. In this manner, the recombination of the free radicals and the formation of toxic pollutant intermediates can be eliminated because they are subsequently exposed to the photocatalytic oxidation as soon as they come out of the HC reactor, where these intermediates are degraded further using photocatalytic oxidation. Also, the suspension of photocatalyst and reactivation of the catalyst surface due to HC enhances the degradation

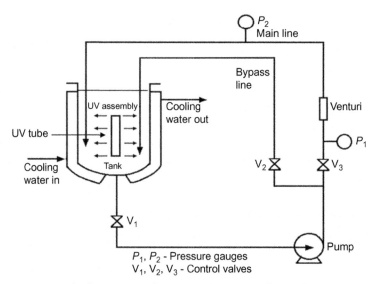

Figure 3.10 Typical schemes used for the combinatorial effects of UV and HC.

efficiency of photocatalytic oxidation. On the other hand, HC has the potential for scale-up and gives a higher cavitational yield if larger volumes are to be treated as compared to acoustic cavitation. Thus, overall it appears, that combining hydrodynamic cavitation with photocatalysis would be more effective as compared to the combination of ultrasound and photocatalysis.

3.5.4 Photo-Fenton (Fenton Process in the Presence of UV Light)

A combination of H_2O_2 and UV radiation with Fe(II) or Fe(III) oxalate ion, the so-called photo-Fenton process, produces more hydroxyl radicals in comparison to the conventional Fenton method (Fe(II) with H_2O_2) or photolysis, thus promoting the rates of degradation of organic pollutants (Kusic et al., 2006; Ma et al., 2005; Papic et al., 2009; Song et al., 2006). It is well known that under acidic conditions some ferrous ions would exist in the form of $Fe(OH)^{2+}$, which can easily get converted into $^{\bullet}OH$ radicals and Fe^{2+} under UV irradiation as indicated in Equation (3.42). This can establish a cycle of reactions generating additional hydroxyl radicals and regenerating catalyst because Fe^{2+} ions are restored. In addition to this, H_2O_2 can also dissociate into more reactive $^{\bullet}OH$ radicals in the presence of UV light. Hence, in the combined process of Fenton and UV radiation, $^{\bullet}OH$ radicals are generated in three ways and hence the concentration of $^{\bullet}OH$ radicals in the solution enhances significantly. The higher concentration of $^{\bullet}OH$ radicals means higher chances of pollutant molecules reacting with $^{\bullet}OH$ radicals and hence a higher rate of mineralization of the pollutant molecules. The following are the major reactions taking place during the photo-Fenton process.

$$Fe^{2+} + H_2O_2 \rightarrow \,^{\bullet}OH + OH^- + Fe^{3+} \qquad (3.40)$$

$$Fe^{3+} + H_2O \rightarrow Fe(OH)^{2+} + H^+ \qquad (3.41)$$

$$Fe(OH)^{2+} + (UV) \rightarrow Fe^{2+} + \,^{\bullet}OH + H^+ \qquad (3.42)$$

$$H_2O_2 + (UV) \rightarrow 2\,^{\bullet}OH \qquad (3.43)$$

Figure 3.11 shows the reaction pathways for the degradation starting with the primary photo-reduction of the dissolved Fe(III) complexes to Fe(II) ions followed by the Fenton reaction and the subsequent oxidation of organic compounds. Additional hydroxyl radicals generated in the first step also take part in the oxidation reaction. For better efficiency of this process, the recombination of the free radicals and scavenging by H_2O_2, if present in excess, should be minimized.

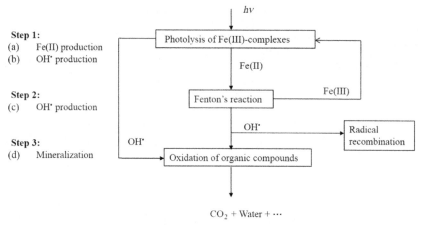

Figure 3.11 Reaction pathways of the photo-Fenton process.

The efficacy of the photo–Fenton technique is significantly higher than that with Fenton oxidation or photolysis alone. The use of sunlight in the case of commercial applications would be significantly cheaper and is also suitable due to the formation of the $Fe(OH)^{2+}$ complex in the case of the photo–Fenton technique, but it may result in lower rates of degradation as compared to the use of UV light. Hence, for the treatment of complex effluents, the loading of Fe(III) as well as H_2O_2 needs to be adjusted. A detailed analysis will be required for investigating the effect of oxidant concentrations on the extent of degradation before adjustment in the dose of oxidants. For the treatment of highly loaded effluents, appropriate dilution factors must be used before the oxidation treatment. Other important parameters that affect the efficiency of the photo–Fenton process are the intensity of the UV light, loading of Fe^{2+} ions, H_2O_2 loading, the presence of ions, and solution pH, which needs to be optimized before it is scaled up to an industrial level. All of these parameters depend on the type and concentration of the pollutant molecules to be oxidized and hence require laboratory-scale experiments to be conducted to establish the optimum operating conditions.

3.5.5 Cavitation Coupled with Fenton

The Fenton process utilizes the reactivity of $^{\bullet}$OH radicals generated in acidic conditions by iron catalyzed decomposition of H_2O_2 (Equation 3.44) for the degradation of organic pollutants (Masomboon et al., 2009).

$$Fe^{2+} + H_2O_2 \rightarrow Fe^{3+} + OH^- + {}^{\bullet}OH \qquad (3.44)$$

The resulting Fe^{3+} ions can react with H_2O_2 (Equation 3.45) to produce the intermediate complex ($Fe-OOH^{2+}$), which can easily get converted into Fe^{2+} and $HO_2^{\bullet}$ under cavitating conditions (Equation 3.46) (Merouani et al., 2010; Pang et al., 2011).

$$Fe^{3+} + H_2O_2 \rightarrow Fe - OOH^{2+} + H^+ \qquad (3.45)$$

$$Fe - OOH^{2+} \xrightarrow{\text{cavitation}} Fe^{2+} + HO_2^{\bullet} \qquad (3.46)$$

Generated Fe^{2+} ions can again react with H_2O_2 to generate even more ${}^{\bullet}OH$ radicals. In addition to this, some part of H_2O_2 directly decomposes to hydroxyl radicals in the presence of cavitation (as explained earlier) as shown in the following equation.

$$H_2O_2 \xrightarrow{\text{cavitation}} {}^{\bullet}OH + {}^{\bullet}OH \qquad (3.47)$$

Thus, the combination of cavitation and Fenton process accelerates the rate of generation of hydroxyl radicals and the degradation of organic pollutants. Various reports suggest that the efficiency of the Fenton process can be enhanced significantly with the use of cavitation. Wang et al. (2008) have shown that the degradation rate of reactive brilliant red K-BP is substantially enhanced by the addition of Fenton reagent to the ultrasound system. The degradation rate through the Fenton reaction was increased by almost sixfold when it was used in combination with ultrasound, and the synergistic coefficient of 2.27 suggest that combination of Fenton and cavitation gives higher efficiency than when used individually. Similar trends have been reported in the literature for the degradation of some other organic pollutants by applying a combination of HC and Fenton. Joshi and Gogate (2012) have investigated the degradation of dichlorvos using a combination of HC and the Fenton process and have reported that the rate of degradation of dichlorvos increases significantly with increasing concentrations of Fe^{2+}, giving a maximum degradation of 91.5% at 3:1 loading of $FeSO_4:H_2O_2$.

The efficiency of this combined process is very much dependent on the solution pH, rate of consumption of generated ${}^{\bullet}OH$ radicals by the pollutant molecules, presence of other ionic species, loading of Fe ions and H_2O_2, and cavitational intensity. The effect of all these parameters on the degradation efficiency has to be studied on a laboratory scale for the case of specific effluents.

3.6 CASE STUDIES

3.6.1 Intensification of Degradation of Imidacloprid in Aqueous Solutions using Combination of HC with Various AOPs

Pesticides are introduced into the environment mainly through industrial effluents, agricultural runoff, and chemical spills. The pesticides and insecticides that find their way into natural bodies of water cause eutrophication and perturbations in aquatic life and pose a major threat to the surrounding ecosystems, mainly due to the documented health hazards caused by toxicity and the potentially carcinogenic nature of such organic pollutants. Imidacloprid belongs to the new class of insecticide called neonicotinoid and has high activity against sucking pests, including rice-hoppers, aphids, thrips, and white flies. It has raised concerns particularly because of its possible impact on bee populations and its potential harmful effects on the aquatic environment. The presence of imidacloprid in water streams causes potential environmental problems due to its high solubility (0.58 g/L), nonbiodegradability, and persistent nature. Due to the toxic nature of imidacloprid, it is more resistant to degradation using conventional biological processes and hence requires the use of advanced oxidation methods that can degrade/mineralize the imidacloprid molecule.

Raut-Jadhav et al. (2013) investigated the degradation of imidacloprid in aqueous solutions using the combination of HC with various other AOPs such as Fenton, photo-Fenton, H_2O_2, and photocatalytic processes. The chemical structure of imidacloprid is shown in Figure 3.12.

In the reported study, all the experiments were performed by treating 5 L aqueous solution of imidacloprid with an initial concentration of 25 ppm. A circular venturi was used as a means to generate the cavitation conditions (Figure 3.13). The typical HC setup used in this study consisted of a closed loop including a feed tank, positive displacement pump ($P = 1.1$ kW),

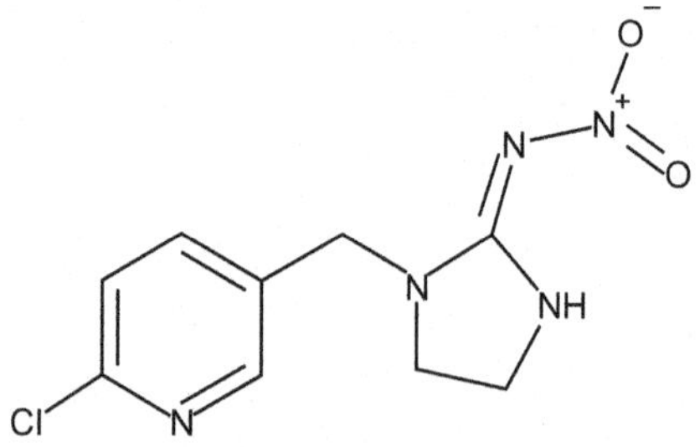

Figure 3.12 Chemical structure of imidacloprid.

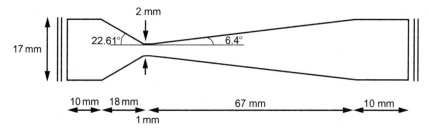

Figure 3.13 Schematic of cavitating device (circular venturi).

pressure gauges, and valves (see Figure 3.10). The degradation of imidacloprid using HC was first optimized in terms of operating inlet pressure and solution pH by carrying out experiments at different inlet pressures and pHs. Further experiments were then conducted at the optimum inlet pressure and solution pH of 2.7. The experiments using the combined process of HC and H_2O_2 were carried out at a different molar ratio of imidacloprid:H_2O_2 to establish the optimum loading of H_2O_2 for the degradation of imidacloprid using combined processes of HC and H_2O_2. Fenton chemistry was employed along with HC at the optimum loading of imidacloprid:H_2O_2 as 1:40 with the concentration of H_2O_2 as 3.91 mmol/L and a varying molar ratio of ferrous sulfate:H_2O_2 as 1:50, 1:40, 1:30, and 1:20. Photo–Fenton alone and HC + photo-Fenton processes were employed at the molar ratio of imidacloprid:H_2O_2 as 1:40 and a molar ratio of ferrous sulfate:H_2O_2 as 1:40. Photocatalytic process alone and HC + photocatalytic processes were employed using niobium pentoxide as a photocatalyst, due to its adequate band gap of 3.4 eV and stability in acidic conditions. The amount of catalyst used was 200 mg/L without any pretreatment. UV assembly was placed centrally in a feed tank for UV irradiation as per the requirement.

3.6.1.1 Degradation of Imidacloprid Using HC-Based Hybrid Method

The degradation of imidacloprid using HC was carried out at different operating inlet pressures and solution pHs. The results obtained for different processes are shown in Table 3.1. It was observed that the rate of degradation of imidacloprid increases with an increase in the inlet pressure until 15 bar, after which an increase in the inlet pressure does not have a significant effect on the rate of degradation of imidacloprid. Also a lower pH of 2.7 is favorable for the degradation of imidacloprid using HC. The maximum extent of degradation of 6.15% was obtained with a reaction rate constant of

Table 3.1 Initial rate constants and extents of degradation obtained during various processes

Process	Degradation Rate Constant $(k) \times 10^3$ (min^{-1})	% Degradation After 15 min	Synergetic Coefficient $= k$ (Combined Process)/ (Addition of k for Individual Process)
HC	2.565	6.15	–
HC + H_2O_2	81.815	74.00	22.79
Fenton	66.71	64.43	–
Photo-Fenton	99.372	81.60	1.4
HC + Fenton	250.749	97.77	3.6
HC + photo-Fenton	297.012	99.23	2.9
Photocatalytic	4.871	7.5	–
HC + photocatalytic	6.837	11.7	0.91

2.565×10^{-3} min^{-1} at the inlet pressure of 15 bar, at a cavitation number of 0.067, and with solution pH of 2.7. The results obtained are attributed to the fact that, at higher inlet pressures such as 15 bar, the number of cavities produced, the intensity of turbulence created, and the intensity of cavity collapse would be higher. This condition would enhance the concentration of $^{\bullet}OH$ radicals and their effective utilization, which subsequently increases the degradation of imidacloprid. Under acidic conditions the state of the imidacloprid molecule changes from ionic to molecular, thereby causing it to locate at the cavity water interface where the concentration of the $^{\bullet}OH$ radicals is high and hence the degradation rate increases.

The degradation of imidacloprid using HC can be further increased by the addition of H_2O_2, and about 74% degradation can be achieved using a combination of HC and H_2O_2 at a much faster rate. A high synergetic coefficient of 22.79 suggests that both processes should be used in combination rather than individually. The efficiency of the combined HC and H_2O_2 can be further enhanced by the addition of ferrous ion into the system, making it more energy efficient. A very high degradation rate constant of around 250.7×10^{-3} min^{-1} was obtained using a combined process of HC and Fenton, with almost 100% degradation achieved in 15 min. A substantial synergetic effect was observed using HC + Fenton process, since the value of the reaction rate constant of 2.565×10^{-3} min^{-1} and 66.711×10^{-3} min^{-1}

obtained in case of HC alone and the conventional Fenton process (without HC) substantially increased to 250.749×10^{-3} min^{-1} when HC was combined with the Fenton process. The inclusion of UV radiation in the process (HC + Fenton) can result into a higher degradation rate constant (297×10^{-3} min^{-1}), but it can be seen that the increase in the degradation rate is marginal and also the additional energy costs make this option (HC + Photo-Fenton) less economical than a combined process of HC and Fenton (as shown in Figure 3.14). In the case of the combined process of HC and photocatalysis, the synergetic effect was not seen. It is observed that the first order reaction rate constant of 2.565×10^{-3} min^{-1} and 4.871×10^{-3} min^{-1} obtained by applying HC alone and photocatalytic process alone increases marginally to 6.837×10^{-3} min^{-1} using the combined HC + photocatalytic process. Further, the efficiency of all these processes was evaluated on the basis of cavitational yield (moles of imidacloprid degraded/energy supplied). Figure 3.14 shows the cavitational yield of the different processes. It can be seen from the figure that the combined process of HC and Fenton is most energy efficient.

Overall, it can be concluded that hybrid methods (combination of different AOPs) are more energy efficient than the individual AOPs because the drawback(s) in one process can be eliminated through the use of other processes.

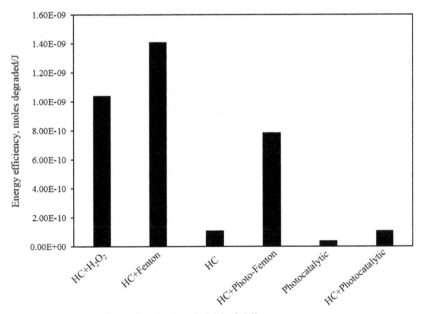

Figure 3.14 Comparison of cavitational yield of different processes.

3.6.2 Biodegradability Enhancement of Distillery Wastewater Using HC

Distilleries are considered as one of the most highly polluting industries worldwide. The wastewater generated from a distillery unit is dark brown in color and contains very high biological oxygen demand (BOD: 50,000–60,000 ppm), chemical oxygen demand (COD: 110,000–190,000 ppm), and high BOD/COD ratio in the range of 0.3–0.6. The amount of inorganic substances such as nitrogen, potassium, phosphates, calcium, and sulfates is also very high. Even after the conventional anaerobic digestion, the biomethanated distillery wastewater (B-DWW) still retains around 40,000 mg/L of COD and significant color and becomes recalcitrant (BI: $BOD_5/COD \approx 0.14$) to further treatment by conventional biological methods. With more stringent rules and regulations imposed by government, distillery industries have been forced to look for more effective treatment technologies. Such technologies would be beneficial to the environment, but also needs to be cost effective. This problem has emphasized the need for further research on developing effective treatment/pretreatment methods for safe disposal of anaerobically digested distillery wastewater.

Padoley et al. (2012) reported the use of HC as a pretreatment option for the complex/recalcitrant B-DWW. The effect of various process parameters such as inlet pressure, dilution, and reaction time on reduction of COD/TOC and enhancement in the biodegradability index (BI: BOD_5:COD ratio) of the B-DWW has been reported with an aim of maximizing the biodegradability index and achieving lower toxicity of the distillery wastewater. To check the efficacy of HC in treating such effluents, distillery wastewater treated in a conventional anaerobic digester from a distillery near Nagpur, India (source not given due to confidentiality issues), was used for the studies. The physicochemical properties of this waste are given in Table 3.2.

Table 3.2 Characteristics of complex wastewater (biomethanated distillery wastewater)

Parameters	Value
pH	7.61
Color	Brown
COD (mg/L)	35,000
BOD (mg/L)	5000
TOC (mg/L)	10,000
Total solids (mg/L)	31,000
Total suspended solids (mg/L)	1600
Biomass (%)	1
BOD:COD ratio	0.168

3.6.2.1 Treatment of B-DWW Using HC

The B-DWW was subjected to HC pretreatment for which 6 L of wastewater was taken in cavitation reactor. Cavitation was achieved using a circular venturi (shown in Figure 3.13). The experiments were conducted at two different inlet pressures (low as well as high) and at different dilutions over the time range of 50–150 min. At the end of each defined time interval, the samples were withdrawn from the reactor through a sampling port, centrifuged, and analyzed for pH, COD, BOD, and TOC. The obtained results are reproduced in Table 3.3.

It has been reported that HC is capable of reducing the COD and TOC of the B-DWW by about 34% and 33%, respectively, at 5 bar inlet pressure; no further improvement was observed with increase in pressure from 5 to 13 bar. It can be observed from the values of COD and TOC that the dilution has no significant effect on the mineralization of distillery wastewater. Although the percentage reduction is marginally higher at 50% dilution, the total quantum (milligram of COD/TOC per unit volume) of COD and TOC reduction is lower at 25% and 50% dilution as compared to undiluted wastewater. But for the enhancement in the biodegradability of B-DWW, the observed conditions are different. It was observed that at lower pressure (5 bar) and zero dilution, the ratio of BOD/COD increases to the tune of 0.24, becoming 0.25 when the sample dilution is 25%. Thus dilution is again

Table 3.3 Effect of cavitation pretreatment on biodegradability index of biomethanated distillery wastewater

Reaction Condition	Time (min)	COD (mg/L)	BOD (mg/L)	BI: BOD$_5$/ COD
Pressure = 5 bar, No dilution	0	34391.00	4853.00	0.14
	50	23723.00	5120.00	0.22
	100	23442.00	5500.00	0.23
	150	23302.00	5500.00	0.24
Pressure = 5 bar, 25% dilution	0	28208.00	3666.00	0.13
	50	19539.00	4170.00	0.21
	100	18300.00	4250.00	0.23
	150	18163.00	4500.00	0.25
Pressure = 13 bar, No dilution	0	33973.00	4756.00	0.14
	50	24128.00	5830.00	0.24
	100	22325.00	6400.00	0.29
	150	22325.00	6400.00	0.29
Pressure = 13 bar, 25% dilution	0	28754.00	3738.00	0.13
	50	18795.00	6000.00	0.32
	100	18374.00	5910.00	0.32
	150	18363.00	5800.00	0.32

not aiding in BOD enhancement significantly. At higher pressure (13 bar), the ratio is enhanced to a value of 0.29 at zero dilution and to 0.32 at a spent wash concentration of 25%. As compared to lower pressure, higher pressure yields a slightly better BOD/COD ratio on treatment. From the obtained results, it can be inferred that lower inlet pressure (5 bar) is suitable for a reduction in toxicity (COD/TOC reduction), whereas for enhanced biodegradability (higher BI), higher inlet pressure operation would be preferred. Thus, it is shown that HC is capable of reducing the toxicity of distillery wastewater, and pretreatment increases the biodegradability of the B-DWW. Hence, depending on the final objective of the pretreatment, HC can be effectively utilized for the treatment of complex wastewater pollutants such as B-DWW.

The biodegradability of the cavitationally pretreated B-DWW was further evaluated by subjecting it to the conventional biological treatment process. The amount of methane generated and reduction in COD was compared for the cavitationally pretreated B-DWW as against the untreated B-DWW. It was observed that in the cavitationally pretreated B-DWW (13 bar, 25% dilution, 50 min, BI: 0.32), 400 mL of gas volume was generated after a total duration of 40 days (including the lag period of 6 days), along with a net 70% COD reduction, whereas in the untreated system (BI: 0.168), the gas volume was observed to be only 60 mL with around 12% COD reduction under similar experimental conditions.

Hence, it can be concluded that HC is capable of enhancing the efficiency of conventional biological processes in terms of a reduction in the toxicity as well as an increase in biogas generation, along with a significantly higher reduction in COD and color. Due to HC pretreatment, the efficiency of the conventional biological process is increased by almost sixfold in terms of COD removal and biogas formation.

3.7 SUMMARY

Different AOPs such as cavitation (acoustic and hydrodynamic), Fenton, photocatalysis, and other combined processes have been tested on the laboratory scale and proven to be effective in degrading various organic pollutants that are biorefractory in nature. The efficacy of these processes depends strongly on the rate of generation of free radicals along with the extent of contact of the generated radicals and/or chemical oxidants with the contaminant molecules. An efficient design should aim at maximizing both these attributes of the process. Although highly successful on a laboratory scale, there are many issues that need to be addressed for the successful implementation of these AOPs on an industrial scale. Economic considerations and effectiveness of these processes in treating a real industrial complex effluent on a larger scale are the major challenges to be overcome for the successful implementation of these technologies. As in the case of acoustic cavitation, the material and fabrication costs for the transducers are very high, making it an uneconomical operation to be tried on industrial scale. In the case of photocatalytic processes the engineering design and fabrication considerations for providing uniform distribution of UV radiation throughout the reactor adds to the total costs. In addition, the maintenance cost for catalyst regeneration and UV lamp life makes this process even more expensive. Similarly, for the Fenton process, the cost of chemical reagent (ferrous sulphate and H_2O_2) is very high. Further, the separation of sludge formed during the Fenton treatment also adds significant cost including that required for effective disposal of sludge after separation. These problems associated with Fenton process makes it an uneconomical process for large-scale operation. Among all the AOPs studied, HC has the potential for application on a larger scale because of its capability for generating hydroxyl radicals under ambient conditions, its ease of scale-up, and its lower material costs, making it more economically feasible. The biodegradability enhancement of distillery wastewater using HC as reported in the case study proves that this method can be used effectively as a pretreatment method in relation to existing biological treatment such that the efficiency of a conventional process can be improved many fold. Further efforts are required on the experimental front as well as in terms of kinetic modeling to establish this method as an effective pretreatment method for different industrial effluents with a high degree of efficiency and lower treatment costs.

On the other hand, hybrid methods offer higher efficiency over individual AOPs. The similarity between the mechanism of destruction and optimum operating conditions also points toward the synergism between these methods. Indeed, combinations of these AOPs should give better results when compared to individual techniques. Moreover, some of the drawbacks of the individual techniques can be eliminated by the characteristics of other techniques (e.g., mass transfer limitations and fouling of the catalyst in the case of photocatalytic oxidation will be eliminated by the turbulence created by cavitation). The expected synergism between different hybrid methods is mainly due to an identical controlling reaction mechanism, i.e., the free radical attack. Generally, the combination of two or more AOPs, such as cavitation/ozone, cavitation/H_2O_2, cavitation/photocatalysis, and cavitation/Fenton, leads to an enhanced generation of the hydroxyl radicals, which eventually results in higher oxidation rates. The efficacy of the process and the extent of synergism depend not only on the enhancement in the number of free radicals but also on the alteration of reactor conditions, leading to a better contact of the generated free radicals with the pollutant molecules and also better utilization of the oxidants and catalytic activity.

The depiction of case study related to the degradation of imidacloprid using the hybrid methods based on HC proves that the combination of different AOPs gives higher extents of degradation per unit supplied energy as compared to individual methods. Combining H_2O_2 with HC leads to better utilization of the oxidant and hence higher degradation rates due to the dissociation of H_2O_2 under the action of cavitation. The mass transfer resistance, which is a major limiting factor for the application of ozone or H_2O_2 alone, is also eliminated due to the enhanced turbulence generated by cavitation. Further, the inclusion of ferrous ions and UV light improves process efficiency. In the combined process of HC and photocatalysis, it is important to run the processes of HC and UV light simultaneously rather than sequentially. The major factor controlling the overall efficiency of destruction is, however, the stability of the photocatalyst under the effect of cavitation. Efforts are required in terms of new designs that will protect the catalyst but at the same time yield enhanced effects.

Overall, it can be said that AOPs are an effective method for degrading complex biorefractory pollutants. However, further research work needs to be done in the direction of developing AOPs for the treatment of real industrial effluents, and focus should be in the direction of solving scale-up problems such as challenges in the design and obtaining higher degradation possibly at lower levels of energy requirements.

REFERENCES

Abbasi, M., Asl, N.R., 2008. Sonochemical degradation of Basic Blue 41 dye assisted by nano TiO_2 and H_2O_2. J. Hazard. Mater. 153 (3), 942–947.

Adewuyi, Y.G., 2001. Sonochemistry: environmental science and engineering applications. Ind. Eng. Chem. Res. 40 (22), 4681–4715.

Adewuyi, Y.G., 2005. Sonochemistry in environmental remediation 2. Heterogeneous sonophotocatalytic oxidation processes for the treatment of pollutants in water. Environ. Sci. Technol. 39 (22), 8557–8570.

Babuponnusami, A., Muthukumar, K., 2012. Advanced oxidation of phenol: a comparison between Fenton, electro-Fenton, sono-electro-Fenton and photo-electro-Fenton processes. Chem. Eng. J. 183, 1–9.

Bashir, T.A., Soni, A.G., Mahulkar, A.V., Pandit, A.B., 2011. The CFD driven optimization of a modified venturi for cavitation activity. Can. J. Chem. Eng. 89 (6), 1366–1375.

Bhatkhande, D.S., Pangarkar, V.G., Beenackers, A.A., 2002. Photocatalytic degradation for environmental applications—a review. J. Chem. Technol. Biotechnol. 77 (1), 102–116.

Bigda, R.J., 1996. Fenton's chemistry: an effective advanced oxidation process. Nat. Env. J. 6 (3), 36–39.

Braeutigam, P., Wu, Z.L., Stark, A., Ondruschka, B., 2009. Degradation of BTEX in aqueous solution by hydrodynamic cavitation. Chem. Eng. Technol. 32 (5), 745–753.

Bremner, D.H., Carlo, S.D., Chakinala, A.G., Cravotto, G., 2008. Mineralization of 2,4 dichlorophenoxyacetic acid by acoustic or hydrodynamic cavitation in conjunction with the advanced Fenton process. Ultrason. Sonochem. 15 (4), 416–419.

Cao, Y., Yi, L., Huang, L., Hou, Y., Lu, Y., 2006. Mechanism and pathways of chlorfenapyr photocatalytic degradation in aqueous suspension of TiO_2. Environ. Sci. Technol. 40 (10), 3373–3377.

Chakinala, A.G., Gogate, P.R., Burgess, A.E., Bremner, D.H., 2009. Industrial wastewater treatment using hydrodynamic cavitation and heterogeneous advanced Fenton processing. Chem. Eng. J. 152 (2–3), 498–502.

Cheng, Z., Quan, X., Xiong, Y., Yang, L., Huang, Y., 2012. Synergistic degradation of methyl orange in an ultrasound intensified photocatalytic reactor. Ultrason. Sonochem. 19 (5), 1027–1032.

Choi, W., Hoffmann, M.R., 1997. Novel photocatalytic mechanisms for $CHCl_3$, $CHBr_3$, and $CCl_3CO_2^-$ degradation and the fate of photogenerated trihalomethyl radicals on TiO_2. Environ. Sci. Technol. 31 (1), 89–95.

Daneshvar, N., Salari, D., Khataee, A.R., 2003. Photocatalytic degradation of azo dye acid red 14 in water: investigation of the effect of operational parameters. J. Photochem. Photobiol. A: Chem. 157 (1), 111–116.

Davis, A.P., Huang, C.P., 1991. The photocatalytic oxidation of sulfur containing organic compounds using cadmium sulfide and the effect on CdS photo-corrosion. Water Res. 25 (10), 1273–1278.

Didenko, Y.T., McNamara, W.B., Suclick, K.S., 1999. Hot spot conditions during cavitation in water. J. Am. Chem. Soc. 121 (24), 5817–5818.

Fındık, S., Gündüz, G., 2007. Sonolytic degradation of acetic acid in aqueous solutions. Ultrason. Sonochem. 14 (2), 157–162.

Fox, M.A., Duley, M.T., 1993. Heterogeneous photocatalysis. Chemical Reviews 93 (1), 341–357.

Francony, A., Petrier, C., 1996. Sonochemical degradation of carbon tetrachloride in aqueous solution at two frequencies: 20 kHz and 500 kHz. Ultrason. Sonochem. 3 (2), 77–82.

Gogate, P.R., Pandit, A.B., 2000a. Engineering design methods for cavitational reactors I: sonochemical reactors. AIChE J. 46 (2), 372–379.

Gogate, P.R., Pandit, A.B., 2000b. Engineering design methods for cavitation reactors II: hydrodynamic cavitation reactors. AIChE J. 46 (8), 1641–1649.

Gogate, P.R., Pandit, A.B., 2004a. A review of imperative technologies for wastewater treatment II: hybrid methods. Adv. Environ. Res. 8 (3–4), 553–597.

Gogate, P.R., Pandit, A.B., 2004b. A review of imperative technologies for wastewater treatment I: oxidation technologies at ambient conditions. Adv. Environ. Res. 8 (3–4), 501–551.

Gogate, P.R., Pandit, A.B., 2005. A review and assessment of hydrodynamic cavitation as a technology for the future. Ultrason. Sonochem. 12 (1–2), 21–27.

Gogate, P.R., Shirgaonkar, I.Z., Sivakumar, M., Senthilkumar, P., Vichare, N.P., Pandit, A. B., 2001. Cavitation reactors: efficiency analysis using a model reaction. AIChE J. 47 (11), 2526–2538.

Gogate, P.R., Mujumdar, S., Pandit, A.B., 2003. Sonochemical reactors for waste water treatment: comparison using formic acid degradation as a model reaction. Adv. Environ. Res. 7 (2), 283–299.

Gogate, P.R., Sutkar, V.S., Pandit, A.B., 2011. Sonochemical reactors: important design and scale up considerations with a special emphasis on heterogeneous systems. Chem. Eng. J. 166 (3), 1066–1082.

Hamadaoui, Q., Naffrechoux, E., 2008. Sonochemical and photosonochemical degradation of 4-chlorophenol in aqueous media. Ultrason. Sonochem. 15 (6), 981–987.

Han, J., Liu, Y., Singhal, N., Wang, L., Gao, W., 2012. Comparative photocatalytic degradation of estrone in water by ZnO and TiO_2 under artificial UVA and solar irradiation. Chem. Eng. J. 213, 150–162.

He, Z., Song, S., Ying, H., Xu, L., Chen, J., 2007. p-Aminophenol degradation by ozonation combined with sonolysis: operating conditions influence and mechanism. Ultrason. Sonochem. 14 (5), 568–574.

Herrmann, J., Matos, J., Disdier, J., Guillard, C., Laine, J., Malato, S., 1999. Solar photocatalytic degradation of 4-chlorophenol using the synergistic effect between titania and activated carbon in aqueous suspension. Catalyst Today 54 (2–3), 255–265.

Hua, I., Hoffmann, M.R., 1997. Optimization of ultrasonic irradiation as an advanced oxidation technology. Environ. Sci. Technol. 31 (8), 2237–2243.

Hung, H.M., Hoffmann, M.R., 1999. Kinetics and mechanism of the sonolytic degradation of chlorinated hydrocarbons: frequency effects. J. Phys. Chem. A 103 (15), 2734–2739.

Ince, N.H., Tezcanlí, G., 2001. Reactive dyestuff degradation by combined sonolysis and ozonation. Dyes Pigments 49 (3), 145–153.

Joshi, R.K., Gogate, P.R., 2012. Degradation of dichlorvos using hydrodynamic cavitation based treatment strategies. Ultrason. Sonochem. 19 (3), 532–539.

Kang, J.W., Hoffmann, M.R., 1998. Kinetics and mechanism of the sonolytic destruction of methyl tert-butyl ether by ultrasonic irradiation in the presence of ozone. Environ. Sci. Technol. 32 (20), 3194–3199.

Kang, Y.W., Hwang, K.Y., 2000. Effects of reaction conditions on the oxidation efficiency in the Fenton process. Water Res. 34 (10), 2786–2790.

Kang, N., Lee, D.S., Yoon, J., 2002. Kinetic modeling of Fenton oxidation of phenol and monochlorophenols. Chemosphere 47 (9), 915–924.

Karci, A., Arslan-Alaton, I., Olmez-Hanci, T., Bekbölet, M., 2012. Transformation of 2,4-dichlorophenol by H_2O_2/UV-C, Fenton and photo-Fenton processes: oxidation products and toxicity evolution. J. Photochem. Photobiol. A: Chem. 230 (1), 65–73.

Koda, S., Kimura, T., Kondo, T., Mitome, H.A., 2003. Standard method to calibrate sonochemical efficiency of an individual reaction system. Ultrason. Sonochem. 10 (3), 149–156.

Konstantinou, I.K., Albanis, T.A., 2004. TiO_2-assisted photocatalytic degradation of azo dyes in aqueous solution: kinetic and mechanistic investigations: a review. Appl. Catal. B: Environ. 49 (1), 1–14.

Kormann, C., Bahnemann, D.W., Hoffmann, M.R., 1991. Photolysis of chloroform and other organic molecules in aqueous suspensions. Environ. Sci. Technol. 25 (3), 494–500.

Kusic, H., Koprivanac, N., Srsan, L., 2006. Azo dye degradation using Fenton type processes assisted by UV irradiation: a kinetic study. J. Photochem. Photobiol. A: Chem. 181 (2–3), 195–202.

Kusvuran, E., Erbatur, O., 2004. Degradation of aldrin in adsorbed system using advanced oxidation processes: comparison of the treatment methods. J. Hazard. Mater. 106B (2–3), 115–125.

Lin, S.H., Lo, C.C., 1997. Fenton process for treatment of desizing wastewater. Water Res. 31 (8), 2050–2056.

Lin, S.H., Lin, C.M., Leu, H.G., 1999. Operating characteristics and kinetic studies of surfactant wastewater treatment by Fenton oxidation. Water Res. 33 (7), 1735–1741.

Lin, F., Zhang, Y., Wang, L., Zhang, Y., Wang, D., Yang, M., et al., 2012. Highly efficient photocatalytic oxidation of sulphur containing organic compounds and dyes on TiO_2 with dual co-catalysts Pt and RuO_2. Appl. Catal. B: Environ. 127, 363–370.

Lorimer, J.P., Mason, T.J., 1987. Sonochemistry part 1—the physical aspects. Chem. Soc. Rev. 16, 239–274.

Ma, J., Song, W., Chen, C., Ma, W., Zhao, J., Tang, Y., 2005. Fenton degradation of organic compounds promoted by dyes under visible irradiation. Environ. Sci. Technol. 39 (15), 5810–5815.

Madhavan, J., SathishKumar, P.S., Anandan, S., Zhou, M., Grieser, F., Ashokkumar, M., 2010. Ultrasound assisted photocatalytic degradation of diclofenac in an aqueous environment. Chemosphere 80 (7), 747–752.

Martinez, N., Fernandez, J., Segura, X., Ferrer, A., 2003. Pre-oxidation of an extremely polluted industrial wastewater by the Fenton's reagent. J. Hazard. Mater. B 101 (3), 315–322.

Masomboon, N., Ratanatamskul, C., Lu, M.-C., 2009. Chemical oxidation of 2,6-dimethylaniline in the fenton process Env. Sci. Tech. 43 (22), 8629–8634.

Mazzarino, I., Piccinini, P., 1999. Photocatalytic oxidation of organic acids in aqueous media by a supported catalyst. Chem. Eng. Sci. 54 (15–16), 3107–3111.

Merouani, S., Hamdaoui, O., Saoudi, F., Chiha, M., 2010. Sonochemical degradation of Rhodamine B in aqueous phase: effects of additives. Chem. Eng. J. 158 (3), 550–557.

Moholkar, V.S., Pandit, A.B., 1997. Bubble behavior in hydrodynamic cavitation: effect of turbulence. AIChE J. 43 (6), 1641–1648.

Moholkar, V.S., SenthilKumar, P., Pandit, A.B., 1999. Hydrodynamic cavitation for sonochemcial effects. Ultrason. Sonochem. 6 (1–2), 53–65.

Nagata, Y., Nakagawa, M., Okuno, H., Mizukoshi, Y., Yim, B., Maeda, Y., 2000. Sonochemical degradation of chlorophenols in water. Ultrason. Sonochem. 7 (3), 115–120.

Ollis, D.F., Pellizzetti, E., Serpone, N., 1991. Photocatalytic destruction of water contaminants. Environ. Sci. Technol. 25, 1522–1529.

Padoley, K.V., Saharan, V.K., Mudliar, S.N., Pandey, R.A., Pandit, A.B., 2012. Cavitationally induced biodegradability enhancement of distillery waste water. J. Hazard. Mater. 219–220, 69–74.

Pang, Y.L., Abdullah, A.Z., Bhatia, S., 2011. Review on sonochemical methods in the presence of catalysts and chemical additives for treatment of organic pollutants in wastewater. Desalination 277 (1–3), 1–14.

Papic, S., Vujevic, D., Koprivanac, N., Sinko, D., 2009. Decolourization and mineralization of commercial reactive dyes by using homogeneous and heterogeneous Fenton and UV/Fenton processes. J. Hazard. Mater. 164 (2–3), 1137–1145.

Patil, P.N., Gogate, P.R., 2012. Degradation of methyl parathion using hydrodynamic cavitation: effect of operating parameters and intensification using additives. Sep. Purif. Technol. 95, 172–179.

Petrier, C., David, B., Laguian, S., 1996. Ultrasonic degradation at 20 and 500 kHz of atrazine and pentachlorophenol in aqueous solutions: preliminary results. Chemosphere 32 (9), 1709–1718.

Pradhan, A.A., Gogate, P.R., 2010. Removal of p-nitrophenol using hydrodynamic cavitation and Fenton chemistry at pilot scale operation. Chem. Eng. J. 156 (1), 77–82.

Ragaini, V., Selli, E., Bianchi, C.L., Pirola, C., 2001. Sono-photocatalytic degradation of 2-chlorophenol in water: kinetic and energetic comparison with other techniques. Ultrason. Sonochem. 8 (3), 251–258.

Raut-Jadhav, S., Saharan, V.K., Pinjari, D.V., Saini, D.R., Sonawane, S.H., Pandit, A.B., 2013. Intensification of degradation of imidacloprid in aqueous solutions by combination of hydrodynamic cavitation with various advanced oxidation processes (AOPs) J. Env. Chem. Eng. 1 (4), 850–857.

Ray, A.K., Beenackers, A.A.C.M., 1997. Novel swirl-flow reactor for kinetic studies of semiconductor photocatalysis. AIChE J. 43 (10), 2571–2578.

Rodriguez, M.L., Timokhin, V.I., Contreras, S., Chamarro, E., Esplugas, S., 2003. Rate equation for the degradation of nitrobenzene by 'Fenton-like' reagent. Adv. Environ. Res. 7 (2), 583–595.

Saharan, V.K., Badve, M.P., Pandit, A.B., 2011. Degradation of Reactive Red 120 dye using hydrodynamic cavitation. Chem. Eng. J. 178, 100–107.

Saharan, V.K., Pandit, A.B., SatishKumar, P.S., Anandan, S., 2012. Hydrodynamic cavitation as an advanced oxidation technique for the degradation of Acid Red 88 dye. Ind. Eng. Chem. Res. 51 (4), 1981–1989.

Saharan, V.K., Rizwani, M.A., Malani, A.A., Pandit, A.B., 2013. Effect of geometry of hydrodynamically cavitating device on degradation of orange-G. Ultrason. Sonochem. 20 (1), 345–353.

Saien, J., Delavari, H., Solymani, A.R., 2010. Sono-assisted photocatalytic degradation of styrene-acrylic acid copolymer in aqueous media with nano titania particles and kinetic studies. J. Hazard. Mater. 177 (1–3), 1031–1038.

Senthilkumar, P., Sivakumar, M., Pandit, A.B., 2000. Experimental quantification of chemical effects of hydrodynamic cavitation. Chem. Eng. Sci. 55 (9), 1633–1639.

Shrestha, R.A., Pham, T.D., Sillanpää, M., 2009. Effect of ultrasound on removal of persistent organic pollutants (POPs) from different types of soils. J. Hazard. Mater. 170 (2–3), 871–875.

Sivakumar, M., Pandit, A.B., 2002. Wastewater treatment: a novel energy efficient hydrodynamic cavitational technique. Ultrason. Sonochem. 9 (3), 123–131.

Sivakumar, M., Tatake, P.A., Pandit, A.B., 2002. Kinetics of p-nitrophenol degradation: effect of reaction conditions and cavitational parameters for a multiple frequency system. Chem. Eng. J. 85 (2–3), 327–338.

Song, W., Cheng, M., Ma, J., Ma, W., Chen, C., Zhao, J., 2006. Decomposition of hydrogen peroxide driven by photochemical cycling of iron species in clay. Environ. Sci. Technol. 40 (15), 4782–4787.

Song, S., Xia, M., He, Z., Ying, H., Lu, B., Chen, J., 2007. Degradation of p-nitrotoluene in aqueous solution by ozonation combined with sonolysis. J. Hazard. Mater. 144 (1–2), 532–537.

Subramanian, V., Pangarkar, V.G., Beenackers, A.A.C.M., 2000. Photocatalytic degradation of para-hydroxybenzoic acid: relationship between substrate adsorption and photocatalytic degradation. Clean Prod. Processes 2 (3), 149–156.

Sun, J.H., Sun, S.P., Fan, M.H., Guo, H.Q., Qiao, L.P., Sun, R.Z., 2007. A kinetic study on the degradation of p-nitroaniline by Fenton oxidation process. J. Hazard. Mater. 148 (1–2), 172–177.

Suslick, K.S., 1990. The chemical effects of ultrasound. Science 247, 1439–1445.

Sutkar, V.S., Gogate, P.R., 2009. Design aspects of sonochemical reactors: techniques for understanding cavitational activity distribution and effect of operating parameters. Chem. Eng. J. 155 (1–2), 26–36.

Tang, W.Z., Huang, C.P., 1996. 2,4 Dichlorophenol oxidation kinetics by Fenton's reagent. Environ. Technol. 17 (12), 1371–1378.

Thiruvenkatachari, R., Kwon, T.O., Jun, J.C., Balaji, S., Matheswaran, M., Moon, S., 2007. Application of several advanced oxidation processes for the destruction of terephthalic acid (TPA). J. Hazard. Mater. 142 (1–2), 308–314.

Vecitis, C.D., Lesko, T., Colussi, A.J., Hoffmann, M.R., 2010. Sonolytic decomposition of aqueous bioxalate in the presence of ozone. J. Phys. Chem. A 114 (14), 4968–4980.

Venkatadri, R., Peters, R.V., 1993. Chemical oxidation technologies: ultraviolet light/ hydrogen peroxide, Fenton's reagent and titanium dioxide assisted photocatalysis. Hazard. Waste Hazard. Mater. 10 (2), 107–149.

Wang, X., Zhang, Y., 2009. Degradation of alachlor in aqueous solution by using hydrodynamic cavitation. J. Hazard. Mater. 161 (1), 202–207.

Wang, J., Ma, T., Zhang, Z., Zhang, X., Jiang, Y., Dong, D., et al., 2006. Investigation on the sonocatalytic degradation of parathion in the presence of nanometer rutile titanium dioxide (TiO_2) catalyst. J. Hazard. Mater. 137 (2), 972–980.

Wang, J., Wang, X., Guo, P., Yu, J., 2011a. Degradation of reactive brilliant red K-2BP in aqueous solution using swirling jet-induced cavitation combined with H_2O_2. Ultrason. Sonochem. 18 (2), 494–500.

Wang, X., Yao, Z., Wang, J., Guo, W., Li, G., 2008. Degradation of reactive brilliant red in aqueous solution by ultrasonic cavitation. Ultrason. Sonochem. 15 (1), 43–48.

Wang, X., Jia, J., Wang, Y., 2011b. Degradation of C.I. Reactive Red 2 through photocatalysis coupled with water jet cavitation. J. Hazard. Mater. 185 (1), 315–321.

Weavers, L., Ling, F., Hoffmann, M., 1998. Aromatic compound degradation in water using a combination of sonolysis and ozonolysis. Environ. Sci. Technol. 32 (18), 2727–2733.

Wei, T.Y., Wang, Y.Y., Wan, C.C., 1990. Photocatalytic oxidation of phenol in the presence of hydrogen peroxide and titanium dioxide power. J. Photochem. Photobiol. A: Chem. 55, 115–126.

Wu, C.H., Kuo, C.Y., Chang, C.L., 2008. Decolorization of C.I. Reactive Red 2 by catalytic ozonation processes. J. Hazard. Mater. 153 (3), 1052–1058.

Xu, N., Shi, Z., Fan, Y., Dong, J., Shi, J., Hu, M.Z.C., 1999. Effects of particle size of TiO_2 on photocatalytic degradation of methylene blue in aqueous suspensions. Ind. Eng. Chem. Res. 38 (2), 373–379.

Xue, X., Hanna, K., Deng, N., 2009. Fenton-like oxidation of Rhodamine B in the presence of two types of iron (II, III) oxide. J. Hazard. Mater. 166 (1), 407–414.

Yan, Y., Thorpe, R.B., 1990. Flow regime transitions due to cavitation in the flow through an orifice. Int. J. Multiphase Flow 16 (6), 1023–1045.

Yawalkar, A.A., Bhatkhande, D.S., Pangarkar, V.G., Beenackers, A.A.C.M., 2001. Solar assisted photochemical and photocatalytic degradation of phenol. J. Chem. Technol. Biotechnol. 76 (4), 363–370.

Zhonga, Y., Jin, X., Qiao, R., Qi, X., Zhuang, Y., 2009. Destruction of microcystin-RR by Fenton oxidation. J. Hazard. Mater. 167 (1–3), 1114–1118.

CHAPTER 4

Advanced Treatment Technology and Strategy for Water and Wastewater Management

Haresh Bhuta
XH2O Solutions Pvt. Ltd., Ahmedabad, India

4.1 INTRODUCTION

Advanced oxidation systems are not widely used in industry. This is not because of their inability to treat wastewater but rather because of general lack of awareness and understanding about the current advanced oxidation technologies and their applications to different cases. It is widely known that conventional wastewater treatment systems have serious shortcomings that can be addressed by advanced oxidation processes.

4.1.1 Principal Bottlenecks of Present Wastewater Treatment Systems

Effective management of wastewater generated from chemical production, processing of textiles, and other industrial activities is becoming a serious cause for concern. With continuous growth in sectors such as dyestuff manufacturing, pharmaceuticals, and textiles, as well as the simultaneous increase in the number of small-scale industries, old methods of treating and managing wastewater are becoming obsolete.

The old methods of treatment commonly comprise primary treatment, followed by biological treatment, and finally polishing by carbon and sand filters. The key to this approach is effective biological treatment. However, it is nearly impossible to have a biological treatment method that is effective for the existing wastewater streams from modern complex chemical manufacturing units. Principal limitations of this conventional method are high volume/detention time, high floor area required, inability to treat complex and unpredictable wastewater, nature of organic load, and the requirement of trained manpower. Some of these limitations are explained below.

- *High volume of wastewater.*
  - The volume of wastewater is one of the principal criteria for the design of effective treatment plants. Most conventional treatment processes are designed for specific flow rates and are not flexible to meet the changing needs of industry.

An industry has to make a highly predictive guess on its rate of expansion and design a treatment plant to cater to the needs for the next decade, which is a waste of critical industrial resources.

- *Very scarce space* available for in-plant treatment.
  - Space for treatment is another major bottleneck. Our experience with industries has indicated that the nonavailability of effective treatment areas becomes a hurdle in implementing a treatment process.
  - Today it is felt that 95 industries out of 100 have severe space limitations for expansion of wastewater treatment plants. There is an *urgent* need for a technology/process that can deliver treatment results with an ultra low footprint.
- *Complex and unpredictable nature of wastewater* with dyes, auxiliaries, and biorefractory compounds.
  - The days of one company–one product are long gone. Today even a small and a medium scale enterprise (SME) chemical industry is forced to produce a large variety of products if it is to stay competitive.
  - Unfortunately, treatment technologies designed decades ago were for a uniform wastewater stream with minimal variations in its composition.
  - Today's industries, such as chemicals and pharmaceuticals, produce such a wide variety of products that no single technology can meet their treatment requirements.

In a nutshell, it can be said that developments in treatment technology have failed to keep pace with the developments in process and synthesis technology.

- *Biological treatment* is unable to handle organic and hydraulic loads.
- *Larger treatment plants based on biological systems are not viable* from engineering and economic point of views.
  - As mentioned earlier, space is a very critical commodity today. An industry would choose production capacity expansion over treatment capacity.
- *Skilled manpower is not readily available* to handle conventional plants.

- Indian industries today face a severe shortage of skilled staffing, and whatever is available has to be used for production and waste management. It requires lot of time to learn to operate biological systems, especially when the characteristics are unpredictable.

A conventional system generates sludge/solid waste at different points, i.e., during primary treatment and from secondary treatment. Management and disposal of the solid waste resulting from these stages is an issue because it contains lot of active organic chemicals, bacteria, and other pollutants that prevent its use as fertilizer and make it difficult to dispose of in landfills because of the hazardous leachate generation.

Common effluent treatment plants (CETPs), which were established a decade ago for treatment and management of wastewater from chemical industries, have been ineffective in treating wastewaters using conventional methods. The inability to treat wastewater has impacted industrial growth and has adversely affected industrial expansion. There is an urgent need to identify new commercially applicable technologies that can treat complex industrial wastewaters.

Advanced oxidation technologies may resolve some of these issues. Key aspects of advanced oxidation treatment are discussed in the following section.

4.2 ADVANCED OXIDATION TREATMENT

The advanced oxidation process operates through different catalytic forms, using oxidizing agents for degradation/mineralization of pollutants. The Fenton process and the advanced electrochemical oxidation process are at the foundation of advanced oxidation treatment processes. The principle of the Fenton process is the catalytic cycle of the reaction between iron (catalyst) and hydrogen peroxide (oxidant) to produce hydroxyl radicals. The hydroxyl radical is produced according to the following reaction:

$$Fe^{2+} + H_2O_2 \rightarrow Fe^{3+} + OH^- + {}^{\bullet}OH$$

Although the Fenton reagent has been known for more than a century, its application in an oxidizing process for destroying hazardous organics was not realized until the late 1960s. The Fenton reagent is one of the most effective methods for oxidizing organic pollutants. The efficiency of the Fenton reaction depends mainly on H_2O_2 concentration, the Fe^{2+}/H_2O_2 ratio, pH, and reaction time. In addition, the initial concentration of the pollutant and its character, as well as temperature, have a substantial influence on final efficiency. The Fenton reagent destroys a wide variety of organic compounds without the formation of toxic by-products. Among the different technologies reported in the literature for the treatment of highly contaminated effluents, Fenton's reagent is characterized by its cost effectiveness, simplicity, and suitability for treating aqueous wastes showing variable compositions. This method offers a cost-effective source of highly oxidizing species, using easy-to-handle reagents. The important advantage of the Fenton process is that oxidation and coagulation take place simultaneously (Barbusiński, 2009).

The electrochemical oxidation process has been known for many years as a potential technology for water treatment using *in situ* generated oxidants such as chlorine and sodium hypochlorite. Commercialization of boron-doped diamond (BDD) technology has brought a revolutionary change in the electrochemical oxidation process. It has been proved that BDD, when used as an anode, is able to generate hydroxyl radicals (as generated in the Fenton process) without the use of any chemicals. BDD anodes are able to generate other more stable oxidants such as hydrogen peroxide, peroxides, and hypochlorite along with hydroxyl radicals for effective wastewater treatment.

4.3 FENTON PROCESS: ADVANCED OXIDATION TECHNOLOGIES

The Fenton process as an advanced oxidation technology has several advantages:
1. Simple to implement
2. Low capital cost
3. Effective across a broad range of organic compounds
4. Nonselective in oxidation of organics
5. Effective as pretreatment to a biological process.

The Fenton process is the simplest of all the advanced oxidation processes. It can be executed by mixing readily available oxidants and catalysts (see Figure 4.1; www.xh2osolutions.com). The reaction can be carried out using either a batch mode or a continuous mode with simple reactor design. The capital cost of the Fenton process is relatively low due to the simple nature of the reactor design involved. Hydroxyl radicals generated from the Fenton process are the most powerful radicals that unselectively destroy organic compounds. One of the biggest benefits of Fenton oxidation is that it has

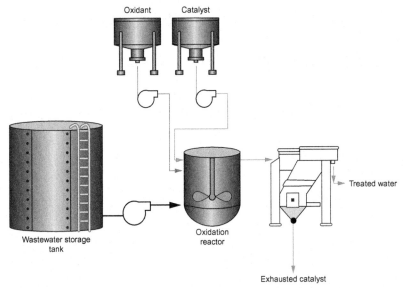

Figure 4.1 Layout of Fenton oxidation process (http://www.xh2osolutions.com).

the ability to increase the BOD/COD ratio of wastewater, thus improving the treatability of the wastewater by conventional biological means.

As indicated in Figure 4.1, the Fenton process is the simplest of all the advanced oxidation processes. The process requires a continuous stirred tank reactor (CSTR), oxidant and catalyst holding tanks, a neutralization vessel, and a settler (filter) for sludge removal. Any small- or medium-scale industry would have the abovementioned infrastructure, which makes Fenton oxidation the most appropriate process for taking the first step toward advanced oxidation treatment. The Fenton process can be used for both destruction of COD and color reduction.

$$\begin{array}{c} Fe^{3+} + O_2^{\bullet-} \rightarrow Fe^{2+} + O_2 \\ Fe^{2+} + H_2O_2 \rightarrow Fe^{3+} + OH^- + {}^{\bullet}OH \\ \hline O_2^{\bullet-} + H_2O_2 \underset{Fe^{3+}}{\rightarrow} OH^- + {}^{\bullet}OH + O_2 \end{array}$$

The Fenton oxidation process follows these simple steps:
1. Acidifying the wastewater to necessary/optimum acidic pH
2. Addition of ferrous solution
3. Addition of hydrogen peroxide
4. Reaction with effective mixing
5. Neutralization of reacted water
6. Sludge removal.

Principal advantages of Fenton oxidation are the following:

1. Simple process, can be conducted in either batch or continuous mode
2. Readily available chemicals for treatment
3. Completely green, environmentally friendly process
4. Destroys a very broad range of organic compounds
5. Can be a pre- or posttreatment step for a biological process
6. Sludge generated is completely nonhazardous and can be safely disposed of.

The disadvantages of the Fenton process are the following:

1. Large quantity of sludge generated
2. Difficulty of hydrogen peroxide management
3. The problem of identifying optimum dosages for cost-effective treatment.

The Fenton oxidation process can be effectively applied to treat dye wastewaters. An example is shown in Figure 4.2a, which shows a photograph of the results obtained from treating direct dye wastewater with Fenton oxidation.

The above results are highly significant because the wastewater contains solvents such as nitrobenzene and amines that make biological treatment difficult. The Fenton process is very successful in COD reduction from around 5000–6000 ppm COD to <1000 ppm COD value with a reaction time of less than an hour. The process is also attractive because it can handle huge fluctuations in COD of incoming wastewater and hydraulic load. The cost of treatment ranges from about Rs. 300 to Rs. 600 m^3. The capital cost of

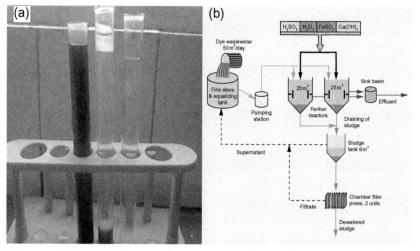

Figure 4.2 (a) Fenton oxidation of direct dye wastewater. (b) Scheme for treatment of textile wastewater using the Fenton process.

treatment is minimal. Another major advantage of this process is a lower footprint, especially when it is carried out in the CSTR process.

The schematic in Figure 4.2b shows actual layout and sizing of a system for treatment of textile dyeing wastewater using the Fenton oxidation process (Barbusiński, 2009).

4.4 ELECTRO-FENTON ADVANCED OXIDATION TREATMENT

Electro-Fenton is a process that is specifically designed to overcome some of the drawbacks of the Fenton process. Two major drawbacks of the Fenton process are the quantity of sludge produced and the imbalance in the Fe/H_2O_2 ratio, which leads to a lower rate of oxidation.

Electro-Fenton (Figures 4.3 and 4.4) (http://www.xh2osolutions.com) is a process that addresses the issues pertaining to the two abovementioned drawbacks by recycling ferrous ions in the Fenton reactor. The advantages of recycling ferrous ions are the following:

1. Less sludge formation compared to advanced oxidation processes (AOPs), because of lower use of ferrous catalyst.
2. Effective against high COD wastewater because there is better control of molar ratio of Fe/H_2O_2.
3. Nonselective oxidation of organics as in Fenton oxidation.
4. Effective both as pretreatment for biological processes and as a standalone treatment for low volume wastewater.

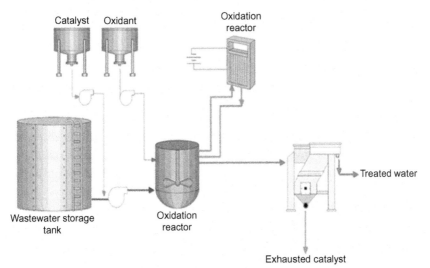

Figure 4.3 Layout of electro-Fenton process (http://www.xh2osolutions.com/).

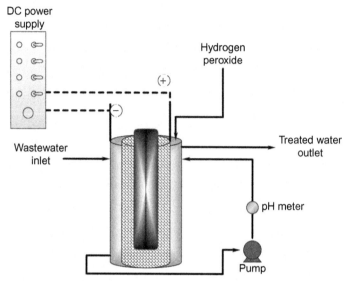

Figure 4.4 Schematic of electro-Fenton process.

As indicated in Figures 4.3 and 4.4, the electro–Fenton process aims to remove one of the principal disadvantages of the Fenton process, i.e., excessive sludge generation and sludge management. Basically, in the Fenton process, sludge is generated when ferrous (Fe^{2+}) is converted into ferric (Fe^{3+}) during the course of the oxidation process. On neutralization, ferric hydroxide is formed, which is separated as sludge.

The electro–Fenton operation overcomes this shortcoming of Fenton oxidation by electrochemically converting the ferric back into ferrous salt. This converted ferrous salt then again takes part in the oxidation process. This is depicted schematically in Figure 4.5 (http://www.xh2osolutions.com).

Recycling of ferrous can, in principle, be done in two ways: (1) *in situ* recycling and (2) *ex situ* recycling. In the case of *in situ* recycling, the dosing of the ferrous catalyst is done in a proportion lower than that required in conventional Fenton process. During the course of the reaction, the ferric salt formed post-Fenton process is converted back into ferrous salt at the cathode as indicated in Figure 4.5. This reaction process is therefore called a Fered Fenton process. A typical setup of the Fered Fenton process is indicated schematically in Figure 4.6 (Anotai et al., 2011). The reaction occurring at the anode is:

$$H_2O \rightarrow 2H^+ + \tfrac{1}{2}O_2 + 2e^-$$

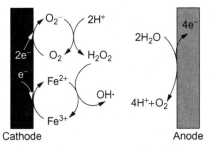

Figure 4.5 Possible reactions in electrochemical cell of electro-Fenton system (http://www.xh2osolutions.com).

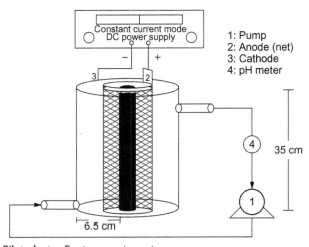

Figure 4.6 Pilot electro-Fenton reactor setup.

while the reaction taking place at cathode is:

$$Fe^{3+} + e^- \rightarrow Fe^{2+}$$

The principal advantage of this process is that the molar ratio of ferrous ions and hydrogen peroxide available at any time during the reaction in the reactor can be controlled and maintained at an optimal level. It has been seen that the Fered Fenton process can not only reduce the quantity of ferrous salt used for oxidation but that it can also accelerate, and achieve a higher percentage of, organic destruction compared to the conventional Fenton process (Haung et al., 2000; Kin et al., 2005; Yao-Hui et al., 2000).

The *ex situ* electro–Fenton process applies a sludge recycling process post-conventional Fenton process. The basic principle of sludge recycling in this process remains the same as in the Fered Fenton process. While in

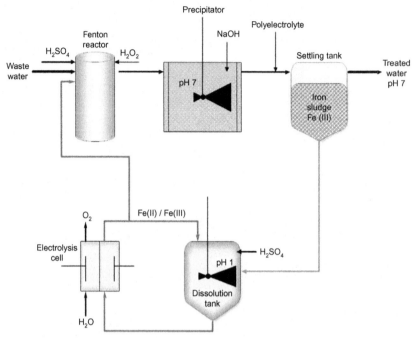

Figure 4.7 *Ex situ* electro-Fenton layout.

the Fered Fenton process, the conversion takes place in the reactor, in the *ex situ* process, the sludge post-neutralization stage is again acidified to the necessary pH and then passed through an electrochemical cell where the ferric is converted to ferrous. Thereafter, this ferrous salt is reused in the process for reaction. This Fenton sludge recycling has been reported in the past (Swinnen et al., 1999). A typical process schematic is shown in Figure 4.7.

Other variants of the electro–Fenton process can be found in the literature, but the two mentioned in this chapter are possible commercially viable process options that can be used in industries at the small, medium, and large scale.

4.5 FENTON CATALYTIC REACTOR ADVANCED OXIDATION TREATMENT

The Fenton process generally is not an ideal choice for treatment of high-volume wastewater because of the limitations associated with it, as discussed earlier. The Fenton catalytic reactor aims to address the limitations of the classical Fenton process using a special reactor design wherein the exhausted ferrous catalyst gets converted into a heterogeneous catalyst that further aids the oxidation process, without getting converted into sludge. The principal advantages of the Fenton catalytic reactor process are less sludge formation, suitability for treatment of high volumes of wastewater, nonselective oxidation of organics, and its ability to be an effective polishing step for CETPs, municipal sewage treatment, and other uses.

Almost all the efforts in developing the Fenton process have aimed at the reduction of sludge formation during the course of the reaction. The Fenton catalytic reaction process, also known as the fluidized-bed Fenton process, was developed by ITRI (Haung et al., 2000; Yao–Hui et al., 2000). Although the Fenton process has been successful in the degradation of organic contaminants present in wastewater, the production of ferric hydroxide sludge in the form of $Fe(OH)_3$ is considered to be a disadvantage of this process, requires further separation and disposal. One of the alternatives for dealing with this problem is the use of the fluidized-bed Fenton reactor. The carriers in the fluidized-bed Fenton reactor can initiate the iron precipitation and/or crystallization process; therefore, the production of sludge can be reduced. Several reactions occur during the operation of the fluidized-bed Fenton reactor, including (1) homogeneous chemical oxidation (H_2O_2/Fe^{2+}), (2) heterogeneous chemical oxidation $(H_2O_2/iron\ oxide)$, (3) fluidized-bed crystallization, and (4) reductive dissolution of iron oxides (Muangthai et al., 2010).

The schematic drawing in Figure 4.8 shows a typical fluidized Fenton process (http://www.xh2osolutions.com), and processes occurring inside

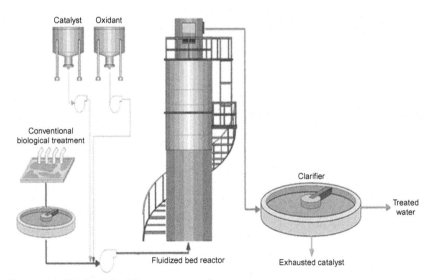

Figure 4.8 Fluidized-bed Fenton system layout.

the reactor are shown schematically in Figure 4.9. Such industrial installations have been reported in the literature (http://www.tridenti.com.my/chemical-oxidation.php). A pilot plant setup for the treatment process for wastewaters of CETPs using this technology has been established. The fluidized-bed Fenton process reduces the possible sludge generated to a minimum. The process is a combination of homogeneous and heterogeneous Fenton processes. Possible reactions happening during a fluidized Fenton process can be as follows (Muangthai et al., 2010):

$$Fe^{2+} + H_2O_2 \rightarrow {}^{\bullet}OH + OH^- + Fe^{3+}$$
$$Fe^{3+} + H_2O_2 \rightarrow Fe - HOO^{2+} + H^+$$
$$Fe - HOO^{2+} + H^+ \rightarrow Fe^{2+} + HO_2{}^{\bullet}$$
$${}^{\bullet}OH + organics \rightarrow products$$
$$H_2O_2 + {}^{\bullet}OH \rightarrow H_2O + HO_2{}^{\bullet}$$
$$Fe^{2+} + {}^{\bullet}OH \rightarrow Fe^{3+} + {}^{\bullet}OH$$
$$Fe^{2+} + HO_2{}^{\bullet} \rightarrow Fe^{3+} + HO_2$$
$$Fe^{3+} + HO_2{}^{\bullet} \rightarrow Fe^{2+} + O_2 + H^+$$

The results from a pilot plant trial conducted for a CETP are presented in Figures 4.10 and 4.11. The purpose of the trials was to reduce COD post-biological treatment of mixed wastewater from the CETP.

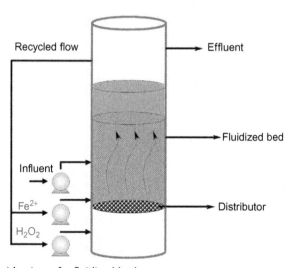

Figure 4.9 Inside view of a fluidized-bed reactor.

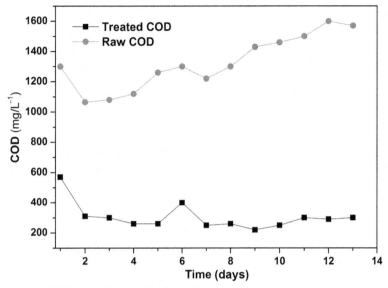

Figure 4.10 COD result for pilot fluidized Fenton reactor.

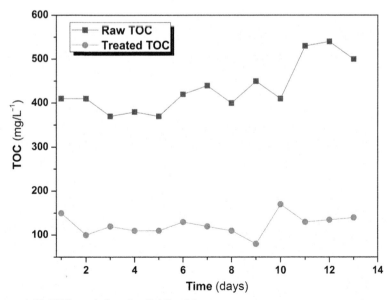

Figure 4.11 TOC result for pilot fluidized Fenton reactor.

The results obtained during the trials clearly show that during the process both COD and TOC are reduced. It was also seen that the consumption of peroxide is lower than that predicted from stoichiometric data.

4.6 ELECTROCHEMICAL ADVANCED OXIDATION TREATMENT WITH BDD

Unlike Fenton-based AOP, where oxidants are generated using external chemicals, the electro-oxidation process generates these oxidants *in situ* using a nonsacrificial electrode set (electrochemical reactor). These nonsacrificial electrodes are selected based on their capacity to generate hydroxyl radicals and other secondary oxidants such as chlorine. Principal advantages of this technology are:

1. It is a zero sludge process, i.e., during the process of oxidation no sludge is formed.
2. It can work as a standalone treatment process for wastewaters that are very difficult to treat because it requires no chemical input and generates no sludge.
3. Unlike the Fenton process, oxidation of organics in electro-oxidation can be selective or unselective depending on the oxidant that is generated in the electrochemical cell.

However,

1. Electro-oxidation is a capital intensive process when compared to the Fenton family of processes.
2. Because different electrodes have varying capacities for generation of oxidants, the selection of electrode is critical for an efficient electro-oxidation process.

In recent years, electrochemical oxidation with conductive-diamond anodes has appeared as one of the most promising technologies in the treatment of industrial wastes polluted with organics. Compared with other electrode materials, conductive-diamond has shown a higher stability and efficiency. During recent years, conductive-diamond electrochemical oxidation has been widely studied with synthetic industrial wastes in lab- and bench-scale plants (Cañizares et al., 2006a).

Recently, it has been demonstrated that hydroxyl radicals are formed during the electrolysis of aqueous electrolytes on conductive-diamond anodes (Aquino et al., 2012; Cañizares et al., 2006b). This has enabled classification of this technology as an AOP. Besides this mechanism, the global oxidation process in conductive diamond anodes is complemented by direct electro-oxidation on the surface and also mediated oxidation by other oxidants electrogenerated on the surface from the electrolyte salts. The combination of these oxidation mechanisms increases the current efficiency of this technique as compared with other electrochemical technologies.

The arrangement of BDD electrochemical cell is well reported (http://www.diaccon.de/english/electrode.htm). The principal advantages of the electrochemical oxidation processes are the following:

1. The process produces less sludge compared to Fenton oxidation processes.
2. Because there is no addition of chemicals, the process does not require any storage of hazardous chemicals.
3. The process can be operated in batches or continuously.
4. Secondary oxidants generated during the process can improve the overall oxidation efficiency.

There are certain disadvantages of the electrochemical oxidation process that can be listed as follows:

1. BDD cells are highly cost prohibitive.
2. The total area required for effective wastewater treatment can make actual application of BDD in industries very difficult.
3. The process is not easily scalable and thus requires extensive pilot studies.
4. Because the process is driven by current, industries with limited power availability cannot use this technology.
5. More research is required across different wastewaters to generate a database of organic/molecules that can be destroyed using an electrochemical process.

A sample of results from laboratory trials conducted on wastewaters using electrochemical oxidation/BDD electrodes is shown in Figures 4.12 and 4.13.

4.7 IMPLEMENTATION OF ADVANCED OXIDATION TECHNOLOGIES

Considering the nature of wastewater, it is very difficult to extrapolate an advanced oxidation technology solution from one industry to another. And it is sometimes difficult to apply the same technology in two different wastewater streams of the same industry. A structured approach is required to identify the right oxidation treatment technology for each industry. Design thinking can be one such approach. The design thinking process comprises seven stages: *define, research, ideate, prototype, choose, implement,* and *learn* (http://en.wikipedia.org/wiki/Design_thinking). Problems can be framed

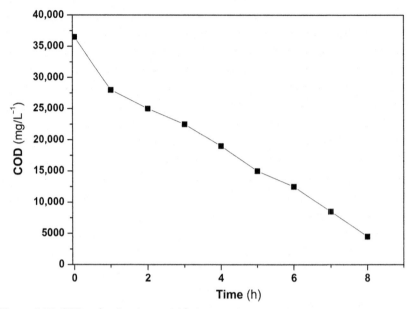

Figure 4.12 COD reduction in pesticide wastewater.

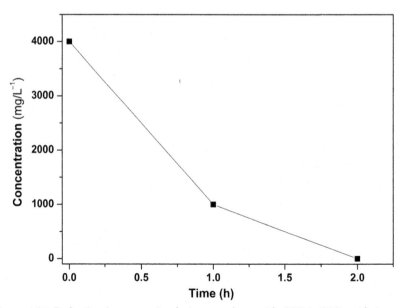

Figure 4.13 Reduction in ammoniacal nitrogen along with COD in BDD oxidation.

using these seven steps, the right questions can be asked, more ideas can be created, and the best answers can be chosen. The steps are not sequential; they can occur simultaneously and can be repeated.

Besides design thinking, it is important to clearly understand the nature of a wastewater and the pollutants in it. Typically, COD and BOD levels are considered as key deliverables of the treatment. However, parameters like COD and BOD may be just the tip of the iceberg. The real problem in wastewater treatment can be anything from pH, ineffective primary treatments, biorefractory compounds, presence of solvents, heavy metals, phenols, lack of space, or other issues. Unless the nature of problem is clearly defined, the correct AOP can never be selected.

For example, textile-processing industries are well known for being associated with wastewater with high volume and low COD. It is also well known that wastewater from textile processing is readily treatable biologically, but the real problem is availability of space for putting a biological treatment system in place. Thus the right solution for such an industry can be an AOP, which can handle a huge volume of wastewater of low COD value. Fluidized-bed Fenton can serve the purpose in this case.

Another example is the case of an SME dyestuff industry that manufactures a broad range of products/dyestuffs that generate wastewater whose nature is difficult to predict. In this scenario, the Fenton and electro-Fenton (Fered Fenton) processes can be the right technology for treatment because they can address a wide variety of wastewater characteristics.

After defining the problem, the next task is to select an appropriate combination of treatment technologies. Figure 4.14 shows the available spectrum of treatment technologies for wastewater treatment, mainly from the point of view of advanced oxidation processes. It is important to know the appropriate technology for application to wastewater treatment for technoeconomic selection. Figure 4.14 indicates one extreme technological selection in the form of biological treatment, while the other extreme is incineration. The pollutants that are most refractory in nature cannot be

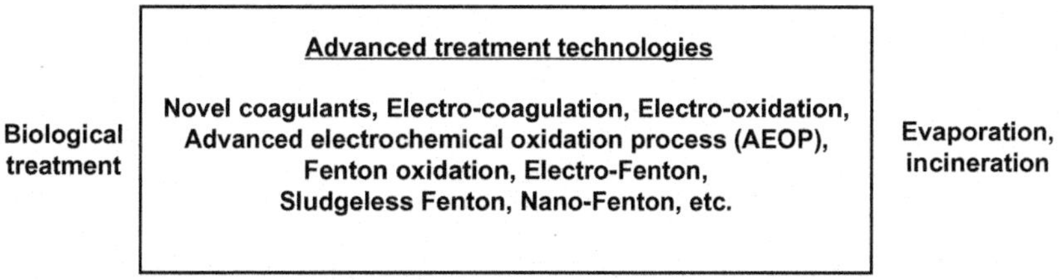

Advanced treatment technologies

Biological treatment | Novel coagulants, Electro-coagulation, Electro-oxidation, Advanced electrochemical oxidation process (AEOP), Fenton oxidation, Electro-Fenton, Sludgeless Fenton, Nano-Fenton, etc. | Evaporation, incineration

Figure 4.14 Spectrum of wastewater treatment technologies.

treated using biological methods; they even cause difficulty in effectively being degraded using advanced oxidation processes and have to be destroyed using an incineration process. Most of wastewaters usually need treatment technologies ranging from biological to incineration, and many times a combination of different treatments.

Possible strategies for implementing AOP are briefly outlined in the following sections.

4.7.1 Advanced Oxidation Process as an End-of Pipe Solution

Conventional wastewater treatment offers limited options regarding postbiological treatment. There is a vacuum in terms of technological options for meeting ultimate discharge and reuse norms postbiological treatment. Advanced oxidation processes can be used as a last resort/treatment to meet stringent discharge norms as shown in Figure 4.15, specifically in light of the following:
• Ideally suited for removal of residual toxicity from wastewater.
• Can reduce COD <1000 ppm.
• Can meet norms for reuse of wastewater.
• Can be applied in CETPs, chemical complexes, municipal sewage systems.

4.7.2 Advanced Oxidation Process as Standalone Treatment

Small- and medium-scale industries have limited options for treatment of high COD and biorefractory wastewater. Limited expertise and infrastructure for wastewater treatment limits the ability of industry to meet even simple CETP inlet norms. This therefore necessitates a different strategy (as depicted in Figure 4.16):
• Versatile treatment for a wide range of wastewater streams.
• A low footprint technology that is well suited for small- and medium-scale industries.
• Easily expandable and automatic in operation.

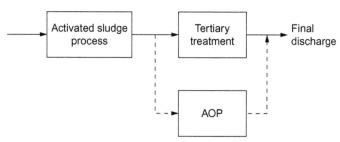

Figure 4.15 AOP as a polishing technology.

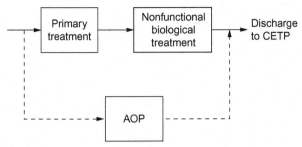

Figure 4.16 AOP as an alternate to conventional treatment process.

4.7.3 Advanced Oxidation Process as a Buffer for Biological Treatment

A single biorefractory wastewater stream can destabilize the entire treatment process for days. It can be stated that 80% of wastewater treatment issues are due to 20% of wastewater streams. Present technologies are unable to address such streams separately. A rethinking concerning the form of application (see Figures 4.17 and 4.18) and using the following aspects can be highly useful.

• Several AOPs can be combined for treatment of each specific stream.
• AOPs can be installed at the point of wastewater generation.
• Stream-specific AOPs can improve the performance of biological treatment systems.

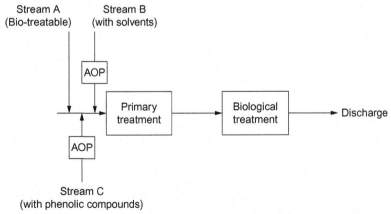

Figure 4.17 AOP as a buffer to conventional technology.

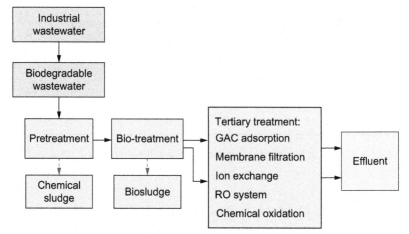

Figure 4.18 Rethinking wastewater treatment methodology and application schemes.

4.8 SUMMARY AND CONCLUSIONS

It is clear from the discussion above that the advanced oxidation process is highly useful for the treatment of industrial wastewaters, especially for the removal of biorefractory pollutants. The important advantages can be summarized as:

- The process is modular.
- The process has a very low footprint.
- The process is scalable.
- The process can be fully automated.
- The process is flexible.
- The process is complementary to conventional processes such as biological processes.

Advanced oxidation technologies are poised to play an important role in making our industrial growth more sustainable. As detailed in this chapter, even the simple Fenton process can significantly improve treatment of wastewaters from various industries because it is complementary to existing biological treatment methodologies. The advanced oxidation process is in a true sense a green technology because it uses chemicals that do not cause secondary pollution, unlike evaporation, which only transfers the pollutant from one phase to another. For an industry, there is broad spectrum of technologies to choose from, ranging from the sludge-generating Fenton process to sludgeless electrochemical oxidation. It has to be noted that research and development in the advanced oxidation process has not kept pace with the introduction of new products into the waste stream. Thus there is an urgent need for R & D in advanced oxidation processes, possibly in the direction of:

- Pilot plant studies and design of new systems.
- Preparation of database of successful AOP cases at lab, pilot, and plant level.
- Identification of heterogeneous catalysts that can address the issue of sludge generation.
- Combining AOP and conventional treatment process such as biological.
- Investment in model AOP systems and plants that can be replicated.
- Use of modern reaction engineering technologies for more efficient and effective AOP reactors and systems.

REFERENCES

Anotai, J., Sairiam, S., Lu, M.C., 2011. Enhancing treatment efficiency of wastewater containing aniline by electro-Fenton process. Sustain. Environ. Res. 21, 141–147.

Aquino, J.M., Rodrigo, M.A., Rocha-Filho, R.C., Sáez, C., Cañizares, P., 2012. Influence of the supporting electrolyte on the electrolyses of dyes with conductive-diamond anodes. Chem. Eng. J. 184, 221–227.

Barbusiński, K., 2009. The full-scale treatment plant for decolourisation of dye wastewater. Arch. Civil Eng. Environ. 2, 89–94.

Cañizares, P., Paz, R., Lobato, J., Sáez, C., Rodrigo, M.A., 2006a. Electrochemical treatment of the effluent of a fine chemical manufacturing plant. J. Hazard Mater. B138, 173–181.

Cañizares, P., Gadri, A., Lobato, J., Nasr, B., Paz, R., Rodrigo, M.A., Saez, C., 2006b. Electrochemical oxidation of azoic dyes with conductive-diamond anodes. Ind. Eng. Chem. Res. 45, 3468–3473.

http://www.diaccon.de/english/electrode.htm (Viewed on 12 March, 2014).

Haung, G.H., Huang, Y.H., Lee, S.N., Lin, S.M. 2000. *Method of Wastewater Treatment by Electrolysis and Oxidization*, US Patent 6126838 A.

Kin, K.T., Tang, H.S., Chan, S.F. 2005. *Water Treatment Reactor for Simultaneous Electrocoagulation and Advanced Oxidation Processes*, US Patent 20050224338 A1.

Muangthai, I., Ratanatamsakul, C., Lu, M.C., 2010. Removal of 2,4-dichlorophenol by fluidized-bed Fenton process. Sustain. Environ. Res. 20 (5), 325–331.

Swinnen, N., Gregor, K.H., Renders, A., TAPPI, 1999. Reuse of wastewater streams in the P&P industry using a physico-chemical treatment process. In: Proceedings of TAPPI International Environmental Conference, vol. 1, pp. 91–98.

http://www.tridenti.com.my/chemical-oxidation.php (Viewed on 12 March, 2014).

http://en.wikipedia.org/wiki/Design_thinking (Viewed on 12 March, 2014).

http://www.xh2osolutions.com/ (Viewed on 12 March, 2014).

Yao-Hui, H., Gaw-Hao, H., Shan-Shan, C., Huey-Song, Y., Shwu-Huey, P. 2000. *Process for Chemically Oxidizing Wastewater with Reduced Sludge Production*, US Patent 6143182 A.

CHAPTER 5

Novel Technologies for the Elimination of Pollutants and Hazardous Substances in the Chemical and Pharmaceutical Industries

Johannes Leonhauser[1], Jyoti Pawar[2], Udo Birkenbeul[3]
[1]Filtration, Membrane Technology and Waste Water, Bayer Technology Services GmbH, Leverkusen, Germany
[2]Bayer Technology Services, Thane, India
[3]Waste water treatment, Bayer Technology Services GmbH, Germany

5.1 INTRODUCTION

Industrial wastewater, especially from chemical and pharmaceutical production, often contains substances that need to be treated before being discharged into a biological treatment plant and subsequent water bodies. Generally, this can be done close to the production site itself, in selected wastewater streams before reaching a central treatment plant as shown in Figure 5.1.

Each of these approaches has certain advantages and disadvantages, and the boundaries between them are fluid. Furthermore, a variety of wastewater treatment processes exist that can be applied at each stage, making it a challenging task to choose the best one in economic and ecological terms.

To eliminate micropollutants and hazardous substances, oxidation processes such as wet air oxidation, Fenton reaction, or ozone are often applied, all of which have proven many times to be reliable in practice. At the same time, new processes are emerging (e.g., electrochemical oxidation with boron-doped diamond electrodes, UV-oxidation) that in certain cases are superior (higher efficiency, lower costs) to the established ones. On the other hand, nondestructive processes such as adsorption or extraction are often feasible alternatives to oxidation and have to be considered, too. Figure 5.2 shows a (non-exhaustive) overview of processes that can be applied.

To eliminate certain pollutants usually not only one process is appropriate, but a multitude of processes could suffice from a technical point of view. Therefore, the task is to find the one that fits best in the complex interaction

215

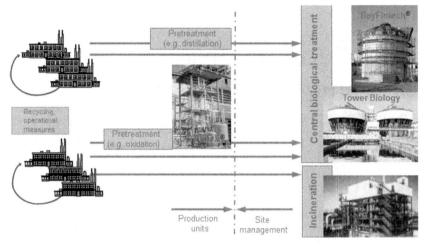

Figure 5.1 Treatment options to eliminate pollutants from wastewater.

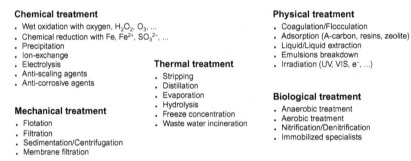

Figure 5.2 Processes that can be applied to eliminate micropollutants and hazardous substances from wastewater in the chemical and pharmaceutical industries.

between wastewater composition, production requirements, and site management, leading to highest efficiency at acceptable costs.

Therefore, the first step in any task of eliminating pollutants and hazardous substances should be through the evaluation of possible processes. Typically, this is done based on experience, theoretical considerations, and selected experimental investigations. Criteria that need to be considered in that evaluation are characteristics of the wastewater stream (further components, salt content, suspended solids, flow-rate, variation in water quality or quantity), desired removal or purification rate (emission limits, process requirements), technical feasibility (removal rates, fouling, corrosion, scale-up experience), local economic background (energy and chemical costs), infrastructure (site condition, availability of energy, chemicals and personnel, existing central ETP/solid waste disposal/off-gas treatment), local engineering and operation staff, investment and operating budget, and time schedule.

Figure 5.3 further shows steps that should be followed to identify, develop engineering, and implement processes to eliminate pollutants and hazardous substances from wastewater in the chemical and pharmaceutical industries. In practice, often even during the very first step, it turns out that not only does the one wastewater stream in question need to be addressed but broader issues of wastewater management and control as well.

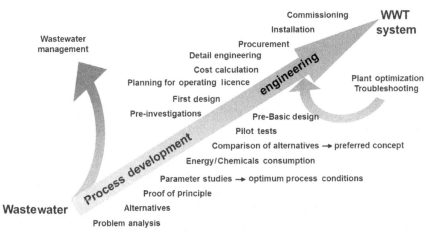

Figure 5.3 Steps to identify and develop processes to eliminate pollutants and hazardous wastes from wastewater in the chemical and pharmaceutical industries.

The next section describes the two in-house technologies developed by Bayer to benefit the industry for eliminating pollutants and hazardous wastes from wastewater.

5.2 THE BAYER LOPROX PROCESS (Holzer et al., 1992)

In the early 1970s, research teams at Bayer began taking a closer look at the wet oxidation process. It was found that, for a number of compounds with poorer biodegradability, less extreme reaction conditions can be sufficient at least to initiate the oxidation process, making them considerably easier to degrade during subsequent biological treatment.

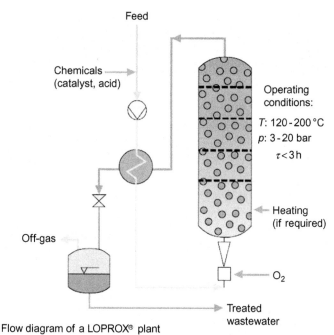

Flow diagram of a LOPROX® plant

Figure 5.4 Flowchart of a LOPROX plant.

This led to the development of Bayer's LOPROX process (low pressure wet oxidaton), a system that operates at temperatures lower than 200 °C (392 °F) and pressure levels of between 3 and 20 bar (45–280 psi).

The principle of the process is shown in Figure 5.4. The untreated wastewater is heated in a countercurrent heat exchanger and is then pumped into the oxidation reactor, a bubble column. Pure oxygen serves as the oxidizing agent. It is injected into the oxidation reactor in fine bubbles.

Oxidation takes place in the acidic range and is catalyzed by a combination of Fe^{2+} ions and organic quinone-forming substances. The residence time in the oxidation reactor is around 1–3 h. The effluent to be treated should have a chemical oxygen demand (COD) of 5000–100,000 ppm.

Oxidation is exothermic, that is, the temperature rises during the course of the reaction. At a reaction temperature of 150 °C (300 °F), a concentration of around 10,000 ppm COD in the effluent is sufficient for the process to take place autothermally. This means that, once the LOPROX plant has been heated up, no additional energy need be supplied.

The treated wastewater flows from the head of the reactor column to the countercurrent heat exchanger where it is cooled. It is then expanded and transferred to a biological treatment plant. This generates a small quantity of flue gas containing CO_2 and small part of CO, which is then treated in a separate process.

Depending on the reaction and the type of organic compound being treated, 60–90% of the organic carbon is converted to CO_2 during wet oxidation. What remains in the wastewater generally consists of easily biodegradable fragments of the aromatic core (acetic acid, acetone, etc.). If the substituting groups include S, CI, or P, then H_2SO_4, HCl, or H_3PO_4 are respectively formed. Most of the organic nitrogen is converted to ammonia, which can be removed from the wastewater by stripping or in the biological plant by nitrification/dinitrification.

Figure 5.5 shows the simple logarithmic plot of a typical total organic carbon (TOC) versus time curve. The values plotted were measured during the wet oxidation at 140 °C (284 °F) of a 2000 ppm phenol solution with a pH of 2. The reaction was catalyzed with Fe^{2+} and quinone-forming substances.

The plot shows a first-order reaction with a TOC elimination of about 70% after 90 minutes. After just 45 minutes, phenol can no longer be detected chromatographically. The phenol was not oxidized in control experiments run under identical conditions but without the addition of the catalysts.

Quinone-forming substances are aromatic compounds in which at least two OH- or NH_2- groups are substituted on each ring. It can be assumed that hydrogen peroxide is formed as an intermediate stage during the

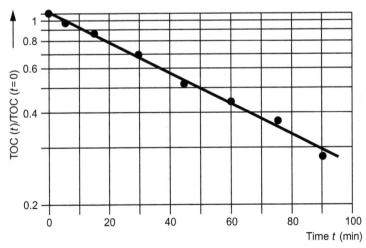

Figure 5.5 TOC elimination in a phenol solution by LOPROX process at 140 °C.

formation of quinone and that this hydrogen peroxide, in the presence of iron, forms OH radicals in the same way as Fenton's reagent. It is radicals that are the actual oxidizing agents (Figure 5.6).

Hydroxylated and aminated phenol and naphthol compounds are the only aromatic compounds that can be oxidized at temperatures as low as 120–150 °C. Most other aromatic compounds require a reaction temperature of 180–190 °C.

Figure 5.7 shows a typical LOPROX reaction in an effluent stream that has very poor biodegradability. The COD is taken as a measure of the level of contamination. The COD decreases with increasing temperature. At the same time, the BOD_5 (biological oxygen demand) increases slightly, resulting in a considerable drop in the COD/BOD_5 ratio from its original level above 10 to about 2. The COD/BOD_5 ratio provides an initial indication of the biodegradability of an effluent stream. Effluents with a BOD_5/COD ratio above 0.5 are generally suitable for immediate treatment in a biological treatment plant.

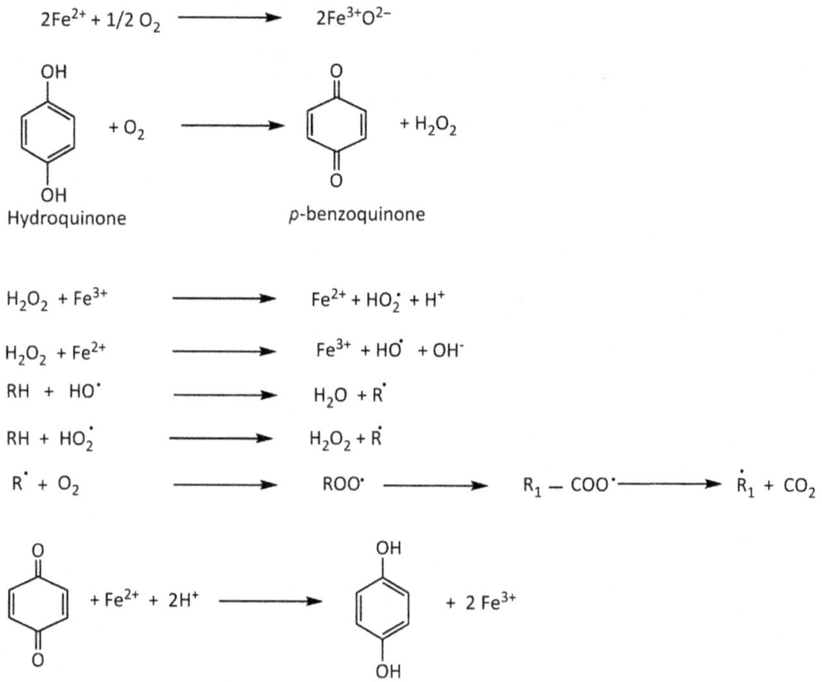

Figure 5.6 Postulated reaction mechanism in the LOPROX process.

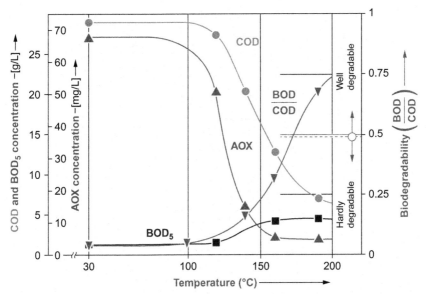

Figure 5.7 Typical behavior of pollutants in the LOPROX process.

The LOPROX process has proven ideal for breaking down organic halogen compounds adsorbable organically bound halogens (AOX). In most cases, the oxidation of toxic wastewater also results in a decrease of fish, daphnia, and bacteria toxicity.

Aliphatic compounds, which are poorly biodegradable and have to be removed from wastewater, include amino compounds. Wet oxidation usually only splits off the amino groups, which in itself results in a considerable improvement in the biodegradability of the wastewater.

It is particularly clear at this point that the catalytic wet oxidation process developed by Bayer is especially suitable for treating effluent streams before they enter a biological treatment plant.

The question about what degradation rates can be achieved in individual cases using a combination of the LOPROX process and biological treatment processes will have to be answered in a laboratory and pilot plant trial.

5.2.1 Examples of the Use of the Loprox Process
(Holzer et al., 1992)

The new wastewater treatment process has already proven successful in practical operation. At Bayer's Leverkusen factory, four plants are in operation to pretreat effluent streams of 6–60 m^3/h. Since 1982 Bayer has been gathering considerable expertise in this process.

Figure 5.8 Two-step LOPROX plant for the pretreatment of 20 m³/h of effluent from the production of dyestuff at Bayer's Leverkusen factory.

Figure 5.8 shows a two-step LOPROX plant for the pretreatment of 20 m³/h (88 gpm) of effluent from the production of dyestuffs at Bayer's Leverkusen factory.

The first step primarily involves the oxidation of inorganic reducing sulfur compounds in an alkaline medium at a temperature of 120 °C (250 °F) and a pressure of 6 bar (85 psi). The pH of the wastewater is then adjusted to between 1 and 2 by the addition of sulfuric acid.

After addition of the catalyst, the organic substances in the wastewater are oxidized at 140 °C (284 °F). The second step takes place in a one-piece, glass-lined steel column, which was the first in the world to be manufactured with dimensions of 10.5 m (34.4 ft) in height and 1.8 m (6 ft) in diameter. The off-gas generated by the reaction is released into the central off-gas treatment unit via a pressure-maintaining valve. The pretreated wastewater is transferred to the central treatment plant for final purification.

Before this plant was commissioned, it was tested for about 2 years in the form of a pilot plant. The capital expenditure for this totalled approximately $3 million, while annual operating costs ran at around $0.9 million.

A further LOPROX plant for treating 6 m³/h (36 gpm) of effluent from the production of anthraquinone dyes was planned and built in 1999. Test results show that one-step operation in an acid medium at a temperature of

190 °C (374 °F) and a pressure of 18 bar (256 psi) results in COD elimination by 85%. The remaining COD is reduced aerobically by a further 90% in the biological wastewater treatment plant, thus resulting in a total COD elimination of 98%. Since the effluent contains up to 5% chlorides, the column has to be lined with titanium to prevent corrosion.

A further example from the pulp industry shows how, using the LOPROX process and relatively mild reaction conditions, long-chain molecules can be broken down and the resultant fragments then degraded easily. The untreated wastewater has a COD of 20,000–30,000 ppm and contains a maximum 50% of easily degradable substances. In trials with the LOPROX process, it was possible to reduce the COD to 8000–9000 ppm (66%) in about one hour in a one-step operation at a temperature of only 150 °C (~300 °F). In addition, there was a drastic improvement in the biodegradability of the effluent. Subsequent treatment in a one-step aerobic test unit reduced the COD by a further 92%, resulting in total COD elimination of more than 97%.

5.3 BAYER TOWER BIOLOGY (Holzer et al., 1992)

Bayer Tower Biology has been used successfully for more than 20 years for the biological treatment of industrial effluents (Figure 5.9). This technology represents a new generation of wastewater treatment plants. Bio-oxidation is carried out in closed tanks holding a water column between 10 and 25 m (33 and 82 ft) high rather than in conventional open basins.

The activation tanks, with a volume of up to 15,000 m^3 (4 million gal), are usually made from steel and are aerated with specially developed Bayer injectors. These are two-phase injectors in which air is dispersed into very fine bubbles by the kinetic energy of a pressure water stream.

A large number of these injectors are installed at regular intervals on the bottom of the activation tank. The wastewater itself is used as the pressure water for the operation of the injectors.

The injectors distribute the wastewater evenly over the whole of the bottom of the Tower Biology unit, simultaneously and intensively mixing it with atmospheric oxygen. The water/air stream emitted by the injector is aimed at the bottom of the tank to prevent sedimentation. This system also means that the air bubbles can rise through the full height of the tank, remaining in the liquid for the longest possible time.

The large gas/water interface resulting from the fineness of the bubbles, the long residence time of the bubbles as they rise through the tank, and the

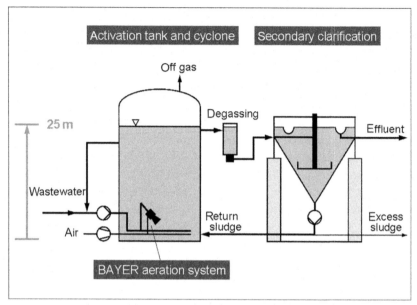

Figure 5.9 Bayer Tower Biology unit with separate secondary clarifier.

pressure of the water column combine to produce optimum oxygen transfer and utilization. Up to 80% of the oxygen in the air can be dissolved in the water in the Tower Biology process. This means that, in comparison with surface aeration in open basins, only one-fifth to one-seventh as much air is needed to introduce the same quantity of oxygen into the process.

This reduces the amount of energy required and the amount of off-gas produced. The closed tank design significantly reduces stripping effects and the formation of aerosols that result in unpleasant odors in open basins.

By building tall tanks instead of flat basins, the amount of ground space required for a Tower Biology unit is some 30–50% less than is needed for a conventional treatment unit. It is also much easier to check for leaks, an important consideration for the protection of the groundwater.

The sludge/water mixture that leaves the activation tank is de-aerated in a special cyclone before it is transferred to the secondary clarifiers. This also results in further flocculation of the sludge, which helps improve sedimentation during secondary clarification.

Secondary clarification usually takes place in funnel-shaped tanks with vertical through-flow known as Dortmund cones. These secondary clarifiers are made from steel and rest on a concrete column under which there is room to construct soundproofed storage facilities for the necessary machinery.

There are no moving parts in either the activation tank or the secondary clarifier, which makes Tower Biology easy to maintain and reliable to operate.

Figure 5.10 shows a Bayer injector. The Bayer injector is a two-phase injector that utilizes the kinetic energy of the liquid with a high degree of efficiency to disperse the gas into very fine bubbles. The injector is made of a special plastic material, polypropylene, and is noncorrosive. Because of its smooth and non-wettable surface, it will not become encrusted. Because of its excellent performance, it is often used in the modernization of old clarification basins. Usually the injectors are grouped together in clusters of four units arranged in rows at the bottom of the bio-volume in such a

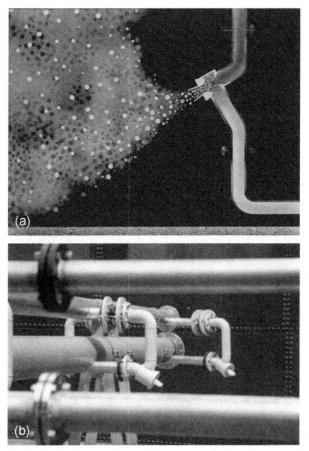

Figure 5.10 (a) Bayer slot injector water/air stream entering the tower. (b) Bayer slot injector with installation.

way that one injector covers 10–20 sq ft of the floor area. The injectors in the cluster are directed downward, so that their jets reach the bottom and prevent solid particles from settling. This arrangement has yet another advantage: The gas–liquid jet that leaves the injector fans out into a bubble swarm after about 80 characteristic nozzle diameters. With this arrangement of the injectors, the bubble swarm starts to rise to the water surface right from bottom of the tank, and the total height of the liquid column is utilized. This gives maximum possible residence time for the gas bubbles in the water. The injector works under optimum conditions when the required oxygen input (kg or lb O_2/h) is achieved with a minimum of mechanical energy (kW or hp) for the pumps and compressors of liquid and gas, that is, when the efficiency (kg O_2/kWh or lb O_2/hp-h) is at a maximum.

It operates economically from a depth of about 7 m. The specific data of an injector, for example, quantities of air and pressure water, pressure drops, level of efficiency, and residual oxygen content in the off-gas, are calculated by computer on the basis of the effluent data for each project.

5.3.1 Process Design Characteristics (Bayer, n.d.)

For the determination of the respective optimum operating conditions, the following process characteristics are necessary:
- The absorption characteristics of the injector combines the O_2 input of the injector with the height of the liquid column, the characteristic concentration difference, and the two independent process parameters, gas and liquid, throughout. In addition, information about the influence of liquids and/or salts dissolved in the wastewater on the O_2 absorption has to be inherent in this characteristic.
- The pressure drop characteristic of the injector is needed in order to size the pump and the compressor correctly. Knowledge of the pressure drops permits calculation of the "kinetic" energy of the water jet that is utilized for the dispersion of the gas into gas bubbles.
- The relationship for the adiabatic compression of the gas throughput takes into account the influence of the height of the liquid column on the power to compress the gas.

Disregarding this last relationship, which can be taken from an appropriate textbook, all the other characteristics were determined by Bayer using a Bayer injector of original size. For greater experimental accuracy, the O_2 input was measured under stationary conditions, using the "hydrazine method," which was specially developed for this purpose.

In addition to the above-mentioned scope, absorption measurements were carried out with aqueous solutions of salts and soluble liquids. They have shown that the O_2 input is strongly influenced by these additives.

When pure liquids are aerated, the primarily generated fine gas bubbles coalesce into bigger ones as soon as they leave the region of high shear stresses. This gas coalescence can be prevented by adding salts or miscible liquids to the system. The higher the concentration of the additives, the higher the degree of non-coalescence.

In a coalescence system (pure liquid) the average diameter of the gas bubbles is 3–5 mm, and it is almost independent of the gas dispersion device. In a non-coalescent system with approximately 20 g NaCl/L, gas bubble coalescence is already significantly prevented, and the average bubble diameter is only 0.5–1.0 mm. Because of this partial preservation of the primary bubble size, it is very important in such systems to use a gas dispersion device that readily generates very fine gas bubbles.

The decrease in the final bubble diameter in the non-coalescence system results in a considerable increase in interfacial area and a longer bubble rising time. Both effects considerably enhance mass transfer, and this has to be taken into account by using different absorption characteristics.

5.3.2 Optimum Design of Injectors (Bayer, n.d.)

For the optimization of the Bayer injector or the Bayer Tower Biology on the basis of the characteristics and relationship described above, the following parameters are usually given:

- O_2 input
- Average concentration of dissolved O_2 in the wastewater
- Average temperature of the wastewater, and hence the degree of O_2 saturation in the wastewater
- Concentration of additives (salts, etc.) in the wastewater

For these conditions, the required O_2 input can be achieved by an infinite number of pairs of values for the gas and liquid throughput, but there is only one pair of values where the required O_2 is transferred into the wastewater with a minimum of mechanical power. Under these process conditions efficiency is at a maximum.

5.3.3 Examples of Tower Biology (Zlokarnik, 1985)

The Bayer Tower Biology process is generally more economical than other treatment systems in terms of both investment and operating costs for high organic load wastewater. To date, a number of Tower Biology units have been constructed to treat between 200 and 90,000 m³/day of effluent. Five of these units are operated by Bayer at its production facilities in Germany, Belgium, India, and the United States. The oldest came on stream in 1979. Tower Biology has also proved effective in other sectors of industry including brewing, sugar processing, and paper production.

A. Bayer AG, Leverkusen Works (Germany)
The industrial effluent from more than 100 chemical production facilities at Bayer AG's Leverkusen works is conveyed through a double sewer more than 2.5 km long to the "joint sewage treatment plant" in the Leverkusen suburb of Burrig. A "basin biology" plant with surface aeration has been in operation there since 1971. This first treatment facility built at Burrig was able to handle part of the industrial effluent from the Bayer factory as well as the municipal sewage from the Wupper water authority area. The Bayer Tower Biology facilities were erected during the second phase of construction, which began in 1976. This facility went onstream in December 1980. Since then it has served as the first stage in the biological treatment of the all the effluent from Bayer's Leverkusen works; the outlet leads to the basin biology, the second stage in biological treatment, which deals not only with the pretreated Bayer effluent but also with the municipal sewage from the Wupper water authority area (dry weather flow).

The design data for the whole plant is shown in Figure 5.11 and includes the following attributes:

Intake: Industrial effluent: 90,000 m³/day with 95 t/day BOD_5; plus municipal sewage: 70,000 m³/day with 14 t/day BOD_5.

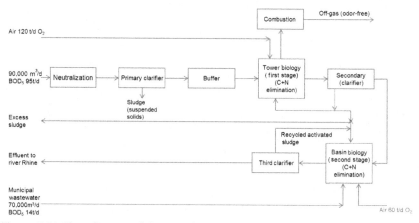

Figure 5.11 Flow diagram of the Leverkusen plant.

Figure 5.12 Tower Biology at Bayer Leverkusen site.

Primary clarification: Residence time 2.5 h.

Tower Biology (Figure 5.12): Four rubber-lined steel tanks (ø26 × 30 m) in parallel; liquid height 26.5 m, tank volume 13,600 m^3, total volume 54,400 m^3. Aeration via 4 × 72 slot injectors. O_2 uptake 120 t/day, 22,000 m^3/h air (standard conditions) being required for an O_2 utilization of up to 4% O_2 by volume in the off-gas. Volume load 1.8 kg/m^{3}d BOD$_5$, sludge load 0.32 kg BOD$_5$ per kg dry mass per day, residence time 14.5 h.

Off-gas purification: The off-gas is preheated in a counter-current to about 580 °C and mixed with approximately 160 m^3/h of natural gas. "Non-flaming combustion" is induced by infrared radiation. This oxidation takes place in a reaction chamber at 750 °C. The flue gas is subsequently cooled in a counter-current heat exchanger to 200 °C, and the HCl resulting from the oxidation of chlorinated hydrocarbons is washed down by caustic scrubber. The off-gas purified in this way is discharged into the atmosphere through a short chimney at a temperature of about 180 °C.

Intermediate clarification: Each tower has a "collar" of 16 sedimentation funnels of ø9 m with a total volume of 16 × 420 m^3 = 6720 m^3 per tower, residence time 7 h, superficial velocity 0.95 m/h.

Basin biology (second stage in the biological treatment of Bayer effluent): Four basin cascade biology with total volumes of 34,000 m^3 for C and N elimination. Residence time 3.4 + 7.5 h, O_2 uptake 60 t/day.

Secondary clarification (sedimentation of biomass): Basin with scraper (t = 3 h) and sedimentation funnels (t = 6.3 h)

B. Bayer AG, Brunsbuttel Works (Germany)

Following several years of experience with the 8 m-high biotanks, the second phase of construction was completed in 1979. Six tanks then went into operation (ø10 × 15 m, volume 1200 m^3) having a total volume of 7200 m^3; these carry out the biological purification of 5600 m^3/day of waste water with a BOD_5 load of 4.2 t/day. This requires an O_2 uptake of 7.2 t/day. The secondary clarifiers (sedimentation funnels) are situated between the tower and are also covered.

C. Bayer Factory at Thane (India) (not in operation from year 2007)

Tower Biology in concrete towers (2 tanks of ø15 × 17 m, water level 15 m) with adjacent sedimentation funnels. Wastewater throughput 3600 m^3/day with a BOD_5 load of 2 t/day; O_2 uptake 4 t/day. This treatment plant commenced operation in 1981.

D. Konigsbacher Brewery, Koblenz (Germany)

One of the main reasons for the choice of Tower Biology in this case was the lack of space available. The treatment plant consists of a tower (ø20 × 20 m) with an adjacent funnel-shaped secondary clarifier. 2000 m^3/day of effluent are treated; the BOD_5 load is 3 t/day. O_2 uptake is 4.2 t/day. The organisms' one-sided diet (carbohydrates) encourages the formation of filamentous activated sludge, which tends to rise to the surface during de-aeration of the Tower Biology outlet. To deal with this, a new type of flotation unit was installed, consisting of a single flotation cell ($D=2.5$ m; $V=3$ m^3) with a funnel-shaped nozzle (induced air flotation). This gives a recycled sludge with about 10 g/L dry matter and a water-clear outflow of the purified wastewater. The Tower Biology went onstream in 1981.

E. Petrochemical Plant in Wesseling, North Rhine–Westphalia (Germany)

Here, there is 8000 m^3 of wastewater to be treated daily, containing an ammonium (NH_4^+) load of 8 t/day as well as BOD_5 load of 6 t/day. Nitrification and denitrification processes therefore have to be integrated into the biological treatment for purposes of nitrogen elimination. Following extensive trials on a semi-technical scale, the industrial-scale plant shown in the sketch in Figure 5.13 was designed. The plant comprises a two-step nitrification, the first step also incorporating the aerobic oxidation of carbon as well as 70% nitrification (from 800–1000 ppm to 100–150 ppm). In the second step, the remaining NH_4^+ is degraded to 5–30 ppm. The nitrites and nitrates thus formed then have to be reduced to nitrogen with the aid of hydrogen donors (e.g., methanol). To save on methanol, the outflow from the second nitrification step is divided into two streams;

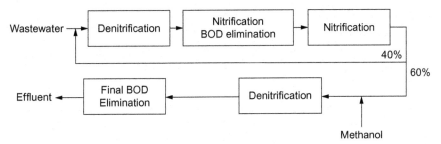

Figure 5.13 Flow diagram of combined BOD elimination and nitrification/denitrification.

40% of the total is fed to a denitrification unit upstream from the first nitrification unit, and 60% passes to the downstream denitrification unit to which is added an effluent free of N but rich in hydrocarbons (methanol). An activated sludge tank is then required downstream from the denitrification unit to degrade the residual BOD_5 resulting from these additional streams.

Since the nitrification unit has a high O_2 demand (4.57 kg oxygen per kg of ammonia nitrogen), the entire plant was designed in the form of a multistep Tower Biology; the O_2 uptake is 50 t/day. It comprises seven towers of between 24 and 17 m height plus six sedimentation funnels. It commenced operation in July 1981 and has proved to be entirely satisfactory.

F. Dynamit Nobel, Lulsdorf Works (Germany)

In this case the effluent (15,000 m^3/day) carries not only the BOD_5 load (14 t/day) but also the NO_3^- load (3 t/day) from the nitrogen process. The plant consists of two parallel units (ø12 × 22 m; $V = 2500$ m^3) and an activated sludge unit (ø17 × 20 m, $V = 4500$ m^3). This treatment plant went onstream in July 1983.

G. Lehrter Zucker AG, Lehrte, near Hannover (Germany)

The special feature of sugar factories as far as their effluents are concerned is that effluents only occur during a particular season (after the sugar beet harvest) and also that the effluent contains a high carbohydrate concentration and is therefore best pretreated in an anaerobic unit. This is also true for Lehrter Zucker AG. Here, only the downstream aerobic treatment unit takes the form of a Tower Biology plant (ø15 × 18 m; $V = 3200$ m^3); the O_2 uptake is 5–7 t/day (max). This plant handles 3000 m^3 of wastewater per day, which contains 7–10 g/L of COD and 4.5–7.2 g/L BOD_5. The plant went into operation in the autumn of 1982.

H. Bitburger Brewery, Bitburg (Germany)

The excellent results obtained with the combination of Tower Biology and induced air flotation at the Konigsbacher Brewery in Koblenz prompted the Bitburger brewery to choose the same system. The plant at Bitburg comprises a tower (ø20 × 24 m; $V = 7300$ m^3) and a downstream flotation unit consisting of two parallel flotation cells, each with a volume of 50 m^3. The plant commenced operation in January 1984 and treats 7200 m^3 of effluent daily.

The Tower Biology utilizes mild steel, rubber-lined towers (two tanks of ø15 × 17 m, water level 15 m).

I. Tower Biology in the United States, Bushy Park, Texas

Figure 5.14 shows two-tower biology followed by gravity clarifiers. It is a treatment plant of capacity 5600 m^3/day.

J. Tower Biology in Mexico, Mexico City

Figure 5.15 shows two-tower biology at a chemical plant in Mexico City.

5.4 SUMMARY OF LOPROX AND TOWER BIOLOGY

The LOPROX process (low pressure wet oxidation) was developed by Bayer for the separate pretreatment of selected effluent streams containing high concentrations of organic substances that degrade too slowly in

Figure 5.14 Tower Biology at the Bushy Park site in the USA.

Figure 5.15 Tower Biology at a site in Mexico City.

conventional biological treatment plants or adversely affect the degradation of other substances.

The LOPROX process is a method of chemical oxidation. Using catalysts, organic substances in effluents are partially oxidized with pure oxygen under pressure and at elevated temperatures. Under relatively mild reaction conditions at temperatures below 200 °C (390 °F) and pressure from 5 to 20 bar (70 to 280 psi), a large number of chemical compounds that otherwise biodegrade very slowly can be oxidized to such an extent that the residual substances generated can be more easily degraded by subsequent aerobic biological treatment.

The new process has proved itself time and again in Bayer's waste management system.

Eight plants are in operation pretreating an effluent stream of up to 60 m^3/h. Since 1982 Bayer has gathered considerable expertise in the use of this process.

In the chemical industry, effluent is disposed of in an integrated wastewater management system. The central biological treatment plant is complemented and supported by a series of decentralized measures, including LOPROX. Bayer Tower Biology has been used successfully for more than 10 years for the biological treatment of industrial effluents. This technology represents a new generation of biological waste water treatment plants. Bio-oxidation is carried out in closed tanks holding a water column between 10 and 25 m high rather than in conventional open basins. To date, 14 Tower Biology units have processed between 299 and 90,000 m^3/day of effluent.

REFERENCES

Bayer Tower Biology: E 589–777/68619.

Holzer, K., Horak, O., Lawson, J.F., 1992. LOPROX: a flexible way to pretreat poorly biodegradeable effluents. In: 46th Prude Industrial Waste Conference Proceedings. Lewis Publishers, Chlesa, MI.

Holzer, K., 1999. AVT-Handbuch, Industriabwasser, Grundlagen, 4. Auflage, p 183 ff.

Zlokarnik, M., 1985. Tower-shaped reactors for aerobic biological waste water treatment. In: Biotechnology, vol. 2. VCH Verlagsgesellschaft, Weinheim.

CHAPTER 6

Reorienting Waste Remediation Towards Harnessing Bioenergy: A Paradigm Shift

S. Venkata Mohan
Bioengineering and Environmental Science (BEES), CSIR-Indian Institute of Chemical Technology (CSIR-IICT), Hyderabad, India

6.1 INTRODUCTION

A huge quantity of wastewater/waste is continuously being generated from various industrial and domestic activities and the volume has increased over time because of rapid and sustained development. Remediation of waste, being an energy-intensive process, increases the economic burden on industry. Reducing the treatment cost of waste/wastewater and finding ways to produce value-added products from treatment has gained importance. Waste material generated from anthropogenic activities has the potential to meet a good fraction of the world's energy demand if it could be economically converted to useful forms of energy. More recently, waste/wastewater is being considered as a potential feedstock/substrate for harnessing various forms of bioenergy and for recovering value-added products because of the biodegradable organic fraction present in wastewater (Venkata Mohan, 2008, 2009; Venkata Mohan and Pandey, 2013; Venkata Mohan et al., 2013a–d). The regulatory need for wastewater treatment before its disposal makes it an even more ideal commodity for producing bioenergy through anaerobic processes. The generation of bioenergy from renewable wastewater, along with a simultaneous treatment of the wastewater that reduces the overall cost, makes the whole process environmentally sustainable. The availability of large quantities of wastewater, the presence of degradable carbon material, and the cost and need for treatment makes wastewater a potential substrate.

Environmental scientists are gradually shifting focus from "pollution control" to "resource exploitation from waste." Using waste/wastewater as a substrate for its value addition coupled with its simultaneous remediation can lead to new opportunities for the exploitation of renewable and inexhaustible energy sources. Recovering energy from treatment benefits the effluent treatment plant (ETP) operators by recouping revenue, simultaneously reducing the overall effluent treatment cost. Recent energy and environment scenarios visualize a paradigm shift from waste remediation to energy generation in order to combine both aspects into a unified and biorefinery approach that is environmentally sustainable. Biological processes are preferred to treat waste because of their simple, economical, and eco-friendly nature. These processes facilitate the conversion of negatively valued waste to useful forms of energy. Simultaneously, they achieve the objective of pollution control. Using waste as a potential source for value addition through biological routes has instigated considerable interest because of its sustainable nature and has further opened up a new avenue for the use of renewable and inexhaustible energy sources. The integration of bioenergy generation and the ETP is the futuristic goal envisioned, with wastewater as the primary feedstock. A comprehensive attempt is made in this chapter to illustrate various routes employed for generating diverse forms of value addition from negative-valued wastewater/waste such as biohydrogen, bioelectricity, bioplastics and algal-based lipids (Figure 6.1).

6.2 ANAEROBIC FERMENTATION

Anaerobic metabolism is a favored process for harnessing bioenergy (Venkata Mohan et al., 2013c). If oxygen is not available as an electron acceptor, bacteria use other terminal electron acceptors (TEAs) such as oxidized organic compounds or pollutants resulting in the generation of reduced energy-rich compounds with simultaneous treatment. However, based on thermodynamic hierarchy of the reactions, the microbes will switch over to other biochemical pathways. Anaerobic fermentation helps in generating energy-rich metabolic intermediates and maintains the carbon flow for longer periods of time without releasing the end products (Venkata Mohan et al., 2013c). Complete degradation of certain persistent chemicals cannot be catalyzed by a single microbe but a syntrophic association of physiologically distinctive organisms is required for organic carbon-driven reduction of iron, manganese, sulfate etc, along with acetogenesis and methanogenesis. Microbial fermentation helps to generate energy-rich reducing factors (e.g., NADH, FADH) that subsequently are reoxidized during respiration with simultaneous generation of biological energy molecules (ATP) in the presence of a terminal electron acceptor (TEA). Both aerobic and anaerobic metabolisms have a common glycolytic pathway.

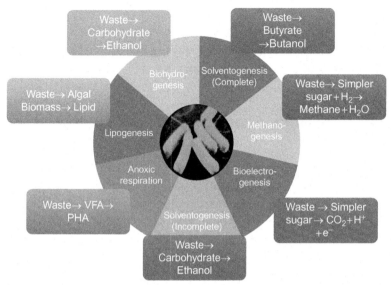

Figure 6.1 Various possible routes of bioenergy generation from waste through simultaneous remediation.

After glycolysis, the aerobic metabolism proceeds with the tricarboxylic acid cycle and oxidative phosphorylation, whereas the anaerobic process continues with interconversion (dehydrogenation), decarboxylation, solventogenesis, methanogenesis, and other mechanisms. During glycolysis, glucose molecule is converted to pyruvate, which is the key molecule of microbial fermentation. During the aerobic process, pyruvate transforms to CO_2 and H_2O. Pyruvate has a different fate during anaerobic fermentation under different environmental conditions. If it enters the acidogenic pathway, it generates volatile fatty acids (VFAs) in association with the generation of H_2 and CO_2. During aerobic respiration, the electron molecules pass through a redox cascade of the respiratory/electron-transport chain where their energy is gradually transformed to ATP through oxidative phosphorylation and get reduced in the presence of an externally available TEA (oxygen). However, ATP generation is not assured during these processes because the energy from the reducing equivalents (protons and electrons) will be used to complete the terminal reduction reaction with TEA but not necessarily transferred to the bonding between ADP and inorganic phosphate (Pi) to generate ATP at the ATPase complex. On the contrary, oxygen does not act as a TEA in an anaerobic metabolism and hence the electrons flow occurs via a series of interconversion reactions that can lead to the formation of energy-rich reduced end products. During anaerobic respiration, bacteria have the ability to utilize a wide range of compounds, such as NO_3^-, SO_4^{2-}, organic and inorganic compounds as electron acceptors for ATP generation with their simultaneous reduction.

6.3 BIOHYDROGEN PRODUCTION FROM WASTE REMEDIATION

H_2 has long been recognized as a promising, green, and ideal energy carrier of the future due to its cleaner efficiency, high energy yield (122 kJ/g), and renewability. The H_2 gas generated either by biological machinery or thermo-chemical treatment of biomass is normally termed biohydrogen production (Figure 6.2). The research fraternity has shown immense interest in biological routes of H_2 production. The past decade has witnessed significant research on biohydrogen. Biological H_2 production processes can broadly be classified into light-independent (dark)-fermentation and light-dependent photosynthetic processes. The photo-biological process can again be classified either into photosynthetic or fermentation processes depending on the carbon source and the biocatalyst used. Biophotolysis of water using green algae and cyanobacteria or photo-fermentation mediated by photosynthetic bacteria (PSB) are light-dependent processes. Cyanobacteria and microalgae undergo direct and indirect biophotolysis to produce H_2 by utilizing inorganic CO_2 in the presence of sunlight and water, while PSB manifest H_2 production by consuming a wide variety

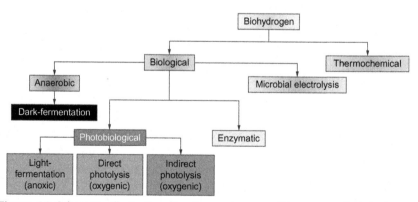

Figure 6.2 Schematic illustration showing various possible routes of biohydrogen production.

of substrates ranging from inorganic to organic acids in the presence of light (Allakhverdiev et al., 2010; Beer et al., 2009; Ntaikou et al., 2010; Venkata Mohan, 2008, 2009, 2010; Venkata Mohan and Pandey, 2013). The dark-fermentation process proceeds to the anaerobic process, where acidogenic bacteria (AB) metabolically generate H_2 along with VFA and CO_2 through acetogenesis. Synthetic enzymes mediate *in vitro* H_2 production, which is one of the most fascinating routes envisaged by scientists, albeit still at the laboratory scale. Microbial electrolysis is a hybrid strategy wherein external potential is applied to the microbial fuel cell (MFC) to enhance biological H_2 production. At present, H_2 is being produced mainly from fossil sources and the electrolysis of water.

6.3.1 Dark-Fermentation

Fermentative conversion of organics to their end products involves a series of biochemical reactions, such as hydrolysis, acidogenesis, acetogenesis, and methanogenesis manifested by five physiologically distinct groups of micro-organisms. The complex organic compounds get degraded to monomers during hydrolysis by hydrolytic microorganisms. Further, these monomers will be fermented by AB in order to generate a mixture of low molecular weight volatile organic acids (e.g., acetic acid, propionic acid, butyric acid, malic acid) associated with H_2 and CO_2 production (Equations 6.1–6.5). The reversible interconversion of acetate from H_2 and CO_2 by acetogens and homoacetogens can also be considered for H_2 production. Acetoclastic methanogens convert organic acids to CH_4 and CO_2 through methanogenesis. Dark-fermentation by anaerobic (acidogenic) bacteria is the most widely understood process for biohydrogen production. AB grow in syntrophic association with the hydrogenotrophic methanogens (H_2 consuming MB) and keep H_2 partial pressure low enough to allow acidogenesis so that the reaction thermodynamically favorable by interspecies H_2 transfer. Methanogenic activity needs to be restricted to make H_2 a metabolic end-product.

$$C_6H_{12}O_6 + 2H_2O \rightarrow 2CH_3 \cdot COOH + 2CO_2 + 4H_2 \text{ (acetic acid)} \quad (6.1)$$
$$C_6H_{12}O_6 \rightarrow CH_3 \cdot CH_2 \cdot CH_2 \cdot COOH + 2CO_2 + 2H_2 \text{ (butyric acid)} \quad (6.2)$$
$$C_6H_{12}O_6 + 2H_2 \rightarrow 2CH_3 \cdot CH_2 \cdot COOH + 2H_2O \text{ (propionic acid)} \quad (6.3)$$
$$C_6H_{12}O_6 + 2H_2 \rightarrow COOH \cdot CH_2 \cdot CH_2 \cdot COOH + CO_2$$
$$\text{(malic acid)} \quad (6.4)$$
$$C_6H_{12}O_6 \rightarrow CH_3 \cdot CH_2OH + CO_2 \text{ (ethanol)} \quad (6.5)$$

The proton-reducing reactions aid in the formation of H_2, a common fermentation by-product when electron-acceptor is limited (Madsen, 2008). Both obligate and facultative AB can catalyze H_2 production from organic substrates (Hallenbeck and Benemann, 2002; Vardar-Schara et al., 2008; Venkata Mohan, 2009, 2010; Venkata Mohan and Pandey, 2013). Pyruvate enters the acidogenic pathway and generates H_2 along with VFA (Equations 6.1–6.4). Obligate anaerobes convert pyruvate to acetyl coenzyme A (acetyl-CoA) and CO_2 through pyruvate ferredoxin oxidoreductase by the reduction of ferredoxin (Fd) (Kraemer and Bagley, 2007; Vardar-Schara et al., 2008). Pyruvate is converted to acetyl-CoA and formate by the action of pyruvate formate lyase by facultative anaerobes that produce H_2 by formate hydrogen lyase (Vardar-Schara et al., 2008). Hydrogenase and nitrogenase are the two most important enzymes involved in fermentative H_2 production. They catalyze the reversible reduction of H^+ to H_2 (Hallenbeck and Benemann, 2002) while [Fe-Fe]-hydrogenase removes the excess reducing equivalents.

6.3.1.1 Selective Enrichment of Biocatalyst

Diverse groups of microorganisms—anaerobic, photosynthetic (heterotrophic and autotrophic), and microalgae—are capable of producing H_2 by taking advantage of their specific metabolic route under defined conditions. Obligate anaerobes, thermophiles, methanogens, and a few facultative anaerobes can produce H_2 through the dark-fermentation mechanism. After wastewater started to be used as a feedstock, particularly in the last decade, the application of mixed consortia as a biocatalyst received a great deal of attention and is considered to be a practical options for scaling up. Mixed cultures facilitate operational flexibility, restrict the requirement of sterile conditions, can use a broad range of substrates, have good stability, and infuse diverse biochemical functions (Angenent et al., 2004; Venkata Mohan, 2008, 2010; Venkata Mohan et al., 2013b; Wang and Wan, 2009). Therefore, producing H_2 with mixed consortia offers lower operational costs and an ease of control in concurrence with the possibility of using waste as a feedstock.

Mixed microbes encompass various physiological groups of bacteria have diverse metabolic functions that are not necessarily specific to H_2 production. The mixed consortia support proton reduction during methanogenesis rather than its shuttling between intermediates during the interconversion of metabolites, which is presumed to be necessary for H_2 to form as an end-product (Venkata Mohan and Goud, 2012; Venkata Mohan et al.,

2008a). H_2 production with typical anaerobic consortia is limited due to its consumption by methanogens. Pretreating biocatalyst plays a vital role in the selective enrichment of mixed consortia and shifting the metabolic function to acidogenesis (Goud and Venkata Mohan, 2012a,b; Srikanth et al., 2010a; Venkata Mohan et al., 2008b; Zhu and Beland, 2006). Physiological differences between H_2 producing bacteria (AB) and H_2 uptake bacteria (MB) form the fundamental basis for the methods used in the preparation of H_2 producing inoculums (Goud and Venkata Mohan, 2012a,b; Sarkar et al., 2013; Venkata Mohan and Goud, 2012; Venkata Mohan et al., 2008b; Zhu and Beland, 2006). Different pretreatment methods, such as heat-shock, chemical, acid-shock, alkaline-shock, oxygen-shock, load-shock, infrared irradiation, microwave irradiation, and freezing and thawing are reported for the selective enrichment of H_2 producing inoculums (Goud and Venkata Mohan, 2012a,b; Venkata Mohan and Goud, 2012; Venkata Mohan et al., 2008b; Wang and Wan, 2008). Efficacy of the pretreatment method depends on the nature and composition of the parent inoculum, conditions adopted for pretreatment, and the nature of substrate apart from other operating conditions (Srikanth et al., 2010a; Venkata Mohan et al., 2008a,b). Pretreatment facilitates selective enrichment of microbial populations specific to acidogenic function and therefore leads to relatively less diversity. It also prevents competitive growth and coexistence of other H_2 consuming bacteria. *Clostridium* sp. became the dominant genus after acid-shock pretreatment (Goud and Venkata Mohan, 2012a; Lee et al., 2009; Venkata Mohan et al., 2011b). This composition of the microbial community in the long-term-operated acidogenic reactor with combined pretreatment (repeated applications) showed the dominance of the *Clostridia* followed by Bacteroidetes, Deltaproteobacteria, and Flavobacteria (Venkata Mohan et al., 2010a) and Aquificales (Goud et al., 2012). On the contrary, application of pretreatment showed marked reduction in substrate degradation efficiency due to specific inhibition of MB, which are essential to metabolize the acid intermediates generated along with the H_2 (Goud and Venkata Mohan, 2012a).

6.3.1.2 Factors Influencing Biohydrogen Production
Conceptually, the physiological and physicochemical conditions under which the microorganisms give optimal H_2 yield are important and need to be established by taking into account other associated aspects. Some of the operating conditions reported in the literature concerning H_2 production are depicted in Table 6.1. Optimization of process parameters is

Table 6.1 Factors affecting fermentative H_2 production with corresponding optimum values

Parameter	Optimum values
Operating pH range	4.5–6.5
Operating temperature	Mesophilic (30–40 °C) and thermophilic (55–70 °C)
Retention time	4–14 h
Organic loading rate	3-50 g of COD/L
Biocatalyst	Anaerobic consortia
Pretreatment	Combined treatment
Reactor configuration	Biofilm/suspended
Mode of operation	Batch mode
Substrate	Organic-based waste/wastewater with good biodegradability
Nitrogen	0.01–0.10 g/L
Phosphorus	0.60 g/L

essential for the up-scaling of this technology. Changing the temperature influences H_2 production as well as metabolite distribution, substrate degradation, and bacterial growth. H_2 production by dark-fermentation was reported under ambient (15–27 °C), mesophilic (30–45 °C), moderate thermophilic (50–60 °C), and extreme thermophilic (over 60 °C) conditions (Yokoyama et al., 2009). The optimal temperature for pure cultures was found in the range of 37–45 °C, while diverse optimum temperatures were reported for mixed consortia (Tang et al., 2008). The optimum temperature for H_2 production usually depends on the nature of the biocatalyst and the type of wastewater used. Temperature control is especially important for regaining the spore-forming acidogenic culture during the reactor operation.

The acidic microenvironment facilitates pyruvate conversion to fatty acids associated with H_2 production by AB. Neutral operations facilitate CH_4 formation by MB. However, the basic operation leads to solventogenesis. The activity of AB is crucial and is the rate limiting step during the dark-fermentative H_2 production process (Venkata Mohan, 2008; Venkata Mohan et al., 2008a, 2010b). AB function well below pH 6, while the optimum pH range for MB is between 6.0 and 7.5. Good H_2 production was observed by maintaining the system's pH in and around 6.0 (Van Ginkel et al., 2001; Venkata Mohan et al., 2008b). However, highly acidic pH (<4.5) is considered to be detrimental for H_2 production because it deactivates AB (Venkata Mohan, 2010; Zhu and Beland, 2006). Hydrogenase enzyme activity is inhibited at a low or high pH beyond the optimum range. The fate of pyruvate depends on the operating pH. Under acidic conditions pyruvate is converted into VFA along with H_2 by AB. Neutral pH operation leads to the formation of CH_4 and CO_2 by MB. Under basic pH, anaerobic digestion leads to solventogenesis. Hydrogenase activity is higher at an acidic pH, but with an increase in pH, the metabolic pathway proceeds to the next step of anaerobic digestion where H^+ get reduced to CH_4 (methanogenesis) or ethanol (solventogenesis). Accumulation of acid metabolites (VFA) during dark-fermentation causes a marked drop in the system's pH, which reduces the buffering capacity, thereby inhibiting the H_2 production (Devi et al., 2010; Lin and Lay, 2004b). The pH range of 5.5–6.0 is optimum to avoid both methanogenesis and solventogenesis and can be considered as a manipulated variable for the process control, especially for dark-fermentation process (Venkata Mohan et al., 2007d, 2010b). A longer fermentation period induces a metabolic shift from acidogenesis to methanogenesis, which is considered to be unfavorable for H_2 production. Maintaining a shorter retention time, between 8.0 and 14 h, therefore helps to restrict the MB growth as well as activity (Hawkes et al., 2007; Venkata Mohan, 2010, Venkata Mohan et al., 2007a–d, 2008d,f, 2011a; Vijaya Bhaskar et al., 2008). Methanogens can be suppressed by maintaining short retention time (2–10 h) as AB grow faster (Fang and Liu, 2002; Nakamura et al., 1993; Ren et al., 2005; Venkata Mohan, 2009; Zhu and Beland, 2006).

Nitrogen at optimal concentration is beneficial for H_2 production, while at higher concentrations it can inhibit the process performance by affecting the intracellular pH of the bacteria or by inhibiting specific enzymes related to H_2 production (Bisaillon et al., 2006; Chen et al., 2008; Salerno et al., 2006). The optimal nitrogen concentration of 0.1 g N/L was found to have a positive effect on H_2 production and substrate degradation (Wang et al., 2009). An ideal carbon-to-nitrogen (C/N) ratio helps in bacterial growth and affects H_2 production by both mixed and pure cultures. A C/N ratio of 47 has been shown to affect fermentative H_2 production by mixed microorganisms (Lin and Lay, 2004a). Wastewater that is low in carbon content can be combined with materials high in N to attain the desired C:N ratio of 30:1 (Yadvika et al., 2004). Wastewater with excess nitrogenous compounds and ammonia inhibits nitrogenase activity (Redwood and Macaskie, 2006). Iron (Fe^{2+}) is very important for the function of hydrogenase and also acts as an active site component for the ferridoxin protein, which is a carrier of electrons to the hydroge-nase. A higher concentration of Fe^{2+} showed an enhancement in H_2 pro-duction efficiency due to its role as a component of hydrogenase and Fd (Karadag and Puhakka, 2010; Lee et al., 2001; Wang and Wan, 2008; Zhang et al., 2005). Optimum Fe^{2+} concentration varied from 25 to 100 mg/L (Srikanth and Venkata Mohan, 2012a).

6.3.1.3 Bioreactor Configuration and Operational Mode

The operation mode of the reactor along with its configuration influences the reactor microenvironment, hydrodynamic behavior, wastewater-biocatalyst contact, and survivability of the microbial population. Diverse reactor configurations, such as suspended growth, biofilm/packed-bed/fixed bed, fluidized bed, expanded bed, upflow anaerobic sludge blanket, granular sludge, membrane-based systems, and immobilized systems, were reported for biohydrogen production. Biofilm systems are resistant to shock-loads, and facilitate an improved reaction potential, leading to stable and robust systems that are also more resilient to changes in the process parameters. More importantly, biofilm systems are well suited for treating highly variable wastewater (Lalit Babu et al., 2009; Venkata Mohan et al., 2007a,b).

Batch, fed-batch, semi-batch/continuous, periodic discontinuous batch (sequencing batch operation), and continuous modes of reactor operation have all been evaluated for H_2 production. Batch mode operation coupled with biofilm configuration helps to maintain stable and robust cultures suit-able for treating highly variable wastewater due to the dual operational advantages of both the systems (Lalit Babu et al., 2009; Luo et al., 2010; Venkata Mohan et al., 2007a–d, 2008a–c; Yokoi et al., 1997). Fed-batch mode operation reduces poor biomass retention/cell washout (Yokoi et al., 1997) and accumulation of soluble metabolic intermediates due to fill-draw mode operation (Venkata Mohan et al., 2007a,b, 2011a).

6.3.2 Renewable Wastewater as Feedstock

The last decade witnessed significant progress in the dark-fermentation pro-cess, owing to its feasibility of utilizing a broad range of wastewaters as a sub-strate with mixed cultures as biocatalysts. In conjunction with wastewater treatment, this process is capable of solving two issues: reduction of pollut-ants in waste and the generation of a clean alternative fuel. The simplicity of the process, its efficiency, and its smaller footprint are some of the striking features of the dark-fermentation process that make it particularly more feasible for the mass production of H_2. Municipal and industrial wastewaters along with the waste generated from agriculture and food-processing industries contains enough organic load that can be appropriately tapped (Venkata Mohan et al., 2013b). Biologically derived organic material and their residue, such as agricultural crops and their waste by-products, wood and wood waste, food-processing waste, aquatic plants, and algae constitute a large source of biomass, which can also be used as a fermentable substrate (Saratale et al., 2008). Cellulosic material or solid waste require an initial pretreatment step to make the organic fraction soluble and bioavailable to the microorganism for metabolic reactions. Its highly crystalline and water insoluble nature makes cellulose recalcitrant to hydrolysis (Saratale et al., 2008). Table 6.2 illustrates the details of wastewater used as substrate to generate H_2. Fermentative H_2 production is relatively less energy-intensive and more environmentally sustainable due to utilization of waste material as substrate.

6.3.3 Thermochemical Process

Thermochemical treatment of biomass is a nonbiological biohydrogen production processes that produces a H_2 rich stream of gas known as *syngas* (a blend of hydrogen and carbon monoxide) by gasification and pyrolysis (heating biomass in the absence of oxygen) (Lipman, 2011). These processes involve a series of thermally assisted chemical reactions that release H_2 from a broad range of feedstocks (Yildiz and Kazimi, 2006). Gasification is a process operated at temperatures above 1000 K in the presence of oxygen and/or steam where the feedstock undergoes partial oxidation and/or steam-reforming reactions yielding gas and char product (Jong, 2009; Navarro et al., 2009). Low-temperature ($<1000\,^{\circ}C$) gasification yields a significant amount of hydrocarbon while at higher temperature the syngas is without any hydrocarbons (Navarro et al., 2009). Pyrolysis facilitates thermal decomposition of biomass at a temperature of 650–800 K (1–5 bar) in the absence of air to yield oils, charcoal, and gaseous compounds (Navarro et al., 2009). Hydrogen can be produced directly through both slow and fast flash pyrolysis if both high temperature and sufficient volatile phase residence time are provided (Agarwal et al., 2013; Navarro et al., 2009). Pyrolysis followed by reforming of bio-oil and gasification of char has received significant interest as this provides an improved quality fuel product (Saxena et al., 2008). Gasification followed by reformation of the syngas and fast pyrolysis, in turn followed by the reformation of the carbohydrate

Table 6.2 Wastewater used for H_2 production by dark-fermentation

Type of wastewater	References
Food-based waste	Van Ginkel et al. (2005); Sentürk et al. (2010); Zhu et al. (2009); Kobayashi et al. (2012); Chakkrit et al. (2011); Kim et al. (2011a,b); Venkata Mohan et al. (2012a); Venkateswar Reddy et al. (2011); Zhu et al. (2011); Siddiqui et al. (2011); Im et al. (2012); Sen and Suttar (2012); Piyawadee and Reungsang (2011); and Goud and Venkata Mohan (2011)
Solid waste from various origins	Li et al. (2011); Chou et al. (2011); Licata et al. (2011); Cheng et al. (2012); Sittijunda and Reungsang (2012); Zhu et al. (2008); Yokoi et al. (2001); Yang et al. (2006); Rozendal et al. (2006); Venkata Mohan et al. (2009c); Sagnak and Kargi (2011); Venkata Mohan et al. (2009b); Gopalakrishnan et al. (2012); Guo et al. (2012); and Saratale et al. (2008)
Dairy and cheese processing	Gustavo et al. (2008); Ren et al. (2007); Venkata Mohan et al. (2007a, 2008a); Julia et al. (2012); Kargi et al. (2012); Ferchichi et al. (2005); Yang et al. (2007); and Rai et al. (2012)
Alcohol-based wastewater	Yu et al. (2002); Froylán et al. (2009); Vatsala et al. (2008); Venkata Mohan et al. (2008d, 2011b); Han et al. (2012a,b); Qiu et al. (2011); Morsy (2011); and Pawinee et al. (2011)
Paper mill	Idania et al. (2005) and Lakshmi Devi and Muthukumar (2010)
Cassava	Luo et al. (2010); Cao et al. (2012); and Sompong et al. (2011)
Chemical wastewater	Venkata Mohan et al. (2007b–d); Vijaya Bhaskar et al. (2008); Tai et al. (2010); and Sivaramakrishna et al. (2009)
Palm oil mill effluent	Vijayaraghavan and Ahmad (2006); Wu et al. (2009); and Yossan et al. (2012)
Olive mill wastewater	Ntaikou et al. (2009) and Kargi and Catalkaya (2011)
Landfill leachate	Liu et al. (2010) and Hafez et al. (2010)
Other types	Liu and Fang (2007); Li et al. (2012a,b); Kongjan et al. (2011); Kim et al. (2012); Kim and Lee (2010); Kim et al. (2011a,b); Gomez et al. (2006); and Tenca et al. (2011)

fraction of bio-oil are also used to produce H_2 (Balat, 2010). Gasification is the most mature technology for large-scale H_2 production, but it needs integrated demonstration plants at sufficiently large scale including catalytic gas upgrading (Jong, 2009). Pyrolysis processes for H_2 production are at present in the smaller-scale demonstration phase while supercritical water gasification is in its early stage of development (Agarwal et al., 2013). Hydrogen production by reformation produces large amounts of CO_2, which contributes to global warming (Navarro et al., 2009).

6.3.4 Process Limitations

Major limitations observed in the dark-fermentative H_2 production process are low substrate conversion efficiency, a drop in system redox conditions, and residual substrate originating from the process as acid-rich wastewater (Venkata Mohan, 2010). About 40–70% residual organic carbon remains in the effluent after dark-fermentation even under optimal operating conditions. The persistent accumulation of VFA causes a sharp drop in the pH, resulting in inhibition of the process (Venkata Mohan et al., 2011a; Wang and Wan, 2009). Biological limitations such as H_2-end-product inhibition, acid or solvent accumulation, and H_2 partial pressure limits process efficiency. Environmental and economic concerns suggest that it is advisable to use the residual carbon fraction of the acidogenic outlet for additional energy generation in the process of its treatment (Mohanakrishna et al., 2010b).

Various integrated approaches were studied to overcome the persistent limitation of the acidogenic process to a certain extent in the context of biorefinery. Various secondary processes, such as methanogenesis (AF) for methane production (Mohanakrishna et al., 2010c), acidogenic fermentation for additional H_2 production (Mohanakrishna et al., 2012), photo-biological process for additional H_2 production (Chandra and Venkata Mohan, 2011; Srikanth et al., 2009), microbial electrolysis cell (MEC) for additional H_2 production (Lenin Babu et al., 2013a,b), anoxygenic nutrient-limiting process for bioplastics production (Venkata Mohan et al., 2010c), hetrotrophic-algae cultivation for lipid accumulation (Venkata Mohan and Devi, 2012), and MFC for bioelectricity generation (Chandra et al., 2012; Subhash et al., 2013), were studied by integration with the primary dark-fermentative H_2 production process with relatively good degrees of success in the case of product recovery and wastewater treatment. Integration approaches facilitate a reduction in wastewater load along with the advantage of value addition to the existing process in the form of product recovery, making the whole process economically and environmentally viable (Mohanakrishna and Venkata Mohan, 2013).

6.4 MFCs FOR HARVESTING BIOELECTRICITY FROM WASTE REMEDIATION

The energy gain in microbes is driven by oxidizing an electron donor and reducing an electron acceptor (Venkata Mohan, 2012; Venkata Mohan et al., 2013a). Variation in the electron acceptor conditions facilitates in harvesting energy. As a part of microbial respiration, electrons move to an exocellular medium towards the available electron acceptor, such as metals, nutrients, minerals, and solid electrodes, in the absence of oxygen (Franks and Nevin, 2010; Logan, 2008, 2010; Venkata Mohan, 2012; Venkata Mohan et al., 2013c). When the microbes use solid electrode as an electron acceptor, the setup is called an MFC, and the electrons can be harvested for different applications. Linking the microbial metabolism to the anode and then transferring the electrons to the cathode generate a net electrical charge from the degradation of the available electron donor (Venkata Mohan et al., 2013a; Figure 6.3). More precisely, MFC is a microbial-catalyzed electrochemical system that facilitates the direct conversion of substrate to electricity through a cascade of redox reactions in the absence of oxygen.

6.4.1 Applications of MFC

MFC have attracted significant interest in the contemporary research arena over the last 10 years (Venkata Mohan, 2012; Venkata Mohan et al., 2013a).

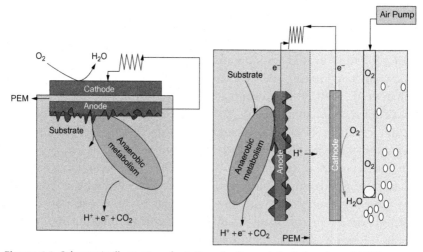

Figure 6.3 Schematic illustration depicting single- and dual-chambered MFC.

In particular, its application for waste remediation attracted considerable attention along with energy harvesting in the form of bioelectricity. Broadly, applications of MFC can be classified as a power generator, wastewater treatment unit, and system for the recovery of value-added products (Venkata Mohan et al., 2013a). Reducing equivalents get reduced in the presence of an electron acceptor at a physically distinct cathode, which results in power generation. When the waste/wastewater functions as an electron donor or acceptor, its remediation is manifested either through anodic oxidation or cathodic reduction. Alternatively, when oxidized metabolites act as electron acceptors during operation, they form reduced end products having commercial importance. The application of MFC was also extended to the production of commercially viable end products such as organic acids, aldehydes, and alcohols (Logan, 2010; Rabaey and Rozendal, 2010; Srikanth et al., 2012). Apart from these, several other distinct applications are reported that also fall in either one or all of these three categories. Microalgae were used as a biocatalyst in the anode chamber of a fuel cell to harness bioelectricity through oxygenic (Subhash et al., 2013) and anoxygenic (Chandra et al., 2012) microenvironments through a photomixotrophic mechanism. An ecologically engineered submerged and emergent macrophyte-based system was studied with an integrated eco-electrogenic design for harnessing power with simultaneous wastewater treatment (Chiranjeevi et al., 2013). Rhizosphere-based fuel cells were also studied for harnessing bioenergy through CO_2 sequestration (Chiranjeevi et al., 2012).

6.4.1.1 Bioelectricity Production

The microbial-catalyzed oxidation of a substrate takes place at the anode, generating reducing equivalents, while their reduction takes place at cathode (Equations 6.6–6.8).

$$C_6H_{12}O_6 + 6H_2O \rightarrow 6CO_2 + 24H^+ + 24e^- \quad \text{(anode)} \tag{6.6}$$
$$4e^- + 4H^+ + O_2 \rightarrow 2H_2O \quad \text{(cathode)} \tag{6.7}$$
$$C_6H_{12}O_6 + 6H_2O + 6O_2 \rightarrow 6CO_2 + 12H_2O \quad \text{(overall)} \tag{6.8}$$

The proton exchange membrane (PEM) between the fermentation (anode) and respiration (cathode) mimics the function of an external membrane, generating a potential gradient, while the electrodes act as redox components of the cell, assisting in the electron flow towards TEA. Electron transfer from its source (metabolism) to the sink (TEA; terminal electron acceptor) will be driven by the potential difference between the redox components of the microbe and the fuel cell. The membrane potential across the cascade of membrane components is called the proton motive force. The electrons reach the anode, creating a negative anodic potential, while the protons go to the cathode, creating a positive potential. The difference between positive cathodic and negative anodic potentials is expressed as cell voltage. The cell voltage drives electrons from the anode to the cathode (electron motive force). Overall, the electron transfer from its source to sink is purely based on the differences in the redox potentials of the components of the fuel cell, irrespective of their nature (biological, chemical, or physical). The MFC's function as a power generator was well established by using a wide range of substrates such as electron donors and acceptors in the anode and cathode chambers, respectively.

6.4.1.1.1 Factors Influencing Bioelectrogenic Activity of MFC

The bioelectrogenic activity of the MFC is governed by several physical, biological, and operational factors (Venkata Mohan et al., 2013a). Physical components such as fuel cell design and configuration, electrode materials, membrane, and stacking regulate the bioelectrogenic activity significantly. Based on the physical separation of fermentation (anode) and respiration (cathode), fuel cell configuration can be either dual chambered or single chambered (Figure 6.3). Anode and cathode chambers are separated by a PEM in a dual-chamber system, while in a single-chamber configuration, only the anode chamber is present. The anode is immersed in liquid medium (anolyte) and the cathode is exposed to air. The electrogenic efficiency of the MFC varies with electron acceptor conditions. Oxygen (O_2) is considered to be the best known electron acceptor in the biological redox system. Apart from O_2, Fe^{3+} (potassium ferricyanide), and Mn^{2+} (potassium permanganate) are the most studied electron acceptors in dual chamber MFC as catholytes (You et al., 2006). However, using metals as electron acceptors has drawbacks such as replenishment after exhaustion and their discharge. Performance of MFC was independent of anolyte volume because the possible theoretical potential is around 1.2 V [NAD^+ (-0.32 V) and O_2 ($+0.816$ V)]. Stacking smaller MFCs in a series result in a cumulative voltage output. The nature of solid electron acceptors (electrodes) also influences the power generation efficiency because of their role as intermediary electron shuttlers. Electrodes should be electrically conductive, biocompatible, and chemically stable in an anolyte as well as efficient electron discharge agents. They should sustain their properties with time and be of a nonfouling nature. Apart from carbon-based materials, the usage of platinum, titanium, vanadium, nickel, stainless steel, aluminum, brass, and copper was also studied (Srikanth et al., 2011). Initial studies with MFC were carried out with a salt-bridge and later the research was extended to use PEM. The presence of negatively charged sulfonate groups facilitates good proton conductivity. However, the use of PEM increases the cost of MFC construction.

The extracellular electron transfer is influenced by the potential difference between the final electron carrier and the anode (Lee et al., 2009; Marsili et al., 2008; Newman and Kolter, 2000). The cell compartmentalization and the highly complicated architecture of cell respiratory chains facilitate energy harvesting from biocatalysts. Biocatalysts play an important role in generating the reducing powers as well as transferring them to electrodes. Initial research on MFCs was specific to pure cultures, but later shifted to the use of mixed consortia that facilitates a synergistic interaction between the individual strains, resulting in a higher power output. Electrochemically active microbes are facultative and metal-reducing in nature. These organisms have an outer membrane cytochrome oxidase type c protein that allows e$^-$ to be transferred from the interior of the bacterial cell membrane to the exterior of the cell membrane. The operational factors, such as organic load, retention time, redox condition (pH), surface area of electrode, and anode and cathode microenvironment, also influence MFC performance (Lenin Babu and Venkata Mohan, 2012; Mohanakrishna et al., 2012; Raghavulu et al., 2011, 2012; Srikanth and Venkata Mohan, 2012b,c; Srikanth et al., 2010b; Velvizhi and Venkata Mohan, 2013a; Velvizhi et al., 2012; Venkata Mohan et al., 2008e–h; Venkateswar Reddy et al., 2010). Biofilm formation on the surface of the anode will significantly influence the current generation in MFC irrespective of the nature of the microbe (Venkata Mohan et al., 2008i).

Wastewater (anolyte fuel) acts as an electron donor to generate reducing equivalents. Various types of waste/wastewater from simple domestic sewage to complex industrial wastewater have been considered as an anodic fuel in MFC (Srikanth and Venkata Mohan, 2012b,c; Venkata Mohan et al., 2013a). Highly biodegradable waste can be loaded at higher loading rates, while low biodegradable waste will interfere with the metabolism of microbes at higher loading rates. Higher loadings can continue the generation of reducing equivalents for longer periods, while lower loading rates can retain the electrogenesis for less time (Velvizhi and Venkata Mohan, 2011; Venkata Mohan and Srikanth, 2011). Neutral pH is the optimum condition for bacterial growth and metabolic activities. Most of the enzymes function near a neutral pH, and biomolecules of the cell become unstable at an extreme pH. Hence, the internal pH of the microbe is maintained at a neutral redox microenvironment irrespective of the external pH (Kim and Gadd, 2008). The external pH can bring alterations in the microbial activities, including synthesis of biomolecules and ion transport across membranes. MFC performance also depends on the redox conditions of fermentation (anolyte). External acidophilic pH is defended by the excessive production of acid-shock proteins that consume the H^+ entering into the cell and bring the internal pH to near neutral. Extreme alkalophilic pH can defend the function of an Na^+/H^+ antiport pump, which converts the proton motive force to the sodium motive force (Kim and Gadd, 2008). The exocellular electron transfer to the anode is dependent on the operating pH, which creates a proton gradient between the cell interior and surrounding environment. Higher performance was reported at an acidophilic pH over neutral and basic pH in diverse configurations of the MFC (Raghavulu et al., 2009a,b). The electron transfer is independent of external pH when the circuit is connected against resistance. The closed circuit creates a strong proton motive force on to the anode due to the continuous reduction of protons with the electron acceptor in the terminal reaction. Moreover, the metabolic activities of biocatalyst will be higher, and waste remediation also favors neutral pH.

Most of the MFC research is confined to the operation of an anodic chamber in anaerobic microenvironments, but a few reports are also available on the application of aerobic metabolic functions at the anode (Ringeisen et al., 2007; Rodrigo et al., 2007; Venkata Mohan et al., 2008f). If a low level of oxygen is allowed in the anode chamber, such that it cannot neutralize all the electrons generated in the system, the remaining electrons can be harnessed. This has an added advantage of higher treatment efficiency, including some toxic compounds (dyes and colored compounds) that need sequential alternative microenvironments. Overall, high carbon concentration, low oxygen levels, and the least possible distance between the anode and the cathode are the prerequisites for power generation from the aerobic MFC. However, detailed studies pertaining to understanding the process are required to establish the advantages of oxygen presence at the anode.

6.4.2 Bioelectrochemical Treatment

MFC operation has documented efficient waste remediation compared to conventional anaerobic treatment process (Aelterman, 2009; Bond et al., 2002; Chae et al., 2009; Chandrasekhar and Venkata Mohan, 2012;

Lee et al., 2008; Liu et al., 2005; Luo et al., 2010; Mohanakrishna et al., 2010a; Rabaey et al., 2003; Sun et al., 2009; Velvizhi and Venkata Mohan, 2011, 2013a,b; Venkata Mohan and Chandrasekhar, 2011b; Venkata Mohan and Srikanth, 2011; Venkata Mohan et al., 2008h,i, 2009a). Hence, MFC can also be termed as a bioelectrochemical treatment system (BET). Enhanced treatment efficiency along with simultaneous electrogenesis has made BET an effective process particularly for complex wastewater treatment. Protons (H^+) and electrons (e^-) released during substrate metabolism by electrochemically active microorganisms through various redox reactions travel from a reduced electron donor to an electrode and finally to an oxidized electron acceptor before generating power. This is the basis for the function of an MFC or a BET. BET operation triggers multiple reactions such as biochemical, physical, physicochemical, electrochemical and oxidative, which cohesively are denoted as bioelectrochemical reactions. The potential difference between anodic oxidation and cathodic reduction reactions can have a positive influence on the pollutant removal in MFCs. Anodic oxidation generates *in situ* biopotential by which potential reactive species like OH^-, O^-, and others are generated at the anode surface. These reactive species help to break the complex chemical structures present in wastewater and also aid in the degradation of different pollutants. The anode chamber of the MFC resembles a conventional anaerobic bioreactor and mimics the functions of a conventional electrochemical cell used for wastewater treatment, where the redox reactions help in the degradation of organic matter and toxic/xenobiotic pollutants (Mohanakrishna et al., 2010a; Venkata Mohan et al., 2009a). Pollutant removal during BET operation is possible mainly due to direct anodic oxidation (DAO) and indirect anodic oxidation (IAO) mechanisms. The pollutants are adsorbed on the anode surface and get destroyed by the anodic electron transfer reactions in the DAO. During the IAO, pollutants will be oxidized by the oxidants formed electrochemically on the anode surface under *in situ* biopotential. DAO allows the formation of primary oxidants that further react with the anode, yielding secondary oxidants such as chlorine dioxide and ozone, both of which might have significant positive effects on the color removal efficiency throughout the oxidation process.

In some reports pollutants present in the wastewater themselves act as mediators in electron transfer. For example, elemental sulfur present in the wastewater acts as a mediator for electron transfer to the anode and converts itself to sulfate in the MFC, which is easier for degradation (Dutta et al., 2009). Azo dyes also act as mediators for the electron transfer in the MFC and decolorize during reduction (Mu et al., 2009a). Biohazardous toxic compounds such as endocrine-disrupting estrogens can also be considered as mediator molecules in the MFC (Kiran Kumar et al., 2012). A few reports are available regarding the cathodic function in the effective removal of pollutants. Azo dyes (Mu et al., 2009a), nitrobenzene (Mu et al., 2009b), and nitrate (Lefebvre et al., 2008) are some of the pollutants studied in the cathode chamber of the BET. Hypothetically, it can be assumed that in the cathode chamber under anaerobic conditions most of the pollutants act as TEAs for power generation. Pollutants in the anodic chamber also act as mediators for electron transfer to the anode, which can increase the power generation efficiency with a simultaneous reduction of pollutants, but very little work has been reported in the use of pollutants as mediators. Apart from substrate removal, considerable reduction of toxicity, color, and TDS in wastewater was also observed (Mohanakrishna et al., 2010a; Velvizhi and Venkata Mohan, 2011). The application of an MFC was also extended to treat solid waste and toxic aromatic hydrocarbons, taking advantage of the *in situ* biopotential and by considering the anode as the electron acceptor (Venkata Mohan and Chandrasekhar, 2011a,b). Studies related to the mechanisms of pollutant reduction and their role in electron transfer or acceptance will give a spectrum of practical feasibility of this technology for the removal of toxic pollutants.

6.4.3 Electrically Driven Biohydrogenesis

MEC is one of the recent advancements and a promising technology for the production of renewable and sustainable hydrogen gas. Use of the MEC provides a completely new approach for hydrogen generation from a wide variety of biomass, such as biowaste and wastewater and accomplishing waste treatment at the same time (Call and Logan, 2008; Cheng and Logan, 2007; Ditzig et al., 2007; Liu et al., 2005; Logan, 2008; Rozendal et al., 2006, 2007). The principle of MEC is similar to the water electrolysis process, except that microbes (biocatalyst) are used in the anodic reaction to degrade organic matter into CO_2, electrons (e^-), and protons (H^+) (Figure 6.4). The e^- and H^+ travel through the external circuit and membrane and are reduced to H_2 at the cathode. MEC operation is similar to the existing MFC operation except that the cathode is sealed in the anolyte in order to avoid oxygen, and an additional voltage is added to the circuit (Venkata Mohan et al., 2013c). An external potential is required because under standard conditions, H_2 formation is not spontaneous. The H^+ migrate to the cathode and get reduced to form H_2 in presence of e^- coming

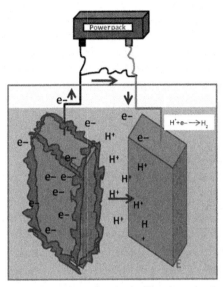

Figure 6.4 Schematic representation of a single-chambered microbial electrolysis cell for biohydrogen production.

from the anode under small applied voltages required to cross the endothermic barrier to form H_2 gas. Low energy consumption compared to the conventional water electrolysis and more substrate degradation than the dark-fermentation process are some of the potential benefits that make MEC an alternate process.

The performance of MEC is influenced by several factors, such as reactor configuration, electrode materials, membrane, and nature of substrate. Initially, double-chambered MECs were used with different types of membranes for the production of H_2 (Cheng and Logan, 2007; Rozendal et al., 2006). The performance of MEC was found to be most effective in a single-chamber without using a membrane (Call and Logan, 2008; Hu et al., 2008a,b; Lee and Rittmann, 2010a,b; Rozendal et al., 2006, 2007). The removal of membrane economizes the construction, operation, and maintenance of the MEC, but it also decreases the internal resistance of the system. Various electrode materials have been used in lab-scale MECs, including carbon felt (Parameswaran et al., 2009), stainless steel brushes (Call et al., 2009), stainless steel and nickel alloys (Selembo et al., 2009), graphite granules (Clauwaert and Verstraete, 2009), graphite fibers (Lee and Rittmann, 2010b), and noncatalyzed graphite electrodes (Venkata Mohan and Lenin Babu, 2011). The MEC has been operated with different substrates based on the substrate H_2, production varied (Cusick et al., 2010; Lenin Babu et al., 2013a,b; Lu et al., 2010; Selembo et al., 2009; Venkata Mohan and Lenin Babu, 2011; Wagner et al., 2009). Integrating the two processes showed good improvement in the treatment as well as in the additional product recovery (Lenin Babu et al., 2013a; Tuna et al., 2009; Wang et al., 2011). Nevertheless, a number of challenges still exist that need to be addressed before MEC can be applied at a practical level.

6.4.4 Microbial Electrosynthesizer

Reduction reactions at the cathode can effectively be used not only for pollutant removal but also for the generation of reduced end products that have commercial value (Figure 6.5). The product formation is mainly based on the electron acceptor and the redox potential of the MFC (Hu et al., 2008a; Rabaey and Rozendal, 2010; Venkata Mohan et al., 2013a). For example, ethanol can be formed at the cathode by using acetate as an electron acceptor under a redox potential of -0.28 V. Likewise a diverse range of value-added products can be harnessed from the MFC, especially at the cathode in the absence of O_2 as an electron acceptor, along with power generation (Rabaey and Rozendal, 2010). Based on the electron-accepting

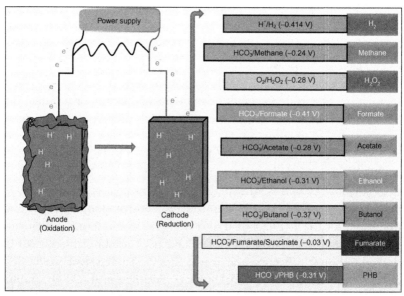

Figure 6.5 Schematic representation of the bioelectrochemical system, including its contribution towards the formation of reduced end products of commercial importance.

conditions at the cathode, different compounds can be synthesized; therefore, the MFC can also be used for product recovery in addition to treatment unit functionality. Bio or microbial electrosynthesis is a process that involves microbially catalyzed synthesis of chemical compounds in an electrochemical cell (Rabaey and Rozendal, 2010). The term *microbial electrosynthesis* has recently been used for the electrically driven reduction of carbon dioxide (Nevin et al., 2010). In recent times, considerable interest has been centered on the application of a biocathode in combination with a catalyzed cathode for bioelectrosynthesis (Rozendal et al., 2006). With the aid of a small input of electric power, many value-added compounds are formed at the cathode.

6.5 BIOPLASTICS

Majority of plastics are based on fossil fuels that are durable and degrade slowly and upon incineration they release gaseous pollutants (Jin et al., 2013). In this area of interest, bioplastics such as polyhydroxyalkanoates (PHAs) have attracted great interest due to their biodegradable nature (Liu et al., 2011). PHAs (mass of 50–100 kDa) are synthesized by bacteria from renewable resources produced as cellular reserve storage products under excess carbon and nutrient-deprived conditions (Madsen, 2008; Steinbüchel, 1992). The most common type of PHA is polyhydroxybutyrate (PHB). PHAs have special physical traits, such as elasticity, a high crystallization rate, and a high degree of polymerization (Madsen, 2008; Satoh et al., 1999). They can be converted to a wide range of finished products for application in industry, agriculture, and medicine (Akaraonye et al., 2010). Major drawbacks for their commercialization are the high production costs in which feedstock accounts for 50% of the overall price (Choi and Lee, 1999). Hence, there is a need for cheap renewable resources, wherein wastewater fits well.

6.5.1 Bioplastics Synthesis from Wastewater

Wastewater contains different types of organic compounds. After hydrolysis these complex organic compounds are degraded into simple sugars (Figure 6.6). These simple sugars through glycolysis are converted into VFAs (acetate, propionate, butyrate, and valerate) (Anderson and Dawes, 1990; Dawes, 1986). VFAs are the key precursors for PHA production. VFAs generated during fermentation are transported across the cell membrane and then activated to the corresponding acyl-CoA. Bacteria are cultivated on carbohydrates, pyruvate, or acetate, and the PHA is synthesized in a three-step

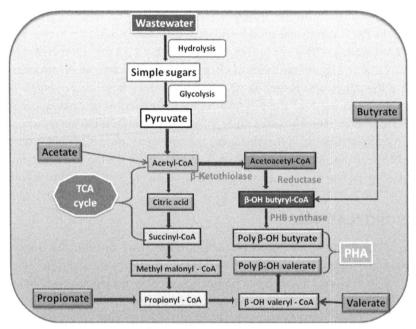

Figure 6.6 Metabolic pathway involved in the production of bioplastics from wastewater.

reaction starting with the formation of acetyl–CoA. Two molecules of acetyl-CoA are condensed to acetoacetyl-CoA, then reduced to 3-hydroxybutyryl-CoA by utilizing the reduced NADPH. Propionate is converted to propionyl-CoA, and both acetyl-CoA and propionyl-CoA can produce PHB and poly-β-hydroxy valerate, respectively. Butyrate is activated by CoA and forms butyryl-CoA, which is then extended to the PHA production pathway (Anderson and Dawes, 1990). ATP is required for the activation of VFA, while reducing equivalents (NADPH) are required for the formation of hydroxyacyl-CoA, the precursor for PHA production. ATP and reducing equivalents are acquired from respiration under aerobic operation, while they are generated in the glycogen metabolism under anaerobic operation.

PHB, a homo-polymer, is highly crystalline, stiff, and brittle with low impact strength. [P(3HB-co-3HV)], a copolymer, has better physical and thermal properties that depend on the hydroxyvalerate (HV) unit (Figure 6.7). If the HV fraction increases, the melting temperature decreases without affecting the degradation temperature, thus providing a polymer

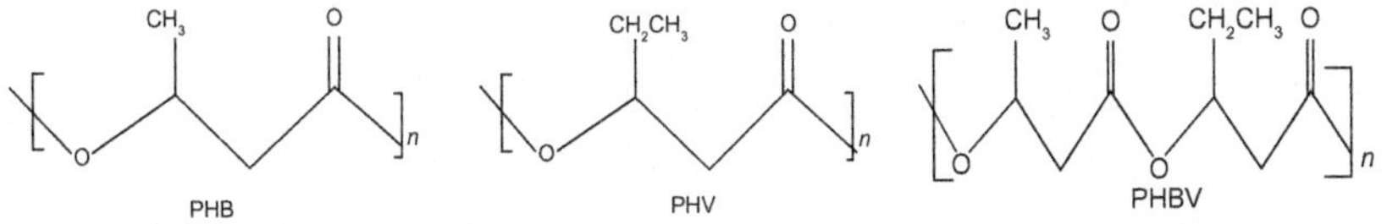

Figure 6.7 General structure of bioplastics.

with structural properties and processability similar to that for conventional plastics (Lee, 1996).

6.5.2 Bioplastics Production from Wastewater and CO_2

The development of pure culture fermentation and the commercialization of PHAs increases the cost of bioplastics to about four to nine times higher than that of conventional plastics (Moita and Lemos, 2012). To overcome this problem, considerable effort has gone into the production of PHAs using mixed cultures and wastewaters such as olive oil mill effluent (Beccari et al., 2009), sugarcane molasses (Albuquerque et al., 2011; Bengtsson et al., 2010), distillery spent wash (Amulya et al., 2014; Khardenavis et al., 2007), paper mill wastewater (Bengtsson et al., 2008), tomato cannery wastewater (Liu et al., 2008), designed synthetic wastewater (Srikanth et al., 2012; Venkata Mohan and Venkateswar Reddy, 2013; Venkateswar Reddy and Venkata Mohan, 2012a), food waste (Venkateswar Reddy and Venkata Mohan, 2012b), pyrolysis by-products (Moita and Lemos, 2011), pea shells (Patel et al., 2012). Integrating PHA production with biohydrogen reactor effluents is also being pursued as it offers the dual benefit of PHA recovery and enhanced waste treatment (Venkateswar Reddy et al., 2013; Amulya et al 2014., Venkata Mohan et al., 2010c; Venkateswar Reddy et al., 2012a, b). The advantages of mixed culture approaches are the use of a cheap carbon source and lower equipment cost (Salehizadeh and Van Loosdrecht, 2004). Disadvantages include low product yields, impurity of the produced PHA, and high PHA recovery costs. Furthermore, the use of open mixed cultures, such as activated sludge, has been developed by employing anaerobic conditions (17% DCW), aerobic conditions (36% DCW), and microaerophilic-aerobic process (oxygen limitation) (62% DCW) for the production of PHA. *Rhodospirillum rubrum*, a purple nonsulfur photosynthetic bacterium is capable of producing PHA when fed with CO_2 as a substrate. PHA can be extracted from bacterial cells and used as a biodegradable plastic with material properties similar to those of polypropylene (Choi et al., 2010). The PHA produced from *R. rubrum* grown on CO_2 gas was a copolymer composed of 86% β-HB and 14% β-HV (Young et al., 2007).

6.6 MICROALGAE CULTIVATION TOWARDS BIODIESEL PRODUCTION

Recently, microalgae have drawn considerable attention as an alternative source of biomass that is capable of generating fuel from the sequestration of carbon from both organic and inorganic sources by photosynthetic machinery. Algae are rapidly growing species whose carbon-fixing rates are much higher than those of terrestrial plants. It is estimated that algae yield is about 60,000 L per hectare compared to 300 ± 100 L from soya and canola crops (Brown et al., 2010). Microalgae commonly double their biomass within 24 h, and this time period during the exponential growth phase (GP) can be as short as 3.5 h (Chisti, 2007; Harrison et al., 2012). The cultivation of algae does not require arable land because algae can be grown in artificial ponds, on land unsuitable for agriculture, surface of lakes, coastal waterways, or in vats on wasteland (Duan and Savage, 2010). Algae-based biofuels have become focal point in current research due to, carbon neutrality, renewability, abundant availability, higher combustion efficiency, and higher biodegradability (Zhang et al., 2003). Algae-based fuels address the major constraints of first- and second-generation biofuels i.e., food vs fuel and food security issues. Microalgae can grow very fast and is capable of producing several fold higher biomass compared to terrestrial crops and trees, requires low and marginal land and produces higher lipid and carbohydrate yields (Singh et al., 2011). Production of biofuels from microalgae is gaining acceptance because of its higher economic feasibility and environmental sustainability compared to agro-based fuels. According to an estimate, the productivity of algae-derived biofuels is predicted to be on the order of 5000 gal/acre/year, which is approximately two orders of magnitude greater than the yield from terrestrial oil seed crops such as soybeans (Demirbas, 2008; Weyer et al., 2009). Algae-derived biodiesel is currently being promoted as a third-generation biofuel feedstock as it does not compete with food crops, it can be produced on non-arable land (Dragone et al., 2010; Hu et al., 2008b), and its scalability has potential (Harrison et al., 2012).

6.6.1 Mode of Nutrition

Biosynthesis of triglycerides by utilizing CO_2 (by biofixation) or wastewater under stress conditions by a photoautotrophic, heterotrophic (photo/dark), or mixotrophic mechanism enumerates the potential of microalgae for generating renewable biodiesel (Devi and Venkata Mohan, 2012; Venkata Mohan et al., 2014). Algae fix CO_2 during the day via photophosphorylation (thylakoid) and produce carbohydrate during the Calvin cycle (stroma), which are converted to various products, including triacyl glycerides (TAGs) depending on the species of algae or specific conditions pertaining to cytoplasm and plastids (Liu and Benning, 2013). The biosynthetic pathway of lipids in algae occur in four steps: (1) carbohydrates accumulate inside the cell, (2) acetyl-CoA is formed followed by malonyl-CoA, (3) palmitic acid is synthesized, and (4) finally, higher fatty acids are synthesized by chain elongation (Venkata Mohan et al., 2014).

Microalgae in the photoautotrophic mode use sunlight as the energy source and inorganic carbon (CO_2) as the carbon source to form biochemical energy through photosynthesis (Huang et al., 2010). This is the most favorable environmental conditions for the growth of microalgae (Chen et al., 2011). The autotrophic nutritional mode at the expense of atmospheric CO_2 generates algal oil. In the heterotrophic mode of nutrition, microalgae utilize the external carbon source in the form of simpler carbohydrates that enter into the cell and participate in other metabolic pathways such as respiration. Heterotrophic nutrition takes place both in the presence and absence of light. In photo-heterotrophic nutrition, light acts as an energy source, but the source of carbon remains organic. Oxidative assimilation takes place by two pathways: (1) the Embden Meyerhof pathway and (2) the pentose phosphate pathway (Neilson and Lewin, 1974). The heterotrophic nutritional mode avoids the limitations of light dependency, one of the major obstruction for gaining high cell density in large-scale photo-bioreactors (Huang et al., 2010). The heterotrophic nutritional mode facilitates wastewater treatment along with organic carbon removal, substrate degradation and lipid productivity. Cost effectiveness and relative simplicity of operations and easy maintenance are the main attractions of the heterotrophic mode of operation (Olguín et al., 2012; Perez-Garcia et al., 2011).

Microalgae can also function under mixotrophic nutrition by combining both the autotrophic and heterotrophic mechanisms that fix atmospheric CO_2 as well as consume the organic molecules and micronutrients from the growing environment. The mixotrophic growth regime is a variant of the heterotrophic growth regime, where CO_2 and organic carbon are simultaneously assimilated and both respiratory and photosynthetic metabolism operate concurrently (Kaplan et al., 1986; Lee, 2004; Perez-Garcia et al., 2011). Mixotrophs have the ability to utilize organic carbon, and therefore light energy is not a limiting factor for the biomass growth (Chang et al., 2011). The acetyl-CoA pool will be maintained from both the carbon source by CO_2 fixation (Calvin cycle) and intake from the outside of cell, which can further make malonyl-CoA. The photosynthetic metabolism utilizes light and CO_2 for growth and organic photosynthate production, while respiration uses the organic photosynthates produced during photosynthesis. Mixotrophic cultures showed reduced photo inhibition and improved growth rates over autotrophic and heterotrophic cultures (Devi et al., 2013). Mixotrophic cultivation was shown to be a good strategy to obtain a large biomass and high growth rates (Lee and Lee, 2002; Ogawa and Aiba, 1981), with the additional benefit of producing photosynthetic metabolites and lipid productivity (Chen and Johns, 1996; Perez-Garcia et al., 2011; Chandra et al., 2014). Algae have the flexibility to switch their nutritional mode based on substrate availability and light conditions. If simpler carbohydrates are present in the system, the algae shift towards heterotrophic nutrition from autotrophic mode to save their energy expenditure (Venkata Mohan et al., 2014).

Subjecting microalgae to stress microalgae causes the photosynthetic mechanism to switch from biomass growth to lipid synthesis. The different stress conditions that trigger lipid synthesis are temperature, light, pH, salinity and nutrients (Venkata Mohan and Devi, 2014). The intracellular lipid granules stored under stress conditions act as precursors for fatty acid biosynthesis (Devi et al., 2012; Venkata Mohan et al., 2014). The triglyceride composition of algae upon transesterification with an alcohol produce algae-derived biodiesel (alkyl esters). Depending on the species, growth conditions, and growth stages, microalgae have been shown to produce various types of lipids including triacylglycerides, phospholipids, glycolipids, and betaine lipids (Greenwell et al., 2010). The influence of various nutritional modes on the algal biomass growth and subsequent lipid production was studied by employing a two-phase operation, including a GP and a stress-induced starvation phase (SP) (Devi et al., 2013). The mixotrophic mode of operation showed higher biomass growth during the GP, while higher lipid productivity was observed with the nitrogen-deprived autotrophic mode followed by heterotrophic and mixotrophic operations. Relative increment in lipid productivities were noticed in the SP operation from GP in mixotrophic operation (2.45) followed by autotrophic (2.2) and heterotrophic (2.14) mode of operations. Effect of salinity stress was evaluated at varying salt concentrations for inducing maximum lipid production (Venkata Mohan and Devi, 2014). Microalgae-derived lipids and biomass can be converted into alcohols, methyl esters, and alkanes for use in spark-ignited engines, compression-ignition engines, and aircraft gas turbine engines (Harrison et al., 2012). Under specific cultivation conditions, algal oil content can exceed 50% by weight of dry biomass (Chisti, 2007).

6.6.2 Carbon Sequestration for Microalgae Growth

Microalgae have the capability to grow in nutrient-rich environments and accumulate nutrients and metals from wastewater (de-Bashan and Bashan, 2010; Devi et al., 2012; Hoffmann, 1998; Mallick, 2002) employing heterotrophic cultivation. Microalgae cultivation with wastewater treatment is a potential option for environmental sustainability and carbon neutrality. Five characteristically different ecological water bodies (mixotrophic) were evaluated to assess the biodiesel production capability of their native microalgae (mixed) (Venkata Mohan et al., 2014). The lipid yield varied between 4% and 26%, which depends mostly on the nature and function of the water body. Carpet mill effluent as feedstock showed its potential for algal biomass growth associated with biodiesel production (Chinnasamy et al., 2010). Cultivation of *Scenedesmus* sp. in fermented swine wastewater yielded lipids and other value-added products in association with nutrient removal (Kim et al., 2007). Nitrogen and phosphorus assimilation associated with lipid production was studied with freshwater microalgae using industrial wastewater (Li et al., 2012a,b). The functional role of macro/micro nutrients, such as carbon, nitrogen, phosphorus, and potassium, in the heterotrophic cultivation of microalgae (mixed) in domestic wastewater was studied on biomass growth and lipid productivity employing sequential GP and SP (Devi et al., 2012). Nutrient limitation during the SP showed a positive influence on the lipid productivity. Nitrogen limitation can also activate diacylglycerol acyl transferase, which converts acyl CoA to TAG (Takagi et al., 2000). Acid-rich effluents from fermentative H_2 producing reactors were evaluated as potential substrate for lipid accumulation by hetrotrophic microalgae cultivation with simultaneous treatment (Venkata Mohan and Devi, 2012). Microalgae can grow heterotrophically by utilizing VFA and results in lipid accumulation. Acetate can be easily assimilated by the algal cell as a part of the acetyl CoA metabolism in a single-step reaction catalyzed by acetyl CoA synthetase (Boyle and Morgan, 2009). For achieving highly dense microalgal cultures in mixotrophic mode, glucose along with inorganic CO_2 can be utilized. Different concentrations of glucose as organic carbon source was optimized for biomass growth and lipid accumulation (Chandra et al., 2014).

Microalgae are considered to be photosynthetically more efficient than terrestrial plants to fix CO_2 (Chiu et al., 2008; Indra et al., 2010). Algae can fix CO_2 from the atmosphere as well as industrial emissions (Brennan and Owende, 2010). In the process of fixation, CO_2 is utilized by microalgae as an inorganic carbon source, while water acts as an electron donor for the storage of reserve food material such as carbohydrates, which further get transformed to lipids under certain stress conditions (Devi and Venkata Mohan, 2012). Algae and cyanobacteria have different CO_2 concentrating mechanisms and act as enhancers for higher growth (Ramanan et al., 2010).

6.6.3 Preparation of Algal Fuel

Selecting the appropriate inoculum and mode of cultivation are the key aspects involved in preharvesting. Followed by preharvesting, a series of sequentially integrated postharvesting steps like harvesting, drying, cell disruption, extraction and transesterification are carried out for converting the algae biomass to biodiesel (Venkata Mohan et al., 2014). Drying the biomass prior to extraction is a prerequisite so as to avoid the interference of moisture with the solvents. After drying, cell disruption, oil extraction, and transesterification of oil to fuel are carried out sequentially. Transesterification facilitates a reaction of triglyceride molecules with alcohol in the presence of a catalyst to produce glycerol and mono-alkyl fatty acid esters (Harrison et al., 2012). Biodiesel is typically transesterified using methanol; therefore, the fatty acid alkyl esters that are produced are fatty acid methyl esters (FAME). In this process, glycerol is formed as a by-product. The transesterification reaction proceeds in short span of time (<5 min) and reduces the viscosity of the FAME compared to the parent oil, while the fatty acid composition will not get altered. The properties of the microalgae oil are mostly dependent on the feedstock and the conversion method used. Key aspects to evaluate the properties of microalgae oil are acid number, iodine number, specific gravity, density, kinematic viscosity, flash point, pour point, heating value, and cetane number. The lower viscosity and higher energy values recorded for the algae oil denotes its comparable features with standard norms and conventional fuel (Demirbas, 2008). Algal lipids contain a substantial quantity of long-chain polyunsaturated fatty acids (LC-PUFA), including eicosapentaenoic acid and docosahexaenoic acid (Chisti, 2007). The algal lipids have greater quantities of LC-PUFA compared to typical feedstocks associated with higher quantities of fully saturated fatty acids (C14:0, C16:0, and C18:0), which have implications in terms of fuel properties (Harrison et al., 2012). The most important characteristics affected by the level of unsaturation are oxidative stability, ignition quality (i.e., cetane number), and cold flow properties (Graboski and McCormick, 1998; Knothe et al., 1997; Ramos et al., 2009). Fully saturated methyl esters have high oxidative stability and a high cetane number, but suffer from poor cold flow properties (Harrison et al., 2012). Algal-based oil is of significant importance due to its renewable and carbon neutral nature.

6.7 SUMMARY

Currently, the processes discussed in this chapter are being evaluated either at the laboratory scale or pilot scale. The low end-product yield is a major stumbling block that needs to be overcome prior to industrial escalation. At present, basic and applied research is on the way to provide more insight into establishing optimized conditions. Process optimization of operational factors needs to be addressed with a multidisciplinary approach. Process integration will definitely have a positive influence on the overall efficiency and paves way for a biorefinery. Process engineering is of paramount importance for the establishment of sustainable remediation technologies along with value addition. Technical feasibility, simplicity, economics, societal needs, and political priorities are some of the vital aspects that can differentiate the bioprocesses used to treat waste in the future. Utilizing wastewater/waste for value addition through its remediation in the future will open a new avenue for the utilization of renewable and inexhaustible feedstock.

ACKNOWLEDGMENTS

The author wishes to thank the Director, CSIR-IICT, Hyderabad, for the encouragement. Funding from Council of Scientific and Industrial Research (CSIR), Government of India, in the form of 12 five-year plan projects (CSC-0113, CSC-0116, ESC-0108), Department of Biotechnology (DBT), Government of India, in the form of EU-FP7-KBBE project on Strengthening Networking on Biomass Research and Biowaste Conversion-Biotechnology for Europe-India Integration (SAHYOG) (No. BT/IN/EU/07/PMS/2011) and Ministry of New and Renewable Energy (MNRE), Government of India, in the form of Mission Mode Project on Hydrogen Production through Biological Routes (No. 103/131/2008-NT) are greatly acknowledged.

REFERENCES

Aelterman, P., 2009. Microbial fuel cells for the treatment of waste streams with energy recovery. Ph.D. Thesis. Gent University, Belgium.

Agarwal, M., Tardio, J., Venkata Mohan, S., 2013. Biohydrogen production from kitchen based vegetable waste: effect of pyrolysis temperature and time on catalysed and non-catalysed operation. Bioresour. Technol. 130, 502–509.

Akaraonye, E., Keshavarz, T., Roy, I., 2010. Production of polyhydroxyalkanoates: the future green materials of choice. J. Chem. Technol. Biotechnol. 85, 732–743.

Albuquerque, M.G.E., Martino, V., Pollet, E., Avérous, L., Reis, M.A.M., 2011. Mixed culture polyhydroxyalkanoate (PHA) production from volatile fatty acid (VFA)-rich streams: effect of substrate composition and feeding regime on PHA productivity, composition and properties. J. Biotechnol. 151, 66–76.

Allakhverdiev, S.I., Thavasi, V., Kreslavski, V.D., Zharmukhamedov, S.K., Klimov, V.V., Ramakrishna, S., et al., 2010. Photosynthetic hydrogen production. J. Photochem. Photobiol. 11, 87–99.

Amulya, K., Venkateswar Reddy, M., Venkata Mohan, S., 2014. Acidogenic spent wash valorization through polyhydroxyalkanoate (PHA) synthesis coupled with fermentative biohydrogen production. 158, 336–342.

Anderson, A.J., Dawes, E.A., 1990. Occurrence, metabolism, metabolic role, and industrial uses of bacterial polyhydroxyalkanoates. Microbiol. Rev. 54 (4), 450–472.

Angenent, L.T., Karim, K., Al-Dahhan, M.H., Wrenn, B.A., Domíguez-Espinosa, R., 2004. Production of bioenergy and biochemicals from industrial and agricultural wastewater. Trends Biotechnol. 22, 477–485.

Balat, M., 2010. Thermochemical routes for biomass-based hydrogen production. Energ. Source. Part A 32 (15), 1388–1398.

Beccari, M., Bertin, L., Dionisi, D., Fava, F., Lampis, S., Majone, M., et al., 2009. Exploiting olive oilmill effluents as a renewable resource for production of biodegradable polymers through a combined anaerobic–aerobic process. J. Chem. Technol. Biotechnol. 84, 901–908.

Beer, L.L., Boyd, E.S., Peters, J.W., Posewitz, M.C., 2009. Engineering algae for biohydrogen and biofuel production. Curr. Opin. Biotechnol. 20, 264–271.

Bengtsson, S., Werker, A., Christensson, M., Welander, T., 2008. Production of polyhydroxyalkanoates by activated sludge treating a paper mill wastewater. Bioresour. Technol. 99, 509–516.

Bengtsson, S., Pisco, A.R., Johansson, P., Lemos, P.C., Reis, M.A.M., 2010. Molecular weight and thermal properties of polyhydroxyalkanoates produced from fermented sugar molasses by open mixed cultures. J. Biotechnol. 147, 172–179.

Bisaillon, A., Turcot, J., Hallenbeck, P.C., 2006. The effect of nutrient limitation on hydrogen production by batch cultures of Escherichia coli. Int. J. Hydrogen Energ. 31, 1504–1508.

Bond, D.R., Holmes, D.E., Tender, L.M., Lovley, D.R., 2002. Electrode-reducing microorganisms harvesting energy from marine sediments. Science 295, 483–485.

Boyle, N.R., Morgan, J.A., 2009. Flux balance analysis of primary metabolism in Chlamydomonas reinhardtii. BMC Syst. Biol. 3, 4.

Brennan, L., Owende, P., 2010. Biofuels from microalgae—a review of technologies for production, processing, and extractions of biofuels and co-products. Renew. Sustain. Energ. Rev. 14, 557–577.

Brown, T.M., Duan, P., Savage, P.E., 2010. Hydrothermal liquefaction and gasification of Nannochloropsis sp. Energ. Fuel. 24, 3639–3646.

Call, D., Logan, B.E., 2008. Hydrogen production in a single chamber microbial electrolysis cell lacking a membrane. Environ. Sci. Technol. 42 (9), 3401–3406.

Call, D., Merrill, M.D., Logan, B.E., 2009. High surface area stainless steel brushes as cathodes in microbial electrolysis cells (MECs). Environ. Sci. Technol. 43 (6), 2179–2183.

Cao, G.L., Guo, W.Q., Wang, A.J., Zhao, L., Xu, C.J., Zhao, Q., Ren, N., 2012. Enhanced cellulosic hydrogen production from lime-treated cornstalk wastes using thermophilic anaerobic microflora. Int. J. Hydrogen Energ. 37, 13161–13166.

Chae, K.J., Choi, M.J., Lee, J.W., Kim, K.Y., Kim, I.S., 2009. Effect of different substrates on the performance, bacterial diversity, and bacterial viability in microbial fuel cells. Bioresour. Technol. 100, 3518–3525.

Chakkrit, S., Plangklang, P., Imai, T., Reungsang, A., 2011. Co-digestion of food waste and sludge for hydrogen production by anaerobic mixed cultures: statistical key factors optimization. Int. J. Hydrogen Energ. 36 (21), 14227–14237.

Chandra, R., Venkata Mohan, S., 2011. Microalgal community and their growth conditions influence biohydrogen production during integration of dark-fermentation and photofermentation processes. Int. J. Hydrogen Energ. 36 (19), 12211–12219.

Chandra, R., Venkata Subhash, G., Venkata Mohan, S., 2012. Mixotrophic operation of photo-bioelectrocatalytic fuel cell under anoxygenic microenvironment enhances the light dependent bioelectrogenic activity. Bioresour. Technol. 109, 46–56.

Chandra, R., Rohit, M.V., Swamy, Y.V., Venkata Mohan, S., 2014. Regulatory function of organic carbon supplementation during growth and nutrient stress phases of mixotrophic microalgae cultivation. Bioresour. Technol. 36, 1221–1221.

Chandrasekhar, K., Venkata Mohan, S., 2012. Bio-electrochemical remediation of real field petroleum sludge as an electron donor with simultaneous power generation facilitates biotransformation of PAH: effect of substrate concentration. Bioresour. Technol. 110, 517–525.

Chang, R.L., Ghamsari, L., Manichaikul, A., Hom, E.F.Y., Balaji, S., Fu, W., et al., 2011. Metabolic network reconstruction of *Chlamydomonas* offers insight into light-driven algal metabolism. Mol. Syst. Biol. 7, 1–13.

Chen, F., Johns, M.R., 1996. Heterotrophic growth of *Chlamydomonas reinhardtii* on acetate in chemostat culture. Process Biochem. 31, 601–604.

Chen, S.D., Lee, K.S., Lo, Y.C., Chen, W.M., Wu, J.F., Lin, C.Y., 2008. Batch and continuous biohydrogen production from starch hydrolysate by *Clostridium* species. Int. J. Hydrogen Energ. 33 (7), 1803–1812.

Chen, C.Y., Yeh, K.L., Aisyah, R., Lee, D.J., Chang, J.S., 2011. Cultivation, photobioreactor design and harvesting of microalgae for biodiesel production: a critical review. Bioresour. Technol. 102, 71–81.

Cheng, S., Logan, B.E., 2007. Sustainable and efficient biohydrogen production via electrohydrogenesis. Proc. Natl Acad. Sci. USA. 104, 18871–18873.

Cheng, H.H., Whang, L.M., Wu, C.W., Chung, M.C., 2012. A two-stage bioprocess for hydrogen and methane production from rice straw bioethanol residues. Bioresour. Technol. 113, 23–29.

Chinnasamy, S., Bhatnagar, A., Hunt, R.W., Das, K.C., 2010. Microalgae cultivation in a wastewater dominated by carpet mill effluents for biofuel applications. Bioresour. Technol. 101, 3097–3105.

Chiranjeevi, P., Mohanakrishna, G., Venkata Mohan, S., 2012. Rhizosphere mediated electrogenesis with the function of anode placement for harnessing bioenergy through CO_2 sequestration. Bioresour. Technol. 124, 364–370.

Chiranjeevi, P., Chandra, R., Venkata Mohan, S., 2013. Ecologically engineered submerged and emergent macrophyte based system: an integrated eco-electrogenic design for harnessing power with simultaneous wastewater treatment. Ecol. Eng. 51, 181–190.

Chisti, Y., 2007. Biodiesel from microalgae. Biotechnol. Adv. 25, 294–306.

Chiu, S.Y., Kao, C.Y., Chen, C.H., Kuan, T.C., Ong, S.C., Lin, C.S., 2008. Reduction of CO_2 by a high-density culture of *Chlorella* sp. in a semi-continuous photobioreactor. Bioresour. Technol. 99, 3389–3396.

Choi, J.I., Lee, S.Y., 1999. High-level production of poly(3-hydroxybutyrate-co-3-hydroxyvalerate) by fed-batch culture of recombinant *Escherichia coli*. Appl. Environ. Microbiol. 65, 4363–4368.

Choi, D.W., Chipman, D.C., Bents, S.C., Brown, R.C., 2010. A techno-economic analysis of polyhydroxyalkanoate and hydrogen production from syngas fermentation of gasified biomass. Appl. Biochem. Biotechnol. 160, 1032–1046.

Chou, C.H., Han, C.L., Chang, J.J., Lay, J.J., 2011. Co-culture of *Clostridium beijerinckii* L9, *Clostridium butyricum* M1 and *Bacillus thermoamylovorans* B5 for converting yeast waste into hydrogen. Int. J. Hydrogen Energ. 36 (21), 13972–13983.

Clauwaert, P., Verstraete, W., 2009. Methanogenesis in membraneless microbial electrolysis cell. Appl. Microbiol. Biotechnol. 82 (5), 829–836.

Cusick, R.D., Kiely, P.D., Logan, B.E., 2010. A monetary comparison of energy recovered from microbial fuel cells and microbial electrolysis cells fed winery or domestic wastewaters. Int. J. Hydrogen Energ. 35, 8855–8861.

Dawes, E.A., 1986. Microbial energy reserve compounds. In: Dawes, E.A. (Ed.), Microbial Energetics. Blackie & Son, London, Glasgow, United Kingdom, pp. 145–165.

de-Bashan, L.E., Bashan, Y., 2010. Immobilized microalgae for removing pollutants: review of practical aspects. Bioresour. Technol. 101, 1611–1627.

Demirbas, A., 2008. Comparison of transesterification methods for production of biodiesel from vegetable oils and fats. Energ. Convers. Manage. 49, 125–130.

Devi, M.P., Venkata Mohan, S., 2012. CO_2 supplementation to domestic wastewater enhances microalgae lipid accumulation under mixotrophic microenvironment: effect of sparging period and interval. Bioresour. Technol. 112, 116–123.

Devi, M.P., Venkata Mohan, S., Mohanakrishna, G., Sarma, P.N., 2010. Regulatory influence of CO_2 sparging on fermentative hydrogen production. Int. J. Hydrogen Energ. 35, 10701–10709.

Devi, M.P., Venkata Subhash, G., Venkata Mohan, S., 2012. Heterotrophic cultivation of mixed microalgae for lipid accumulation and wastewater treatment during sequential growth and starvation phases: effect of nutrient supplementation. Renew. Energy 43, 276–283.

Devi, M.P., Swamy, Y.V., Venkata Mohan, S., 2013. Nutritional mode influences lipid accumulation in microalgae with the function of carbon sequestration and nutrient supplementation. Bioresour. Technol. 142, 278–286.

Ditzig, J., Liu, H., Logan, B.E., 2007. Production of hydrogen from domestic wastewater using a bioelectrochemically assisted microbial reactor (BEAMR). Int. J. Hydrogen Energ. 32, 2296–2304.

Dragone, G., Fernandes, B., Vicente, A.A., Teixeira, J.A., 2010. Third generation biofuels from microalgae. In: Méndez-Vilas, A. (Ed.), Applied Microbiology and Microbial Biotechnology. Current Research, Technology and Education. Formatex Research Centre, Spain.

Duan, P., Savage, P.E., 2010. Hydrothermal liquefaction of a microalge with heterogeneous catalysts. Ind. Eng. Chem. Res. 50, 52–61.

Dutta, P.K., Keller, J., Yuan, Z., Rozendal, R.A., Rabaey, K., 2009. Role of sulfur during acetate oxidation in biological anodes. Environ. Sci. Technol. 43, 3839–3845.

Fang, H.H.P., Liu, H., 2002. Effect of pH on hydrogen production from glucose by a mixed culture. Bioresour. Technol. 82, 87–93.

Ferchichi, M., Crabbe, V., Gil, G.-H., Hintz, W., Almadidy, A., 2005. Influence of initial pH on hydrogen production from cheese whey. J. Biotechnol. 120, 402–409.

Franks, A.E., Nevin, K.P., 2010. Microbial fuel cells, a current review. Energies 3, 899–919.

Froylán, M.E.-E., Carlos, P.-O., Jose, N.-C., Yolanda, G.-G., André, B., Humberto, G.-P., 2009. Anaerobic digestion of the vinasses from the fermentation of Agave tequilana Weber to tequila: the effect of pH, temperature and hydraulic retention time on the production of hydrogen and methane. Biomass Bioenerg. 33, 14–20.

Gomez, X., Moran, A., Cuetos, M.J., Sanchez, M.E., 2006. The production of hydrogen by dark fermentation of municipal solid wastes and slaughterhouse waste: a two-phase process. J. Power. Sources 157, 727–732.

Gopalakrishnan, K., Lay, C.H., Chu, C.Y., Wu, J.H., Lee, S.C., Lin, C.Y., 2012. Seed inocula for biohydrogen production from biodiesel solid residues. Int. J. Hydrogen Energ. 37 (20), 15489–15495.

Goud, R.K., Venkata Mohan, S., 2011. Pre-fermentation of waste as a strategy to enhance the performance of single chambered microbial fuel cell (MFC). Int. J. Hydrogen Energ. 36, 13753–13762.

Goud, R.K., Venkata Mohan, S., 2012a. Acidic and alkaline shock pretreatment to enrich acidogenic biohydrogen producing mixed culture: long term synergetic evaluation of microbial inventory, dehydrogenase activity and bio-electro kinetics. RSC Adv. 2 (15), 6336–6353.

Goud, R.K., Venkata Mohan, S., 2012b. Regulating biohydrogen production from wastewater by applying organic load-shock: change in the microbial community structure and bio-electrochemical behavior over long-term operation. Int. J. Hydrogen Energ. 37, 17763–17777.

Goud, R.K., Raghavulu, S.V., Mohanakrishna, G., Naresh, K., Venkata, Mohan S., 2012. Predominance of Bacilli and Clostridia in microbial community of biohydrogen producing biofilm sustained under diverse acidogenic operating conditions. Int. J. Hydrogen Energ. 37, 4068–4076.

Graboski, M.S., McCormick, R.L., 1998. Combustion of fat and vegetable oil derived fuels in diesel engines. Prog. Energ. Combust. 24, 125–164.

Greenwell, H.C., Laurens, L.M.L., Shields, R.J., Lovitt, R.W., Flynn, K.J., 2010. Placing microalgae on the biofuels priority list: a review of the technological challenges. J. R. Soc. Interface 7, 703–726.

Guo, L., Zhao, J., She, Z., Lu, M., Zong, Y., 2012. Effect of S-TE (solubilization by thermophilic enzyme) digestion conditions on hydrogen production from waste sludge. Bioresour. Technol. 117, 368–372.

Gustavo, D.V., Felipe, A.M., Antonio de, L.R., Elías, R.F., 2008. Fermentative hydrogen production in batch experiments using lactose, cheese whey and glucose: influence of initial substrate concentration and pH. Int. J. Hydrogen Energ. 33, 4989–4997.

Hafez, H., Nakhla, G., Naggar, HEl, 2010. An integrated system for hydrogen and methane production during landfill leachate treatment. Int. J. Hydrogen Energ. 35, 5010–5014.

Hallenbeck, P.C., Benemann, J.R., 2002. Biological hydrogen production: fundamentals and limiting processes. Int. J. Hydrogen Energ. 27, 1185–1193.

Han, W., Chen, H., Jiao, A., Wang, Z., Li, Y., Ren, N., 2012a. Biological fermentative hydrogen and ethanol production using continuous stirred tank reactor. Int. J. Hydrogen Energ. 37 (1), 843–847.

Han, W., Wang, B., Zhou, Y., Wang, D.X., Wang, Y., Yue, L., et al., 2012b. Fermentative hydrogen production from molasses wastewater in a continuous mixed immobilized sludge reactor. Bioresour. Technol. 110, 219–223.

Harrison, B.B., Marc, E.B., Anthony, J.M., 2012. Chemical and physical properties of algal methyl ester biodiesel containing varying levels of methyl eicosapentaenoate and methyl docosahexaenoate. Algal Res. 1, 57–69.

Hawkes, F.R., Hussy, I., Kyazze, G., Dinsdale, R., Hawkes, D.L., 2007. Continuous dark fermentative hydrogen production by mesophilic microflora: principles and progress. Int. J. Hydrogen Energ. 32, 172–184.

Hoffmann, J.P., 1998. Wastewater treatment with suspended and non suspended algae. J. Phycol. 34, 757–763.

Hu, H., Fan, Y., Liu, H., 2008a. Hydrogen production using single chamber membrane-free microbial electrolysis cells. Water Res. 42 (15), 4172–4178.

Hu, Q., Sommerfeld, M., Jarvis, E., Ghirardi, M., Posewitz, M., Seibert, M., Darzins, A., 2008b. Microalgal triacylglycerols as feed stocks for biofuel production: perspectives and advances. Plant J. 54, 621–639.

Huang, G., Chen, F., Wei, D., Zhang, X., Chen, G., 2010. Biodiesel production by microalgal biotechnology. Appl. Energ. 87, 38–46.

Idania, V.V., Sparling, R., Risbey, D., Noemi, R.S., Hec, M., Poggi Varaldo, H.M., 2005. Hydrogen generation via anaerobic fermentation of paper mill wastes. Bioresour. Technol 96, 1907–1913.

Im, W.T., Kim, D.H., Kim, K.H., Kim, M.S., 2012. Bacterial community analyses by pyrosequencing in dark fermentative H_2-producing reactor using organic wastes as a feedstock. Int. J. Hydrogen Energ. 37 (10), 8330–8337.

Indra, S., Halldor, G.S., Sigurbjörn, E., Ása, B., Grzegorz, M., 2010. Geothermal CO_2 biomitigation techniques by utilizing microalgae at the blue lagoon, iceland. In: Thirty-Fourth Workshop on Geothermal Reservoir Engineering Stanford University, Stanford, California.

Jin, Y.-X., Shi, L.-H., Kawata, Y., 2013. Metabolomics-based component profiling of *Halomonas* sp. KM-1 during different growth phases in poly(3-hydroxybutyrate) production. http://dx.doi.org/10.1016/j.biortech.2013.04.059.

Jong, W., 2009. Sustainable hydrogen production by thermochemical biomass processing. In: Gupta, R.B. (Ed.), Hydrogen Fuel: Production, Transport, and Storage. Taylor & Francis Group, LLC, Boca Raton, Florida, USA.

Julia, C.R., Celis, L.B., Felipe, A.M., Elías, R.F., 2012. Different start-up strategies to enhance biohydrogen production from cheese whey in UASB reactors. Int. J. Hydrogen Energ. 37 (7), 5591–5601.

Kaplan, D., Richmond, A.E., Dubinsky, Z., Aaronson, S., 1986. Algal nutrition. In: Richmond, A. (Ed.), Handbook for Microalgal Mass Culture. CRC Press, Boca Raton, FL, pp. 147–198.

Karadag, D., Puhakka, J.A., 2010. Enhancement of anaerobic hydrogen production by iron and nickel. Int. J. Hydrogen Energ. 35, 8554–8560.

Kargi, F., Catalkaya, E.C., 2011. Hydrogen gas production from olive mill wastewater by electrohydrolysis with simultaneous COD removal. Int. J. Hydrogen Energ. 36 (5), 3457–3464.

Kargi, F., Eren, N.S., Ozmihci, S., 2012. Bio-hydrogen production from cheese whey powder (CWP) solution: comparison of thermophilic and mesophilic dark fermentations. Int. J. Hydrogen Energ. 37 (10), 8338–8342.

Khardenavis, A.A., Kumar, M.S., Mudliar, S.N., Chakrabarti, T.N., 2007. Biotechnological conversion of agro-industrial wastewaters into biodegradable plastic, poly b-hydroxybutyrate. Bioresour. Technol. 98, 3579–3584.

Kim, B.H., Gadd, G.M., 2008. Bacterial Physiology and Metabolism. Cambridge University Press, Cambridge.

Kim, M.S., Lee, D.Y., 2010. Fermentative hydrogen production from tofu-processing waste and anaerobic digester sludge using microbial consortium. Bioresour. Technol. 101, S48–S52.

Kim, M.K., Park, J.W., Park, C.S., Kim, S.J., Jeune, K.H., Chang, M.U., Acreman, J., 2007. Enhanced production of Scenedesmus spp. (green microalgae) using a new medium containing fermented swine wastewater. Bioresour. Technol. 98, 2220–2228.

Kim, D.H., Kim, S.H., Jung, K.W., Kim, M.S., Shin, H.S., 2011a. Effect of initial pH independent of operational pH on hydrogen fermentation of food waste. Bioresour. Technol. 102 (18), 8646–8652.

Kim, M.S., Lee, D.-Y., Kim, D.-H., 2011b. Continuous hydrogen production from tofu processing waste using anaerobic mixed microflora under thermophilic conditions. Int. J. Hydrogen Energ. 36 (14), 8712–8718.

Kim, S.H., Cheona, H.C., Lee, C.Y., 2012. Enhancement of hydrogen production by recycling of methanogenic effluent in two-phase fermentation of food waste. Int. J. Hydrogen Energ. 37, 13777–13782.

Kiran Kumar, A., Venkateswar Reddy, M., Chandrasekhar, K., Srikanth, S., Venkata Mohan, S., 2012. Endocrine disruptive estrogens role in electron transfer: bio-electrochemical remediation with microbial mediated electrogenesis. Bioresour. Technol. 104, 547–556.

Knothe, G., Dunn, R., Bagby, M., 1997. Biodiesel: the use of vegetable oils and their derivatives as alternative diesel fuels. Fuels and Chemicals from Biomass. American Chemical Society, Washington, DC, pp. 172–208.

Kobayashi, T., Xu, K.Q., Li, Y.Y., Inamori, Y., 2012. Effect of sludge recirculation on characteristics of hydrogen production in a two-stage hydrogenemethane fermentation process treating food wastes. Int. J. Hydrogen Energ. 37 (7), 5602–5611.

Kongjan, P., O-Thong, O., Angelidaki, I., 2011. Biohydrogen production from desugared molasses (DM) using thermophilic mixed cultures immobilized on heat treated anaerobic sludge granules. Int. J. Hydrogen Energ. 36, 14261–14269.

Kraemer, J.T., Bagley, D.M., 2007. Improving the yield from fermentative hydrogen production. Biotechnol. Lett. 29, 685–695.

Lakshmi Devi, R., Muthukumar, K., 2010. Enzymatic saccharification and fermentation of paper and pulp industry effluent for biohydrogen production. Int. J. Hydrogen Energ. 35, 3389–3400.

Lalit Babu, V., Venkata Mohan, S., Sarma, P.N., 2009. Influence of reactor configuration on fermentative hydrogen production during wastewater treatment. Int. J. Hydrogen Energ. 34, 3305–3312.

Lee, S.Y., 1996. Bacterial polyhydroxyalkanoates. Biotechnol. Bioeng. 49, 1–14.

Lee, Y.K., 2004. Algal nutrition: Heterotrophic carbon nutrition. In: Richmond, A. (Ed.), Handbook of Microalgal Culture. Biotechnology and Applied Phycology. Blackwell Publishing, Oxford, pp. 116–124.

Lee, H.S., Rittmann, B.E., 2010a. Characterization of energy losses in an upflow single-chamber microbial electrolysis cell. Int. J. Hydrogen Energ. 35, 920–927.

Lee, H.S., Rittmann, B.E., 2010b. Significance of biological hydrogen oxidation in a continuous single-chamber microbial electrolysis cell. Environ. Sci. Technol. 44, 948–954.

Lee, K., Lee, C.G., 2002. Nitrogen removal from wastewaters by microalgae without consuming organic carbon sources. J. Microbiol. Biotechnol. 12, 979–985.

Lee, Y.J., Miyahara, T., Noike, T., 2001. Effect of iron concentration on hydrogen fermentation. Bioresour. Technol. 80, 227–231.

Lee, H.S., Parameswaran, P., Kato-Marcus, A., Torres, C.I., Rittman, B.E., 2008. Evaluation of energy-conversion efficiencies in microbial fuel cells (MFCs) utilizing fermentable and non-fermentable substrates. Water Res. 42, 1501–1510.

Lee, H.S., Torres, C.I., Rittmann, B.E., 2009. Effects of substrate diffusion and anode potential on kinetic parameters for anode-respiring bacteria. Environ. Sci. Technol. 43, 7571–7577.

Lefebvre, O., Mamun, A., Ng, H.Y., 2008. A microbial fuel cell equipped with a biocathode for organic removal and denitrification. Water Sci. Technol. 58, 881–885.

Lenin Babu, M., Venkata Mohan, S., 2012. Influence of graphite flake addition to sediment on electrogenesis in a sediment-type fuel cell. Bioresour. Technol. 110, 206–213.

Lenin Babu, M., Sarma, P.N., Mohan, S.V., 2013a. Microbial electrolysis of synthetic acids for biohydrogen production: influence of biocatalyst pretreatment and pH with the function of applied potential. J. Microb. Biochem. Technol. S6, http://dx.doi.org/10.4172/1948-5948.

Lenin Babu, M., Venkata Subhash, G., Sarma, P.N., Venkata Mohan, S., 2013b. Bioelectrolytic conversion of acidogenic effluents to biohydrogen: an integration strategy for higher substrate conversion and product recovery. Bioresour. Technol. 133, 322–331.

Li, Y.C., Wu, S.Y., Chu, C.Y., Huang, H.C., 2011. Hydrogen production from mushroom farm waste with a two-step acid hydrolysis process. Int. J. Hydrogen Energ. 36 (21), 14245–14251.

Li, F.W., Pei, C., Ai, P.H., Chi, M.L., 2012a. The feasibility of biodiesel production by microalgae using industrial wastewater. Bioresour. Technol. 113, 14–18.

Li, Y.C., Chu, C.Y., Wu, S.Y., Tsai, C.Y., Wang, C.C., Hung, C.H., Lin, C.Y., 2012b. Feasible pretreatment of textile wastewater for dark fermentative hydrogen production. Int. J. Hydrogen Energ. 37, 15511–15517.

Licata, B.L., Sagnelli, F., Boulanger, A., Lanzini, A., Leone, P., Zitella, P., Santarelli, M., 2011. Bio-hydrogen production from organic wastes in a pilot plant reactor and its use in a SOFC. Int. J. Hydrogen Energ. 36 (13), 7861–7865.

Lin, C.Y., Lay, C.H., 2004a. Carbon/nitrogen-ratio effect on fermentative hydrogen production by mixed microflora. Int. J. Hydrogen Energ. 29, 41–45.

Lin, C.Y., Lay, C.H., 2004b. Effects of carbonate and phosphate concentrations on hydrogen production using anaerobic sewage sludge microflora. Int. J. Hydrogen Energ. 29, 275–281.

Lipman, T., 2011. An overview of hydrogen production and storage systems with renewable hydrogen case studies. Energy Efficiency and Renewable Energy Fuel Cell Technologies Program (US DOE Grant DE-FC3608GO18111 A000).

Liu, F., Fang, B., 2007. Optimization of biohydrogen production from biodiesel wastes by *Klebsiella pneumoniae*. Biotech. J. 2, 374–380.

Liu, B., Benning, C., 2013. Lipid metabolism in microalgae distinguishes itself. Curr Opin in Biotechnol 24 (2), 300–309.

Liu, H., Grot, S., Logan, B.E., 2005. Electrochemically assisted microbial production of hydrogen from acetate. Environ. Sci. Technol. 39, 4317–4320.

Liu, H.Y., Hall, P.V., Darby, J.L., Coats, E.R., Green, P.G., Thompson, D.E., Loge, F.J., 2008. Production of polyhydroxyalkanoate during treatment of tomato cannery wastewater. Water Environ. Res. 80, 367–372.

Liu, Q., Zhang, X., Yu, L., Zhao, A., Tai, J., Liu, J., et al., 2010. Fermentative hydrogen production from fresh leachate in batch and continuous bioreactors. Bioresour. Technol. 102 (9), 5411–5417.

Liu, Y., Wang, Y., He, N., Huang, J., Zhu, K., Shao, W., et al., 2011. Optimization of polyhydroxybutyrate (PHB) production by excess activated sludge and microbial community analysis. J. Hazard. Mater. 185, 8–16.

Logan, B.E., 2008. Microbial fuel cells. John Wiley & Sons, Inc., Hoboken, NJ.

Logan, B.E., 2010. Scaling up microbial fuel cells and other bioelectrochemical systems. Appl. Microbiol. Biotechnol. 85, 1665–1671.

Lu, L., Xing, D., Xie, T., Ren, N., Logan, B.E., 2010. Hydrogen production from proteins via electrohydrogenesis in microbial electrolysis cells. Biosens. Bioelectron. 25, 2690–2695.

Luo, G., Xie, L., Zou, Z., Wang, W., Zhou, Q., 2010. Evaluation of pretreatment methods on mixed inoculum for both batch and continuous thermophilic biohydrogen production from cassava stillage. Bioresour. Technol. 101, 959–964.

Madsen, E.L., 2008. Environmental Microbiology: from Genomes to Biogeochemistry. Blackwell Publishing, Malden, MA, USA.

Mallick, N., 2002. Biotechnological potential of immobilized algae for wastewater N, P and metal removal: a review. BioMetals 15, 377–390.

Marsili, E., Baron, D.B., Shikhare, I.D., Coursolle, D., Gralnick, J.A., Bond, D.R., 2008. *Shewanella* secretes flavins that mediate extracellular electron transfer. Proc. Natl Acad. Sci. USA. 105, 3968–3973.

Mohanakrishna, G., Venkata Mohan, S., 2013. Multiple process integrations for broad perspective analysis of fermentative H_2 production from wastewater treatment: technical and environmental considerations. Appl. Energ. 107, 244–254.

Mohanakrishna, G., Venkata Mohan, S., Sarma, P.N., 2010a. Bio-electrochemical treatment of distillery wastewater in microbial fuel cell facilitating decolorization and desalination along with power generation. J. Hazard. Mater. 177, 487–494.

Mohanakrishna, G., Goud, R.K., Venkata Mohan, S., Sarma, P.N., 2010b. Enhancing biohydrogen production through sewage supplementation of composite vegetable based market waste. Int. J. Hydrogen Energ. 35 (2), 533–541.

Mohanakrishna, G., Venkata Mohan, S., Sarma, P.N., 2010c. Utilizing acid-rich effluents of fermentative hydrogen production process as substrate for harnessing bioelectricity: an integrative approach. Int. J. Hydrogen Energ. 35, 3440–3449.

Mohanakrishna, G., Mohan, S.K., Venkata Mohan, S., 2012. Carbon based nano-tubes and nanopowder as impregnated electrode structures for enhanced power generation: evaluation with real field wastewater. Appl. Energ. 95, 31–37.

Moita, R., Lemos, P.C., 2012. Biopolymers production from mixed cultures and pyrolysis by-products. J. Biotechnol. 157 (4), 578–583.

Morsy, F.M., 2011. Hydrogen production from acid hydrolyzed molasses by the hydrogen overproducing *Escherichia coli* strain HD701 and subsequent use of the waste bacterial biomass for biosorption of Cd(II) and Zn(II). Int. J. Hydrogen Energ. 36 (22), 14381–14390.

Mu, Y., Rozendal, R.A., Rabaey, K., Yuan, Z., Keller, J., 2009a. Nitrobenzene removal in bioelectrochemical systems. Environ. Sci. Technol. 43, 8690–8695.

Mu, Y., Rabaey, K., Rene, A., Zhiguo, R., Yuan, Z., Keller, J., 2009b. Decolorization of azo dyes in bioelectrochemical systems. Environ. Sci. Technol. 43, 5137–5143.

Nakamura, M., Kanbe, H., Matsumoto, J., 1993. Fundamental studies on hydrogen production in the acid-forming phase and its bacteria in anaerobic treatment processes—the effects of solids retention time. Water Sci. Technol. 28, 81–88.

Navarro, R.M., Sanchez-Sanchez, M.C., Alvarez-Galvan, F., Valle, D., Fierro, J.L.G., 2009. Hydrogen production from renewable sources: biomass and photocatalytic opportunities. Energy Environ. Sci. 2, 35–54.

Neilson, A.H., Lewin, R.A., 1974. The uptake and utilization of organic carbon by algae: an essay in comparative biochemistry. Phycologia 13, 227–264.

Nevin, K.P., Woodard, T.L., Franks, A.E., Summers, Z.M., Lovley, D.R., 2010. Microbial electrosynthesis: feeding microbes electricity to convert carbon dioxide and water to multicarbon extracellular organic compounds. MBio 1 (2), e00103–001010.

Newman, D.K., Kolter, R.A., 2000. A role of excreted quinones in extracellular electron transport. Nature 405, 94–97.

Ntaikou, I., Kourmentza, C., Koutrouli, E.C., Stamatelatou, K., Zampraka, A., Kornaros, M., 2009. Exploitation of olive oil mill wastewater for combined biohydrogen and biopolymers production. Bioresour. Technol. 100 (15), 3724–3730.

Ntaikou, I., Antonopoulou, G., Lyberatos, G., 2010. Biohydrogen production from biomass and wastes via dark fermentation: a review. Waste Biomass Valor. 1, 21 39.

Ogawa, T., Aiba, S., 1981. Bioenergetic analysis of mixotrophic growth in *Chlorella vulgaris* and *Scenedesmus acutus*. Biotechnol. Bioeng. 23, 1121–1132.

Olguín, E.J., Giuliano, G., Porro, D., Tuberosa, R., Salamin, F., 2012. Biotechnology for a more sustainable world. Biotechnol. Adv. 30, 931–932.

Parameswaran, P., Torres, C.I., Lee, H.S., Krajmalnik-Brown, R., Rittmann, B.E., 2009. Syntrophic interactions among anode respiring bacteria (ARB) and non-ARB in a biofilm anode: electron balances. Biotechnol. Bioeng. 103 (3), 513–523.

Patel, S.K.S., Singh, M., Kumar, P., Purohit, H.J., Kalia, V.C., 2012. Exploitation of defined bacterial cultures for production of hydrogen and polyhydroxybutyrate from pea-shells. Biomass Bioenerg. 36, 218–225.

Pawinee, S., Rangsunvigit, P., Leethochawalit, M., Chavadej, S., 2011. Hydrogen production from alcohol distillery wastewater containing high potassium and sulfate using an anaerobic sequencing batch reactor. Int. J. Hydrogen Energ. 36 (20), 12810–12821.

Perez-Garcia, R.O., Bashan, Y., Puente, M.E., 2011. Organic carbon supplementation of municipal wastewater is essential for heterotrophic growth and ammonium removing by the microalgae *Chlorella vulgaris*. J. Phycol. 47 (1), 190–199.

Piyawadee, S., Reungsang, A., 2011. Biological hydrogen production from sweet sorghum syrup by mixed cultures using an anaerobic sequencing batch reactor (ASBR). Int. J. Hydrogen Energ. 36 (14), 8765–8773.

Qiu, C., Wen, J., Jia, X., 2011. Extreme-thermophilic biohydrogen production from lignocellulosic bioethanol distillery wastewater with community analysis of hydrogen-producing microflora. Int. J. Hydrogen Energ. 36 (14), 8243–8251.

Rabaey, K., Rozendal, R.A., 2010. Microbial electrosynthesis—revisiting the electrical route for microbial production. Nat. Rev. Microbiol. 8, 706–716.

Rabaey, K., Lissens, G., Siciliano, S.D., Verstraete, W., 2003. A microbial fuel cell capable of converting glucose to electricity at high rate and efficiency. Biotechnol. Lett. 25, 1531–1535.

Raghavulu, S.V., Venkata Mohan, S., Goud, R.K., Sarma, P.N., 2009a. Anodic pH microenvironment influence on microbial fuel cell (MFC) performance in concurrence with aerated and ferricyanide catholytes. Electrochem. Commun. 11, 371–375.

Raghavulu, S.V., Venkata Mohan, S., Venkateswar Reddy, M., Mohanakrishna, G., Sarma, P.N., 2009b. Behavior of single chambered mediatorless microbial fuel cell (MFC) at acidophilic, neutral and alkaline microenvironments during chemical wastewater treatment. Int. J. Hydrogen Energ. 34, 7547–7554.

Raghavulu, S.V., Sarma, P.N., Venkata Mohan, S., 2011. Bioelectrochemical behavior of Pseudomonas aeruginosa and Escherichia coli with the function of anaerobic consortia during biofuel cell operation. J. Appl. Microbiol. 110, 666–674.

Raghavulu, S.V., Babu, P.S., Goud, R.K., Subhash, G.V., Srikanth, S., Venkata Mohan, S., 2012. Bioaugmentation of an electrochemically active strain to enhance the electron discharge of mixed culture: process evaluation through electro-kinetic analysis. RSC Adv. 2, 677–688.

Rai, K.P., Singh, S.P., Asthana, R.K., 2012. Biohydrogen production from cheese whey wastewater in a two-step anaerobic process. Appl. Biochem. Biotechnol. 167, 1540–1549.

Ramanan, R., Kannan, K., Deshkar, A., Yadav, R., Chakrabarti, T., 2010. Enhanced algal CO_2 sequestration through calcite deposition by Chlorella sp. and Spirulina platensis in a mini-raceway pond. Bioresour. Technol. 101, 2616–2622.

Ramos, M.J., Fernandez, C.M., Casas, M., Rodriguez, L., Perez, A., 2009. Influence of fatty acid composition of raw materials on biodiesel properties. Bioresour. Technol. 100, 261–268.

Redwood, M.D., Macaskie, L.E., 2006. A two-stage, two-organism process for biohydrogen from glucose. Int. J. Hydrogen Energ. 31, 1514–1521.

Ren, N., Chen, Z., Wang, A., Hu, D., 2005. Removal of organic pollutants and analysis of MLSS-COD removal relationship at different HRTs in a submerged membrane bioreactor. Int. Biodeter. Biodegrad. 55, 279–284.

Ren, N.Q., Chua, H., Chan, S.Y., Tsang, Y.F., Wang, Y.J., Sin, N., 2007. Assessing optimal fermentation type for bio-hydrogen production in continuous flow acidogenic reactors. Bioresour. Technol. 98, 1774–1780.

Ringeisen, B.R., Ray, R., Little, B., 2007. A miniature microbial fuel cell operating with an aerobic anode chamber. J. Power Sources 165, 591–597.

Rodrigo, M.A., Canizares, P., Lobato, J., Paz, R., Saez, C., Linares, J.J., 2007. Production of electricity from the treatment of urban waste water using a microbial fuel cell. J. Power Sources 169, 198–204.

Rozendal, R., Hamelers, H.V.M., Euverink, G.J.W., Metz, S.J., Buisman, C.J.N., 2006. Principle and perspectives of hydrogen production through biocatalyzed electrolysis. Int. J. Hydrogen Energ. 31, 1632–1640.

Rozendal, R.A., Hamelers, H.V.M., Molenkamp, R.J., Buisman, C.J.N., 2007. Performance of single chamber biocatalyzed electrolysis with different types of ion exchange membranes. Water Res. 41, 1984–1994.

Sagnak, R., Kargi, F., 2011. Photo-fermentative hydrogen gas production from dark fermentation effluent of acid hydrolyzed wheat starch with periodic feeding. Int. J. Hydrogen Energ. 36 (7), 4348–4353.

Salehizadeh, H., Van Loosdrecht, M.C.M., 2004. Production of polyhydroxyalkanoates by mixed culture: recent trends and biotechnological importance. Biotechnol. Adv. 22, 261–279.

Salerno, M.B., Park, W., Zuo, Y., Logan, B.E., 2006. Inhibition of biohydrogen production by ammonia. Water Res. 40, 1167–1172.

Saratale, G.D., Chen, S., Lo, Y., Saratale, J.L.G., Chang, J., 2008. Outlook of biohydrogen production from lignocellulosic feedstock using dark fermentation—a review. J. Sci. Ind. Res. 67, 962–979.

Sarkar, O., Goud, R.K., Subash, G.V., Venkata Mohan, S., 2013. Relative effect of different inorganic acids on selective enrichment of acidogenic biocatalyst for fermentative biohydrogen production from wastewater. Bioresour. Technol. 147, 321–331.

Satoh, H., Mino, T., Matsuo, T., 1999. PHA production by activated sludge. Int. J. Biol. Macromol. 25, 105–109.

Saxena, R.C., Seal, D., Kumar, S., Goyal, H.B., 2008. Thermo-chemical routes for hydrogen rich gas from biomass: a review. Renew Sust. Energ. Rev. 12, 1909–1927.

Selembo, P.A., Merrill, M.D., Logan, B.E., 2009. The use of stainless steel and nickel alloys as low-cost cathodes in microbial electrolysis cells. J. Power Sources 190, 271–278.

Sen, B., Suttar, R.R., 2012. Mesophilic fermentative hydrogen production from sago starch-processing wastewater using enriched mixed cultures. Int. J. Hydrogen Energ. 37, 15588–15597.

Sentürk, E., Ince, M., Engin, O.G., 2010. Treatment efficiency and VFA composition of a thermophilic anaerobic contact reactor treating food industry wastewater. J. Hazard. Mater. 176 (1-3), 843–848.

Siddiqui, Z., Horan, N.J., Salter, M., 2011. Energy optimisation from co-digested waste using a two-phase process to generate hydrogen and methane. Int. J. Hydrogen Energ. 36 (8), 4792–4799.

Singh, A.P., Singh, N., Murphy, J.D., 2011. Renewable fuels from algae: an answer to debatable land based fuel. Bioresour. Technol. 102 (1), 10–16.

Sittijunda, S., Reungsang, A., 2012. Biohydrogen production from waste glycerol and sludge by anaerobic mixed cultures. Int. J. Hydrogen Energ. 37, 13786–13796.

Sivaramakrishna, D., Sreekanth, D., Himabindu, V., Anjaneyulu, Y., 2009. Biological hydrogen production from probiotic wastewater as substrate by selectively enriched anaerobic mixed microflora. Renew. Energ. 34, 937–940.

Sompong, O.T., Hniman, A., Prasertsan, P., Imai, T., 2011. Biohydrogen production from cassava starch processing wastewater by thermophilic mixed cultures. Int. J. Hydrogen Energ. 36 (5), 3409–3416.

Srikanth, S., Venkata Mohan, S., 2012a. Regulatory function of divalent cations in controlling the acidogenic biohydrogen production process. RSC Adv. 2, 6576–6589.

Srikanth, S., Venkata Mohan, S., 2012b. Change in electrogenic activity of the microbial fuel cell (MFC) with the function of biocathode microenvironment as terminal electron accepting condition: influence on overpotentials and bio-electro kinetics. Bioresour. Technol. 119, 241–251.

Srikanth, S., Venkata Mohan, S., 2012c. Influence of terminal electron acceptor availability to the anodic oxidation on the electrogenic activity of microbial fuel cell (MFC). Bioresour. Technol. 123, 471–479.

Srikanth, S., Venkata Mohan, S., Devi, M.P., Lenin Babu, M., Sarma, P.N., 2009. Effluents with soluble metabolites generated from acidogenic and methanogenic processes as substrate for additional hydrogen production through photo-biological process. Int. J. Hydrogen Energ. 34, 1771–1779.

Srikanth, S., Venkata Mohan, S., Lalit Babu, V., Sarma, P.N., 2010a. Metabolic shift and electron discharge pattern of anaerobic consortia as a function of pretreatment method applied during fermentative hydrogen production. Int. J. Hydrogen Energ. 35, 10693–10700.

Srikanth, S., Venkata Mohan, S., Sarma, P.N., 2010b. Positive anodic poised potential regulates microbial fuel cell performance with the function of open and closed circuitry. Bioresour. Technol. 101, 5337–5344.

Srikanth, S., Pavani, T., Sarma, P.N., Venkata Mohan, S., 2011. Synergistic interaction of biocatalyst with bio-anode as a function of electrode materials. Int. J. Hydrogen Energ. 36, 2271–2280.

Srikanth, S., Venkateswar Reddy, M., Venkata Mohan, S., 2012. Microaerophilic microenvironment at biocathode enhances electrogenesis with simultaneous synthesis of polyhydroxyalkanoates (PHA) in bioelectrochemical system (BES). Bioresour. Technol. 125, 291–299.

Steinbüchel, A., 1992. Biodegradable plastics. Curr. Opin. Biotechnol. 3, 291–297.

Subhash, G.V., Chandra, R., Venkata Mohan, S., 2013. Microalgae mediated bioelectrocatalytic fuel cell facilitates bioelectricity generation through oxygenic photomixotrophic mechanism. Bioresour. Technol. 133, 322–331.

Sun, M., Sheng, G.P., Mu, Z.X., Liu, X.W., Chen, Y.Z., Wang, H.L., Yu, H.Q., 2009. Manipulating the hydrogen production from acetate in a microbial electrolysis cell-microbial fuel cell-coupled system. J. Power Sources 191, 338–343.

Tai, J., Adav, S.S., Su, A., Lee, D.J., 2010. Biological hydrogen production from phenol-containing wastewater using Clostridium butyricum. Int. J. Hydrogen Energ. 35, 13345–13349.

Takagi, M., Watanabe, K., Yamaberi, K., Yoshida, T., 2000. Limited feeding of potassium nitrate for intracellular lipid and triglyceride accumulation of Nannochloris sp. UTEX LB1999. Appl. Microbiol. Biotechnol. 54, 112–117.

Tang, G.-L., Huang, J., Sun, Z.-J., Tang, Q.-Q., Yan, C.-H., Liu, G.-Q., 2008. Biohydrogen production from cattle wastewater by enriched anaerobic mixed consortia: influence of fermentation temperature and pH. J. Biosci. Bioeng. 106, 80–87.

Tenca, A., Schievano, A., Perazzolo, F., Adani, F., Oberti, R., 2011. Biohydrogen from thermophilic co-fermentation of swine manure with fruit and vegetable waste: maximizing stable production without pH control. Bioresour. Technol. 102 (18), 8582–8588.

Tuna, E., Kargi, F., Argun, H., 2009. Hydrogen gas production by electrohydrolysis of volatile fatty acid (VFA) containing dark fermentation effluent. Int. J. Hydrogen Energ. 34, 262–269.

Van Ginkel, S., Sung, S.W., Lay, J.J., 2001. Biohydrogen production as a function of pH and substrate concentration. Environ. Sci. Technol. 35, 4726–4730.

Van Ginkel, S., Oh, S., Logan, B.E., 2005. Biohydrogen gas production from food processing and domestic wastewaters. Int. J. Hydrogen Energ. 30, 1535–1542.

Vardar-Schara, G., Maeda, T., Wood, T.K., 2008. Metabolically engineered bacteria for producing hydrogen via fermentation. Microb. Biotechnol. 1, 107–125.

Vatsala, T.M., Mohan Raj, S., Manimaran, A., 2008. A pilot-scale study of biohydrogen production from distillery effluent using defined bacterial co-culture. Int. J. Hydrogen Energ. 33, 5404–5415.

Velvizhi, G., Venkata Mohan, S., 2011. Biocatalyst behavior under self-induced electrogenic microenvironment in comparison with anaerobic treatment: evaluation with pharmaceutical wastewater for multi-pollutant removal. Bioresour. Technol. 102, 10784–10793.

Velvizhi, G., Venkata Mohan, S., 2013a. In situ system buffering capacity dynamics on bioelectrogenic activity during the remediation of wastewater in microbial fuel cell. Environ. Prog. Sustain. Energ.. http://dx.doi.org/10.1002/ep.11809.

Velvizhi, G., Venkata Mohan, S., 2013b. Electrogenic activity and electron losses under increasing organic load of recalcitrant pharmaceutical wastewater. Int. J. Hydrogen Energ. 37, 5969–5978.

Velvizhi, G., Babu, P.S., Mohanakrishna, G., Srikanth, S., Venkata Mohan, S., 2012. Evaluation of voltage sag-regain phases to understand the stability of bioelectrochemical system: electro-kinetic analysis. RSC Adv. 2 (4), 1379–1386.

Venkata Mohan, S., 2008. Fermentative hydrogen production with simultaneous wastewater treatment: influence of pretreatment and system operating conditions. J. Sci. Ind. Res. 67 (11), 950–961.

Venkata Mohan, S., 2009. Harnessing of biohydrogen from wastewater treatment using mixed fermentative consortia: process evaluation towards optimization. Int. J. Hydrogen Energ. 34, 7460–7474.

Venkata Mohan, S., 2010. Waste to renewable energy: a sustainable and green approach towards production of biohydrogen by acidogenic fermentation. In: Singh, O.V., Harvey, S.P. (Eds.), Sustainable Biotechnology: Sources of Renewable Energy. Springer, the Netherlands, pp. 129–164.

Venkata Mohan, S., 2012. Harnessing bioelectricity through microbial fuel cell from wastewater. Renew. Energ. 5 (5), 25–29.

Venkata Mohan, S., Chandrasekhar, K., 2011a. Solid phase microbial fuel cell (SMFC) for harnessing bioelectricity from composite food waste fermentation: influence of electrode assembly and buffering capacity. Bioresour. Technol. 102, 7077–7085.

Venkata Mohan, S., Chandrasekhar, K., 2011b. Self-induced bio-potential and graphite electron accepting conditions enhances petroleum sludge degradation in bio-electrochemical system with simultaneous power generation. Bioresour. Technol. 102, 9532–9541.

Venkata Mohan, S., Devi, M.P., 2012. Fatty acid rich effluent from acidogenic biohydrogen reactor as substrate for lipid accumulation in heterotrophic microalgae with simultaneous treatment. Bioresour. Technol. 123, 627–635.

Venkata Mohan, S., Goud, R.K., 2012. Pretreatment of biocatalyst as viable option for sustained production of biohydrogen from wastewater treatment. In: Mudhoo, A. (Ed.), Biogas Production: Pretreatment Methods in Anaerobic Digestion. John Wiley & Sons, Inc., Hoboken, NJ.

Venkata Mohan, S., Lenin Babu, M., 2011. Dehydrogenase activity in association with poised potential during biohydrogen production in single chamber microbial electrolysis cell. Bioresour. Technol. 102, 8457–8465.

Venkata Mohan, S., Pandey, A., 2013. Biohydrogen. In: Pandey, A., Chang, J., Hallenbeck, P.C., Larroche, C. (Eds.), Biohydrogen Production: An Introduction. Elsevier, Burlington, pp. 1–24.

Venkata Mohan, S., Srikanth, S., 2011. Enhanced wastewater treatment efficiency through microbial catalyzed oxidation and reduction: synergistic effect of biocathode microenvironment. Bioresour. Technol. 102, 10210–10220.

Venkata Mohan, S., Venkateswar Reddy, M., 2013. Optimization of critical factors to enhance polyhydroxyalkanoates (PHA) synthesis by mixed culture using Taguchi design of experimental methodology. Bioresour. Technol. 128, 409–416.

Venkata Mohan, S., Lalit Babu, V., Sarma, P.N., 2007a. Anaerobic biohydrogen production from dairy wastewater treatment in sequencing batch reactor (AnSBR): effect of organic loading rate. Enzyme Microb. Technol. 41 (4), 506–515.

Venkata Mohan, S., Bhaskar, Y.V., Sarma, P.N., 2007b. Biohydrogen production from chemical wastewater treatment by selectively enriched anaerobic mixed consortia in biofilm configured reactor operated in periodic discontinuous batch mode. Water Res. 41, 2652–2664.

Venkata Mohan, S., Mohanakrishna, G., Raghavulu, S.V., Sarma, P.N., 2007c. Enhancing biohydrogen production from chemical wastewater treatment in anaerobic sequencing batch biofilm reactor (AnSBBR) by bioaugmenting with selectively enriched kanamycin resistant anaerobic mixed consortia. Int. J. Hydrogen Energ. 32, 3284–3292.

Venkata Mohan, S., Bhaskar, Y.V., Krishna, T.M., Rao, N.C., Lalit Babu, V., Sarma, P.N., 2007d. Biohydrogen production from chemical wastewater as substrate by selectively enriched anaerobic mixed consortia: influence of fermentation pH and substrate composition. Int. J. Hydrogen Energ. 32, 2286–2295.

Venkata Mohan, S., Lalit Babu, V., Srikanth, S., Sarma, P.N., 2008a. Bio-electrochemical evaluation of fermentative hydrogen production process with the function of feeding pH. Int. J. Hydrogen Energ. 33 (17), 4533–4546.

Venkata Mohan, S., Lalit Babu, V., Sarma, P.N., 2008b. Effect of various pretreatment methods on anaerobic mixed microflora to enhance biohydrogen production utilizing dairy wastewater as substrate. Bioresour. Technol. 99, 59–67.

Venkata Mohan, S., Mohankrishna, G., Sarma, P.N., 2008c. Integration of acidogenic and methanogenic processes for simultaneous production of biohydrogen and methane from wastewater treatment. Int. J. Hydrogen Energ. 33, 2156–2166.

Venkata Mohan, S., Mohanakrishna, G., Ramanaiah, S.V., Sarma, P.N., 2008d. Simultaneous biohydrogen production and wastewater treatment in biofilm configured anaerobic periodic discontinuous batch reactor using distillery wastewater. Int. J. Hydrogen Energ. 33 (2), 550–558.

Venkata Mohan, S., Mohanakrishna, G., Reddy, B.P., Sarvanan, R., Sarma, P.N., 2008e. Bioelectricity generation from chemical wastewater treatment in mediatorless (anode) microbial fuel cell (MFC) using selectively enriched hydrogen producing mixed culture under acidophilic microenvironment. Biochem. Eng. J. 39, 121–130.

Venkata Mohan, S., Mohanakrishna, G., Sarma, P.N., 2008f. Effect of anodic metabolic function on bioelectricity generation and substrate degradation in single chambered microbial fuel cell. Environ. Sci. Technol. 42, 8088–8094.

Venkata Mohan, S., Mohanakrishna, G., Srikanth, S., Sarma, P.N., 2008g. Harnessing of bioelectricity in microbial fuel cell (MFC) employing aerated cathode through anaerobic treatment of chemical wastewater using selectively enriched hydrogen producing mixed consortia. Fuel 87, 2667–2676.

Venkata Mohan, S., Saravanan, R., Veer Raghuvulu, S., Mohankrishna, G., Sarma, P.N., 2008h. Bioelectricity production from wastewater treatment in dual chambered microbial fuel cell (MFC) using selectively enriched mixed microflora: effect of catholyte. Bioresour. Technol. 99, 596–603.

Venkata Mohan, S., Veer Raghuvulu, S., Sarma, P.N., 2008i. Influence of anodic biofilm growth on bioelectricity production in single chambered mediatorless microbial fuel cell using mixed anaerobic consortia. Biosens. Bioelectron. 24, 41–47.

Venkata Mohan, S., Veer Raghuvulu, S., Dinakar, P., Sarma, P.N., 2009a. Integrated function of microbial fuel cell (MFC) as bio-electrochemical treatment system associated with bioelectricity generation under higher substrate load. Biosens. Bioelectron. 24, 2021–2027.

Venkata Mohan, S., Lenin Babu, M., Mohanakrishna, G., Sarma, P.N., 2009b. Harnessing of biohydrogen by acidogenic fermentation of Citrus limetta peelings: effect of extraction procedure and pretreatment of biocatalyst. Int. J. Hydrogen Energ. 34 (15), 6149–6156.

Venkata Mohan, S., Mohanakrishna, G., Goud, R.K., Sarma, P.N., 2009c. Acidogenic fermentation of vegetable based market waste to harness the biohydrogen. Bioresour. Technol. 100 (12), 3061–3068.

Venkata Mohan, S., Mohanakrishna, G., Velvizhi, G., Lalit Babu, V., Sarma, P.N., 2010a. Bio-catalyzed electrochemical treatment of real field dairy wastewater with simultaneous power generation. Biochem. Eng. J. 51, 32–39.

Venkata Mohan, S., Mohanakrishna, G., Sarma, P.N., 2010b. Composite vegetable waste as renewable resource for bioelectricity generation through non-catalyzed open-air cathode microbial fuel cell. Bioresour. Technol. 101, 970–976.

Venkata Mohan, S., Venkateswar Reddy, M., Venkata Subhash, G., Sarma, P.N., 2010c. Fermentative effluents from hydrogen producing bioreactor as substrate for poly (β-OH) butyrate production with simultaneous treatment: an integrated approach. Bioresour. Technol. 101, 9382–9386.

Venkata Mohan, S., Mohanakrishna, G., Srikanth, S., 2011a. Biohydrogen production from industrial effluents. In: Pandey, A., Larroche, C., Ricke, S.C., Dussap, C., Gnansounou, E. (Eds.), Biofuels: Alternative Feedstocks and Conversion Processes. Academic Press, Burlington, pp. 499–524 (Chapter 22).

Venkata Mohan, S., Agarwal, L., Mohanakrishna, G., Srikanth, S., Kapley, A., Purohit, H.J., Sarma, P.N., 2011b. Firmicutes with iron dependent hydrogenase drive hydrogen production in anaerobic bioreactor using distillery wastewater. Int. J. Hydrogen Energ. 36, 8234–8242.

Venkata Mohan, S., Chiranjeevi, P., Mohanakrishna, G., 2012. A rapid and simple protocol for evaluating biohydrogen production potential (BHP) of wastewater with simultaneous process optimization. Int. J. Hydrogen Energ. 37, 3130–3141.

Venkata Mohan, S., Srikanth, S., Velvizhi, G., Lenin Babu, M., 2013a. Microbial fuel cells for sustainable bioenergy generation: principles and perspective applications. In: Gupta, V.K., Tuohy, M.G. (Eds.), Biofuel Technologies: Recent Developments. Spinger, Berlin.

Venkata Mohan, S., Chandrasekhar, K., Chiranjeevi, P., Suresh Babu, P., 2013b. Biohydrogen. In: Pandey, A., Chang, J., Hallenbeck, P.C., Larroche, C. (Eds.), Biohydrogen Production from Wastewater. Elsevier, Burlington.

Venkata Mohan, S., Reddy, M.V., Chandra, R., Venkata Subhash, G., Devi, M.P., Srikanth, S., 2013c. Bacteria for bioenergy: a sustainable approach towards renewability. In: Gaspard, S., Ncibi, M.C. (Eds.), Biomass for Sustainable Applications: Pollution, Remediation and Energy. RSC Publishers, London.

Venkata Mohan, S., Devi, M.P., Venkata Subhash, G., Chandra, R., 2014. Algae oils as fuels. In: Pandey, A., Lee, D.J., Christi, Y. (Eds.), Biofuels from Algae. CJL School, Elsevier, pp. 155-187.

Venkata, Mohan S., Devi, M.P., 2014. Salinity stress induced lipid synthesis during dual mode cultivation of mixotrophic microalgae. Bioresour. Technol. http://dx.doi.org/10.1016/j.biortech.2014.02.103.

Venkateswar Reddy, M., Venkata Mohan, S., 2012a. Effect of substrate load and nutrients concentration on the polyhydroxyalkanoates (PHA) production using mixed consortia through wastewater treatment. Bioresour. Technol. 114, 573–582.

Venkateswar Reddy, M., Venkata Mohan, S., 2012b. Influence of aerobic and anoxic microenvironments on polyhydroxyalkanoates (PHA) production from food waste and acidogenic effluents using aerobic consortia. Bioresour. Technol. 103, 313–321.

Venkateswar Reddy, M., Srikanth, S., Venkata Mohan, S., Sarma, P.N., 2010. Phosphatase and dehydrogenase activities in anodic chamber of single chamber microbial fuel cell (MFC) at variable substrate loading conditions. Bioelectrochemistry 77, 125–132.

Venkateswar Reddy, M., Chandrasekhar, K., Venkata Mohan, S., 2011. Influence of carbohydrates and proteins concentration on fermentative hydrogen production using canteen based waste under acidophilic microenvironment. J. Biotechnol. 155, 387–395.

Venkateswar Reddy, M., Nikhil, G.N., Venkata Mohan, S., Swamy, Y.V., Sarma, P.N., 2012a. *Pseudomonas otitidis* as a potential biocatalyst for polyhydroxyalkanoates (PHA) synthesis using synthetic wastewater and acidogenic effluents. Bioresour. Technol. 123, 471–479.

Venkateswar Reddy, M., Nikhil, G.N., Venkata Mohan, S., Swamy, Y.V., Sarma, P.N., 2012b. *Pseudomonas otitidis* as a potential organism for polyhydroxyalkanoates (PHA) production using synthetic wastewater and acidogenic effluents. Bioresour. Technol. 123, 480–487.

Venkateswar Reddy, M., Amulya, K., Rohit, M.V., Sarma, P.N., Venkata Mohan, S., 2013. Valorization of fatty acid waste for bioplastics production using *Bacillus tequilensis*: integration with dark-fermentative hydrogen production process. Int. J. Hydrogen. Energy. 39, 7616–7626.

Vijaya Bhaskar, Y., Venkata Mohan, S., Sarma, P.N., 2008. Effect of substrate loading rate of chemical wastewater on fermentative biohydrogen production in biofilm configured sequencing batch reactor. Bioresour. Technol. 99, 6941–6948.

Vijayaraghavan, K., Ahmad, D., 2006. Biohydrogen generation from palm oil mill effluent using anaerobic contact filter. Int. J. Hydrogen Energ. 31, 1284–1291.

Wagner, R.C., Regan, J.M., Oh, S.E., Zuo, Y., Logan, B.E., 2009. Hydrogen and methane production from swine wastewater using microbial electrolysis cells. Water Res. 43 (5), 1480–1488.

Wang, J., Wan, W., 2008. Comparison of different pretreatment methods for enriching hydrogen-producing bacteria from digested sludge. Int. J. Hydrogen Energ. 33 (12), 2934–2941.

Wang, J., Wan, W., 2009. Factors influencing fermentative hydrogen production: a review. Int. J. Hydrogen Energ. 34, 799–811.

Wang, B., Wan, W., Wang, J., 2009. Effect of ammonia concentration on fermentative hydrogen production by mixed cultures. Bioresour. Technol. 100, 1211–1213.

Wang, A., Gao, L., Ren, N., Xu, J., Liu, C., Gao, G., et al., 2011. Isolation and characterization of *Shigella flexneri* G3 for effective cellulosic saccharification under mesophilic conditions. Appl. Environ. Microbiol. 77, 517–523.

Weyer, K.M., Bush, D.R., Darzins, A., Willson, B.D., 2009. Theoretical maximum algal oil production. Bioenerg. Res. 3, 204–213.

Wu, T.Y., Mohammad, A.W., Md Jahim, J., Anuar, N., 2009. A holistic approach to managing palm oil mill effluent (POME): biotechnological advances in the sustainable reuse of POME. Biotechnol. Adv. 27, 40–52.

Yadvika, S., Sreekrishnan, T.R., Kohli, S., Rana, V., 2004. Enhancement of biogas production from solid substrates using different techniques: a review. Bioresour. Technol. 95 (1), 1–10.

Yang, H., Shao, P., Lu, T., Shen, J., Wang, D., Xu, Z., Yuan, X., 2006. Continuous biohydrogen production from citric acid wastewater via facultative anaerobic bacteria. Int. J. Hydrogen Energ. 31, 1306–1313.

Yang, P., Zhang, R., McGarvey, J.A., Benemann, J.R., 2007. Biohydrogen production from cheese processing wastewater by anaerobic fermentation using mixed microbial communities. Int. J. Hydrogen Energ. 32, 4761–4771.

Yildiz, B., Kazimi, M.S., 2006. Efficiency of hydrogen production systems using alternative nuclear energy technologies. Int. J. Hydrogen Energ. 31, 77–92.

Yokoi, H., Tokushige, T., Hirose, J., Hayashi, S., Takasaki, Y., 1997. Hydrogen production by immobilized cells of aciduric *Enterobacter aerogenes* strain HO-39. J. Ferment. Bioeng. 83, 481–484.

Yokoi, H., Saitsu, A.S., Uchida, H., Hirose, J., Hayashi, S., Takasaki, Y., 2001. Microbial hydrogen production from sweet potato starch residue. J. Biosci. Bioeng. 91, 58–63.

Yokoyama, H., Ohmori, H., Waki, M., Ogino, A., Tanaka, Y., 2009. Continuous hydrogen production from glucose by using extreme thermophilic anaerobic microflora. J. Biosci. Bioeng. 107 (1), 64–66.

Yossan, S., Sompong, O.T., Prasertsan, P., 2012. Effect of initial pH, nutrients and temperature on hydrogen production from palm oil mill effluent using thermotolerant consortia and corresponding microbial communities. Int. J. Hydrogen Energ. 37, 13806–13814.

You, S., Zhao, Q., Zhang, J., Jiang, J., Zhao, S., 2006. A microbial fuel cell using permanganate as the cathodic electron acceptor. J. Power Sources 162, 1409–1415.

Young, S.D., Smeenk, J., Broer, K.M., Kisting, C.J., Brown, R., Heindel, T.J., et al., 2007. Growth of *Rhodospirillum rubrum* on synthesis Gas, conversion of CO to H_2 and Poly-*b*-hydroxyalkanoate. Biotechnol. Bioeng. 97, 279–286.

Yu, H., Zhu, Z., Hu, W., Zhang, H., 2002. Hydrogen production from rice winery wastewater in an upflow anaerobic reactor by using mixed anaerobic cultures. Int. J. Hydrogen Energ. 27, 1359–1365.

Zhang, Y., Dub, M.A., McLean, D.D., Kates, M., 2003. Biodiesel production from waste cooking oil: 2. Economic assessment and sensitivity analysis. Bioresour. Technol. 90, 229–240.

Zhang, Y.F., Liu, G.Z., Shen, J.Q., 2005. Hydrogen production in batch culture of mixed bacteria with sucrose under different iron concentrations. Int. J. Hydrogen Energ. 30, 855–860.

Zhu, H., Beland, M., 2006. Evaluation of alternative methods of preparing hydrogen producing seeds from digested wastewater sludge. Int. J. Hydrogen Energ. 31, 1980–1988.

Zhu, H., Stadnyk, A., Beland, M., Seto, P., 2008. Co-production of hydrogen and methane from potato waste using a two-stage anaerobic digestion process. Bioresour. Technol. 99, 5078–5084.

Zhu, H., Parker, W., Basnar, R., Proracki, A., Falletta, P., Bélanda, M., Seto, P., 2009. Buffer requirements for enhanced hydrogen production in acidogenic digestion of food wastes. Bioresour. Technol. 100 (21), 5097–5102.

Zhu, H., Parker, W., Conidi, D., Basnar, R., Seto, P., 2011. Eliminating methanogenic activity in hydrogen reactor to improve biogas production in a two-stage anaerobic digestion process co-digesting municipal food waste and sewage sludge. Bioresour. Technol. 102 (14), 7086–7092.

FURTHER READING

Becker, E.W., 1994. Microalgae—Biotechnology and Microbiology. Cambridge University Press, Cambridge.

Becker, W., 2004. Microalgae in human and animal nutrition. In: Richmond, A. (Ed.), Handbook of Microalgal Culture: Biotechnology and Applied Phycology. Blackwell Science, Oxford, pp. 312–351.

Blackader, W., Rensfelt, E., 1984. Synthesis gas from wood and peat: the mino process. In: Bridgwater, A.V. (Ed.), Thermochemical Processing of Biomass. Butterworth, London, pp. 137–149.

Hallenbeck, P.C., Abo-Hashesh, M., Ghosh, D., 2012. Strategies for improving biological hydrogen production. Bioresour. Technol. 110, 1–9.

Logan, B.E., Call, D., Cheng, S., Hamelers, H.V.M., Sleutels, T.J.A., Jeremiasse, A.W., Rozendal, R.A., 2008. Microbial electrolysis cells for high yield hydrogen gas production from organic matter. Environ. Sci. Technol. 42 (23), 8630–8640.

Mohanakrishna, G., Subhash, G.V., Venkata Mohan, S., 2011. Adaptation of biohydrogen producing reactor to higher substrate load: redox controlled process integration strategy to overcome limitations. Int. J. Hydrogen Energ. 36, 8943–8952.

Momirlan, M., Veziroglu, T.N., 2002. Current status of hydrogen energy. Renew. Sust. Energ. Rev. 6, 141–179.

Venkata Mohan, S., Devi, M.P., Mohanakrishna, G., Amarnath, N., Lenin Babu, M., Sarma, P.N., 2011c. Potential of mixed microalgae to harness biodiesel from ecological water-bodies with simultaneous treatment. Bioresour. Technol. 102, 1109–1117.

CHAPTER 7

Urban Wastewater Treatment for Recycling and Reuse in Industrial Applications: Indian Scenario

R. Saravanane[1], Vivek V. Ranade[2], Vinay M. Bhandari[2], A. Seshagiri Rao[3]

[1]Environmental Engineering Laboratory, Department of Civil Engineering, Pondicherry Engineering College, Pondicherry, India
[2]Chemical Engineering and Process Development Division, CSIR-National Chemical Laboratory, Pune, India
[3]Department of Chemical Engineering, National Institute of Technology, Warangal, Andhra Pradesh, India

7.1 INTRODUCTION

The availability and quality of water resources is one of the major environmental challenges facing the increasing demands of domestic and industrial sectors of India. The stress for water resources come from various levels and can impact the environment in diverse ways. A rapid increase in population together with urbanization, industrialization, and agricultural development has increased the impact on the quantity and quality of water resources. The situation has resulted in many risks to the quality of water in various forms. The availability of fresh water resources in India has been declining over the years because of the increase in population growth and improper management without recycling and reusing wastewater (Bharadwaj, 2005). On average, the combination of three resources—rainfall, surface water, and groundwater—have been sufficient to provide adequate water to the Indian population. However, many modern Indian cities are beginning to experience water shortages because of the various effects of urbanization, industrialization, and agricultural growth. These shortages are expected to manifest more explicitly and rapidly in the coming years. They can be mitigated only if the wastewater treatment, reuse, and recycling are adopted in appropriate sectors.

This chapter assesses the potential applicability of recycling and reuse of urban wastewater and explores options for industrial applications in the Indian scenario. The chapter is organized into various subtopics covering significant details of recycling and reuse, namely, the current water status

283

of India, sewage treatment options, industrial water reuse via recycling, urban–industry joint venture, and urban–industrial water sustainability in 2030. This presents a major advance in the current preparedness to reach a long-term sustainability goal of the water sector at large. It also highlights the cost effectiveness of recycling and reusing urban wastewater for industrial water production through a few case studies that are in current operation. This would be useful to practicing engineers in industry, industrialists using various water-intensive industrial manufacturing and production operations, pollution control authorities and monitoring engineers, field engineers of the urban and municipal water sector, water regulatory authorities/boards, and academia, and, as a whole, to the water resources engineering domain.

7.2 URBAN WATER SECTOR: INDIAN SCENARIO

Sewage wastewater is generated mainly by domestic use such as household water usage. The generation of such wastewater is highly dependent on people and varies significantly from place to place. It is customary to specify the sewage generation figures on the basis of *per capita* use of water, which is city/ place of habitation specific predominantly on the basis of habits of people, water availability, and the water sanitation system. There are major cities where a large population lives with high levels of *per capita* water consumption along with systematic sewage systems for the collection and disposal of sewage wastewaters. In contrast, there are rural areas that have small populations and a total lack of proper sewage systems. Thus, sewage wastewater characterization can vary substantially; therefore, this discussion applies only to general sewage wastewater from a reasonably populated place with a proper sewage system. Also, it is assumed that the industrial wastewaters are not mixed with the municipal wastewaters.

Typical characteristics of municipal wastewaters are given in Table 7.1 (http://nptel.iitm.ac.in/courses/105105048/M11_L14.pdf; www.cpcb. nic.in/newitems/12.pdf—CPCB Report, 2005).

Sewage wastewaters mostly contain organics, dissolved solids, suspended solids, nitrogen, phosphorus, chlorides/salts, and metals. It is generally observed that the sewage wastewater pH is in the range of 5.5–8, and its appearance ranges from cloudy to dark colored. Many times, there is an appreciable foul odor, which is attributed to the microbial presence in these wastewaters. In Indian systems, the temperatures vary from place to place; however, the typical range for most cities is from 15 to 30 °C for

Table 7.1 Typical characteristics of sewage wastewater in india

pH	5.5–8	
Total solids	~0.1%	
TSS	100–700	mg/L
TDS	200–900	mg/L
BOD	100–400	mg/L
COD	200–700	mg/L
Nitrogen content	20–90	mg/L
Phosphorus content	5–20	mg/L
Chlorides	30–90	mg/L

an appreciable period of the year, barring the summer period when temperatures soar close to 40 °C or even higher. The temperature is an important parameter. Most municipal wastewaters are biologically treated using the conventional aerobic process-activated sludge process (ASP). Temperature here plays crucial role in the treatment because biological activity is strongly temperature dependent, thereby directly affecting the efficiency of the biological processes. The lower the temperature, the lower the efficiency of the biodegradation, thereby adversely affecting the process. The organics present in the wastewaters need to be decomposed suitably and constitute the major load on the biological treatment process. Typically, the organics in the sewage wastewaters are biodegradable and do not require chemical oxidation for making water suitable for discharge. The inorganic fraction, particularly chlorides/salts, can be a major problem if the water is to be further used for farming/irrigation because the salt presence can adversely affect the land/soil quality and fertility. There are some phosphorus compounds present in wastewaters that normally come from the use of soaps and detergents apart from food residues. The nitrogen content in the wastewaters is mainly in the form of proteins and amino acids. Both nitrogen and phosphorus are useful nutrients for the bacteria present in the biological processes. However, salts of these two elements need monitoring, especially if the water is to be reused.

The chemical oxygen demand (COD) for sewage wastewaters is not very high, generally of the order of 200–700 mg/L. This is in stark contrast to industrial wastewaters where typically the COD values range from 2000 to 4000 mg/L for textile wastewaters; 30,000 to 40,000 mg/L for dye wasters, and as high as 150,000 to 200,000 mg/L for distillery wastewaters. The biological oxygen demand (BOD) for sewage wastewaters is typically below 400 mg/L. Further, the ratio of COD to BOD is generally of the order of 1.7, although some variation is found depending on the nature of the sewage. In India, the domestic consumption of water is rising practically at the same rate as that for industrial water and is likely to triple between 2000 and 2050. Further, only ~31% of the sewage wastewater generated from the metro cities is treated, and thus untreated sewage is a major factor contributing to severe water pollution. The available data on the major metro cities indicate sewage wastewater generation of ~38,400 MLD (Million Litres per Day) with sewage treatment capacity of only ~12,000 MLD. It is believed that apart from these metro cities, the treatment facilities for sewage waters in other cities could be as low as 10% or even less.

The untreated sewage waters largely find their way out through the rivers close by. The Central Pollution Control Board (CPCB, 2005; CPCB, 2010), India, has identified polluted regions in 18 major rivers and, not surprisingly, these are located close to urban areas (water bodies with BOD higher than 6 mg/L are considered polluted). The data on river pollution indicate severe pollution in many of the major rivers and nearly 14% of total river length is believed to be severely polluted in India, ~20% being moderately polluted based on the BOD levels. Apart from high BOD and COD levels, high levels of heavy metals, arsenic, fluorides, and hazardous chemicals are also found in many places, especially in groundwater.

In recent years, a number of technologies have been made available, especially using biological treatment methods for sewage wastewaters. These are:
1. Membrane bioreactor (MBR)
2. Fluidized aerobic bed
3. Fluidized aerated bed reactor
4. Submerged aeration fixed-film reactor
5. Biological filter oxygenated reactor
6. Anaerobic filter
7. Expanded granular sludge blanket
8. Sequencing batch reactor
9. Up-flow anaerobic sludge blanket (UASB).

7.2.1 Water Requirements of the Urban Population

It is estimated that about 50 billion liters of municipal water is required every day based on the urban Indian population, which accounts for about 360 million people. By 2050, the urban population will exceed around 850 million, and the estimated water requirements will be around 110 billion liters per day. The rural population, which is estimated to be around 1.1 billion in 2050, will require about 44 billion liters per day (CWC, 2005). Annually, more and more people are moving into the cities, and the figures are expected to reach about 600 million by 2030, making India more semi-urban than rural. Already, there is enormous pressure to provide utility services and industrial use, so water supply is a priority, especially where semi-urban water is exported formally or informally to fulfill city requirements. At the same time, the urban return flow (wastewater) is also increasing, and is usually about 70–80% of the water supply. This study attempted to estimate the current status of wastewater generation, its uses, and livelihood benefits especially in agriculture, based on national data and case studies from specific regions and cities such as Ahmedabad, New Delhi, Hyderabad, Kanpur and Kolkata.

The challenge of the growing Indian economy is that, in many cities, the wastewater processed is a mixture of domestic and industrial wastewater, which makes the system complicated, and resulting reuse remains a challenge. Lack of systematic data on the different discharges makes it difficult to estimate the volume and quality of wastewater discharged and the total area under (usually informal) wastewater irrigation. Data from more than 900 Class I cities and Class II towns (with the population of each over 1 million and between 0.5 and 1 million, respectively) have shown that more wastewater gets collected than eventually treated. In general, wastewater generation is around 60–70% over the established treatment capacity, which varies from city to city.

7.2.2 Urban Water Supply System

Urbanization and rapid industrialization have revamped the present ideas on the reversible nature of water resources. A large population of India does not have access to safe water. Treatment of wastewater, sewage treatment, industrial liquid and chemical waste treatment, and sludge from desalination (Indian Desalination Association INDACON, 2008) requires a cost-effective sustainable solution facilitating a lead on zero liquid discharge (ZLD) with carbon neutral mechanisms. The total water requirements for 2050 and 2065, as estimated by the government in 1999 with revisions based on 2011, predicts that India would face a water demand of 90 bcm in 2050 as compared to 20 bcm by the Water Commission's predictions (Garg and Hassan, 2007). With the large, growing demand for water, India would experience a huge deficit in water and energy in 2030 (Ministry of Water Resources, 2008). A worldwide trend toward acceptance of the concept of reuse is currently visible, because water shortages are intensifying in India

(Ministry of Water Resources, 2003). Almost 80% of the water supplied for urban sectors comes as wastewater, which will lead to pollution in surface and groundwater resources, creating a troublesome situation for the supply systems. A critical assessment of the urban water supply system would reveal ample scope for incorporating recycling and reuse of treated urban wastewater to augment reversing the deficit of water scarcity.

7.2.3 Urban Sewerage System

The current practice of sewerage system in India uses manholes and sewer lines as collection and transport to the treatment plant. However, with year-on-year increase in population, the system results in higher operational and maintenance costs, despite reasonable efficiency under Indian conditions. The lower occupancy of population causes a reduction in the fully-fledged layout systems and gives rise to many hazards to the environment because of poor transport conditions resulting in poor operational schemes. The current *per capita* water supplies are not even sufficient to meet the minimum *per capita* demand of 135 L to satisfy the self-cleaning velocity in the sewer lines, and the conventional sewerage systems are not capable of meeting required efficiency. It is estimated that about 38,254 million liters per day of wastewater is generated in tier 1 and tier 2 cities, which cover about 31% of the total sewage generation considering both sewered and unsewered areas (NCIWRDP/MWR, 1999). An increase in population will also give rise to an increase in water supplies and wastewater generation. About 80% of the water supplied is generated as wastewater. From the GOI 2010 data, it is estimated that around 120 million m^3 of sewage will be generated during the projected year of 2051 with additional wastewater generation around 50 million m^3 of increasing water supplies from rural and community areas. However, the management plan for the sewerage systems lack strategies for handling wastewater in the future. The CPCB India studies show that there are 269 sewage treatment plants (STPs) in India, out of which only 231 plants are in operating condition, and they cover only 21% of the country's total sewage generation (MoUD, 2013).

7.2.4 Wastewater Treatment: Recycling and Reuse Option

The volume of wastewater generated increases with the increase in population, urbanization, and people's improved living standards. The productive use of wastewater has also been increased; in developing countries millions of small-scale farmers use wastewater for high-value crops because there is no alternative irrigation (Asano et al. 1985). In a conventional process setup, sewage is collected in a large network of pipes usually known as sewers. It is then transported to the centralized treatment plants and treated. Efforts are also being directed at treating sewage by a decentralized method, whereby it is treated locally with the aim of reusing the treated wastewater (CPCB, 2009). Depending on the characteristics of the sewage, various techniques have been used in various treatment plants operating under favorable local conditions. In most of the Class I cities, the ASP is commonly used in sewage treatment. This covers around 59.5% of the total installed capacities; upflow anaerobic methods cover about 26% of the installed capacity.

Sector demands for water are reaching new heights where irrigation, household supply, energy, and industry seek increased volumes to meet growing needs. The 2050 projections for India report that there will be a requirement of 1447 cubic kilometers (km^3) of water, of which 74% is identified for irrigation purposes, while the rest is for drinking water (7%), industry (4%), energy (9%), and other uses (6%; CPCB, 2009). However, with rapid urban growth with 498 Class I cities and 410 Class II towns (GOI, 2008), the demand for drinking water is also rising and has a high priority, competing with rural water needs, including irrigation. The current water supply to these cities is estimated at about 48 million m^3 per day and is projected to increase further with the increased demand from different sectors (CPCB, 2009). A large number of these growing cities are located in major river basin catchments, taking fresh water away and discharging wastewater back into the catchments, thus polluting irrigation water. The urban return flow is seen not only as a hazard but also as an asset. From the past experience of urban/municipal wastewater treatment, it is quite evident that the industrial sector requires a huge volume of fresh water for increased productivity (Asano and Levine, 1996). The withdrawal of fresh water by industries further depletes the availability of resources. Realizing this, and for long-term sustainability, urban sectors can join with industry to share its water potential from treated secondary sewage or cooling water for primary/secondary industrial uses. Newer technologies offering significantly higher removal rates should be designed and implemented. Membrane technologies, which were formerly restricted to water desalination applications, are now being tested for the production of high-quality water for indirect potable reuse, and they are expected to lead current treatment technologies in the near future.

7.2.5 Water Balance for India

The water balance for India is estimated on the basis of average annual precipitation of India which is around 4000 km^3. The precipitation contributes nearly 2150 km^3 to ground water and 1150 km^3 to surface runoff. The available total water resources have been estimated to 1953 km^3. Out of 1953 km^3, Ganga-Brahmaputra-Maghna basin shares 1202 km^3 and the remaining basins share 751 km^3. (Hydrology and Water Resources System for India, 2007).

The utilisable part of the water resources is estimated to 1122 km^3 from the total annual water availability in India. An additional demand of 123–169 km^3 is projected to be required by 2050 anticipating the increased usage in irrigation, domestic and industrial sectors. The per capita availability of water was seen to decrease from 3000 m^3 in 1951 to 1100 m^3 in 1998. It is predicted that the per capita water availability will further shrink to 687 m^3 by 2050. The *per capita* water availability is presented in Table 7.2.

7.2.5.1 Atmospheric Water Balance

Under Indian condition, the equation for atmospheric water balance is obtained by equating inflow to the sum of outflow and change in storage and is given below:

$$V_I + E_T + V_{AI} = P + V_0 + V_{AE}$$

where V_I is inflow of water vapor that reaches to the Indian atmosphere from land and sea routes; E_T, total quantity of evapotranspiration; V_{AI}, initial water vapor content of atmosphere; P, total precipitation; V_0, outgoing water vapor content; and V_{AE}, the water vapor content present at the end of period under consideration.

A schematic of the atmospheric water balance is shown in Table 7.3 (*Source*: Hydrology and Water Resources System for India). It is inferred from Table 7.3 that the inflow of water vapour should be large enough to maintain the balance between natural vapourisation and precipitation.

7.2.5.2 Hydrologic Water Balance

The hydrologic water balance of India can be deduced from average annual conditions and the pertinent equation is given below:

Table 7.2 *Per capita* availability of water (http://www.nih.ernet.in/rbis/india_information/Water%C2%A0Budget.htm)

Year	1951	1991	2010	2025	2050
Population (Million)	361	846.3	1157	1333	1581
Average water resources (m^3/capita/year)	3008	1283	938	814	687

Source: Hydrology and Water Resources System for India.

Table 7.3 Atmospheric water balance of India (http://www.nih.ernet.in/rbis/india_information/Water%C2%A0Budget.htm)

Inflow		Outflow	
Parameter	Value	Parameter	Value
V_{AI}	13	P	400
V_I	1440	V_o	1318
E_T	278	V_{AE}	13
Total	1731	Total	1731

All components in (10^{10}) m^3.
Source: Hydrology and Water Resources system for India.

$$P + I = Q_s + E_T + Q_g + \Delta S$$

where P is total precipitation; E_T, total evapotranspiration; I, sum of surface water (I_s) and groundwater (I_g) flow; Q_s surface water outflow to oceans and other countries; (Q_g) ground outflow; and ΔS soil moisture change.

A diagrammatic representation of the hydrologic water balance for India is shown in Figure 7.1. The hydrologic water balance of India is shown in Table 7.4. It is assessed from Table 7.4 that if the precipitation reaches a low value, artificial augmentation of water resources through recharge, recycle and reuse would be an optimal choice for sustaining water balance of India.

7.2.6 Water Balance: Convergence to Recycling and Reuse

The infrastructural development for sewage and wastewater treatment has not kept pace with wastewater generation. As a result, vast amounts of polluted water are being discharged into natural waterways, with pollutants well above the permissible levels (Lorenzen et al., 2010). In many Indian cities, the wastewater discharges comprise domestic and industrial wastewater and are often mixed and not separately accounted for in wastewater reuse. Lack of systematic record-keeping of the different discharges makes it difficult to arrive at reasonable estimates of wastewater discharge and its quality (Heggade, 1998; Misra, 1998). For the period 1947–1997, a sixfold increase in wastewater generation was recorded in Class I cities and Class II towns. Current generation for Class I cities and Class II towns is more than 38,000 million liters per day, out of which only 35% is treated (CPCB, 2009). Conservation, augmentation, and recycling of urban water are major foci in India's national water policy (CIA, 2009). The policy also advocates the reuse of treated sewage in view of the looming future water scarcity. Thus, the policy support for reuse of treated wastewater from STPs is inherently embedded in the overall water policy of India, despite the shortcomings at the implementation level. The Ganga Action Plan was one of the

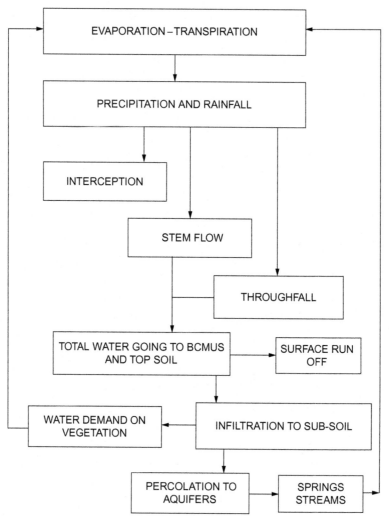

Figure 7.1 Hydrological balance for water in India. *Source: Hydrology and Water Resources System for India.*

first restoration plans for water bodies; it commenced in 1985 and led to a larger program, bringing the entire country under the National River Action Plan. In this program, the identification of pollution sources, interception, or diversions and treatment were planned for 157 major cities along the main rivers. However, fast urbanization and industrialization have outpaced the installation of STPs and regulatory processes with only marginal improvements in the field. Domestic sewage and industrial waste are the major causes of deterioration of water quality and contamination of lakes, rivers, and

Table 7.4 Hydrologic water balance of India (http://www.nih.ernet.in/rbis/india_information/Water%C2%A0Budget.htm)

Inflow		Outflow	
Term	**Value**	**Term**	**Value**
P	400	E_T	278
I_s	20	Q_s	126
I_g	4	Q_g	20
Total	424	Total	424

All components in (10^{10}) m^3 per year.
Source: Hydrology and Water Resources System for India.

groundwater aquifers. Septic tanks constitute one of the most common forms of urban sanitation facilities in India. The major parts of urban India have not been connected to a municipal sewer system, which makes people dependent on the conventional individual septic tank. Access to improved sanitation in urban India has increased, but the management of on-site sanitation systems, such as septic tanks, remains a neglected component of urban sanitation and wastewater management. Sewage generated lacking treatment facilities contributes to 80% of the pollution of the national surface waters (CSE, 2011, 2012). Based on water pollution, five different classes of water quality have been identified. Data show that, from an Indian river 45,000 km in length, 6000 km had a $BOD_{3.27}$ above 3 mg/L, making the water unfit for drinking. Matters relating to sewage treatment as well as the drinking and industrial water supply are dealt with at the state level while the municipal authorities of cities are responsible for providing these services. Currently, only the networked sewage systems are targeted for treatment, while the vast nonpoint source discharges go undetected and untreated. Therefore, the pollution loads in rivers are highly variable, depending on the season, modulated by rainfall, sewage, and solid waste management practices in towns and cities, and types of industry in the proximity. While the regulatory mechanisms have been outlined, uncontrolled industrial discharges contribute to heavy environmental pollution and potential health hazards (Rawat et al., 2009). Hence the order of the day would be to reassess and redesign the whole urban wastewater treatment system with a perspective to recycling and reuse for industrial and other domestic applications.

7.2.7 Urban Sewage Quality and Quantity

The total water requirement for the industries was around 11 km^3/year for the base year 1996–1997. It was observed that the pulp and paper, integrated iron and steel, and textile subsectors account for approximately 60% of the water requirements of industry. For future projections, the water requirements for each subsector have been separately calculated. The overall water requirements for 2010, 2025, and 2050 (Table 7.5) have been estimated as 37, 67, and 81 km^3, respectively. It needs to be stated here that until the 1990s, the Indian economy was on a slow growth path (*Source*: Hydrology and Water Resources System for India, 2007).

7.2.7.1 Sewage Generation and Existing Treatment Capacity

Some critical field-oriented results are presented to understand the urban population effect on urban sewage and are as follows:

- A rapid population growth in terms of urban to total population shows 2.1% increase over the decade from 1991 to 2001 (CPCB, November 2005) (*Source*: Status of Sewage Treatment in India, CPCB, November 2005). In a status report on the STPs of Class I cities and Class II towns, it is seen that out of 269 STPs, 231 are operational and 38 are under construction.
- Pollution of surface water bodies has been realized due to STPs of Class I cities and Class II towns, necessitating a complete secondary treatment for recycling and reuse.

Out of total sewage generation of 29,129 MLD from Class I cities and Class II towns, only 6190 MLD was installed and 22,939 MLD of sewage remains as a gap for further planning and scope. The pertinent details of the sewage generation and treatment capacity in Class I cities and Class II towns are given in Figure 7.2.

A schematic view of the statewise gap in sewage generation and installed treatment capacity in Class I cities is given in Figure 7.3. The following data analysis further explains the scope for the gap in capacity to be in the installed in urban areas.

- Based on the 2001 census population for Class I cities (CUPS/44/1999–2000), the estimation of sewage generation is presented for the Eastern, Northern, Central, Western, and Southern regions of the country. The Western region shows around 8605 MLD of sewage generation, whereas the other regions show a marginal difference ranging from 4000 to 5000 MLD.
- The installed treatment capacity attained a sewage discharge of 3000–4000 MLD in the Western region, 3000 MLD in the North, and around 1000 MLD in East, Central, and South.
- Hence it is evident that the Eastern, Central, and Southern regions leave scope for newer perspectives with modern STPs and to opt for recycling and reuse schemes for industrial water production.

Table 7.5 Water requirement for different industries for 2010, 2025, and 2050

Category of industry	Year 1997 (1000 tons)	Growth rate (%)	Water requirement per unit of production (m³) 1997–2010	Water requirement (km³)		
				2010	2025	2050
Integrated Iron & Steel	1,19,390.00	9.00	22.00	5.84	5.74	10.94
Smelters	174.00	5.80	82.50	2.41×10^{-2}	3.18×10^{-2}	4.30×10^{-2}
Petrochemicals & Refinery	1255.00	3.00	17.00	3.06×10^{-2}	3.55×10^{-2}	4.90×10^{-2}
Textile & Jute	24,730.20	11.57	200.00	19.02	36.52	35.19
Fertilizer	11,155.30	8.00	16.70	6.31×10^{-1}	1.03	1.19
Leather Products	912.50	5.80	40.00	8.76×10^{-1}	8.99×10^{-2}	1.43×10^{-1}
Food	96,450.00	14.00	11.00	5.57	8.04	8.32
Inorganic Chemicals	2200.00	7.00	200.00	1.60	3.35	3.01
Pharmaceuticals	4190.00	12.00	22.00	1.84×10^{-1}	2.43×10^{-1}	3.43×10^{-1}
Distillery	1790.80	4.20	22.00	6.73×10^{-2}	9.80×10^{-2}	1.17×10^{-1}
Paper & Pulp	3100.00	4.00	280.00	2.90	10.19	18.90
Average		7.67	Total	34.93	64.87	77.55

Source: Hydrology and Water Resources System for India.

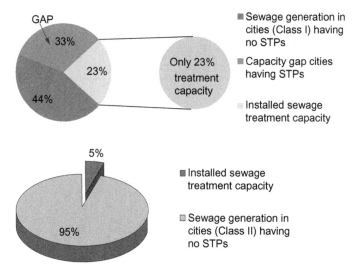

Figure 7.2 Sewage generation and treatment capacity in Class I cities and Class II towns. *Source: Status of Sewage Treatment in India, CPCB, November 2005.*

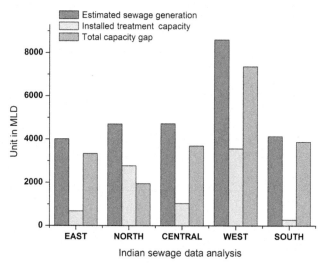

Figure 7.3 Statewise gap in sewage generation and installed treatment capacity in Class I cities. *Source: Status of Sewage Treatment in India, CPCB, November 2005.*

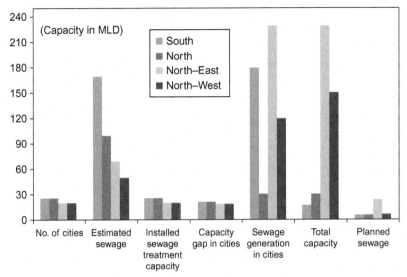

Figure 7.4 Statewise gap in sewage generation and installed treatment capacity in Class II towns. *Source: Status of Sewage Treatment in India, CPCB, November 2005.*

A statewise gap in sewage generation and installed treatment capacity in Class II towns is represented in Figure 7.4. The estimated sewage generation reached a sewage capacity of 280 MLD in the Northern region; 180–200 MLD in the Western region; 150–200 MLD in the Central region; around 100 MLD in the Eastern region, and around 200 MLD in the Southern region. The installed capacity is in the range of 10–12% of the estimated capacity. The data could be visualized in terms of the potential availability of secondary sewage. Hence the STPs in Class II towns could reveal an opportunity to exercise new choices and options for utilizing the potential availability of secondary sewage for reuse applications.

7.2.7.2 Water Quality Requirement for Different Uses

For any water body to function adequately in satisfying the desired use, it must have a corresponding degree of purity. Because the magnitude of demand for water is fast approaching the available supply, the concept of management of water quality is becoming as important as its quantity. Water quality for various uses is listed in Table 7.6.

The price of water, just as for other commodities, can be determined using the demand and supply curves. Figure 7.5 shows the supply curve (*S*) and demand curve (*D*) for water. Their point of intersection gives the

Table 7.6 Designated best uses of water

Designated best use	Class	Criteria
Drinking water source without conventional treatment but after disinfection	A	1. Total coliform organism MPN/100 mL shall be 50 or less 2. pH between 6.5 and 8.5 3. Dissolved oxygen 6 mg/L or more 4. Biochemical oxygen demand 5 days 20 °C, 2 mg/L or less
Outdoor bathing (organized)	B	1. Total coliform organism MPN/100 mL shall be 500 or less 2. pH between 6.5 and 8.5 3. Dissolved oxygen 5 mg/L or more 4. Biochemical oxygen demand 5 days 20 °C, 3 mg/L or less
Drinking water source after conventional treatment and disinfection	C	1. Total coliform organism MPN/100 mL shall be 5000 or less 2. pH between 6 and 9 3. Dissolved oxygen 4 mg/L or more 4. Biochemical oxygen demand 5 days 20 °C, 3 mg/L or less
Propagation of wildlife and fisheries	D	1. pH between 6.5 and 8.5 2. Dissolved oxygen 4 mg/L or more 3. Free ammonia (as N) 4. Biochemical oxygen demand 5 days 20 °C, 2 mg/L or less
Irrigation, industrial cooling, controlled waste disposal	E	1. pH between 6.0 and 8.5 2. Electrical conductivity at 25 °C micro mhos/cm, maximum 2250 3. Sodium absorption ratio maximum 26 4. Boron maximum 2 mg/L
	Below-E	Not meeting any of the A, B, C, D, and E criteria

Source: Hydrology and Water Resources Information System for India & CPCB.

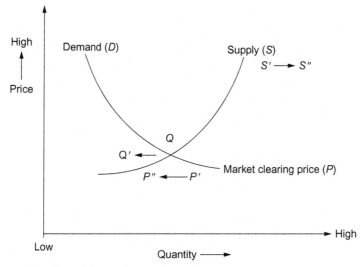

Figure 7.5 Supply and demand curve for water.

quantity of water (Q) that should be supplied at price P under normal market conditions with no government interventions. Whenever there is a scarcity of water because of higher demand (population increase) or lower rainfall, or when the government decides to conserve water for future use, it needs to reduce the present supply (from Q to Q′). Accordingly, P′ gives the scarcity price of water. To reduce the gap between supply and demand, water boards and communities are exploring a number of alternate water options to complement the existing urban water supplies, including wastewater recycling, rainwater harvesting, storm water recycling, exploring new groundwater sources, diverting agriculture water to cities, and the construction of new dams. If any or a combination of these sources are tapped, the supply curve will move toward the right (S″), and urban people can pay a lower price, P′. Wastewater, if treated to appropriate levels, has a huge potential to complement the existing water sources and bring down the price from P′ to P″. However, the costs of treatment need to be deducted from the benefits to realize the net profit from recycling and water markets.

7.2.8 Urban Water Market

The urban water market and pricing have the following components:
(1)   Cost of water supply: In most OECD (Organization for Economic Cooperation and Development) countries and in the United States water pricing is based on average cost pricing or marginal cost pricing.

The consumers are charged at a rate of per kiloliter of water consumed. This rate varies depending on the pricing structure in each city.

(2) Cost of maintenance of sewerage services: In most cities around the world, water boards are also responsible for maintaining the sewerage system and consumers are charged for this service. In Hyderabad, nearly 35% of the water supply charge is charged as sewage cess. Similarly, in New Delhi, 50% of water supply charge is charged as sewage cess.

(3) Cost of treatment of sewage water or wastewater discharged by households and industries: Urban consumers in Indian cities are not charged for treatment of sewage. However, most of the developed countries have introduced the "polluter pays principle" for the amount of water pollution load discharged by companies. Wastewater treatment charges are fully recovered from urban consumers of water.

(4) Service charge: In India, in most cities, a minimum service charge is included in the water bill. In India water is a highly subsidized commodity, leading to market inefficiencies and hence the inefficient use of the already scarce resource. The water subsidy in the urban areas has important consequences for the poor and the environment. The urban water authorities, usually known as Water Supply and Sewerage Boards, are responsible for the city's water supply and sewage services. Since urban water is subsidized, these institutions constantly incur losses and lack funds to invest in repairs and maintenance of existing water supply infrastructure, wastewater treatment, and expansion of their services. Another important consequence of urban water subsidies is that the urban water consumers and polluters are not charged for sewage treatment; hence, in most developing countries, only 20–30% of wastewater is treated to the secondary level. It is also seen that the average cost incurred by the water boards to supply water in most metropolitan cities ranges from Rs. 10 to 35 per m^3 and the price charged to urban domestic consumers ranges from Rs. 6 to 36 per m^3 depending on the volume consumed. The price for non-domestic consumers varies from Rs. 20 to 100 per m^3 depending on the volume consumed and the type of industry.

7.3 URBAN SEWAGE TREATMENT OPTIONS

Based on the information available, an analysis of various treatment technologies employed in different STPs is presented in Tables 7.7 and 7.8, respectively, for Class I and II towns. The following are vital issues:

S. No.	Technology	No. of plants	Combined capacity (MLD)	% Age capacity	Average size (MLD)	Remarks
1.	Activated sludge process (ASP)					The preference for conventional ASP is realized in more installations than the others
	PST + ASP	42	3059.63	52.6	72.8	
	ASP + Ext. Aer	3	63.36	1.1	21.1	
	ASP – Ext. Aer + Ter. Sed.	7	58.04	1.0	8.3	
	High-rate ASP + biofilter	1	181.84	3.1	181.8	
	Aerated lagoon + fish pond	3	49.50	0.9	16.5	
	Facultative lagoon + ASP	1	44.50	0.8	44.5	
2.	Fluidized aerobic bioreactor (attached growth)	5	66.00	1.1	13.2	Obvious objection due to higher energy input
3.	Trickling filter or biofilters	6	192.62	3.3	32.1	Operational difficulties and poor maintenance
4.	UASB + activated sludge process	1	86.00	1.5	86.0	Sensitive to pH and environmental conditions
5.	UASB					Pretreatment is sensitive to consistent efficiency
	Grit channel or PST + UASB + PP	24	1229.73	21.2	51.2	
	UASB + Sedimentation	1	126.00	2.2	126.0	
	Grit channel or PST + UASB	5	158.17	2.7	31.6	
	UASB (sum of all the above process)	30	1513.90	26.0	50.5	
6.	Waste stabilization ponds	42	327.53	5.6	7.8	Poor efficiency and higher land costs (common for S.No. 6 & 7)
7.	Oxidation pond (single stage)	3	69.00	1.2	23.0	
8.	Anaerobic digester + trickling filter	1	4.45	0.1	4.5	Higher cost and maintenance issue
9.	Kamal technology (for plantation)	2	12.46	0.2	6.2	Large land area
10.	Only primary treatment	3	84.00	1.4	28.0	Limited use
	Total	150 (100%)	5812.83 (100%)			

Source: Status of Sewage Treatment in India, CPCB, November 2005.

Table 7.8 Sewage treatment technologies employed in STPs of class II towns

S. No.	Technology	No. of plants	Combined capacity (MLD)	Average size (MLD)	Remarks
1.	ASP (Preceded by primary sedimentation)	1	12.5	12.5	Less area requirement and no option for energy recovery
2.	Grit channel or PST + UASB + PP	3	23.83	7.9	Energy recovery is an option
3.	Waste stabilization ponds	21	161.26	7.7	Adopted where land cost is not a constraint
4.	Trickling filter	2	16.68	8.3	Operation and maintenance
5.	Kamal technology (for plantation)	2	10.13	5.1	Land area for acceptability
	Total	29 (100%)	224.4 (100%)		

Source: Status of Sewage Treatment in India, CPCB, November 2005.

- At present, the field scenario of treatment processes at STPs of Class I cities involves activated sludge process (ASP) being used to an extent of 59.5% of total installed capacity, and UASB technology is used to 26% of total installed capacity. These technologies have been the most commonly used, including primary or tertiary treatment units. Waste stabilization ponds (WSPs), in series, are implemented in up to 28% of the plants, even though the treatment efficiency is less than 60%.
- The series of WSP technology is the most commonly employed technology, covering 71.9% of total installed capacity and 72.4% of STPs in Class II towns. The UASB technology covers 10.6% of total installed capacity and 10.3% of STPs.
- ASP technology is seen as adaptable for large cities because it requires less space as compared to the other two technologies, namely, UASB technology and WSP technology. The WSP technology requires a large area for its implementation. However, the extended-aeration sludge

process is expected to provide a better-quality effluent and require less land area as compared to conventional sludge process because the process is accommodated in substrate-limited conditions and also because of better settling properties of mixed liquor. The secondary excess sludge is also well stabilized and has better drainability for reuse.

- Conventionally, UASB technology is the suitable technology for cities of all sizes. However, all anaerobic treatment processes including UASB technology are very sensitive to environmental changes. Intermittent loading can greatly affect the performance of a UASB reactor, because the anaerobic bacteria are very sensitive to shock loading and power cuts. The polishing pond (PP), which is the terminal unit of the scheme, is also very crucial in deciding overall performance of the plant. The UASB with PP technology may be a suitable option if energy recovery is the priority. Two UASB plants of 86 and 126 MLD have been operating at Vadodara, Gurarat, India; however, these need to be studied in detail to assess the optimal efficiency of the treatment scheme in reducing BOD, COD, TSS, and Fecal & Total Coliform and reuse pattern to comply with pollution norms.

- The application of technology has constantly been upgraded over the years. The use of aerated lagoons was common practice in the early years. The conventional ASP gained a prominent place later and is still in use in many places. In recent years, newer modifications of ASP have gained importance, especially MBR for lowered land and operating cost apart from newer versions of aerobic as well as anaerobic treatment processes. According to the CPCB India (2005) report, most of the STPs in India employ any one of the three technologies, namely, primary settling followed by ASP (PST + ASP), UASB + PP, and a series of WSPs. The first technology has been found capable of providing a final effluent having BOD <20 mg/L and TSS <30 mg/L. The other two technologies may provide final effluent of this quality, if operated properly. The Gujarat State Pollution Control Board has already stipulated 20 mg/L, 100 mg/L, and 30 mg/L limits for BOD, COD, and TSS, respectively, for treated sewage quality (CPCB, 2005). Metal removal in municipal wastewater treatment plants can be accomplished primarily from sorption or physicochemical processes. Because metals are hydrophobic and have a tendency to get sorbed, their removal is easier through sorption by organic material in the wastewater, even in the primary settlement stage. Coprecipitation of metals also occurs with chemical removal of phosphorus. The highest removal efficiencies are therefore expected with both primary treatment and chemical precipitation. Newer technologies offering significantly higher removal rates need to be designed and implemented, e.g., membrane technologies are now increasingly used for the production of high-quality water for indirect potable reuse and are expected to lead current treatment technologies in the near future, either in isolation or in combination in the form of process integration.

In view of the continued development in the area of sewage treatment and also in view of the fact that sewage wastewaters are relatively easy to treat, especially using biological methods, it is recommended that most appropriate technologies be used for proper treatment such that the water can be treated reused. The concept of wastewater treatment, recycling, and reuse is most appropriate in this case, and it is believed that it can substantially alleviate the pains of water scarcity by reducing water requirements significantly—be it for public consumption or for agricultural application. The benefits go beyond water treatment, recycling, and reuse. The successful operation in this regard would relieve the rivers from pollution, reduce health hazards to the surrounding population, and thereby increase sustainability. It will be most appropriate that some industries in the adjoining area also be involved in this scheme so that if the industrial wastewater is not severely polluted, it can also be included in the treatment. Alternatively, industries can also be assigned part of the wastewater treatment, if they can treat wastewater better than the conventional approach for enhanced reuse. This type of cooperation at local and governmental level can certainly help.

7.3.1 Urban Sewage: Primary and Secondary Treatment Options

7.3.1.1 Primary Treatment

The first step in wastewater treatment involves separating large pieces of debris when the wastewater first enters the treatment plant. Debris may include wood, cloth, plastics, glass, metal, sand, and gravel. This is referred to as pretreatment.

The wastewater is held in a large sedimentation tank for several hours. This allows heavier solids to settle to the bottom and form a sludge layer. Lighter solids, fats, oil, and grease float to the top, creating a scum layer. The solids and scum are removed to receive further treatment as sludge. The clarified wastewater flows on to the next stage of wastewater treatment if there is to be further treatment of the effluent. If primary treatment is the only level of treatment, the clarified wastewater is disinfected and then discharged into the receiving water body.

In *enhanced primary treatment* (also called *advanced primary treatment*) chemicals are added to the sedimentation tanks to help waste particles bond together and settle out more readily.

7.3.1.2 Secondary Treatment

Secondary treatment typically involves biological treatment of wastewater usually following the primary treatment stage. Some secondary treatment plants do not include the primary treatment process. Naturally occurring bacteria break down the organic components of wastewater, and additional settling occurs. These solids are either reused in the biological process or removed for further treatment and disposal. If secondary treatment is the final level of treatment, the clarified wastewater is disinfected and then discharged into the receiving water body.

7.3.2 Urban Wastewater: Tertiary Treatment Options

The technical factor is often considered as least important for the success of water reuse projects as today's available technologies make it possible to reach any water quality required by users and for regulatory compliance. The range of technologies that can be applied is broad, from the well-proved standard ones to the newest advanced types (IAEA, 2007). However, the final choice will be strongly dependent upon local conditions, plant size, and water quality standards. An adequate choice, which means the selection of the most appropriate or "the best available" technology in a given situation, plays a key role for its future reliable operation and provides the guarantee of having a suitable water quality at lower operation and maintenance costs.

To achieve an adequate disinfection performance, in which fecal indicators are to be kept below the detection limits, it is recommended to implement a pretreatment to decrease the concentration of suspended solids. This is of special importance if UV light is used as disinfection agent. The combination of UV with other disinfection agents provide a greater reliability and higher efficiency for inactivation of different types of microorganisms.

During the last decades, membrane treatment has been approved as one of the best available technologies for the production of high-quality recycled water for indirect potable reuse and industrial applications. Among the membrane processes used for wastewater treatment, MBR technology is advancing rapidly worldwide. Microfiltration and ultra filtration are becoming the preferred pretreatment options over reverse osmosis (RO).

Increasing concerns on low carbon emissions and sustainability are creating a newer interest in nonconventional technologies such as soil-aquifer treatment, stabilization ponds, and wetlands as secondary treatment and polishing. Decentralized treatment plays an increasing role in water recycling both in urban and rural areas.

7.3.3 Water Recycling and Reuse: Strategy

The selective combination of wastewater treatment methods and practices can lead to successful recycle and reuse scheme. This scheme is very important for agricultural irrigation, for which recycled water is becoming the inevitable alternative resource in dry and urban areas (Biotechnology and Water Security in the 21st Century, 1999). Source quality is beneficial for the diversification of water reuse, e.g., limiting industrial discharges, on-site recycling, urine separation, reduction of salt discharge, or intrusion into sewers.

Water policy and economic water pricing are important to achieve the cost effectiveness in arriving at water reuse project formulations. Water reuse needs to be focussed on the basis of full cost recovery and "polluter pays" principle.

The participation of stackholders' in water recycling and reuse projects and their perception of the water cycle management is unquestionably the crucial factor for the success and the future development of water reuse practices, in particular in urban areas, agriculture, and for indirect potable reuse.

The current issues of *climate change*, the *European Water Framework Directives*, and the *Millennium Development Goals* are expected to effect water recycling and reuse development because water recycling is widely recognized as a proven water scarcity solution, drought-proof alternative resource, and environment sustainability approach.

7.4 INDUSTRIAL WATER PRODUCTION AND REUSE/URBAN-INDUSTRY JOINT VENTURE

From the past experience of urban/municipal wastewater treatment, it is quite evident that the industrial sectors require huge volumes of fresh water for increased productivity. The withdrawal of fresh water by industry further depletes the availability of resources. Realizing this long-term sustainability need, urban sectors are coming forward for joint ventures with industry to share its water potential in the form of treated sewage to industries for cooling water makeup and secondary industrial uses. The urban sectors are benefitted by reduction in cost incurred for their own treatment and in turn obtaining revenue for wastewater volume shared to industries. As per the studies conducted by the Ministry of Environment and Forests (MoEF) published in 2009, the comparison of costs of the various sewage treatment technologies depends upon various factors such as capital costs, operation and maintenance costs, land area costs, reinvestment, energy, and outputs costs. Data collected from the various STPs operating in India indicate that WSPs are operated with lower-cost techniques over the other treatment methods, around Rs. 1 per m^3, whereas WSPs require higher costs for their land area requirements, around 20,000 m^2 per MLD of wastewater, while the sequence batch reactor requires low land area costs, around 600 m^2 per MLD of sewage, but it operates at high costs for treatment process, Rs. 5 per m^3. The conventional ASP has moderate costs for the treatment process and in terms of land requirement for the plants. This is operated at a cost of about Rs. 3.5 per m^3 and requires a land area of 2000 m^2 per MLD (Table 7.9).

Apart from domestic sewage, about 13,468 MLD of wastewater is generated by industries, of which only 60% is treated. In the case of small-scale industries that may not be able to afford the cost of a wastewater treatment plant, common effluent treatment plants (CETPs) have been set up for clusters of small-scale industries (CPCB, 2005). The treatment methods adapted in these plants are dissolved air flotation, dual media filter, activated carbon filter, sand filtration and tank stabilization, flash mixer, clariflocculator, secondary clarifiers, and sludge drying beds. Coarse material and settable solids are removed during primary treatments by screening, grit removal, and sedimentation. Treated industrial wastewater from CETPs is mixed and disposed of in rivers. For example, 10 CETPs from Delhi with capacity of 133 MLD discharge their effluents into the Yamuna River.

The conventional wastewater treatment processes are expensive and require complex operations and maintenance. It is estimated that the total cost for establishing a treatment system for all domestic wastewater is around Rs. 7560 crores (CPCB, 2005), which is about 10 times the amount that the Indian government plans to spend (Kumar, 2003). Table 7.9 illustrates the economics of different levels of treatments through conventional measures (CPCB, 2007). Sludge removal, treatment, and handling are the most neglected areas in the operation of STPs in India. Due to improper design, poor maintenance, frequent electricity breakdowns, and lack of technical

Table 7.9 Economics of different levels of treatment through conventional measures

Salient Features	Primary treatment	Primary treatment + ultra- filtration	Primary treatment + ultra- filtration + RO	Remarks
Capital cost (Rs. lakhs)	30.0	90.64	145	The cost details are for municipal sewage at sewage contribution of 80% of water supply rate
Annual capital cost at 15% p.a. interest and depreciation	5.79	18.06	29.69	
Operation and maintenance (lakhs per year)	5.88	7.04	12.63	
Annual burden (annual and maintenance, Rs. lakhs)	11.85	27.1	42.5	
Treatment cost (Rs./KL) (without interest and depreciation)	34.08	52.40	73.22	

Source: Kaur et al. (2007)

staffing, the facilities constructed to treat wastewater do not function properly and remain closed most of the time (CPCB, 2007). Utilization of biogas generated from UASB reactors or sludge digesters is also not adequate in most cases. In some cases the gas generated is being flared and not being utilized. One of the major problems with wastewater treatment methods is that none of the available technologies has a direct economic return. Because there are no economic returns, local authorities are generally less interested in addressing wastewater treatment. A performance evaluation of STPs carried out by CPCB in selected cities has indicated that out of 92 STPs studied, 26 STPs had not complied with prescribed standards in respect to BOD, thereby making these waters unsuitable for household reuse. As a result, although the wastewater treatment capacity in the country has increased by about 2.5 times since 1978-1979, hardly 10% of the sewage generated is treated effectively, while the rest finds its way into the natural ecosystems and is responsible for large-scale pollution of rivers and groundwaters (Trivedy and Nakate, 2001).

However, it is necessary to evaluate all the treatment options by studying each technique in detail before concluding the treatment process is favorable in all aspects for the local conditions. The additional use of certain methods, such as energy recovery in the treatment process and the use of biomethanation techniques, will change the overall cost aspects of the STPs. The revenues generated by the treatment plants are not sufficient to address their operating costs and require proper funding from the Government sector in order to utilize the systems to maximum extent.

Financial Conditions of Sewage and Sewage Treatment Systems: Almost all the urban bodies in India have difficulty with financial sustainability to meet their requirements and therefore in the discharge of their functions in a proper manner to meet the demands of the society. In India sewage treatment is receiving low priority as far as funding is concerned. As per the study conducted by the National Institute of Urban Affairs (NIUA), the sanitation systems generate very low revenues and are sometimes referred to as non-revenue generators. The current revenues and costs for all urban bodies are very low in India; thus there is no planning for cost-effective policies for sewage collections at treatment systems and disposal facilities. Revenues will be increased by raising the cost tariffs to meet the capital costs, operation, and maintenance costs gradually over a period of time.

Some of the options for use of treated sewage in industrial application are given below. These are evident from current technological attainability.

1. Minimizing cooling water requirements with the superior GenGard technology—Leading Metals Company, Rajasthan.
2. Reducing dependency on freshwater sources by reusing industrial wastewater with the ZeeWeed MBR technology: Global automobile manufacturer, Bangalore.
3. Reducing dependency on freshwater sources by reusing wastewater in an urban setting with the ZeeWeed MBR technology: National IT company, Pune and Mysore.
4. Protecting the environment by treating chemical industry wastewater using ZLD technology: International chemical company, Madhya Pradesh (Ministry of Water Resources, 2008).

Some of the case studies relating industrial water production and reuse to an urban-industry joint venture are given below.

7.4.1 Sewage Reclamation Plant, the Rashtriya Chemicals and Fertilizers Plant, Chembur, Mumbai, India

In its 2011 report (MoEF, 2011), Rashtriya Chemicals and Fertilizers states that it is treating multiplex wastewater for industrial reuse purposes. The wastewater is reported to be extremely complex in nature, consisting of municipal sewage, which is mixed with pollutants of various industrial wastes. The main objective of the recycling and reuse of sewage is for various nonpotable industrial uses. Municipal sewage generated in the vicinity of the Rashtriya Chemicals and Fertilizers (RCF) Plant, Chembur, Mumbai, is heavily contaminated with various streams of industrial wastes and results in complex wastewater. In order to become water self-sufficient and to meet increasing process water requirements, the RCF Plant realizes the importance of recycling and reuse of wastewater for nonpotable industrial use and commissioned a sewage reclamation plant for the industry (MoEF, 2011).

7.4.1.1 Salient Features

The salient features reported for the RCF Plant include a 23 MLD capacity sewage reclamation plant involving RO in 2000 and treating complex wastewater comprising municipal sewage heavily contaminated with various industries' wastes. The sewage reclamation plant broadly consists of the following treatment units, which are schematically shown in Figure 7.6.

- Screening
- Grit removal
- Activated sludge system
- Clarifier

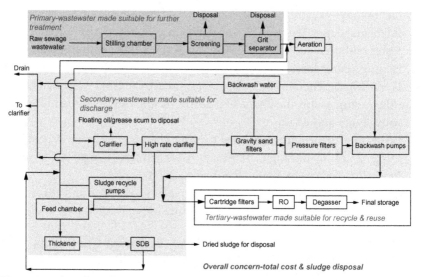

Figure 7.6 Flow sheet for 23 MLD sewage reclamation plant for the Rashtriya Chemicals and Fertilizers (RCF) Ltd., Chembur, Mumbai. *Data from MoEF (2011).*

- Sand filter
- Pressure filter
- Cartridge filters
- Reverse osmosis
- Degasification to remove CO_2
- Reuse in industry.

It is reported that the plant cost was nearly Rs. 40 crores in 1998, and the operating cost as reported in 2005 came to Rs. 39 per m^3. With the passage of time and the success of reuse schemes, the municipal taxes levied also became higher at Rs. 6 per m^3 of raw sewage. Some additional treatment steps, such as the use of ultrafiltration, became necessary in order to improve the quality of the water reaching the RO system (keeping the silt density index, SDI, <3.0), owing to the more polluted nature of the influent wastewater (MoEF, 2011). It is to be noted that the overall scheme here emphasizes primary treatment followed by secondary in the form of conventional ASP and finally tertiary treatment using RO to make water recyclable for reuse. A number of process modifications and/or alternatives can be possible to increase effectiveness and to reduce overall cost of treatment.

7.4.2 Tertiary Treated Municipal Sewage Reuse, Madras Refineries Ltd. (MRL) and Madras Fertilizers Ltd., Chennai, India

A similar case study has been reported under the treatment scheme employed at MRL and Madras Fertilizers Ltd. for recycling and reuse of municipal sewage for nonpotable uses in the refinery and fertilizer plant. MRL and the Madras Fertilizer Ltd. (MFL) are consumers of a very large quantity of water for their processing requirements. The treatment plant encompasses a tertiary treatment plant (TTP) unit for reusing the secondary municipal sewage to minimize the water demand from the source and become self-supporting in water availability to compensate for the increasing process water requirements (MoEF, 2011).

7.4.2.1 Salient Features

It is evident from the report that since 1991, MRL initiated the reuse of municipal sewage, generating 12 MLD of reusable water. MFL produced 16 MLD of reusable water. The operation of these TTPs was facilitated by the Chennai Metro Water and Sewerage Board to ensure a secondary treated sewage with a biochemical oxygen demand of 120 mg/L. The TTPs of MRL and MFL, which received secondary treated wastewater from the Chennai, consisted of the following treatment units (Figure 7.7):

- Additional secondary biological treatment

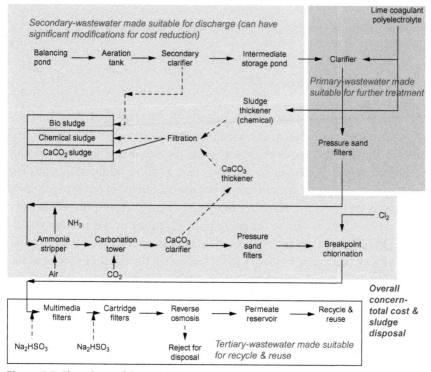

Figure 7.7 Flow sheet of the 12 MLD tertiary treatment plant (TTP) for Madras Refineries Ltd., Chennai. *Data from MoEF (2011).*

- Chemically aided settling + pressure filtration + ammonia stripping, carbonation, clarification, pressure filtration
- Chlorination
- Sodium bisulfate dosing
- Multimedia filtration
- Cartridge filtration
- Reverse osmosis
- Permeate for reuse.

The rejects containing high TDS (Total Dissolved Solids) are disposed of in the sea through a submerged outfall. As per the 1991 estimate, the capital cost for building the MRL plant was around Rs. 24 crores. The treatment costs for the MRL plant are reported to be about Rs. 35 per 1000 L of water, which is much lower in comparison to the charge of Rs. 60 per liters for fresh water supplied to industries. The Chennai Metro Water and Sewerage Board also charges a much higher tariff rate of Rs. 5.2 per 1000 L of water to cover its treatment costs up to secondary stage. It is to be noted that the overall treatment scheme here has more emphasis on primary treatment involving clarification and secondary in the form of biological treatment, with generation of both chemical as well as biological sludge. The final tertiary treatment again is in the form of RO to make water suitable for recycle and reuse. The overall concern remain same and it would be advisable to reduce/eliminate sludge, especially chemical sludge. The overall cost of operation again can be reduced significantly by intelligent combination of various treatment methodologies.

7.4.3 RO Plant for Wastewater Reuse, Vadodara, Gujarat, India

Another case study for industrial wastewater treatment and recycling/reuse is the RO-based plant in Gujarat, which is used for treating the highly polluted wastewater consisting of various industrial effluent streams for recycling and reusing for industrial uses for its nonpotable purposes. The plant uses highly polluted wastewater from an effluent disposal channel into which several industries—refineries, fertilizer, petrochemical—discharge their raw wastes. The successful operation of the plant demonstrated that at least 75% of the wastewater could be made available for reuse at a treatment cost of Rs. 36 per 1000 L as per the 1999 estimates. The remaining 25% consists of the rejects and sludge from the RO plant and needs to be disposed of separately.

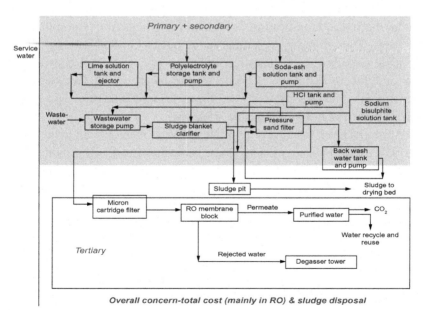

Figure 7.8 Flow diagram of the three MLD reverse osmosis plant for wastewater reuse at Vadodara, Gujarat. *From MoEF (2011).*

The treatment chain for the 3 MLD capacity RO plant for wastewater reuse at Vadodara (Figure 7.8) comprises the following units:
- Wastewater from effluent channel
- Chemical feeding (lime, polyelectrolyte, soda ash)
- Clarification
- HCl addition
- Press filtration
- Sodium bisulfate
- Cartridge filtration
- Reverse osmosis
- Degasser to remove CO_2
- For reuse.

It can be seen that here the major emphasis is on RO process that is cost intensive process. Use of many chemicals can be reduced that will also help in reduction in the sludge generation. Different treatment alternatives/combinations in this regard, need to be searched and used.

7.5 URBAN-INDUSTRIAL WATER SUSTAINABILITY: 2030

7.5.1 Water Management, Policies, and Legislation Related to Water Use in Agriculture

Under the Indian Constitution, water is, by and large, the responsibility of the states. Thus, the states are primarily responsible for the planning, implementation, funding, and management of water resources development. This responsibility in each state is borne by the Irrigation and Water Supply Department. The Inter-State Water Disputes Act of 1956 provides a framework for the resolution of possible conflicts.

At central level, which is responsible for water management in the union territories and in charge of developing guidelines and a policy frame work for all the states, there are three main institutions involved in water resources management, as described below.

It is conceived from the Ministry of Water Resources (MWR) goal that the use of country water resources is governed by the policy guidelines and programs formulated at regular stipulated periods. The Central Water Commission (CWC), being the ministry's technical arm, provides general infrastructural, technical, and research support for water resources development at the state level (CWC, 2006). The CWC is also responsible for the regular assessment of water resources. The Planning Commission is responsible for the allocation of financial resources required for various programs and schemes of water resources development in the states and MWR. It also actively engages in policy formulation and development of water resources at the national level (CWC, 2007).

It is also quite evident that the Ministry of Agriculture promotes irrigated agriculture through its Department of Agriculture and Cooperation. The Indian National Committee of Irrigation and Drainage coordinates with the International Commission on Irrigation and Drainage and promotes research in the relevant areas. The Central Pollution Control Board handles water quality monitoring and the preparation and implementation of action plans to solve pollution problems. In 1996, the Central Groundwater Authority was established to regulate and control groundwater development with a view to preserving and protecting the resource.

7.5.2 Water Management

Water resources planning and augmentation should attract increased potential of water through sustainable agriculture and food grain yield availability (Sarma, 1999). Sustained irrigation has also helped reduce interannual fluctuations in agricultural output and India's vulnerability to drought (Anbumozhi et al., 2001). One of the goals of Indian policy is now to find ways of maintaining the level of food grain availability per inhabitant in a context of population increase. Total water demand is expected to equal water availability by 2025, but in the meantime industrial and municipal water demand are expected to rise unproportionately at the expense of the agricultural sector, which will have to produce more with less water (Sarma, 2002).

The centrally sponsored Command Area Development (CAD) Program was launched in 1974–1975. The main objectives of the program are improving the utilization of the area equipped for irrigation and optimizing agriculture production and productivity from irrigated agriculture. The program involves the implementation of on-farm development works like construction of field channels and field drains, reclamation of waterlogged areas, renovation and rehabilitation of minor irrigation tanks, and correction of irrigation water distribution system deficiencies. The program also involves "software" activities like adaptive trials, demonstrations, training of farmers, and evaluation studies. An amount of Rs. 35,280 million has been released to the states as central assistance under the program from its inception until the end of March 2008. An area of about 18.07 million ha has been covered under the program since inception until the end of March 2007. The CAD Programme has been restructured and renamed the "Command Area Development and Water Management (CADWM) Programme" since 1 April 2004.

The use of effluents via irrigation can make a significant contribution to the integrated management of our water resources. When the water and nutrients in effluents are beneficially utilized through irrigation, some of the water extracted from rivers can be replaced, and the amount of pollutants discharged into our waterways can be reduced. The Department of Environment and Conservation (NSW) has ensured a policy of encouraging the beneficial use of effluents where it is safe and practicable to do so and where it provides the best environmental outcome. This Guideline is educational and advisory in nature. It is not a mandatory or regulatory tool, and it does not introduce new environmental requirements.

The best management practices related to the management of effluents by irrigation, are to be used to design and operate effluent irrigation systems, by reducing the risk on health implications and possible hazards to food chain (Directorate of Economics and Statistics, 2010). The Guideline will assist decision makers and industry members in achieving the best environmental outcome for each site at least cost.

The Guideline is not intended to provide specific guidance on every individual industry's issues. On the other hand, it provides an information base to be used for addressing issues that might arise in the range of situations, circumstances, and industries in which effluent irrigation may be considered or underway. Industry-specific guidelines or site-specific information may need to be taken into account when applying the Guideline.

Approaches to effluent irrigation management other than those outlined in this Guideline will always be considered on their merits, provided that they demonstrate environmental sustainability and are safe from a public health perspective (Sarma, 2006).

Extensive consultation with industry and government are made for formulating National guidelines to reveal the reuse practice. The Guideline reflects the idea that a sustainable effluent irrigation system will be a function of the interactions between the site, soil, agronomic system and effluent characteristics, and diligent operational practices. These interactions should ensure effective management to maximize the resources available in effluents and ensure that the environment is protected. Selecting a suitable site is important for successfully establishing an effluent irrigation system. The Guideline provides criteria for assessing a proposed irrigation site and discusses related issues important to the assessment of a site. The relationship between effluent quality and soil characteristics that should also be considered when selecting a site are also outlined to ensure that soil structure is not likely to be adversely affected and/or pollution is not likely to be caused.

In relation to effluent quality, effluents contain valuable resources (water, organic matter, and nutrients). However, in excessive amounts these can be detrimental to soils or plant growth. Effluents can also contain chemical contaminants, salts, and pathogens that can pose a risk to the wider environment or public health, or may cause pollution. These risks can be minimized by applying the criteria and information provided in the Guideline during the site selection, design, and operation phases of an effluent irrigation system.

Best management practices that optimize the use of the water, nutrients, and organic matter and reduce the potential for harm from other contaminants are also critical. For an effluent irrigation system to be sustainable, the amount of water, nutrients, and chemicals that will be applied should be determined to ensure that it is the optimum for the crop or cultivar, the agronomic system employed, and site-specific factors such as climate, topography, and soil.

Adjustments to the amount of effluent applied or the area over which it is applied can then be made to ensure that irrigated plants and environments are not stressed by water or by the organic material, nutrients, or chemicals applied.

Water and nutrient balances are used to calculate the amount of water and nutrients that should be applied, and at what times, to meet the crop requirements while ensuring increases in runoff and percolation are minimized. The water balance is calculated to determine the maximum volume of effluent that can be sustainably used. The elements to be considered in a water balance are rainfall, evapotranspiration, runoff, and percolation.

For some effluents, the loading rates of nutrients such as nitrogen and phosphorus can limit the quantity of effluent to be used for irrigation in a given area. In a nutrient balance the amount of the specific nutrient (e.g., nitrogen or phosphorus) assumed to be applied in a year is compared with the amount taken up by the biological or physical processes of the crop-soil system. Preirrigation soil nutrient status is also considered.

In some systems the amount of effluent that can be applied is limited by potential adverse impacts of salinity, heavy metals, and persistent organic chemicals. The Guideline suggests that key components in managing these types of limitations include designing the system to avoid any potential impacts and having in place a management and monitoring system to correct any emerging problems and to identify when action needs to be taken to ensure the environmental and agronomic performance of the system.

The National Water Policy 2002 emphasizes a participatory approach in water resources management. It has been recognized that participation of beneficiaries will help greatly for the optimal upkeep of the irrigation system and efficient utilization of irrigation water. The participation of farmers in irrigation management is formulated through the constitution of Water Users' Associations (WUAs). The aims of the WUAs are to (i) promote and secure distribution of water among users, (ii) ensure adequate maintenance of the irrigation systems, (iii) improve efficiency and economic utilization of water, (iv) optimize agricultural production, (v) protect the environment, and (vi) ensure ecological balance by involving the farmers and inculcating a sense of ownership of the irrigation systems in accordance with the water budget and operational plan. The WUAs are formed and work on the basis of executive instructions and guidelines laid down by each state government. There is no central legislation or legal instrument in this regard. However, the one state that has passed legislation exclusively for farmer participation in the management of irrigation systems is Andhra Pradesh. A total of 55,500 WUAs were constituted in India covering an area of 10.23 million ha.

The National Groundwater Recharge Master Plan provides a nationwide assessment of the groundwater recharge potential and outlines the guiding principles for an artificial groundwater recharge program. The plan estimates that through dedicated recharge structures in rural areas and rooftop water harvesting structures in urban areas a total of 36 km^3 can be added to groundwater recharge annually. The master plan follows two criteria for identifying recharge: availability of surplus water and availability of storage space in aquifers. Investments in the program would therefore be driven by the potential available for groundwater recharge, and not by the need for recharge. Thus, the three states-Andhra Pradesh, Rajasthan, and Tamil Nadu, which together account for over half of India's threatened groundwater blocks, receive only ~21% of funds, whereas the states of the Ganges-Brahmaputra basin, which face no groundwater overdevelopment problems, receive ~43% of the funds. If implemented successfully, this recharge program will be able to add a significant quantity of water to India's groundwater storage, but it will not provide much help in the areas that are most in need of help.

7.5.3 Finances

At present there is no uniform set of principles in fixing water rates. The water charges vary from state to state, project to project, and crop to crop. The rates vary widely for the same crop in the same state depending on irrigation season, type of system, and other parameters. Because water rates are abysmally low, not enough funds are generated for proper maintenance of irrigation systems, leading to poor quality of service. It is necessary that the state governments evolve a policy for periodical rationalization and revision of water rates so that the revenue generated by the irrigation sector is able to meet the cost of operation and maintenance. However, in view of unreliable and poor quality services, farmers are reluctant to pay increased water charges. They may not be averse to paying increased water charges if the quality of services is first improved. It is imperative that the tariff structure is reviewed and revised with simultaneous improvement in the quality of services provided so as to restore efficiencies. Rationalization of water rates will also act as a deterrent to the excessive and wasteful use of water. A shift toward fixing water rates on a volumetric basis is desirable. This will encourage farmers to avoid overirrigation and wasteful use of water, thereby increasing water-use efficiencies. A uniform formula of water pricing for the entire country would have no practical value. It may be helpful to recommend setting up an independent State Regulatory Authority for rationalization of water rates by each state.

7.5.4 Policies and Legislation

India adopted a National Water Policy in 1987, which was revised in 2002, for the planning and development of water resources to be governed by national perspectives. It emphasizes the need for river basin-based planning of water use. Water allocation priority has been given to drinking water, followed by irrigation, hydropower, navigation, and industrial or other uses. Because water resources development is a state responsibility, all the states are required to develop their state water policy within the framework of the national water policy and, accordingly, set up a master plan for water resources development.

7.6 SUMMARY AND PATH FORWARD

On comparing an average Indian industry with an optimally performing global industry, the former can be seen to be far more water intensive than its global counterpart. Indian industries are facing huge water crises due to the current allocation level of only ~8% of national water with ~87% allocated for the agricultural sector (National Water Policy, 2002). This led to the devising of a new way of sustainable water use to satisfy the water demand of industrial sectors both at current and at projected demand levels without altering the existing policies. The huge water demand of industry can be balanced by recycling and reuse of urban wastewater. This would create a large opportunity for Indian industry to grow within a closed loop.

A leading-edge assessment of urban wastewater treatment for recycling and reuse in industrial applications reveals a potential and strategic way forward by addressing the following crucial issues:

1. It would be possible to reuse 40–50% of secondary sewage for industrial use and indirect potable water reuse.

2. Adhering to the sustainability of the resources and increasing industrial water demand, the present reuse of less than 8% is expected to increase to 30–40% at least toward estimated water demands of 2030.

3. Industrial water production from secondary sewage of urban and semi-urban areas can increase the water availability to Indian industrial sectors by establishing a joint municipal-industrial collaborative venture operated within a sustainable loop (without altering current national water policy) that would eventually be expected to lead to a water price far less than that of natural fresh water resources with a countdown toward 2030 from today.

REFERENCES

Anbumozhi, V., Matsumoto, K., Yamaji, E., 2001. Towards improved performance of irrigation tanks in semi-arid regions of India: modernization opportunities and challenges. Irrigat. Drainage Syst. 5 (4), 293–309.

Asano, T., Levine, A.D., 1996. Wastewater reclamation, recycling and reuse: past, present, and future. Water Sci. Technol. 33 (10–11), 1–14.

Asano, T., Smit, R.G., Tchobanoglous, G., 1985. Municipal wastewater: treatment and reclaimed water characteristics. In: Pettygrove, G.S., Asano, T. (Eds.), Irrigation with Reclaimed Municipal Wastewater: A Guidance Manual. Lewise Publishers, Inc., Chelsea, MI.

Bharadwaj, R.M., 2005. Central Pollution Control Board, *Status of wastewater generation and treatment in India*, at IWG- Environment International Work Session on Water Statistics, Vienna.

Central Pollution Control Board (CPCB), 2009. Annual Report 2009–10, Ministry of Environment and Forests (MoEF), New Delhi, India.

Central Pollution Control Board (CPCB), 2010. Annual Report 2010–11, Ministry of Environment and Forests (MoEF), New Delhi, India.

Central Water Commission, Ministry of Water Resources, 2005. *Rainfall in different meteorological zones of India*.

Central Water Commission, Ministry of Water Resources, 2006. *Water data pocket book*, http://www.cwc.nic.in/Water_Data_Pocket_Book_2006/T8.3final.pdf.

Central Water Commission, Ministry of Water Resources, 2007. *Water-related statistics 2007*.

Central Intelligence Agency (CIA), 2009. World fact book, http://www.cia.gov/library/publications/the-world-factbook/print/in.html.

Central Pollution Control Board (CPCB), November 2005. *Status of sewage treatment plant in India*.

CPCB, 2007. Advance methods for treatment of textile industry effluents, Resource Recycling Series : RERES/&/2007. Central Pollution Control Board, India.

CSE (Centre for Science and Environment), 2011. Policy paper on septage management in India. New Delhi. available at: http://www.urbanindia.nic.in/programme/uwss/slb/SeptagePolicyPaper.pdf (accessed on October 12, 2012).

CSE. (Centre for Science and Environment), 2012. Excreta matters volume 1: How urban India is soaking up water, polluting rivers and drowning in its own waste. 7th report in the State of India's Environment series. New Delhi.

Directorate of Economics and Statistics, 2010. Department of Agriculture and Cooperation, Ministry of Agriculture, Government of India, *Agricultural statistics at a glance 2010*, http://dacnet.nic.in/eands/latest_2006.htm, accessed on 2011-01-19.

Garg, N.K., Hassan, Q., 2007. Alarming scarcity of water in India. Curr. Sci. 93, 932–941.

Government of India (GoI, 2008), Ministry of Water Resources, 2008. Compressive Mission Document: National Water Mission under National Action Plan on Climate Change', Vol II.

Government of India (GoI, 2010), Central Water Commission, 2010. Water and related statistics. Water Planning and Project Wing. India.

Heggade, O.D., 1998. Urban development in India: Problems, policies and programmes. New Delhi: Mohit publications, pp. 463.

IAEA, 2007. Status of Nuclear Desalination in IAEA Member States. IAEA, Vienna, IAEA TECDOC 1524.

Indian Desalination Association INDACON, 2008. http://www.indaindia.org/intro_desalination.htm.

Jain S. K., Agarwal P. K., Singh V. P. Hydrology and Water Resources of India., 2007, Springer, The Netherlands.

Kaur, R., Wani, S.P., Singh, A.K., La, K., 2007. Wastewater production, treatment and use in India, UN water country report.

Kumar, R.M., 2003. Financing of wastewater treatment projects, Infrastructure Development Finance Corporation and Confederation of Indian Industries, Water Summit, Hyderabad, 4–5 December.

Lorenzen, G., Sprenger, C., Taute, T., Pekdeger, A., Mittal, A., Massmann, G., 2010. Assessment of the potential for bank filtration in a water-stressed megacity (Delhi, India). Env. Earth Sci. 61 (7), 1419–1434.

Biotechnology and water security in the twenty-first Century. 1999. Global water scenario and emerging challenges, Madras Declaration, 1999, M. S. Swaminathan Research foundation, Chennai.

Ministry of Environment & Forest, 2011. *Review of wastewater reuse projects world wide*, National Ganga River Basin Authority (NGRBA), Consortium report.

Ministry of Water Resources, 2003. *3rd Minor irrigation census*. http://wrmin.nic.in/micensus/mi3census/chapter5.pdf.

Ministry of Water Resources, 2008. *Efficient use of water for various purposes*. National Water Mission under National Action Plan on Climate Change.

Misra, S. 1998. Economies of scale in water pollution abatement: A case of small scale factories in an industrial estate in India. Working Paper No. 57. Centre for Development Economics, Delhi School of Economics, University of Delhi. Delhi.

MoEF-Ministry of Environment and Forest, Government of India, 2009. State of Environment Report India-2009.

MoUD, 2013. *Manual on sewerage system and sewage treatment*, CPHEEO, New Delhi, Government of India.

National Commission for Integrated Water Resources Development Plan, Ministry of Water Resources, 1999. *Report of the sub-group No. IV on water logging and drainage of the working group on water management for agriculture and other allied sectors.* http://wrmin.nic.in/writereaddata/linkimages/component%20of%20reclamation7872571015.pdf.

National Water Policy, 2002. Ministry of Water Resources, Government of India. New Delhi, India.

Rawat, M., Ramanathan, A., Subramanian, V., 2009. Quantification distribution of heavy metals from small-scale industrial areas of Kanpur city, India. Journal of Hazardous Materials 172 (2–3), 1145–1149.

Sarma, P.B.S., 1999. Irrigated agriculture, strengths, constraints and opportunities for the future. In: Proceedings of 17th International Congress on Irrigation and Drainage, September 15–19. Granada, Spain, pp. 91–105.

Sarma, P.B.S., 2002. Water resources and their management for sustainable agricultural production in India. In: Prasad, R., Vedulapp, S. (Eds.), Research Perspectives, in Hydraulics and Water Resources Engineering. World Scientific, Singapore, pp. 193–286.

Sarma, P.B.S., 2006. *Strategies for irrigation and drainage sector in India.* Country Policy Support Programme (CPSP) CPSP report #13, International Commission on Irrigation and Drainage (ICID).

Trivedy, R.K., Nakate, S.S., 2001. Treatment of hospital waste and sewage in hyacinth ponds. In: Kaul, S., Trivedy, R.K. (Eds.), Low Cost Wastewater Treatment Technologies. ABD, Jaipur, India.

CHAPTER 8

Phenolic Wastewater Treatment: Development and Applications of New Adsorbent Materials

Laxmi Gayatri Sorokhaibam[1], M. Ahmaruzzaman[2]
[1]Chemical Engineering & Process Development Division, CSIR-National Chemical Laboratory, Pune, India
[2]Department of Chemistry, National Institute of Technology, Silchar, India

8.1 INTRODUCTION

The principal objective of wastewater treatment has conventionally been to dispose of industrial effluents without danger to human health or the environment. There are a number of technologies already in place—both in terms of physicochemical and biological processes for the treatment of various types of wastewaters—that have been discussed at length in this book. However, it is evident that many of these conventional technologies have limitations in terms of overall process economics; therefore, it becomes imperative to devise solutions that are cost effective without compromising process effectiveness. It is expected that the development of newer forms of less expensive adsorbent materials, e.g., those derived from biomass, may have potential to fulfill this objective. Such cost-effective alternatives would enhance treatment abilities, especially for developing countries.

Typical physical characteristics of wastewater are gray color, stale odor, a solids content of 0.1%, and 99.9% water content. The chemical composition of wastewater consists of various organic and inorganic compounds as well as numerous dissolved gases. There are a number of conventional and nonconventional pollutants present in domestic and industrial wastewaters, and these have been discussed in detail in Chapters 1 and 2. One particular class of organic compounds that is of major concern in industrial wastewater treatment due to its toxicity and hazardous nature is phenol and phenolic compounds. They are hydroxyl derivatives of benzene that may occur in domestic and industrial wastewaters. It is desirable to recover these chemicals prior to discharge of waters. In any case, they have to be effectively removed prior to discharge in order to prevent environmental pollution. The limit for phenol in wastewaters for discharge is typically 0.1 mg L^{-1} (USEPA, 1985),

Industrial Wastewater Treatment, Recycling, and Reuse

323

and the World Health Organization has established the maximum level of phenol in drinking water allowable as $1\ \mu g\ L^{-1}$.

In the present chapter, the focus will be mainly on new adsorbent materials mainly derived from biomass, metal oxides, and application in the wastewater treatment for removal of three of the most commonly encountered industrial pollutants, namely, phenol, 4-nitrophenol, and 4-chlorophenol, represented, respectively, by P, 4-NP, and 4-CP hereafter. Wastewaters containing phenolics are generated from various types of industrial processing procedures relating to the production of these chemicals, preparation of intermediates, and industries using these compounds as raw materials. The prominent industries include petroleum refineries; textiles; coke ovens; fuel production facilities; wood preserving plants; manufacturers of plastics, resins, dyes, pesticides, pharmaceuticals; engineering; metallurgy (Das and Sharma, 1998); and industries producing a host of other chemicals. Phenols are also found in wastes of synthetic resin and plastics, rubber proofing, cutting-oil, dye manufacturing, and many chemicals (Davi and Gnudi, 1999). The composition and nature of pollutants vary from industry to industry, and treatment is usually complex. It is reported that the release of phenol, 4-CP, and 4-NP for the year 2000 in the United States (www.epa.gov/tri) was 22.0, 0.046, and 0.007 t/year, respectively.

8.1.1 Technologies for Removal of Phenolic Compounds from Wastewaters

A number of technologies, both destructive and nondestructive, are available for the removal of phenolics from industrial wastewaters. The selection of appropriate treatments for achieving removal and recovery or removal through destruction to the desired degree with cost effectiveness is in most cases a challenging and a complex task. The principal types of technologies available for treatment are briefly discussed in the following sections.

8.1.1.1 Coagulation

Coagulation is a physicochemical process that is often used to remove turbidity and color from materials that are typically colloidal in nature (1–200 μm). Both inorganic and organic coagulants have been used. The optimum time for rapid mixing is often achieved in few minutes. Rapid mixing is frequently followed by flocculation whereby agglomeration of settled turbid particles into larger flocs takes place. The flocs then settle and remove the contaminants in the sludge. Coagulation is often carried out using chemicals such as ferric chloride, ferrous sulfate, ferric sulfate, alum, lime, polymers, or a combination of these chemicals. Sometimes a combination of coagulation and adsorption processes is used (Nowack et al., 1999). pH plays an important role in chemical coagulation. The use of several chemical coagulants and disposal of sludge is a major problem in this technique. Further, desirable limits for the pollutants are rarely achieved using coagulation alone.

8.1.1.2 Ion Exchange

Since phenols are weak acids, ion exchange technology can be employed and can be a potentially attractive process from a recovery point of view, especially at low concentrations. The exchange process is reversible, and resins can be regenerated after the recovery of phenols. Resins and zeolites are widely employed for decontamination of wastewater containing phenol. However, the process has some limitations in terms of capacity/selectivity, and the main disadvantages associated with this method include organic and bacterial fouling and chlorine contamination.

8.1.1.3 Chemical Oxidation

Phenolics can be chemically converted to more oxygenated forms by means of reactions with oxidizing agents. Chemical oxidation can achieve high degradation efficiency for phenols, but it needs more energy and may result in secondary pollution problems. The oxidants that are commonly used in the chemical oxidation method of water treatment include ozone, chlorine, hydrogen peroxide, sodium hypochlorite (NaOCl), and potassium permanganate. 4-NP can be degraded into an environmentally friendly compound by chemical oxidation (Oturan et al., 2000; Stüber et al., 2005), but this method cannot recover 4-NP from industrial effluents. Usually, the operation cost is high and requires trained manpower or advanced reactors (Langlais et al., 1991). Several chemical oxidation techniques, such as catalytic wet air oxidation (CWAO), hydrogen peroxide promoted CWAO, wet peroxide oxidation (WPO), and Fenton WPO, have been tested to remove different aromatic organics belonging to commonly encountered pollutants. Sometimes chemical oxidation leads to the formation of secondary products that are highly acidic, leading to corrosion problems. Although the addition of Fenton reagent may improve the efficiency, it cannot be used for treatment of wastewater containing p-nitrophenol due to the formation of an insoluble complex. The overall oxidation rates are slow, and the use of chemicals is high. Alternatively, such wastewaters can be treated using further advanced oxidation processes (AOPs), where highly reactive free radicals, especially hydroxyl radicals, are mostly utilized to destroy the pollutants in wastewater. The process can involve direct ozonization, wet air oxidation (WAO), hydrogen peroxide oxidation, and UV photolysis (Esplugas et al., 2002; Poznyak and Arazia, 2005; Salah et al., 2004). Ideally, the final products in advanced oxidation technology are simple products such as water and carbon dioxide, which means that there are no secondary pollution problems.

8.1.1.4 Ozonation
Ozonation is widely employed in water treatment or disinfection. Hoigné and co-workers have revealed that ozonation produces hydroxyl (OH) radicals through the decomposition of ozone (O_3) with OH and proposed the Staehelin, Buhler, and Hoigné model for the complex reaction cascade during ozonation (Staehelin et al., 1984). The OH radical is one of the most reactive species in ozonation, and the amount of OH radical determines the efficiency of ozonation (Hoigné, 1988). Recently, some kinetic models were provided to explain the efficiency of OH radicals in AOPs using OH radical scavengers (Andreozzi et al., 1999). The ozonation method of wastewater treatment is applicable to wastewaters at very low phenol concentrations of 1–5 mg L^{-1}.

8.1.1.5 Electrolysis
The electrolytic method can also offer one tool for the treatment of wastewaters. The oxidation of organic matter may be classified into two types—direct oxidation, taking place at the surface of the anode, and indirect oxidation, which is distant from the anode surface. This technique can also remove 4-NP (Cañizares et al., 2004) but has limitations similar to those of chemical oxidation.

8.1.1.6 Biodegradation
Biological treatment of organic chemicals can be accomplished if proper microbial communities are established for degradation. Microbial strains such as *Pseudomonas putida*, *Pseudomonas fluorescens*, *Acinetobacter*, *Trichosporon cutaneum*, and *Candida tropicalis* are capable of degrading phenol at low concentrations. At sufficiently high concentrations, phenol can become sufficiently toxic to inhibit the growth rate of these micro-organisms. Hence, for achieving satisfactory performance, the phenol concentration needs to be maintained well below toxic level, and the acclimatization of these organisms to the wastewater environment is also required. Biodegradation can also

remove 4-NP (Yi et al., 2006). Typically, phenols are biodegradable provided that the concentration is below the toxic level. Chlorinated phenols are not amenable to biodegradation except at low temperatures or specific conditions. pH can have a significant and adverse impact on the process. The biological method of wastewater treatment is generally adopted at concentration levels in the range 5–500 mg L^{-1}.

8.1.1.7 Reverse Osmosis

Reverse osmosis in the form of membrane technology is being used for the large-scale reclamation of wastewater, although the technique depends on proper pretreatment, chemical control, and use of reverse osmosis membranes that are resistant to fouling. It is also an energy-intensive operation. Other disadvantages include risk of bacterial contamination of the membrane, cost of pretreatment of the wastewater to avoid membrane fouling, and sensitivity of the membrane toward chemicals.

8.1.1.8 Solvent Extraction

This method involves separation of the organic phase using solvents that are immiscible with water. Phenol is then extracted by treating the organic layer with alkali or distilling the phenol or the solvent in case there is a large difference between the boiling temperatures of phenol and the extracting agent. Some of the organic solvents that have been used include n-hexane and cyclohexane, benzene, toluene, ethyl benzene, cumene, acetate esters, and methyl-isobutyl ketone (Greminger et al., 1982). Because of the formation of a third phase, solvent extraction is accompanied by solvent loss in wastewater that inevitably creates secondary pollution (Shen et al., 2006; Yang et al., 2006). This secondary contamination of wastewater with solvents is the main drawback of this method over and above the requirement of high energy input and high operating cost. This method is also less efficient for the removal of trace levels of pollutants (Goto et al., 1984).

8.1.1.9 Adsorption

Adsorption is more suitable for treatment of phenolic wastewaters, especially in view of low concentrations and superior process efficiencies compared to many of the conventional treatment methodologies. The advantages of the adsorption technique over several conventional methods of wastewater treatment include:

1. Less space requirement, particularly in comparison to biological treatment.
2. Flexibility in design and operation.

3. Lower sensitivity to process variation.
4. Greater efficiency.
5. Insensitivity to toxic pollutants.
6. No formation of harmful substances.
7. Wider applicability because it can remove or minimize different types of pollutants.

The adsorption process can be either physical or chemical in nature. Physical adsorption involves weak forces that are reversible and occur mostly at low temperatures. On the other hand, chemisorption or chemical activation occurs at high temperatures, involving significant activation energy along with the participation of strong chemical bonds that are usually irreversible in nature. Activated carbon (AC) is one of the most widely used adsorbents for the removal of organic contaminants from wastewater that is refractory or biologically resistant in nature. The large surface area, high adsorption capacity, and high surface reactivity are the positive points. The problem of high initial cost of AC or polymeric adsorbents/specialty materials coupled with the problems of regeneration has necessitated the use/search for alternative adsorbents. Newer materials derived from biomass can be substantially effective in this regard.

The conventional treatment technologies implemented in industrialized nations are expensive to build, operate, and maintain in developing countries. The present research is one such effort to develop affordable treatment technologies particularly important for developing and underdeveloped countries. Thus, in spite of several technologies available, our main focus is on adsorption technology and development of newer low-cost adsorbent materials for application in phenolic wastewater treatment. This chapter reviews developments in this regard with new data and analysis.

8.2 NEWER ADSORBENTS AND THE POTENTIAL FOR THEIR APPLICATION IN PHENOLIC WASTEWATER TREATMENT

Wastewater treatment using the adsorption technique has been a subject of much exciting research and various novel materials. Low-cost adsorbents have been proposed for a myriad variety of toxic chemicals in wastewater. Existing and developed materials can be further modified to increase capacity or activity by means of several pretreatment/activation methods that result in much greater efficacy in their applications. AC, owing to its high surface area and porosity, is certainly a near-perfect candidate for treatment of almost all kinds of contaminants. But the limitation of huge capital investment and loss in regeneration restricts its use for wide-scale application. There has been increased interest in newer biomass-derived adsorbents that are low-cost materials, but have lower efficiency and capacity compared to conventional adsorbents. However, by employing different pretreatment and activation methods, such as charring or chemical treatment with orthophosphoric acid, the performance of these low-cost adsorbents could be increased significantly.

AC has a complex surface and porous structure. Generally, it has 45–1800 $m^2 g^{-1}$ total surface area distributed in micropore (<2 nm), mesopore (2-50 nm), and macropore (>50 nm) regions. AC generally exhibits high adsorption capacity for phenolic compounds and is considerably cost effective. Due to these reasons, there have been numerous studies on these materials; for example, Wang et al. (1997) reported adsorption of phenol, 4-CP, and 4-NP and found the adsorptive removal with granular carbon to be better than that with powder form; Varghese et al. (2004) investigated the use of AC prepared from water hyacinth for the removal of phenol, 4-CP, and 4-NP with maximum adsorption capacity of 1.20, 1.28, and 1.35 $mg L^{-1}$, respectively; Jung et al. (2001) reported adsorption of phenol and chlorophenols on four commercial granular ACs.

As an alternative to conventional adsorbents, a number of low-cost adsorbents were developed from materials such as fertilizer waste, wood, and rice husk by several chemical treatments. Also, several naturally occurring materials having the characteristics of adsorbents can be similarly low-cost adsorbents for the removal of pollutants from waste water, for example, clays, siliceous materials, zeolites, and bentonites. The adsorption capability, here, is a result of net negative charge on the structure of minerals that provides the capacity to adsorb positively charged species. The adsorption properties also come from their high surface area and high porosity. The use of bentonite for phenol adsorption from aqueous solutions was reported by Banat and Al-Asheh (2000). Taha et al. (2003) reported adsorption of phenol in granite residual soil and kaolinite stating that the residual soil possesses a greater adsorption capacity compared to kaolinite. Kaleta (2006) reported modified clarion clay and clinoptylolite as adsorbents for the removal of phenolic compounds from water. Results showed improved capacities by modification of the clay materials. Experimental results were reported on the adsorption characteristics of phenol and *m*-chlorophenol using organobentonites (Lin and Cheng, 2000). Removal of phenol from water was also reported by silica gel, HiSiv 3000, activated alumina, AC, Filtrasorb-400, HiSiv 1000 (Roostaei and Tezel, 2004), and zeolites (Khalid et al., 2004). To adsorb phenol selectively from water, adsorbents are required to possess a hydrophobic nature, i.e., zeolites having a high Si/Al ratio. The surfactant-modified zeolites can also be useful in adsorption of phenol and 4-CP (Kuleyin, 2007).

8.2.1 Biomass as Adsorbent

The use of biomass for wastewater treatment is increasing because of its availability in large quantities and at low prices. Almost any carbonaceous material may be used as precursor for the preparation of carbon adsorbents as against commercial ACs that are usually derived from natural materials such as wood, coconut shell, lignite, or coal. Coal is the most commonly used precursor for producing AC (Carrasco-Marín et al., 1996). The adsorption properties are influenced by the nature of the original vegetation and also by the physical-chemical changes occurring after deposition (Karaca et al., 2004). The presence of surface functional groups is an important factor, especially for adsorption of phenolics due to the chemical nature of the interactions. The presence of acidic functional groups imparts the ability to adsorb phenolic compounds under oxic conditions, and the presence of oxygen-containing basic groups such as chromene-type and pyrone-type can be crucial in promoting irreversible adsorption. Notable examples of modifications in this regard include the presence of molecular oxygen that reportedly increased the adsorptive capacity of GAC for phenolic compounds by threefold (Vidic et al., 1993), high-surface-area ACs prepared by chemical activation of coconut shell with KOH as active agent (Zhonghau and Srinivasan, 1999), AC prepared from agricultural waste by-products such as rubber seed coat (Rengaraj et al., 2002), tamarind nutshell (Goud et al., 2005) that can be used for the adsorption of phenol from aqueous solution, and beet pulp to prepare carbon for phenol adsorption by Dursun et al. (2005). Wu et al. (2005) prepared carbonaceous adsorbents with controlled pore sizes from carbonized fir wood (i.e., char) by KOH and steam activation for the removal of phenols. By changing the KOH/char ratio from 0.5 to 6, the KOH-ACs had surface areas ranging from 891 to 2794 $m^2\,g^{-1}$ and micropore volume of $0.76-0.82\ cm^3\,g^{-1}$. As opposed to this, carbons activated by steam at 900 °C for 5 and 7 h had a surface area of 1016 and 131 $m^2\,g^{-1}$ with micropore volume of 0.51 and 0.48 $cm^3\,g^{-1}$, respectively. Teng et al. (1998) reported surface modification using phosphoric acid impregnation followed by carbonization in a nitrogen atmosphere at 400 to 600 °C to obtain AC from Australian brown coal. Hobday et al. (1994) used a range of Victorian brown coal-based material to remove nitrophenol from an aqueous solution. Das and Sharma (1998) and Ahmaruzzaman and

Sharma (2005) investigated the possible use of coal, residual coal, and residual coal treated with H_3PO_4 for removal of phenol from wastewater.

Biomass such as agricultural solid wastes and waste materials, e.g., sawdust, rice husk, and bark, have been used as adsorbents. The role of sawdust materials in the removal of pollutants from aqueous solutions has been reviewed recently (Shukla et al., 2002). Dutta et al. (2001) carried out studies on the adsorption of p-nitrophenol on charred sawdust. The adsorption of phenol on to sawdust, polymerized sawdust, and sawdust carbon was also studied by Jadhav and Vanjara (2004). The potential of rice husk for phenol adsorption from aqueous solution was studied by Mahvi et al. (2004). Ahmaruzzaman and Sharma (2005) also investigated the potential of rice husk and rice husk char for the removal of phenolic compounds from wastewater.

Phenol removal has also been reported using industry waste such as fly ash by Aksu and Yener (1999), and capacity was found to be 27.9 mg g^{-1} for fly ash and 108.0 mg g^{-1} for granular AC. Application of bagasse fly ash as a low-cost adsorbent for the removal of phenol and p-nitrophenol has also been reported (Gupta et al., 1998; Mukherjee et al., 2007). A detailed account of various low-cost adsorbents for the adsorption of phenols was given in the review by Ahmaruzzaman (2008).

It is evident from the discussion above that a number of efficient adsorbents could be developed through judicious selection of raw precursors and/ or using different methods of activation. In this chapter, we further consolidate the existing knowledge base for synthesis, modification, and characterization of newer adsorbents and their applications in phenolic wastewater treatment. The adsorbents that are reviewed belong to biomass-based sorbents, charred biomass, industrial waste, and synthetic adsorbents. New data is provided on untreated forms of biomass-derived adsorbents such as banana peel (BP) and tea waste (TW); modified adsorbents derived from egg shell; and synthetic adsorbents from the class of binary mixed oxides.

All the materials derived from biomass that are reported here are available in abundant quantities all over the world and attach specific significance for developing countries such as India for utilization of such materials as adsorbents to tackle wastewater treatment problems. TW refers to tea fiber/stalks/ fluff or sweepings not including green tea or green tea stalks. The insoluble cell walls of tea leaves are largely made up of cellulose and hemicelluloses, lignin, condensed tannins, and structural proteins. Hence, due to the presence of a number of functional groups, mainly carboxylate, aromatic carboxylate, and phenolic hydroxyl groups, TW may be considered for the adsorption of phenolic compounds from aqueous solutions. The yearly production of tea in India is approximately 857,000 tonnes, which is 27.4% of the total world production (Wasewar et al., 2008). It is said that 100 kg of green tea leaves produces 22 kg of dry mass on average and ~18 kg tea goes for marketing. The remaining 4 kg of dry tea material is therefore wasted (Çay et al., 2004). The amount of TW produced annually by the factories after processing is believed to be thousands of tonnes in India alone.

8.2.2 Preparation and Activation of the Adsorbent

The standard procedure for preparation of raw untreated samples for utilization as adsorbents involves washing, color removal, drying, powdering, and sieving to the desired sizes. Color removal is required so that interference can be avoided from water-soluble fractions of the sample under study. Surface activation of the adsorbent surface is necessary because it increases the surface area, porosity, and number of functional groups that are known to enhance the adsorption process. The activation procedure can involve either physical or chemical activation. The selection of the time and temperature of activation needs to be optimized by characterizing the surface properties, such as surface area of the adsorbents obtained by treating the raw materials for different intervals of time and temperature. For the studies reported here, the activation temperature of 500 °C was chosen as an optimum temperature because the products obtained at a temperature higher or lower than 500 °C exhibited lower adsorption capacity, which is probably due to the collapse of the surface functional groups. In addition, all adsorbents were sieved to a particle size of 200–300 B.S.S. mesh size for use in adsorption studies.

Physical activation has two forms: (1) carbonization of the material in an inert atmosphere and (2) activation of the char at high temperatures in the presence of carbon dioxide or steam (Ahmad et al., 2007). In the activation process, the density as well as the number of the pores can increase, and amorphous decomposition products such as tars are burned off. However, the yield of the adsorbent is low in this method of activation, and thus the method is less preferred. This method is not included in the present study.

Chemical activation is carried out by impregnation of chemical agents followed by pyrolysis in an inert atmosphere at a high temperature. The purpose of activation here is to enhance the adsorption efficiency besides carbonizing the raw precursor without the evolution of huge amounts of fumes as in the case of physical methodology. Chemical activation is a preferred method over physical activation due to comparatively lower temperatures of 400–700 °C. Also, the yield is higher since carbon burnoff is not required.

The activating agents such as H_3PO_4, $ZnCl_2$, KOH, H_2O_2, NaOH, and MnO_2 can influence the modification and consequently the performance of the adsorbents. In the following discussion, the activated adsorbents were prepared by chemical activation with orthophosphoric acid. Phosphoric acid is a proven effective activating agent for many lignocellulosic materials. It produces adsorbents with high surface area and iodine adsorption value (Patnukao and Pavasant, 2008). Phosphoric acid is preferred over $ZnCl_2$ because of the environmental disadvantage associated with the latter. Phosphoric acid activation has also been applied to a wide variety of cellulose precursors such as coconut shell. The use of chemical reagents also promotes cross linking, resulting in a product that is less prone to volume contraction and volatile loss upon heating.

Different newer adsorbents, such as those derived from BP, TW, activated egg shell (AES), and neem leaf (NL) can be used for phenol adsorption. The preparation of untreated adsorbents from BP and TW involves extensive washing to remove any particulates, cutting into small pieces, drying in sunlight, sizing, and sieving through a 1 mm size before its use in adsorption experiments without any further treatment. Similarly, TW–based adsorbent materials are prepared by removing hydrolysable tannins and other soluble and colored components from the crushed TWs by washing with hot water (80 °C) for 6 h periods until a colorless solution of TW is observed at room temperature, drying the decolorized and cleaned TW in an oven at 105 °C, and sizing to 60–170 mesh particles. In the present chapter, we report another newer material: chicken egg shells. The egg shells were collected and dried for 2 h in an oven after properly washing and treated them with phosphoric acid in 1:1 ratio at 373 K for 2 h followed by heat treatment at 773 K for 2 h in a muffle furnace and subsequent washing with double distilled water to remove any free acids that might have been present on the adsorbent surface until the pH reaches nearly 7. The final product was then kept in an oven for 12 h, maintained at a temperature of 373 K to remove the excess moisture trapped inside the pores of the adsorbent. The adsorbent so prepared is named AESs.

8.2.3 Synthetic Adsorbents

The use of mixtures of metal oxides as catalysts has been known for years. Here, we report on the binary mixed oxide of iron and aluminum for its efficiency as an adsorbent for the removal of phenolic compounds. Preparation of Fe-Al mixed oxide was envisioned with the view that alumina has a specifically high surface area, while Fe_2O_3 is associated with potential applications in catalytic and adsorption studies. Our aim in preparation of Fe-Al mixed oxide is to explore the possibility of synthesizing porous media of a synthetic nature and to provide a comparative view with the rest of the agro-based adsorbents so prepared in the present study.

8.2.3.1 Binary Mixed Oxide of Al-Fe

The hydroxide of the mixed oxide was first prepared by the co-precipitation method. Equimolar volumes of $FeCl_3$ and $AlCl_3$ prepared in 0.01 mol L^{-1} HCl were mixed and heated to 60 °C. Aqueous ammonia was used as a precipitating agent, which was added drop-wise with constant stirring until the pH of the solution was nearly 7. The solution was then aged at 65 °C in a water bath for 24 h, and the mother liquor was decanted. The solid mass was then washed repeatedly with double distilled water in order to remove chloride ions and dried in hot-air oven at 65 °C for 2 days. The corresponding product was obtained by calcination at three different temperatures of 600, 800, and 1000 °C for 2 h to study the effect of calcination temperature in a muffle furnace at a heating rate of 10 °C min^{-1}. For convenience, the mixed oxide so prepared was denoted by the acronyms BMO_L, BMO_M, and BMO_H. The subscripts L, M, and H denote low, medium, and high temperatures for the mixed oxide so prepared.

8.3 ADSORBENT CHARACTERIZATION

Characterization of the adsorbent is a crucial step for adsorption studies. BET (Brunauer-Emmett-Teller) surface area analysis, FTIR (Fourier transform infrared) studies, XRD (X-ray diffraction), SEM (scanning electron micrography), proximate and ultimate analysis, and pH_{ZPC} are some of the methods used to characterize the adsorbent.

8.3.1 Fourier Transform Infrared

FTIR studies can provide useful qualitative information about the surface functionality. The FTIR spectrum helps to reveal the various functional groups present on the adsorbent surface before and after adsorption. The spectra can therefore help in interpretation of which functional groups are responsible for adsorption and whether H bonding is involved in the interaction between the adsorbate and the adsorbent. All these inferences pertaining to chemisorption can also be drawn from FTIR studies. FTIR spectra were collected in the transmission mode from 4000 to 400 cm^{-1}

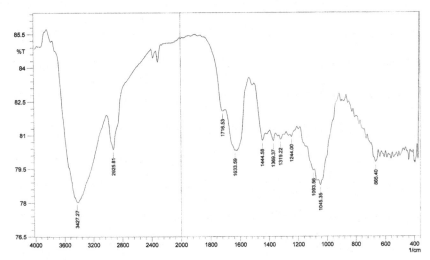

Figure 8.1 FTIR spectrum of untreated banana peel powder.

with a resolution of $2\ \text{cm}^{-1}$. For mixed metal oxides, IR spectra were collected in Far-IR region below $400\ \text{cm}^{-1}$ because a majority of the peaks lie in this region.

The FTIR spectrum of untreated BP (Figure 8.1) depicts a sharp peak at $3427.27\ \text{cm}^{-1}$, which is indicative of H-bonded OH group stretching. Another intense band at $1633.59\ \text{cm}^{-1}$ is ascribed to hydrated hydroxyl groups (—OH). The band centered at $1045.35\ \text{cm}^{-1}$ corresponds to the stretching vibration of the Si—O bond. The band at $1244\ \text{cm}^{-1}$ indicates the presence of a carboxyl group.

The FTIR spectrum of unmodified TW is shown in Figure 8.2. The intense, nearly sharp adsorption band at $3425\ \text{cm}^{-1}$ can be assigned to OH stretching of a hydroxyl group. These vibrations are often associated with H bonds; thus the surface —OH groups interact with water molecules adsorbed (Zawadzki, 1980). The sharp and intense bands at 2925 and $2363\ \text{cm}^{-1}$ belong to H—C—H stretching, while the broad and intense peak at $1638\ \text{cm}^{-1}$ belongs to C=O stretching of amides. The peak position at $1053\ \text{cm}^{-1}$ is compatible with a silicate ion, while the broad band centered at $668\ \text{cm}^{-1}$ may be linked to the C—H group frequencies of acetylenic linkage.

The FTIR spectrum of AES (Figure 8.3) also indicates the presence of hydrogen bonded O—H stretching at 3444.63 and $2925.81\ \text{cm}^{-1}$. The C—O functional group can be envisaged from the presence of intense to minor peaks at 1257.50, 1205.43, 1118.64, and $1062.70\ \text{cm}^{-1}$. The presence of a silicate ion is revealed by the sharp peak at $939.27\ \text{cm}^{-1}$, while the

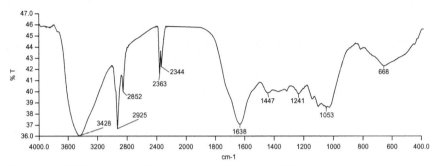

Figure 8.2 FTIR spectrum of raw untreated tea waste (TW).

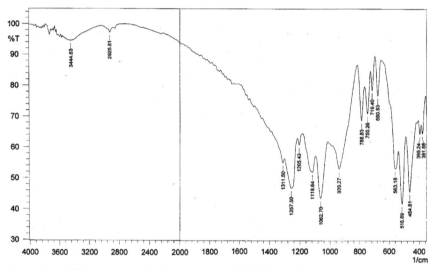

Figure 8.3 FTIR spectrum of activated egg shell (AES) adsorbent.

peaks at 750.26 and 680.83 may be due to the presence of C—Cl stretching. A sharp and an intense peak at 464.81 cm^{-1} is assigned to aryl disulfides (S—S stretching).

Figure 8.4 depicts the FTIR spectrum of binary mixed oxide of Al-Fe (BMO) of Fe and Al, showing bands at low wave numbers ≤ 400.00 cm^{-1}, which can be attributed to vibrations of Fe—O bonds of iron oxide in the case of BMO_L. The disappearance of certain functional groups at higher temperatures of 800 and 1000 °C is vividly displayed in the FTIR spectrum of BMO_M and BMO_H, respectively, clearly showing the sintering effect. As demonstrated by the spectrum, practically no peaks were observed in the regions above 450 cm^{-1} and hence the spectrum was analyzed in the lower wavelength region below 400 cm^{-1} (Figure 8.5). Almost all the peaks in BMO_L were observed in the far-IR region.

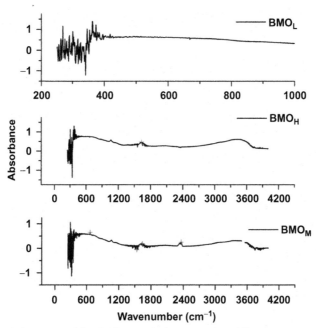

Figure 8.4 FTIR spectra of BMO of Fe and Al prepared at different temperatures.

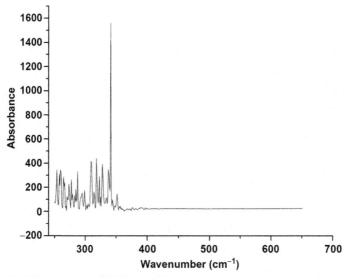

Figure 8.5 FTIR spectrum of BMO_L in elaborated view.

The major IR bands for BMO_L are recorded at 385, 360, 341.37, 327.87, 318.23, and 287.37 cm^{-1}. The bands at 318.23, 360, and 385 cm^{-1} are attributed to $\alpha\text{-}Fe_2O_3$. The peaks above and around 250 cm^{-1} also represent vibrations arising out of Al—O bond. Peaks below 250 cm^{-1} are due to the distortion of the hematite network due to introduction of Al atoms.

8.3.2 Scanning Electron Micrography

SEM is one of the most widely used diagnostic tools. It helps to reveal whether the surface of the adsorbent is heterogeneous or homogeneous and the degree of porosity of the adsorbent for entrapment of the adsorbate. The large depth of field of 30 mm provided by SEM allows a large amount of sample to be studied at a time, resulting in a three-dimensional image representation of the sample. The magnification in SEM is of the range 10×500 K $\times$, and this property of greater magnification, larger depth of field, and higher resolution coupled with information on composition and crystallography delivers SEM as one of the highly used instruments in morphological analysis. SEM is therefore a primary tool for characterizing the surface morphology and fundamental physical properties of various adsorbents.

The SEM images of some of the candidate adsorbents under study reveal the characteristic morphology of the adsorbents. The analysis of the SEM images of AES (Figure 8.6a) shows the irregularity in surface morphology and the porous nature of the adsorbent. It is apparent from the surface texture that AES is composed of rough heterogeneous sites. The magnified image of AES (Figure 8.6b) shows the development of pores as a result of

Figure 8.6 SEM images of AES at different magnifications, (a) micrograph at $200 \times$ magnification and (b) at 2 K$\times$.

Figure 8.7 SEM of ATW at different magnifications, (a) at 1 K × magnification and (b) at 3 K × magnification.

chemical activation. The surface morphology of ATW (activated tea waste) prepared by a similar method of activation as that of AES is shown in Figure 8.7. It shows the presence of several porous structures that may have developed during the activation process with phosphoric acid. The roughness of the surface indicates a high surface area for adhering adsorbate molecules. One can observe the partially porous nature of ATW when the image at lower magnification (Figure 8.7a) and that at higher magnification (Figure 8.7b) are compared. The minute cavities may have been created during the chemical activation process when volatile components and gases escaped through the adsorbent surfaces, creating pores of various sizes.

The micrograph of natural adsorbent BP (Figure 8.8) is uniquely different. It appears to show a nearly uniform arrangement/stacks of flattish structures intercalated by pores of several sizes at 100 × magnification (Figure 8.8a). The arrangement looks compact in nature with pores of several sizes (in higher magnified form at 1000 ×, see Figure 8.8b). Although the precursors are different, one can ascertain the difference in surface morphology comparing SEM images of chemically activated forms as in the earlier two incidences and those of untreated biomass-based adsorbent, BP. The surface morphology of the synthesized binary mixed oxide of aluminum and iron, BMO_L as analyzed by SEM (Figure 8.9a and b), shows a partly homogeneous surface with polygonal-shaped crystals. A relatively ordered 3D pattern with a tendency to form an agglomerate that may be due to shrinkage as a result of calcination. This is very similar to the SEM of zeolite HY. The SEM is a typical of crystallized structure with good microporosity, which is

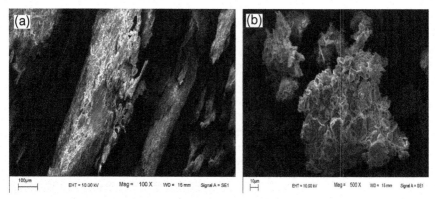

Figure 8.8 SEM image of untreated adsorbent BP at different magnifications, (a) at 100 × and (b) at 1 K×.

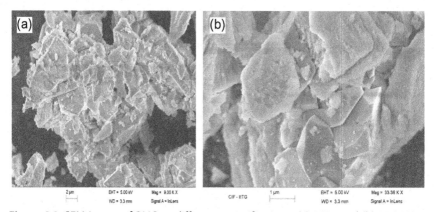

Figure 8.9 SEM image of BMO at different magnifications, (a) 9 K× and (b) 33.36 K×.

further affirmed by the total pore volume determined by N_2 adsorption-desorption isotherm (Table 8.1).

8.3.3 BET Surface Area Analysis

Surface area is one of the most important parameters for adsorbent development. As a general rule, the higher the surface area, the greater the adsorption capacity of the adsorbent (Cooney, 1999), barring specific adsorption. The internal surface area study is confined to the pore structure of the adsorbent material. The pore structure is also a major factor affecting the adsorption process. If the pore diameter is such that the adsorbate molecules are larger, then

Table 8.1 Adsorption parameters of BMO_L by the application of the BET model to nitrogen isotherms at 77 K

Single point surface area $(m^2\,g^{-1})$[a]	S_{BET} $(m^2\,g^{-1})$	Pore volume $(cm^3\,g^{-1})$[b]	Pore size (nm)[c]
98.93	102.63	0.35	13.61

[a] $P/P_0 = 0.20$.
[b] Pores < 1150.48 Å at $P/P_0 = 0.982$.
[c] $4V/A$ by BET.

less adsorption will take place due to steric hindrance. The nitrogen adsorption and desorption tests were used to characterize the surface area and the pore size using Coulter SA3100. The samples were first degassed at 473 K for at least 4 h under vacuum. The BJH (Barret-Joyner-Halenda) desorption method was used to calculate the average pore diameter and cumulative volume of the pores, while the specific surface area was evaluated from the N_2 adsorption isotherms by applying the equation described by Brunauer et al. (1938) in the relative pressure (P/P_0) range. The study of N_2 adsorption and desorption is important because many conclusions concerning the structure of the adsorbents can be drawn on the basis of the adsorption-desorption isotherm.

An important characteristic of good adsorbents is their high porosity and larger surface area with more specific adsorption sites (Boer and Linsen, 1970; Tein, 1994). The specific surface area of the different adsorbents was calculated from nitrogen adsorption data, according to the BET model. The pore structures of the adsorbents were characterized on the basis of low-temperature nitrogen adsorption/desorption isotherms in the micro and mesopore range.

The hysteresis loop at relatively low pressure in ATW (Figure 8.10) indicates small pore size. Comparative data on surface area, pore size, and pore volume of ATW with TW is given in Table 8.2. The specific surface area and pore volume of TW is significantly smaller than those of ATW. Some drastic increases in surface area and pore sizes are observed when the raw biomass adsorbent is chemically activated. After activation, biomass adsorbents such as NL powder improved from an initial surface area of $1.73\ m^2\,g^{-1}$ to a surface area of $890.45\ m^2\,g^{-1}$ (Ahmaruzzaman and Laxmi Gayatri, 2011). The effect of activation is also observed in other pore parameters, such as pore volume ($0.003\ cm^3\,g^{-1}$ in untreated form to $0.89\ cm^3\,g^{-1}$ in chemically activated form) and pore diameter from $5.60\ nm$ in untreated form to $44.53\ nm$ in activated form. Adsorbents with

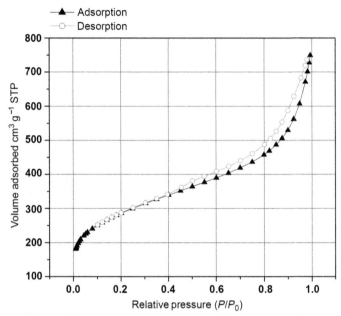

Figure 8.10 Nitrogen isotherm plot of ATW; low pressure dose $= 10.0 \, cm^3 \, g^{-1}$ STP, analysis bath $= 77.74$ K.

Table 8.2 Surface Characteristics of TW and ATW

Property	TW	ATW
1. BET surface area $(m^2 \, g^{-1})$	2.91	1016.26
2. Single point surface area[a] $(m^2 \, g^{-1})$	2.34	994.45
3. Adsorption cumulative surface area of pores[b] $(m^2 \, g^{-1})$	2.19	635.77
4. Desorption cumulative surface area of pores[b] $(m^2 \, g^{-1})$	0.64	792.44
5. Total pore volume[c] $(cm^3 \, g^{-1})$	0.003	1.04
6. Adsorption average pore diameter[d] (nm)	4.56	4.08
7. BJH adsorption average pore (nm)	15.52	6.35
8. BJH desorption average pore diameter (nm)	53.35	5.45

[a]At P/P_0 0.20.
[b]Applying BJH model, pore diameter in range of 1.70–300.00 nm.
[c]Pores less than 73.40 nm at P/P_0 0.97.
[d]BET method.

specifically low surface area in the range of 0.50–$4.00 \, m^2 \, g^{-1}$ include BP and AES. These figures are very low in comparison with other prepared adsorbents. Their surface characterization properties are given in Table 8.3.

The BET surface area of binary mixed oxide, BMO, was $102.64 \, m^2 \, g^{-1}$, while the single point surface area at P/P_0 of 0.20 was $98.93 \, m^2 \, g^{-1}$.

Table 8.3 Specific surface area, pore volume, and average pore diameter of BP and AES

Adsorbent	Specific surface area (m² g⁻¹)			Total pore volume (cm³ g⁻¹)	Pore diameter (nm)		
	BET	BJH$_{ads}$	BJH$_{des}$		BET	BJH$_{ads}$	BJH$_{des}$
1. BP	0.69	0.58	0.25	0.0007^a	4.22	6.09	12.52
2. AES	3.96	3.12	4.01	0.008^b	8.63	17.73	17.50

^{a}Pores < 214.66 nm.
^{b}Pores < 78.81 nm.

The BJH adsorption and desorption cumulative surface area of the pores between 17 and 3000 Å were determined to be 112.96 and 127.47 m² g⁻¹, respectively. The single point adsorption total pore volume of pores less than 1150.49 Å in diameter at P/P_0 0.98 was 0.35 cm³ g⁻¹. The pore size of BMO, as determined by the BET method, was 136.06 Å in diameter, while the adsorption and desorption average pore diameter determined by the BJH method were 126.13 and 111.39 Å, respectively, thus proving the microporous nature of the material. It is seen from Table 8.1 that the single point surface area and BET–specific surface area (S_{BET}) of BMO$_L$ found in this investigation are almost comparable.

The nitrogen adsorption isotherm analysis of BMO indicated that the isotherm of BMO belong to the I-II hybrid shape defined by IUPAC classification, which is exhibited by a microporous solid containing a well-developed mesoporous system. The presence of a hysteresis loop in the desorption branch of the isotherm at relatively high pressure suggested a considerable development of mesoporous structure. Another important characteristic of an adsorbent is pore size distribution (PSD). It determines the fraction of the total pore volume that is accessible to adsorbate molecules of definite size and shape. As discussed earlier, there are three broad classifications of pore dimension: micropore (≤ 2 nm), mesopore (2–50 nm), and macropore (≥ 50 nm). Figure 8.11 shows the PSD curve of BMO during adsorption (Figure 8.11a) and desorption (Figure 8.11b), calculated by using the standard BJH method (Barrett et al., 1951). It shows that there is a sharp increase in the pore volume in the micropore region.

It may be summarized from the data of this work and the literature that adsorbent surface area can be ranked in the following increasing order:

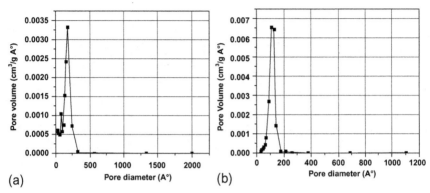

Figure 8.11 Pore size distribution of BMO. (a) BJH adsorption pore volume and (b) BJH desorption pore volume.

Banana Peel < Tea Waste < Neem Leaf < Activated Egg Shell
< Activated Neem Leaf < Activated Tea Waste

Considerable differences in surface area are observed for the natural and chemically activated forms of adsorbents.

8.3.4 X-ray Diffraction

XRD patterns provide information on the particle size and defects, while the peak relative intensities provide insight into the atomic distribution in the unit cell. For the correct interpretation of powder diffractograms, a good peak-to-background ratio is an important issue. The background in powder diffraction can originate from many sources, which can be related to instruments or the sample itself. The analysis of XRD patterns and diffraction peaks will characterize the crystalline phase of the adsorbent. This is particularly important in the case of synthesized metal oxides.

Powder XRD patterns were obtained with PAN analytical, X'pert PRO using 40 kV, 30 mA, Cu Kα radiation ($\lambda = 1.5405$ Å) with a scan speed of $2\theta = 5$ min^{-1} in the scan range from 10° to 130° 2θ. Qualitative analysis was performed by the Powder Diffraction File database. The XRD method appears to be a reliable method for studying surface area, which requires the use of crystal size measurement. The peak broadening is a suitable XRD parameter (Schwertmann and Latham, 1986) used for determining the mean crystallite dimension.

The Debye-Scherrer equation is often used to calculate the mean crystallite size.

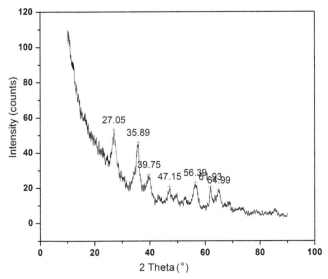

Figure 8.12 X-ray diffractogram for BMO$_L$.

$$d = 0.9\lambda/\beta\cos\theta \qquad (8.1)$$

where d is the grain size of the crystallite, λ is the wavelength of the X-ray used, θ is the angle of reflection, and β is the full width at half maximum or the broadening of the diffraction line in radians. The grain size thus determined by using Equation (8.1) is 15.26 nm. The XRD pattern of BMO$_L$ of Fe–Al (Figure 8.12) shows strong Cu Kα peaks at 27.05° (AlFeO$_3$), 35.89° (FeAlO$_3$), 39.75° (Fe$_2$O$_3$), 47.15° (FeAlO$_3$), 56.39° (Fe$_2$O$_3$), and 61.93° (Al$_2$O$_3$). The crystalline nature of BMO is confirmed by the presence of several peaks in the XRD pattern. The results of XRD analysis at different temperatures (Figure 8.13) showed that the structure of BMO changed with heat treatment.

8.3.5 Physicochemical Analysis of the Adsorbents

Data from proximate, ultimate, and chemical oxygen demand (COD)/biological oxygen demand (BOD) of the candidate adsorbents constitute the physicochemical analysis. Proximate analysis of the adsorbent constitutes determination of moisture, volatile matter, fixed carbon, and ash content. The chemical and structural analyses of the adsorbents were conducted according to the American Standard Testing method. The microanalysis study consisted of the determination of carbon, hydrogen, and nitrogen content of the adsorbent samples. In the present study, the CHNS (O) Analyzer

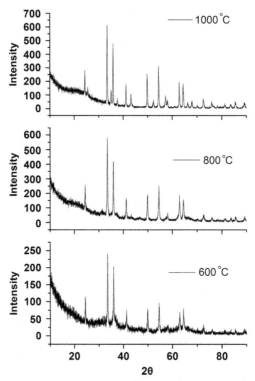

Figure 8.13 Effect of calcination temperature on the structure of BMO.

(model: Perkin-Elmer 2400 series II) is used for determining the percentages of carbon, hydrogen, nitrogen, and sulfur/oxygen in organic compounds. The principle used is the "Dumas method," where complete and instantaneous oxidation of the samples takes place by "flash combustion." The combustion products are segregated by a chromatographic column and detected by the thermal conductivity detector, which gives the concentration of the individual components of the mixture. The precision of the instrument used is 1–1.5%, while the accuracy depends on the quality and the preparation of the sample.

CHN analysis study of TW showed the carbon, hydrogen, and nitrogen content as 45.37%, 6.04%, and 3.05%, respectively, while NL (neem leaf powder) showed a carbon, hydrogen, and nitrogen content of 42.10%, 5.95%, and 3.37%, respectively. Table 8.4 provides the result of macro- and microanalysis tests for raw untreated BP. It depicts a fairly high ash content with low fixed carbon content, which may somehow contribute to the low adsorption capacity of this adsorbent.

Table 8.4 Macro- and microanalysis data for BP

Macroanalysis data (%)		Microanalysis data (%)	
Moisture content	3.31	C	42.60
Ash content	50.24	H	6.56
Volatile matter	40.82	N	1.58
Fixed carbon	5.63		

In the adsorptive removal of 4-NP by ATW, the COD/BOD$_5$ ratio was found to be 28.24, which indicates the nonbiodegradable nature of the waste. The synthetic stock solution had an initial BOD value of 203.70 mg L^{-1}, which decreased to 10.20 mg L^{-1} after treatment with ATW indicating ~95% BOD removal. The reduction in these values indicates the effectiveness of the treatment technique. The initial COD was found to be 288 mg L^{-1}, while the final COD was found to be 128 mg L^{-1} with % COD removal of 55.55%. AES showed a very low content of C, H, and N, explaining its low adsorptivity. The %C, %H, and %N content were 8.18, 0.12, and 0.94, respectively.

8.4 SINGLE-SOLUTE ADSORPTION STUDIES: PERFORMANCE AND EVALUATION

A synthetic solution of phenol-containing wastewater was prepared. The performance of some potential adsorbents such as BP, TW, ATW, AES, and BMO were evaluated, compared with JSC (jute stick char), PPC (potato peel char), AAJSC (acid-activated jute stick char), ANL (activated neem leaf), and APPC (activated potato peel char), and an attempt was made to understand the mechanism of adsorption in these different types of adsorbent. Desorption studies using various eluents in different concentrations were investigated for the regeneration of the adsorbent. Efficiency was also tested in several cycles.

8.4.1 Adsorption on Untreated Adsorbents

The untreated adsorbents showed low adsorption capacity. The studies on utilization of these untreated adsorbents for the effect of several parameters on adsorption were not extended further as such. In an attempt to increase their efficiency, the acid-activated forms of the raw precursors were tested. More extensive study has been done on the adsorption studies of activated forms rather than the raw/untreated form of the adsorbent.

8.4.1.1 Untreated Banana Peel

Earlier researchers have utilized banana pith (Namasivayam and Kanchana, 1992) for the removal of colors from wastewater, but the peel of the fruit itself has not been utilized for the removal of phenolic compounds. BP was collected from the local market in the Silchar area of northeast India in large batches as required.

The adsorption isotherms of BP powder at different temperatures (Figure 8.14) exhibited a decrease in adsorption with increasing temperature. Additionally, the adsorption isotherms also showed increased steepness at higher equilibrium concentration, C_e. The adsorption that occurs follows the type III isotherm, which proceeds when the adsorbate interaction with the adsorbed layer is greater than the interaction with the adsorbent surface. Due to its relatively low adsorption capacity (Langmuir's theoretical adsorption capacity, $a_L = 8.40$ mg g^{-1}), the application of BP powder adsorbent on other low-adsorbing phenols, namely, 4–CP and phenol, was not extended as such. Furthermore, the specific surface area of this adsorbent is the least recorded in the present study; consequently, the adsorption of 4-CP and phenol can be considered to be negligible. The shape of the isotherm as exhibited by this adsorbent (Figure 8.14) indicates that it is not economical enough to be used

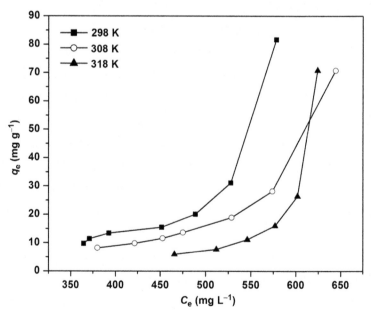

Figure 8.14 Single-solute adsorption of 4-NP on BP adsorbent at different temperatures.

on a large scale. These results are for the untreated form of the banana fruit peel. However, after certain modifications, there may be some potential use. This was indicated by our earlier studies with some charred forms of biomass adsorbents such as jute stick char (Ahmaruzzaman and Laxmi Gayatri, 2010a) and PPC (Laxmi Gayatri and Ahmaruzzaman, 2012), which had good potential for phenol removal.

8.4.1.2 Tea Waste Powder

The degree of adsorption of the three phenolic compounds in single solute using TW (Figure 8.15) was not seen to be satisfactory. Consequently, extensive study for the removal of 4-CP, 4-NP, and phenol from this industrial waste when it is not subjected to any pretreatment is not warranted. In the adsorption experiment, the adsorbent dose was varied within the range of 0.12–1.50 g per 20 mL of the adsorbate. It was observed that the percentage removal of phenols increased with the increase in the adsorbent dose, which was expected in an adsorption equilibrium, while the capacity at equilibrium, q_e (m g^{-1}) decreased, corresponding to the lowered concentrations. There is a possibility of partial aggregation at a high adsorbent dose, leading to a decrease in active sites.

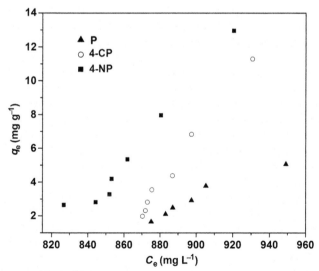

Figure 8.15 Experimental adsorption isotherms at 298 K for the three phenolic systems on TW.

8.4.2 Adsorbent Modification Using Phosphoric Acid Treatment

8.4.2.1 Activated Tea Waste

TW can serve as a potential low-cost adsorbent for the removal of phenolic compounds, especially in a country like India, which is the second largest producer of tea (*Camellia sinensis*) in the world, with the state of Assam alone having a major share. A huge amount of TW is generated by a number of tea industries during the processing of tea leaves, resulting in a solid waste disposal problem. Keeping in mind this large-scale availability of TW from *Camellia assamica* and *C. sinensis* in this Indian region, research on utilization of TW for phenolic compound removal will certainly be one step toward utilizing waste in the form of a cost-effective adsorbent requiring no regeneration.

The experimental equilibrium data for 4-CP adsorption by ATW indicated that the theoretical maximum adsorption capacity was 43.10 mg g^{-1}. The adsorption efficiency of 4-CP by ATW is found to be relatively lower than that of 4-NP, having a capacity of 142.85 mg g^{-1} (Ahmaruzzaman and Laxmi Gayatri, 2010b). In both cases, the Langmuir adsorption was found to be the best fitting isotherm in describing the adsorption process with a coefficient of determination, R^2, value greater than 0.9. On the other hand, the adsorption of phenol by the same adsorbent was found to be best explained by the Freundlich model. This shows that every adsorption system is unique, and the adsorption phenomenon is not only influenced by adsorbent properties but by certain adsorbate properties such as degree of solubility/hydrophobicity, molecular size, and adsorption energy. These adsorption capacity values using ATW give evidence of increased performance of the same biomass with specific modifications.

8.4.2.2 Activated Egg Shell

Chicken egg shells were activated for use as a low-cost adsorbent for the removal of phenolic compounds. Calcite and calcareous soil are known to be good adsorbent sources, and since egg shells are mainly composed of calcium carbonate, they are also expected to show good adsorption properties. Figure 8.16 illustrates the single-solute equilibrium adsorption for the three-phenol system at 298 K using AES. It also indicates an adsorption order (decreasing) of 4-NP, 4-CP, and phenol, respectively. Analysis of the data for both 4-CP and phenol adsorption by AES indicates a physical adsorption process. When isotherm modeling with the five different linearized isotherms was conducted, the Temkin model was found to describe the

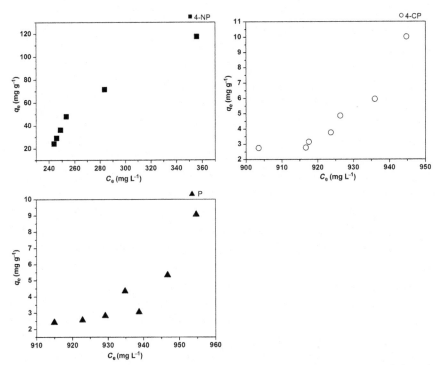

Figure 8.16 Comparative single-solute adsorption isotherms of 4-NP, 4-CP, and phenol using AES as adsorbent.

adsorption of 4-NP on AES, while the Freundlich model was able to describe the adsorption of phenol and 4-CP. Thermodynamic studies on batch adsorption by AES showed the feasibility of the process as indicated by the negative values of $\Delta G°$ for all the three phenol systems. In all the three systems, the process was exothermic as indicated by negative values of $\Delta H°$.

8.4.3 Adsorption by Synthetic Adsorbents

Oxides of iron and aluminum are present quite abundantly in the Earth's crust, while Fe or Al hydroxides occur as a common industrial waste. The oxide forms have higher surface area and are associated with a variety of applications, including catalytic agent, but mixed Fe-Al oxide (abbreviated as BMO in the present study) has not been investigated for the removal of phenols. Its adsorption behavior for the removal of phenols is investigated in the present work. Three types of BMO were synthesized at different temperatures: at 600 °C labeled as BMO_L, 800 °C labeled as BMO_M, and 1000 °C labeled as BMO_H for adsorption of 4-CP, 4-NP, and phenol.

8.4.3.1 Binary Mixed Oxide of Iron and Aluminum

The thermal treatments of these hydroxides lead to the formation of a mixed oxide with high surface area. Preliminary adsorption experiments with a known amount of the sample for the BMO samples prepared at different temperatures showed that BMO_M and BMO_H have relatively low adsorption capacity; thus the batch studies on BMO_M and BMO_H were not investigated for these forms. The lower adsorption capacity of BMO_H and BMO_M is attributed to the destruction of oxygen surface groups at higher temperatures. The equilibrium adsorption experiments studied in detail for BMO_L exhibited a comparatively higher adsorption capacity among the three mixed oxides. BMO_L showed a distinctly stronger adsorption of 4-NP (20.33 mg g^{-1}) than 4-CP (13.13 mg g^{-1}) and phenol (11.23 mg g^{-1}). Only slight differences in removal capacity were observed for the latter two phenols. The relatively low adsorption capacity of BMO from the activated adsorbents may be attributed to the highly crystalline phase as exhibited by the SEM and XRD image. It may be recalled that in wastewater treatment, a higher crystalline phase is less desirable because it provides a comparatively smaller surface area.

8.5 ADSORPTION MECHANISM

Understanding the mechanism of adsorption is useful in order to optimize adsorptive wastewater treatment. In the case of carbon-based adsorbents, the carboxyl groups are mainly responsible for the uptake of phenol by the donor-acceptor complex mechanism (Mattson et al., 1970). Considerable research has been done to identify mechanisms of adsorption of phenolic compounds, and it is suggested that carbonyl oxygens of the carbon surface act as the electron donor and the aromatic ring of the solute as an acceptor. However, the real mechanism in AC adsorption is complex and difficult to explain. The mechanism of irreversible phenol adsorption is proposed based on the analysis of the average hysteresis on adsorption-desorption isotherms. Terzyk (2003) suggested that the irreversibility is caused by two factors: the creation of strong complexes between phenol and surface carbonyl and lactones as well as by the polymerization. Accordingly, the mechanism of phenol adsorption is not only determined by so called "π-π interactions" and "π donor-acceptor complex formation" but also by (strongly depending on temperature) the "solvent effect" balancing the influence of the two mentioned factors on this mechanism.

Adsorbate-adsorbent interactions are governed by dispersion, electrostatic interaction, and chemical bonds. Weak chemical bonds, such as those of π electrons or π-complexation, indicate the potential for fabricating new and highly selective adsorbents, which is very much desirable in the industrial wastewater treatment process with adsorption methodology where selective removal of hazardous chemicals is required. The substitution of electron-withdrawing groups such as the nitro group reduces the electron density of the π-system of the benzene ring, which acts as an electron acceptor; the carboxyl oxygen on the carbon surface acts as an electron donor.

On the basis of the FTIR spectrum of ATW (Ahmaruzzaman and Laxmi Gayatri, 2010b), it is seen that the surface of ATW is enriched with groups such as —COOH, —OH, and —C=C. Consequently, it is assumed that there is a possibility of π—π interaction between C=C double bonds or the benzene ring of phenol molecules and the C=C of ATW surface (Chen et al., 2007; Ma et al., 2011). In addition to this π—π interaction, the presence of a typical electron acceptor group, the —Cl group in the case of 4-CP, can contribute toward electron-donor acceptor interaction between 4-CP and the ATW surface (which can act both as an acceptor and a donor group). Thus, the most likely mechanism for 4-CP adsorption by ATW is electron donor-acceptor interaction. Similarly, the presence of the electron-withdrawing —NO$_2$ group causes the aromatic ring of the solute molecule (4-NP) to act as an electron acceptor. The details of the mechanism of interaction of phenols on the ATW surface may also be supplemented from pH and pH$_{ZPC}$ (value of 2.40) studies where the uptake of phenols was observed to be higher in the acidic pH range with the surface remaining mostly negatively charged with a fall in percent adsorption at higher pH above the pK_a value of the respective dissociated phenol molecules. This therefore suggests the participation of neutral phenol molecules toward the diffusion into the porous surface of ATW.

8.5.1 Adsorption Mechanism in AES

The nature of adsorption in AES may similarly be explained on the basis of FTIR. The presence of carboxylic, carbonyl, and hydroxyl groups in the adsorbent is expected to play a role in the adsorption mechanism. The usual pH factor also plays a significant role in explaining the adsorption mechanism. Again, properties of the adsorbate molecule, like their molecular sizes, solubility factor, pK_a, and nature of the substituents, also influence the adsorption process. The molecular size determines the accessibility of the adsorbate to the pores of carbon, while aqueous solubility is related to the degree of hydrophobic interaction between the adsorbate and the adsorbent surface and pK_a to the dissociation of the adsorbate. The substituents present on the benzene ring of the adsorbate also influence the adsorption process due to their electron withdrawing/releasing abilities, which finally affects the nonelectrostatic interactions between the adsorbate and the adsorbent. Mattson et al. (1970) proposed that the adsorption of phenolic compounds on an AC surface was due to the π-electron system of the aromatic ring, which can act as an electron acceptor in a donor-acceptor complex, while Coughlin and Ezra (1968) suggest dispersion forces acting between the π-electrons of the graphitic basal planes of AC and the phenol molecule. However, further research is needed to gain better insights into the real mechanism of these adsorbents, including the one for untreated adsorbents where there is more likely participation of weak van der Waals forces.

8.6 RESULTS FROM BATCH ADSORPTION

The performances of the prepared adsorbents were compared using the maximum adsorption capacities. Table 8.5 lists data for the adsorption of 4-CP, 4-NP, and phenol by both the charred and activated forms.

A comparatively good adsorption capacity was noted for the prepared adsorbents for which the value of q_{max} is quite notable. Therefore, this assessment provides a suitable option for further commercial exploitation of the product, which is also expected to be sustainable. It is also seen from Table 8.5 that kinetic data in the studied cases followed the pseudo-second order model more favorably. It may be summarized that activated adsorbents with higher surface area and porosity adsorb phenols to a greater extent, whereas untreated adsorbents having low surface area show little or negligible adsorption. Overall, adsorption capacity follows 4-NP > 4-CP > phenol. The solubility factor also attributed to explaining this observed result. It may be noted that the solubility of the three phenols in aqueous media is of the order phenol > 4-CP > 4-NP. A comparison of the solubility and adsorption clearly indicates an inverse relationship between the extent of adsorption and solubility, i.e., phenols with lesser affinity for water have a higher tendency to get adsorbed at the solid-liquid interface.

Adsorption of 4-nitrophenol, 4-CP, and phenol on different activated and nonactivated adsorbents indicated significant performance differences, and the transport mechanism was found to differ widely for different adsorbents. The activated adsorbents were efficient for removal of 4-CP, 4-NP, and phenol. The capacity of the adsorbents increased tremendously after

Table 8.5 Comparative summary of various chemically and thermally activated adsorbents

Adsorbent	System	Max. adsorption capacity (mg g^{-1})	Thermodynamic property
ANL (Ahmaruzzaman and Laxmi Gayatri, 2011)	4-NP	549.01	Exothermic
	4-CP	219.81	Exothermic
	Phenol	69.7	Exothermic
ATW	4-NP (Ahmaruzzaman and Laxmi Gayatri, 2010b)	142.85	Exothermic
	4-CP	44.91	Exothermic
	Phenol	35.53	Endothermic
AES	4-NP	117.57	Exothermic
	4-CP	10.02	Exothermic
	Phenol	9.06	Exothermic
AAJSC	4-NP (Ahmaruzzaman and Laxmi Gayatri, 2010c)	39.38	Exothermic
	4-CP	31.91	Exothermic
	Phenol	27.48	Endothermic
APPSC	4-NP (Laxmi Gayatri and Ahmaruzzaman, 2011)	31.96	Exothermic
	4-CP	23.82	Exothermic
	Phenol	22.82	Exothermic
JSC	4-NP (Ahmaruzzaman and Laxmi Gayatri, 2010a)	40.62	Exothermic
	4-CP	24.06	Exothermic
	Phenol	22.93	Exothermic
PPC	4-NP (Laxmi Gayatri and Ahmaruzzaman, 2012)	27.48	Exothermic
	4-CP	20.66	Exothermic
	Phenol	16.99	Endothermic

chemical activation, while intermediate capacities were observed for the thermally activated or the charred forms. This indicates that the activated adsorbents are most likely to be efficacious for certain wastewater treatment purposes, depending upon effluent requirements. It was noted that the untreated adsorbents have almost identical isotherms for the three-phenol system. The efficiency of ATW and ANL for phenol adsorption is also ascertained by the rapid uptake of adsorbate within the first 1 h of adsorption. The adsorption capacity in all the studied adsorbents showed the order P < 4-CP < 4-NP.

8.7 MULTICOMPONENT ADSORPTION STUDIES

In practice, wastewater consists of a broad range of adsorbates with interaction among the adsorbate molecules that even complicates the theoretical study of multicomponent adsorption. Very little literature or information is available on column performance of multicomponent systems. Some recent reports on the multicomponent adsorption of toxic chemicals from wastewater, which focuses on the removal of mixtures of metals (Srivastava et al., 2008, 2009), metals with phenols (Aksu and Akpinar, 2000; Aksu and Yener, 2001; Quintelas et al., 2006), and mixtures of organic compounds (Seidel and Gelbin, 1988; Sulaymon and Ahmed, 2008; Yang et al., 2008), attempt to predict multisolute adsorption. However, interpretation of the interactive influence of the solute components from experimental data needs to be studied in the context of understanding the real adsorption phenomena.

Binary solute adsorption of mixtures of 4-CP and 4-NP, phenol and 4-NP, and phenol and 4-CP were studied. Also, the single component adsorption was extended to binary study using an ANL adsorbent because ANL gave the highest adsorption capacity. The stock solution of the binary system consisted of equal mass concentrations of the two selected adsorbates under study to make up a binary solute system having an initial concentration of 1000 mg L^{-1}. This ratio was used and equilibration was carried out for 24 h.

It can be inferred from Table 8.6 that the adsorption capacity of 4-NP and 4-CP decreased more during binary adsorption than in single component adsorption. This shows the double antagonistic effect of 4-NP and 4-CP adsorption in the binary mixture of 4-NP and 4-CP. There is extreme reduction in adsorption capacity in the case of 4-NP in the binary mixture compared to that in the single system. The reduction in adsorption capacity in the binary system may be explained because of the presence of strong

Table 8.6 Comparison of individual and total adsorption equilibrium uptake in single and binary systems (4-NP + 4-CP)

Binary system (1:1)

$C_{e(4\text{-}NP)}$ (mg L^{-1})	$C_{e(4\text{-}CP)}$ (mg L^{-1})	$q_{e(4\text{-}NP)}$ (mg g^{-1})	$q_{e(4\text{-}CP)}$ (mg g^{-1})	$Ad_{(4\text{-}NP)}$ %	$Ad_{(4\text{-}CP)}$ %	$Ad_{(total)}$ %
144.61	189.10	156.51	148.37	85.53	81.08	83.31
15.42	24.03	62.24	64.67	98.45	97.59	98.02
4.59	32.58	39.85	38.73	99.54	96.74	98.14
2.43	23.82	28.37	27.76	99.75	97.61	98.68
2.13	18.42	21.97	21.62	99.78	98.15	98.97
2.48	6.24	17.85	17.85	99.75	99.37	99.56

Single system

$C_{e(4\text{-}NP)}$ (mg L^{-1})	$C_{e(4\text{-}CP)}$ (mg L^{-1})	$q_{e(4\text{-}NP)}$ (mg g^{-1})	$q_{e(4\text{-}CP)}$ (mg g^{-1})	$Ad_{(4\text{-}NP)}$ %	$Ad_{(4\text{-}CP)}$ %
107.85	0	549.01	0	89.21	0
70.82	0	261.73	0	92.91	0
46.83	0	177.82	0	95.31	0
10.49	0	98.80	0	98.90	0
1.84	0	66.19	0	99.81	0
0	626.31	0	219.81	0	37.36
0	546.49	0	176.12	0	45.35
0	436.84	0	144.03	0	56.31
0	344.73	0	124.57	0	65.52
0	160.52	0	81.26	0	83.94
0	101.75	0	59.66	0	89.82

competition between the solute molecules for the adsorbent surface. The comparative equilibrium uptake and the adsorption yield obtained for single-component and binary mixtures of 4-NP and 4-CP solution at 298 K are shown in Table 8.6, which inevitably shows the higher adsorption capacity for 4-NP than for 4-CP.

8.7.1 Comparison of Single-Solute and Binary Solute Adsorption

In single-solute adsorption, the maximum adsorption capacity for 4-NP was 549 and for 4-CP, 219.8 mg g^{-1} (Ahmaruzzaman and Laxmi Gayatri, 2011). The adsorption capacity was found to decrease in the bi-solute system as indicated by values of 156.5 mg g^{-1} for 4-NP and 148.4 mg g^{-1} for 4-CP. Obviously, the bi-solute adsorbates have an antagonistic effect on one another with competition for adsorption sites being a prominent mechanism. The presence of 4-NP produces greater hindrance for the adsorption of 4-CP. The overall net effect is the presence of competition, which lowers the adsorption capacity of both the individual components in the binary mixture. The repulsion between the individual binary components resulted in specific heterogeneous sites being less available for adsorption.

Figure 8.17 shows the adsorption of 4-CP under varying concentrations of 4-NP (100–1000 mg L^{-1}). The inverse of that was investigated to determine the optimum mixture concentration under which ANL could effectively remove the mixture. The adsorption of 4-CP was found to be higher at low initial 4-NP concentrations. 4-CP showed a binary adsorption yield of 98.76% at 100 mg L^{-1} of 4-NP. Hence, a binary solution consisting of 4-NP:4-CP in 1:10 ratio was found to be the critical combination for maximum adsorption of the less favorable adsorbate, 4-CP. Similarly, in the case of 4-NP in the binary mixture consisting of various ratios, the uptake of the more favorable adsorbate was higher at a lower 4-CP initial concentration, and a maximum q_e of 99.69 mg g^{-1} was achieved at a 1:10 mixing ratio. This is because of less competition for adsorption sites at lower initial concentration of the less favorable adsorbate and a higher concentration gradient of 4-NP, which becomes freely available to the active sites of ANL.

The binary adsorption of 4-NP and phenol over ANL was similarly investigated to observe the changes in the adsorption behavior when phenol

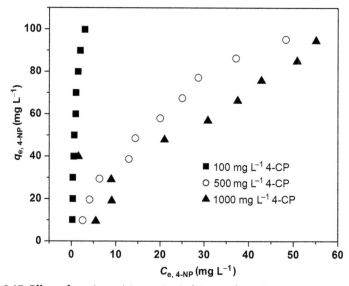

Figure 8.17 Effect of varying mixing ratios in binary adsorption using ANL.

is present instead of 4-CP. The equilibrium uptake showed that there is an increase in adsorption capacity—reaching 96.11 mg g^{-1}—while the single-solute study demonstrated a lower uptake of 69.70 mg g^{-1} (Ahmaruzzaman and Laxmi Gayatri, 2011). On the other hand, there is a decrease in the removal of 4-NP in the binary mixture when the experimental adsorption capacity is decreased to 98.91 mg g^{-1} as compared to 549.01 mg g^{-1} (Ahmaruzzaman and Laxmi Gayatri, 2011) in the single-solute system. Thus, the presence of phenol in the binary solute mixture has an antagonistic effect on the simultaneous adsorption of 4-NP, while phenol itself experiences a synergistic effect. This shows the competition for the components for the specific adsorption sites, which are readily more accessible to phenol rather than to 4-NP due to the smaller size of the phenol molecule, and which has greater affinity toward the micropores of ANL. A close observation of Table 8.7 shows that the q_e values for different components decrease with the increase in initial adsorbence dose. This may be due to the adsorption and desorption occurring simultaneously and overcrowding of the adsorption sites, which may result in some of the layers being not available for adsorption. However, it is observed that the adsorption yield increases with the increase in adsorbence dose in both of the components, reaching a maximum of 99.98% and 99.64% for 4-NP and phenol, respectively. However, the increase in individual adsorption yield with an increase in dose is not very significant. The total maximum adsorption yield was 99.81%, which equates to almost total removal of phenol, thereby proving the efficiency of ANL adsorbent.

Binary adsorption experiments using 4-CP and phenol were conducted to further investigate the interaction effects. Between these two components, it is speculated that 4-CP will have higher adsorption capacity over phenol as

Table 8.7 Summary of individual and total adsorption in binary solute mixture of (4-NP +phenol)[a]

$C_{e(4-NP)}$ (mg L^{-1})	$C_{e(P)}$ (mg L^{-1})	$q_{e(4-NP)}$ (mg g^{-1})	$q_{e(P)}$ (mg g^{-1})	$Ad_{(4-NP)}$ %	$Ad_{(P)}$ %	$Ad_{(total)}$ %
3.447	31.601	98.913	96.118	99.65	96.83	98.24
1.521	19.781	66.52	65.304	99.84	98.02	98.93
1.087	14.478	49.87	49.202	99.89	98.55	99.22
0.639	9.679	39.942	39.581	99.93	99.03	98.48
0.328	7.24	33.189	32.96	99.96	99.27	99.62
0.162	4.296	24.958	24.855	99.98	99.57	99.77
0.207	3.58	22.21	22.13	99.97	99.64	99.81

[a]Conditions: $C_{0,(4-NP)}$: $C_{0,P}$ = 1:1 ratio, $C_{0,i}$ = 1000 mg L^{-1}

Table 8.8 Binary component equilibrium results and total adsorption yield in the solute mixture of 4-CP + phenol[a]

$C_{e(4\text{-}CP)}$ (mg L^{-1})	$C_{e(P)}$ (mg L^{-1})	$q_{e(4\text{-}CP)}$ (mg g^{-1})	$q_{e(P)}$ (mg g^{-1})	Ad$_{(4\text{-}CP)}$ %	Ad$_{(P)}$ %	Ad$_{(total)}$ %
464.87	621.66	53.11	37.55	53.51	37.83	45.67
402.49	531.56	39.81	31.21	59.75	46.84	53.29
314.63	444.1	34.21	27.75	68.53	55.59	62.06
257.3	365.88	29.68	25.34	74.27	63.41	68.84
204.86	307.51	29.39	22.99	79.51	69.25	74.38
117.94	169.12	22.02	20.74	88.21	83.09	85.65
86.28	126.67	20.3	19.42	91.37	87.43	89.4

[a]Conditions: $C_{0,4\text{-}CP} : C_{0,P} = 1{:}1$ ratio, $C_{0,i} = 1000$ mg L^{-1}

obtained from single-solute data. Besides, the surface characteristic of an ANL adsorbent and the presence of the electron-withdrawing —Cl group implicates their role in the higher uptake. For the experimental study of this combination, 4-CP had a similar concentration to phenol. Among the various binary solute investigations, the minimum adsorption yield occurred in the combined of 4-CP + phenol system. Table 8.8 shows a decrease in the adsorbed phase concentration. A total adsorption yield of 89.40% was achieved at a high adsorbent dose. It is again evident from this table that the adsorption yield of 4-CP is higher than for phenol in all the doses of adsorbent. Overall, an antagonistic effect is observed in this binary combination because both capacities decreased drastically.

8.8 DESORPTION STUDIES

Desorption and regeneration studies are important in adsorption. Desorption may occur either by thermal treatment or through suitable desorbing agents. More acidic groups such as carboxyls and lactones can be desorbed as CO_2 at 200–650 °C, while the less acidic (phenol and carbonyl) basic groups may be desorbed as CO or a mixture of CO and CO_2 in the range of 500–1000 °C (Brennan et al., 2001). In the present study, however, for phosphoric acid-activated adsorbents, a chemical method of regeneration was employed.

Figures 8.18 and 8.19 show desorption of 4-NP, 4-CP, and phenol for ATW and AES adsorbents, respectively. The study shows maximum desorption efficiency using NaOH for 4-CP and phenol while HNO_3 also served as a good eluent for desorption of 4-NP-loaded adsorbents. In the case of ATW, maximum desorption of 27.02%, 28.66%, and 52.44% was

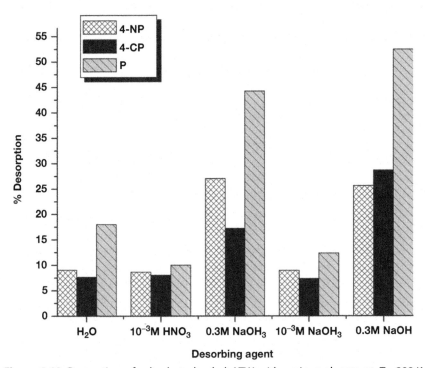

Figure 8.18 Desorption of adsorbate loaded ATW with various eluents at $T = 298$ K using 0.2 g of the adsorbent.

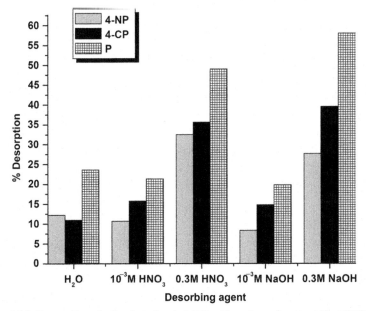

Figure 8.19 Desorption of adsorbate loaded AES with various eluents at $T = 298$ K using 0.2 g of the adsorbent.

obtained for 4-NP, 4-CP, and phenol, respectively. The maximum desorption was higher for AES-loaded adsorbents among the activated groups, reaching 32.53%, 39.55%, and 58.01% for 4-NP, 4-CP, and phenol. For 4-NP, HNO_3 is a better desorbing agent, while for 4-CP and phenol, NaOH was found to be a good desorbing agent. The study indicates that 4-NP desorption is the most difficult among 4-NP, 4-CP, and phenol.

8.9 DISPOSAL AND COST ANALYSIS

The safe disposal of phenol-loaded adsorbents is an environmental problem because leaching of the loaded adsorbents due to rain or percolated water may occur. Some of the methods, with their own inherent advantages and limitations, include elution, incineration, and pyrolysis. In the elution method, there is both recovery of the adsorbate molecules and regeneration of the adsorbent for subsequent use in a number of cycles. Although simple and inexpensive, there is some loss in efficiency after three or four cycles of desorption that may be due to blocking of the pores present on the adsorbent surface. The phenol-loaded adsorbent can be incinerated at high temperatures to yield ash, which can be recycled and reused appropriately. The release of toxic gases is one of the major disadvantages associated with the incineration process, especially in the case of chlorinated phenols. This method of regeneration involves a high cost of energy. However, a lot of research is required to make the adsorption method technoeconomically feasible.

Cost is an important parameter for comparing adsorbent materials. It is usually the cost of the adsorbent that determines the cost of the adsorption process. The capacity/dose of the adsorbent and kinetics control the cost of treatment. The total cost for the preparation of the adsorbent can be calculated in a stepwise manner where the breakup costs of heating, drying, chemicals, and other overhead expenses are taken into consideration. The typical cost of special-grade AC materials is generally above Rs. 1000 per kg and many times runs into several thousand rupees per kg. The costs of the adsorbents used in the present study are less than Rs. 700 per kg.

8.10 Summary

The adsorption technique, especially using new adsorbent materials, in wastewater treatment provides a promising alternative to treat industrial wastewaters, in general, and phenolic industrial effluents, in particular. The study on several activated/nonactivated biomass-derived adsorbents and bimetallic oxides for the removal of three industrially important pollutants, phenol, 4-CP, and 4-NP, indicate:

1. The adsorption capacity of the adsorbents is significantly lower when used in untreated form than in their charred or activated forms. The carbonaceous adsorbents showed greater efficiency than the synthesized binary mixed oxide of Fe-Al because of greater porosity and the presence of certain functional groups that facilitated the adsorption process.
2. The order of the adsorption follows: phenol $<$ 4-CP $<$ 4-NP.
3. Most of the adsorption processes can be described by the Freundlich and Langmuir isotherm model.
4. The activated adsorbents have a high surface area, generally above $1000 \, m^2 \, g^{-1}$. The pore size for most of the adsorbents is in the microporous to mesoporous range.
5. The cost of the adsorbents is believed to be substantially lower than for commercial activated carbon and hence they have the potential to compete with commercial products.

The development of new adsorbents can provide better opportunities for commercial applications. The synthesis of porous materials from zeolite, the preparation of AC fibers, and the fabrication of nanomaterials can also be considered in the future scope of this research. Growth can be seen in the area of the synthesis of new porous materials with tailored structural and surface properties. Thus, there is an immense potential for creating functional structures to meet specific requirements. In this regard, the potential use of the biomass-derived new materials remains largely unexplored. The low-cost adsorbents used in the present study may also be effective in removing other harmful phenolics, organics, and metal ions present in industrial effluents.

Nomenclature

4-CP	4-chlorophenol
4-NP	4-nitrophenol
AAJSC	acid activated jute stick chars
Ad$_i$ %	individual adsorption in %
AES	activated egg shells
a_L	Langmuir's theoretical adsorption capacity
ANL	activated neem leaf
AOPs	advanced oxidation processes
APPC	activated potato peel chars
ATW	activated tea waste
BET	Brunauer-Emmett-Teller
BJH	Barret-Joyner-Halenda

BMO	binary mixed oxide of Al-Fe
BOD	biological oxygen demand
BP	banana peel powder
C_0	initial liquid phase adsorbate concentration
C_e	equilibrium liquid phase concentration
COD	chemical oxygen demand
CWAO	catalytic wet air oxidation
d	grain size of the crystallite
FTIR	Fourier transform infrared
JSC	jute stick chars
NL	neem leaf
P	phenol
PPC	potato peel chars
q_e	quantity of adsorbate per unit mass of the adsorbent at equilibrium
SEM	scanning electron micrograph
TW	tea waste
WPO	wet peroxide oxidation
XRD	X-ray diffraction
β	full width at half maximum
θ	angle of reflection
λ	wavelength of the X-ray used

Subscripts

ads	adsorption
des	desorption
e	equilibrium
H	high
L	low
M	medium
max	maximum
ZPC	zero point charge

REFERENCES

Ahmad, A.L., Loh, M.M., Aziz, J.A., 2007. Preparation and characterization of activated carbon from oil palm wood and its evaluation on methylene blue adsorption. Dyes Pigments 75, 263–272.

Ahmaruzzaman, M., 2008. Adsorption of phenolic compounds on low-cost adsorbents: a review. Adv. Colloid Interface Sci. 143, 48–67.

Ahmaruzzaman, M., Laxmi Gayatri, S., 2010a. Adsorptive removal of p-nitrophenol (p-NP) on charred jute stick. Int. J. Chem. React. Eng. 8, A98.

Ahmaruzzaman, M., Laxmi Gayatri, S., 2010b. Activated tea waste as a potential low-cost adsorbent for the removal of p-nitrophenol from wastewater. J. Chem. Eng. Data 55, 4614–4623.

Ahmaruzzaman, M., Laxmi Gayatri, S., 2010c. Batch adsorption of 4-nitrophenol by acid activated jute stick char: Equilibrium, kinetic and thermodynamic studies. Chem. Eng. J. 158, 173–180.

Ahmaruzzaman, M., Laxmi Gayatri, S., 2011. Activated neem leaf: a novel adsorbent for the removal of phenol, 4-nitrophenol, and 4-chlorophenol from aqueous solutions. J. Chem. Eng. Data 56, 3004–3016.

Ahmaruzzaman, M., Sharma, D.K., 2005. Adsorption of phenols from wastewater. J. Colloid Interface Sci. 287, 14–24.

Aksu, Z., Akpinar, D., 2000. Modelling of simultaneous biosorption of phenol and nickel(II) onto dried aerobic activated sludge. Sep. Purif. Technol. 21, 87–99.

Aksu, Z., Yener, J., 1999. The usage of dried activated sludge and fly ash wastes in phenol biosorption/adsorption: comparison with granular activated carbon. J. Environ. Sci. Health A Tox. Hazard. Subst. Environ. Eng. 34, 1777–1796.

Aksu, Z., Yener, J., 2001. A comparative adsorption/biosorption study of mono-chlorinated phenols onto various sorbents. Waste Manage. 21, 695–702.

Andreozzi, R., Caprio, V., Insola, A., Marotta, R., 1999. Advanced oxidation processes (AOP) for water purification and recovery. Catal. Today 53, 51–59.

Banat, F., Al-Asheh, S., 2000. Biosorption of phenol by chicken feathers. Environ. Eng. Policy 2, 85–90.

Barrett, E.P., Joyner, L.G., Halenda, P.P., 1951. The determination of pore volume and area distributions in porous substances. I. Computations from nitrogen isotherms. J. Am. Chem. Soc. 73, 373–380.

Boer, J.H., Linsen, B.G., 1970. Physical and Chemical Aspects of Adsorbents and Catalyst. Academic Press, London.

Brennan, J.K., Bandosz, T.J., Thomson, K.T., Gubbins, K.E., 2001. Water in porous carbons. Colloids Surf. A Physicochem. Eng. Asp. 187–188, 539–568.

Brunauer, S., Emmett, P.H., Teller, E., 1938. Adsorption of gases in multimolecular layers. J. Am. Chem. Soc. 60, 309–319.

Cañizares, P., Sáez, C., Lobato, J., Rodrigo, M.A., 2004. Electrochemical treatment of 2,4-dinitrophenol aqueous wastes using boron-doped diamond anodes. Electrochim. Acta 49, 4641–4650.

Carrasco-Marín, F., Alvarez-Merino, M.A., Moreno-Castilla, C., 1996. Microporous activated carbons from a bituminous coal. Fuel 75, 966–970.

Çay, S., Uyanik, A., Özasik, A., 2004. Single and binary component adsorption of copper (II) and cadmium (II) from aqueous solutions using tea-industry waste. Sep. Purif. Technol. 38, 273–280.

Chen, W., Duan, L., Zhu, D., 2007. Adsorption of polar and nonpolar organic chemicals to carbon nanotubes. Environ. Sci. Technol. 41, 8295–8300.

Cooney, D.O., 1999. Adsorption Design for Wastewater Treatment. CRC Press, Lewis Publishers, Boca Raton, Florida.

Coughlin, R.W., Ezra, F.S., 1968. Role of surface acidity in the adsorption of organic pollutants on the surface of carbon. Environ. Sci. Technol. 2, 291–297.

Das, A., Sharma, D.K., 1998. Adsorption of phenol from aqueous solutions by oxidized and solvent-extracted residual coal. Energ. Sources 20 (9), 821–830.

Davi, M.L., Gnudi, F., 1999. Phenolic compounds in surface water. Water Res. 14, 3213–3219.

Dursun, G., Çiçek, H., Dursun, A.Y., 2005. Adsorption of phenol from aqueous solution by using carbonised beet pulp. J. Hazard. Mater. 125, 175–182.

Dutta, S., Basu, J.K., Ghar, R.N., 2001. Studies on adsorption of p-nitrophenol on charred saw-dust. Sep. Purif. Technol. 21, 227–235.

Esplugas, S., Giménez, J., Contreras, S., Pascual, E., Rodríguez, M., 2002. Comparison of different advanced oxidation processes for phenol degradation. Water Res. 36, 1034–1042.

Goto, S., Goto, M., Uchiyama, S., 1984. Adsorption equilibria of phenol on anion exchange resins in aqueous solution. J. Chem. Eng. JPN 17, 204–205.

Goud, V.V., Mohanty, K., Rao, M.S., Jayakumar, S., 2005. Phenol removal from aqueous solutions by tamarind nutshell activated carbon: batch and column studies. Chem. Eng. Technol. 28, 814–821.

Greminger, D.C., Burns, G.P., Lynn, S., Hanson, D.N., King, C.J., 1982. Solvent extraction of phenols from water. Ind. Eng. Chem. Process Des. Dev. 21, 51–54.

Gupta, V.K., Sharma, S., Yadav, I.S., Mohan, D., 1998. Utilization of bagasse fly ash generated in the sugar industry for the removal and recovery of phenol and p-nitrophenol from wastewater. J. Chem. Technol. Biotechnol. 170, 180–186.

Hobday, M., Li, P.H.Y., Crewdson, D.M., Bhargava, S.K., 1994. The use of low rank coal-based adsorbents for the removal of nitrophenol from aqueous solution. Fuel 73, 1848–1854.

Hoigné, J., 1988. The chemistry of ozone in water. In: Stucki, S. (Ed.), Process Technologies for Water Treatment. Plenum Press, New York.

Jadhav, D.N., Vanjara, A.K., 2004. Removal of phenol from wastewater using sawdust, poly-merized sawdust and sawdust carbon. Indian J. Chem. Technol. 11, 35–41.

Jung, M., Ahn, K., Lee, Y., Kim, K., Rhee, J., Park, J., Paeng, K., 2001. Adsorption characteristics of phenol and chlorophenols on granular activated carbon. Microchem. J. 70, 123–131.

Kaleta, J., 2006. Removal of phenol from aqueous solution by adsorption. Can. J. Civil Eng. 33, 546–551.

Karaca, S., Gurses, A., Bayrak, R., 2004. Effect of some pre-treatments on the adsorption of methylene blue by Balkaya lignite. Energ. Convers. Manage. 45, 1693–1704.

Khalid, M., Joly, G., Renaud, A., Magnoux, P., 2004. Removal of phenol from water by adsorption using zeolites. Ind. Eng. Chem. Res. 43, 5275–5280.

Kuleyin, A., 2007. Removal of phenol and 4-chlorophenol by surfactant-modified natural zeolite. J. Hazard. Mater. 144, 307–315.

Langlais, B., Reckhow, D.A., Brank, D.R., 1991. Ozone in Water Treament. Lewis Publishers, Chelsea, Michigan.

Laxmi Gayatri, S., Ahmaruzzaman, M., 2011. The use of activated carbon derived from potato peel as an adsorbent for the removal of 4-nitrophenol. J. Sci. Forum II (1), 132–137.

Laxmi Gayatri, S., Ahmaruzzaman, M., 2012. Development of adsorbent from solid waste of potato peel for decontamination of wastewater containing 4-nitrophenol. J. Int. Acad. Phys. Sci. 16, 1–14.

Lin, S.H., Cheng, M.J., 2000. Phenol and chlorophenol removal from aqueous solution by organobentonites. Environ. Technol. 21, 475–482.

Ma, X., Anand, D., Zhang, X.F., Talapatra, S., 2011. Adsorption and desorption of chlorinated compounds from pristine and thermally-treated multi-walled carbon nanotubes. J. Phys. Chem. C 115, 4552–4557.

Mahvi, A.H., Maleki, A., Eslami, A., 2004. Potential of rice husk and rice husk ash for phenol removal in aqueous systems. Am. J. Appl. Sci. 1, 321–326.

Mattson, J.S., Lee, L., Mark Jr., H.B., Weber Jr., W.J., 1970. Surface oxides of activated carbon: internal reflectance spectroscopic examination of activated sugar carbons. J. Colloid Interface Sci. 33, 284–293.

Mukherjee, S., Kumar, S., Misra, A.K., Fan, M., 2007. Removal of phenols from water environment by activated carbon, bagasse ash and wood charcoal. Chem. Eng. J. 129, 133–142.

Namasivayam, C., Kanchana, N., 1992. Waste banana pith as adsorbent for color removal from wastewaters. Chemosphere 25, 1691–1705.

Nowack, K.O., Cannon, F.S., Aroraoc, H., 1999. Ferric chloride plus GAC for removing TOC. J . Am. Water Works Assoc. 91 (2), 65–78.

Oturan, M.A., Peiroten, J., Chartrin, P., Acher, A.J., 2000. Complete destruction of p-nitrophenol in aqueous medium by electro-Fenton method. Environ. Sci. Technol. 34, 3474–3479.

Patnukao, P., Pavasant, P., 2008. Activated carbon from *Eucalyptus camaldulensis* Dehn bark using phosphoric acid activation. Bioresour. Technol. 99, 8540–8543.

Poznyak, T., Arazia, B., 2005. Ozonation of non-biodegradable mixtures of phenol and naphthalene derivatives in tanning wastewaters. Ozone: Sci. Eng. 27, 351–357.

Quintelas, C., Sousa, E., Silva, F., Neto, S., Tavares, T., 2006. Competitive biosorption of ortho-cresol, phenol, chlorophenol and chromium(VI) from aqueous solution by a bacterial biofilm supported on granular activated carbon. Process Biochem. 41, 2087–2091.

Rengaraj, S., Moon, S.H., Sivabalan, R., Arabindoo, B., Murugesan, V., 2002. Removal of phenol from aqueous solution and resin manufacturing industry wastewater using an agricultural waste: rubber seed coat. J. Hazard. Mater. 89, 185–196.

Roostaei, N., Tezel, F.H., 2004. Removal of phenol from aqueous solutions by adsorption. J. Environ. Manage. 70, 157–164.

Salah, N.H., Bouhelassa, M., Bekkouche, S., Boultif, A., 2004. Study of photocatalytic degradation of phenol. Desalination 166, 347–354.

Schwertmann, U., Latham, M., 1986. Properties of iron oxides in some New Caledonian oxisols. Geoderma 39, 105–123.

Seidel, A., Gelbin, D., 1988. On applying the ideal adsorbed solution theory to multicomponent adsorption equilibria of dissolved organic components on activated carbon. Chem. Eng. Sci. 43, 79–89.

Shen, S., Chang, Z., Liu, H., 2006. Three-liquid-phase extraction systems for separation of phenol and p-nitrophenol from wastewater. Sep. Purif. Technol. 49, 217–222.

Shukla, A., Zhang, Y.H., Dubey, P., Margrave, J.L., Shukla, S.S., 2002. The role of sawdust in the removal of unwanted materials from water. J. Hazard. Mater. 95, 137–152.

Srivastava, V.C., Mall, I.D., Mishra, I.M., 2008. Antagonistic competitive equilibrium modeling for the adsorption of ternary metal ion mixtures from aqueous solution onto bagasse fly ash. Ind. Eng. Chem. Res. 47, 3129–3137.

Srivastava, V.C., Mall, I.D., Mishra, I.M., 2009. Competitive adsorption of cadmium(II) and nickel(II) metal ions from aqueous solution onto rice husk ash. Chem. Eng. Process. Process Intensif. 48, 370–379.

Staehelin, J., Buehler, R.E., Hoigné, J., 1984. Ozone decomposition in water studied by pulse radiolysis. 2. Hydroxyl and hydrogen tetroxide (HO4) as chain intermediates. J. Phys. Chem. 88, 5999–6004.

Stüber, F., Font, J., Eftaxias, A., Paradowska, M., Suarez, M.E., Fortuny, A., et al., 2005. Chemical wet oxidation for the abatement of refractory non-biodegradable organic wastewater pollutants. Process Saf. Environ. 4, 371–380.

Sulaymon, A.H., Ahmed, K.W., 2008. Competitive adsorption of furfural and phenolic compounds onto activated carbon in fixed bed column. Environ. Sci. Technol. 42, 392–397.

Taha, M.R., Leng, T., Mohamad, A.B., Kadhum, A.H., 2003. Batch adsorption tests of phenols in soils. Bull Eng. Geol. Environ. 62, 251–257.

Tein, C., 1994. Adsorption Calculations and Modelling. Butterworth-Heinemann, Boston.

Teng, H., Yeh, T., Hsu, L., 1998. Preparation of activated carbon from bituminous coal with phosphoric acid activation. Carbon 36, 1387–1395.

Terzyk, A.P., 2003. Further insights into the role of carbon surface functionalities in the mechanism of phenol adsorption. J. Colloid Interf. Sci. 268, 301–329.

United States Environmental Protection Agency (USEPA), 1985. Technical Support Document for Water Quality Based Toxics Control. EPA/440/485032, USEPA, Washington, DC.

Varghese, S., Vinod, V.P., Anirudhan, T.S., 2004. Kinetics and equilibrium characterization of phenols adsorption onto a novel activated carbon in water treatment. Indian J. Chem. Technol. 11, 825–833.

Vidic, R., Suidan, M., Brenner, R., 1993. Oxidative coupling of phenols on activated carbon: impact on adsorption equilibrium. Environ. Sci. Technol. 27, 2079–2085.

Wang, R.C., Kuo, C.C., Shyu, C.C., 1997. Adsorption of phenols onto granular activated carbon in a liquid–solid fluidized bed. J. Chem. Technol. Biotechnol. 68, 187–194.

Wasewar, K.L., Mohammad, A., Prasad, B., Mishra, I.M., 2008. Adsorption of zinc using tea factory waste: kinetics, equilibrium and thermodynamics. CLEAN: Soil, Air, Water 36, 320–329.

Wu, F.C., Tseng, R.L., Juang, R.S., 2005. Preparation of highly microporous carbons from fir wood by KOH activation for adsorption of dyes and phenols from water. Sep. Purif. Technol. 47, 10–19.

Yang, C., Qian, Y., Zhang, L., Feng, J., 2006. Solvent extraction process development and on-site trial-plant for phenol removal from industrial coal-gasification wastewater. Chem. Eng. J. 117, 179–185.

Yang, K., Wu, W., Jing, Q., Zhu, L., 2008. Aqueous adsorption of aniline, phenol, and their substitutes by multi-walled carbon nanotubes. Environ. Sci. Technol. 42, 7931–7936.

Yi, S., Zhuang, W.Q., Wu, B., Tay, S.T.L., Tay, J.H., 2006. Biodegradation of p-nitrophenol by aerobic granules in a sequencing batch reactor. Environ. Sci. Technol. 40, 2396–2401.

Zawadzki, J., 1980. IR spectroscopic investigations of the mechanism of oxidation of carbonaceous films with HNO_3 solution. Carbon 18, 281–285.

Zhonghau, H., Srinivasan, M.P., 1999. Preparation of high-surface-area activated carbons from coconut shell. Micropor. Mesopor. Mater. 27, 11–18.

FURTHER READING

United States Environmental Protection Agency (2014), http://www2.epa.gov/toxics-release-inventory-tri-program, viewed 4 March.

CHAPTER 9

An Introduction to Biological Treatment and Successful Application of the Aqua EMBR System in Treating Effluent Generated from a Chemical Manufacturing Unit: A Case Study

Nilesh Tantak[1], Nitin Chandan[2], Pavan Raina[2]
[1]Aquatech Systems (Asia) Pvt. Ltd., Pune, India
[2]AMP Technologies (A Division of Aquatech Systems Asia Pvt. Ltd.), Pune, India

9.1 INTRODUCTION

Scarce freshwater resources and rapidly deteriorating quality of water resources because of the discharge of untreated or noncompliant treated wastewater are the prime compelling reasons for industries to adopt new and advanced technologies for treating their effluent, not only to meet stringent discharge standards but also to possibly recycle it for various uses within the plant area. The complex question facing the authorities is up to what levels of treatment must be achieved in a given application and beyond those prescribed by discharge permits to protect public health and environment. Based on detailed analysis of local conditions and needs and the application of scientific knowledge and engineering judgment based on past experience, various methods are employed for treatment of wastewater. These may be physical, chemical, biological, or advanced methods for tertiary treatment. On one hand, it is regulatory drivers that are pushing for maximizing wastewater treatment and reuse, but on the other hand, industry is also taking proactive initiatives to implement wastewater treatment and recycling projects. A wastewater treatment system (as shown in Figure 9.1) is generally comprised of:

1. Primary treatment
2. Secondary treatment
3. Tertiary treatment.

Industrial Wastewater Treatment, Recycling, and Reuse

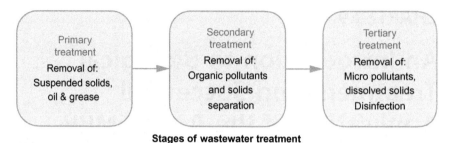

Stages of wastewater treatment

Figure 9.1 Stages of wastewater treatment.

Primary Treatment: Primary treatment generally comprises the removal of inert solids by using screens, a grit removal system, and clarifiers; oil is removed by means of a corrugated plate interceptor, American Petroleum Institute separator, dissolved air flotation, or induced gas flotation.

Secondary Treatment: Secondary treatment is the heart of the wastewater treatment plant and comprises the removal of dissolved organic pollutants. Organic removal can be achieved by biodegradation or chemical oxidation. Generally, biodegradation is preferred over chemical oxidation mainly due to its economical advantage. Biodegradation may be aerobic or anaerobic. This chapter will mainly focus on the aerobic biological treatment method.

Tertiary Treatment: Depending on the end use of treated water, tertiary treatment can have one or more of the following process systems:
1. *Removal of suspended solids to less than 10 mg/L*: Use of filters such as pressure sand filter, dual media filter, multimedia filters, or ultra-filtration (UF) membranes.
2. *Removal of any particular hard-to-degrade contaminant*: Treatment methods can vary from simple chemical oxidation to advanced oxidation processes (mixture of different oxidants).
3. *Removal of bacteria, viruses*: Disinfection by means of chlorination or UV disinfection.
4. *Removal of dissolved solids*: Reverse osmosis.

9.2 SECONDARY WASTEWATER TREATMENT

Generally, biological treatment methods are used as a secondary stage for removal of biodegradable organic matter, as well as removal of nutrients and suspended solids. Biodegradable organics, principally composed of proteins, carbohydrates, and fats, if discharged untreated in to the environment, can lead to the depletion of oxygen and the development of septic conditions. Therefore, biological treatment is an important and integral part of any wastewater treatment plant that treats wastewater from either a municipality or industry having soluble organic impurities or a mix of the two types of wastewater sources. The obvious economic advantage, both in terms of capital investment and operating costs, of biological treatment over other treatment processes such as chemical oxidation or thermal oxidation, has cemented its place in any integrated wastewater treatment plant.

Biological treatment methods can be further divided into different categories (Sutton, 2006): such as aerobic, anaerobic, anoxic processes which are used in specific condition of waste to be purified and constituents to be removed. However we would go in details of aerobic process with reference to the subject of this article.

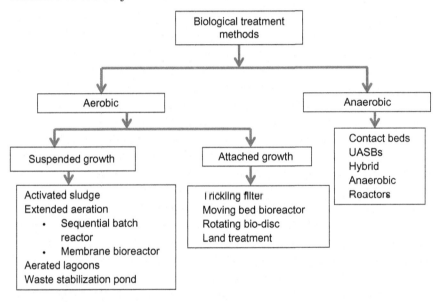

9.2.1 Aerobic and Anaerobic (Mittal, 2011)

Before we go in to the discussions of various aerobic biological treatment processes, it is important to briefly discuss the terms *aerobic* and *anaerobic*. Aerobic, as the title suggests, means in the presence of air (oxygen), while anaerobic means in the absence of air (oxygen). These two terms are directly related to the type of bacteria or microorganisms that are involved in the degradation of organic impurities in given wastewater and the operating conditions of the bioreactor. Therefore, aerobic treatment processes take place in the presence of air and utilize those microorganisms (also called

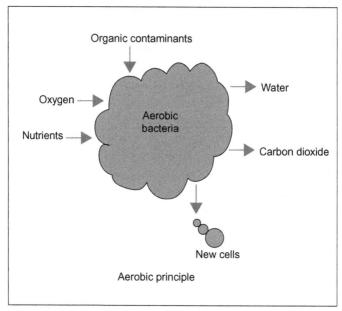

Figure 9.2 Aerobic principle.

aerobes) that use molecular/free oxygen to assimilate organic impurities, that is, convert them into carbon dioxide, water, biomass, and inorganic nitrogen products at ambient temperature without significant onerous by-product formation. The anaerobic treatment processes, on other hand, take place in the absence of air (and thus molecular/free oxygen), and exploit those microorganisms (also called anaerobes) that do not require air (molecular/free oxygen) to assimilate organic impurities. The final products of organic assimilation in anaerobic treatment are methane and carbon dioxide gas and biomass. The illustrations in Figures 9.2 and 9.3 depict simplified principles of the two processes. Biotreatment processes are generally robust to variable organic loads, create little odor (if aerobic), and generate a waste product (sludge).

9.3 AEROBIC TREATMENT PRINCIPLE

As explained earlier, the aerobic treatment process is classified depending on the working principles and also according to the process configuration, feed condition, and oxidation state. Process configuration defines the way in which water is in contact with the biomass, which can form a layer on some

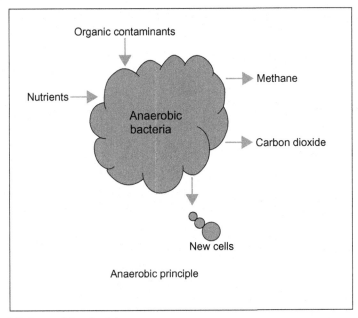

Figure 9.3 Anaerobic principle.

supporting media to form a fixed biofilm or be suspended in a reactor, or sometimes a combination of these. Suspended growth provides higher mass transfer, but the biomass subsequently needs to be separated from the water. Both configurations generate excess biomass, which needs to be disposed of. Feeding condition defines the way in which feed water is introduced, which can either be continuous or batch wise. Feeding batch–wise allows the same vessel to be used for both biodegradation and separation; such bioreactors are known as sequential batch reactors. The oxidation and reduction conditions are maintained by the presence or absence of dissolved oxygen in the system. Different redox conditions favor different microbial communities and are used to affect different types of treatment (Judd, 2011).

Aerobic treatment is used to remove organic compounds (BOD or COD) and to oxidize ammonia to nitrate. Aerobic tanks may be combined with anoxic or anaerobic tanks to provide biological nutrient removal. Total removal of organic nitrogen from feed water can be achieved by recycling the nitrate effluent from activated sludge process (ASP) upstream of the aerobic process in anoxic conditions.

In all biotreatment processes the treated water must be separated from biomass.

9.4 DIFFERENT TYPES OF AEROBIC TREATMENT TECHNOLOGIES

Selection of treatment technology depends on:
1. Type of wastewater and its characteristics
2. Quality of treated wastewater required
3. Footprint available.

9.4.1 Activated Sludge Process

The ASP, a suspended growth process, is used routinely for biological treatment of both municipal and industrial wastewaters. It is the oldest and most widely practiced treatment method for treatment of municipal as well as industrial wastewater; this process is 130 years old and was the result of the work of Dr. Angus Smith. Dr. Smith discovered that the aeration of wastewater resulted in the more rapid oxidation of organic matter. The aeration of wastewater was subsequently studied by a number of investigators, including Black and Phelps in 1910 and then in 1912 and 1913 by Clark and Gage respectively. Arden and Lockett, in further investigations, found that sludge played an important role in the results obtained by aeration. The process was thus named ASP by Arden and Lockett because it involved the production of an activated mass of micro-organisms capable of aerobic stabilization of organic material in the wastewater (Metcalf & Eddy, Inc. 2003). The process finally came into common use in 1920 (Sarin, 2013).

Activated sludge plant involves:
1. Wastewater aeration in the presence of a microbial suspension.
2. Solid-liquid separation following aeration.
3. Discharge of clarified effluent.
4. Wasting of excess biomass.
5. Return of remaining biomass to the aeration tank.

Typically, after primary treatment, suspended impurities are removed from wastewater using an ASP-based biological treatment system comprised of an aeration tank followed by a secondary clarifier (Figure 9.4). The aeration tank is a completely mixed or a plug flow (in some cases) bioreactor where specific concentration of biomass measured as mixed liquor suspended solids (MLSS) or mixed liquor volatile suspended solids (MLVSS) is maintained along with sufficient dissolved oxygen (DO) concentration (typically 2 mg/L) to effect biodegradation of soluble organic impurities measured as biological oxygen demand (BOD_5) or chemical oxygen demand (COD). The aeration tank is provided with fine bubble diffused aeration pipework at the bottom to transfer the required oxygen to the biomass

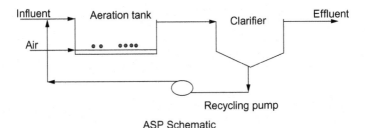

Figure 9.4 Activated sludge process (ASP) schematic.

and also to ensure that product is completely mixed in the reactor. A roots-type air blower supplies air to the diffuser pipework. In several older installations, mechanical surface aerators have been used to meet the aeration requirement. The aerated mixed liquor from the aeration tank overflows by gravity to the secondary clarifier unit to separate out the biomass and allow clarified, treated water into the downstream filtration system for finer removal of suspended solids. The separated biomass is returned to the aeration tank by means of a return activated sludge (RAS) pump. Excess biomass (produced during the biodegradation process) is drawn out to the sludge handling and dewatering facility (Mittal, 2011).

9.4.1.1 Types of ASPs

ASPs can be operated in different modes. Each variation utilizes the basic process of suspended growth in the aeration tank. Three modes of operation for activated sludge are:

A. Conventional activated sludge
B. Extended aeration activated sludge
C. Contact stabilization activated sludge.

The primary difference between these three modes of operation has to do with length of time micro-organisms reside in the treatment system, which is also termed sludge retention time (SRT) (nmenv, 2014). Table 9.1 shows a comparison between the ASPs.

9.4.2 Sequencing Batch Reactor

The sequencing batch reactor (SBR) process is one of the most popular aerobic treatment technologies employed to treat municipal wastewater and wastewater from a variety of industries including refineries and petrochemical plants.

This technology offers several operational and performance advantages over the conventional ASP. The SBR process performs all the functions of a conventional activated sludge plant (biological removal of pollutants,

Table 9.1 Comparison between different activated sludge processes

Process Description	Conventional Activated Sludge	Extended Aeration Activated Sludge	Contact Stabilization Activated Sludge
Hydraulic retention time (HRT)	4–8 h	12–14 h	0.3–3 h in contact tank and 4–8 h in stabilization tank
MLSS in aeration basin	1000–4000 mg/L	2000–5000 mg/L	1000–3000 mg/L contact tank 2–6 times concentration in stabilization tank
System SRT	3.5–10 days	>10 days	<3.5 days
System F/M ratio	0.25–0.5:1	0.05–0.15:1	0.5–>1.0:1

solid/liquid separation, and treated effluent removal) by using a single variable volume basin in an alternating mode of operation that dispenses with the need for final clarifiers and high-RAS pumping capacity. With continued value addition, the performance of the SBR system can be enhanced with a high level of process sophistication in a configuration that is cost and space effective and offers a methodology with operational simplicity, flexibility, and reliability that is not available in conventionally configured activated sludge systems. Also with this advancement, it is possible to control the growth of filamentous sludge bulking, a common problem with conventional processes and other activated sludge systems (Mittal, 2011).

SBR utilizes a simple repeated time-based sequence that incorporates:

- Fill–aeration (for biological reactions)
- Fill–settle (for solid/liquid separation)
- Decant (for treated effluent removal).

9.4.2.1 Advantages of SBR (Mittal, 2011)

- It operates under continuous reduced loading through simple cycle adjustments.
- It operates with feed-starve selectivity, So/Xo operation (control of limiting substrate to micro-organism ratio), and aeration intensity to prevent filamentous sludge bulking and ensures endogenous respiration (removal of all available substrate), nitrification, and de-nitrification together with enhanced biological phosphorus removal.
- It operates simultaneous (co-current) nitrification and de-nitrification by variation of aeration intensity.

- It tolerates shock load caused by organic and hydraulic load variability. The system is easily configured and adjusted for short-term diurnal and long-term seasonal variations.
- It eliminates the need for a secondary clarifier.
- It eliminates separate load equalization. The SBR basin is in itself an equalization basin and a clarifier with a much lower solids flux compared to conventional clarifier design.
- It provides the inherent ability to remove nutrients without chemical addition by controlling the oxygen demand and supply.
- It provides for energy optimization through nutrient removal mechanisms. The feed water carbonaceous BOD used in denitrification and enhanced biological phosphorus removal reduces overall oxygen demand and energy requirements.
- It offers capital and operating cost advantages.

9.4.3 Moving Bed Bio Reactor (Mittal, 2011)

Moving bed bio reactor (MBBR) technology is another advanced aerobic treatment technology used widely for treating different industrial wastewater. MBBR technology is based on the attached growth aerobic system and is more efficient and requires less area than ASP and SBR. The moving media is introduced into the aeration tank, which provides higher surface area for microbes to grow. This helps to increase the microbial biomass in the reactor and eventually reduces the reactor's footprint area. MBBR is also one of the robust aerobic technologies. Compared to conventional ASP, it has better sustainability against hydraulic as well as organic load fluctuations. Unlike conventional ASP, the MBBR process design is not based on bioreactor hydraulic retention time (HRT), but it mainly depends on:

- Active surface area of MBBR media.
- Specific COD loading per unit active surface area of MBBR media.
- Oxygen transfer requirements.

There are various MBBR media suppliers available in the market.

9.5 MEMBRANE BIOREACTOR TECHNOLOGY

Numerous modifications of ASP have evolved over the years aimed at more effective and efficient removal of organics, nutrients, and suspended solids. As the membrane design improved, water treatment applications, mainly, and enhanced solid separation, in particular, improved, thus making water reuse possible. Membranes have been used more recently in suspended growth reactors for wastewater treatment (Sarin, 2013). The membrane bioreactor (MBR) is the latest technology for biological degradation of soluble organic impurities. MBR technology has been in extensive usage for treatment of domestic sewage, but for industrial waste treatment applications, its use has been somewhat limited or selective. The MBR process is very similar to the conventional ASP in that both have mixed liquor solids in suspension in an aeration tank. The difference in the two processes lies in the method of separation of bio-solids. In the MBR process, the bio-solids are separated by means of a polymeric membrane based on a microfiltration or UF unit, as compared to the gravity settling process in the secondary clarifier in the conventional ASP (Mittal, 2011). Therefore, the advantages of an MBR system over a conventional activated sludge system are obvious:

- Membrane filtration provides a positive barrier to suspended bio-solids so that they cannot escape the system, unlike gravity settling in the ASP, where the bio-solids continuously escape the system along with clarified effluent. Sometimes a total loss of solids is also encountered due to process upsets causing sludge-bulking in the clarifier. As a result, the bio-solid concentration measured as MLSS/MLVSS can be maintained at three to four times greater in an MBR process ($\sim$10,000 mg/L) in comparison to the ASP ($\sim$2500 mg/L).

- Due to the above aspect of MBR, aeration tank size in the MBR system can be one-third to one-fourth the size of the aeration tank in an activated sludge system. Further, instead of a gravity settling-based clarifier, a much more compact tank is needed to house the membrane cassettes in the case of submerged MBR along with skid-mounted membrane modules in the case of a non-submerged, external MBR system.

- Thus, an MBR system requires only 40–60% of the space required for an activated sludge system, therefore significantly reducing the construction work/cost and overall footprint.

- Due to membrane filtration (micro/UF), the treated effluent quality in the case of the MBR system is far superior compared to conventional activated sludge, so the treated effluent can be directly reused as cooling tower make-up or for gardening or other uses. Typical treated water quality from the MBR system is:

  $BOD_5 < 5$ mg/L

  Turbidity < 0.2 NTU

The MBR system has three important parts that play an important role in overall performance; the development at each has been moving fast since 1990 for its commercial usage (Frenkel, 2013).

1. Biological reactor
2. Membrane properties
3. MBR configuration.

9.5.1 Biological Reactor

Membrane is merely a filtration device in the entire process, thus the biological process of the effluent needs is the first step of importance. With advancement in understanding of bio-technology, the biological treatment section of MBR is being improved, which makes the bioreactor sturdy and efficient. However, dampening of hydraulic and organic load fluctuations is a key factor to guarantee a consistent bio-treatment operation.

Biological processes are slow and invariably require close monitoring because any change, if it happens, appears as an observation after a lapse of time. There are specific parameters, such as DO, MLSS, SRT, HRT, inlet wastewater quality, and quantity, that need to be checked and logged. The sludge character varies with the variation in feed characteristics However, it is very important to acclimate the MLSS properly so that the system can operate efficiently. The biologically treated wastewater is then fed to the MBR membrane system. Because the membrane performs the filtration, oversized sludge is retained in the feed side and is recirculated back to the aeration tank to maintain the MLSS quantity in the reactors. Bleed (sludge wastage) is maintained from the reactor to remove the dead cells or to maintain MLSS. The deposition of sludge over the membrane surface is bound to take place and so periodic back flush/cleaning is recommend by individual suppliers.

9.5.2 Membrane Properties

Filtering properties depend on the type of membrane, its pore size, its hydrophilic properties, and chemical properties (Judd, 2011). The membrane supplier recommends a process wherein to operate at a defined minimum to maximum flux under different conditions. This is to safeguard the membrane against becoming fouled without notice to avoid reaching a non-recovery point on the membrane and thus the system.

9.5.3 MBR Configuration

Available MBR technologies employed for wastewater treatment can be classified according to membrane configuration: flat sheet (FS), hollow fiber, and multi-tubes.

The FS base submerged MBR was piloted early in 1990 for wastewater application (Judd, 2011). The membrane material used was chlorinated polyethylene over a robust nonwoven support. The pore size of such membrane used was 0.4 μm, which is microfiltration size, but once a biofilm layer was formed over this surface, it worked in the UF range and produced filtrate with good turbidity (Judd, 2011). The PES-based (Polyethersulfone) flat membrane over PP (Polypropylene) support was reportedly used in sewage applications subsequently at a 150 kDa pore size, which falls in the UF range.

Generally, air is used for removal of sludge over the membrane surface, keeping the membranes fixed. In one instance, a moving membrane module has also been developed to reduce the sludge adhering to the membrane surface (Judd, 2011). There was a lot of development in the 1990s when commercial applications were tried. Many players have participated in the development and improvement of membrane performance and also have improved processes to suit their particular products.

The MBR configurations used in industries fall in two general categories: submerged MBR and side-stream MBR, as shown in Figure 9.5. The basic difference between the two schemes is the placement of the membrane module, which is either inside or outside the bioreactor depending on the configuration.

A general comparison between the two configurations is shown in Table 9.2. Each configuration has its advantages and drawbacks with respect to the operating parameters, energy consumption, footprint, and type of feed-water handling.

In 1993, the first reinforced tubular UF fibers were introduced commercially for industrial wastewater applications (Judd, 2011) and the momentum soon grew behind this technology. There has been no looking back since then, and today it is a well-accepted technology in the advanced countries of the world.

As the world has become more environmentally conscious, conserving and reusing water has become important globally, and MBR is likely to achieve greater importance. The global use of MBR requires many players with unique

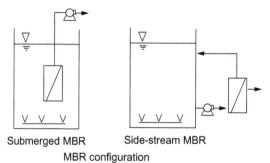

Submerged MBR   Side-stream MBR
MBR configuration

Figure 9.5 Membrane bioreactor (MBR) configuration.

Table 9.2 Comparison between submerged and side stream configurations

Item	Unit	Submerged MBR	Side Stream MBR
Typical configuration	–	Hollow fiber (HF) Flat sheet (FS)	Tubular (TB) Plate & frame (PF)
Mode of operation		Cross-flow	Cross-flow
Type of permeation		Outside-In	TB: inside-out PF: outside-in
Operating pressure	kPa	5–30 (vacuum)	300–600
Long-term average flux	LMH	15–30	50–100
Recycle ratio	m^3 feed/m^3 permeate	–	20–70
Superficial velocity	m/s	0.2–0.3	2–6
Specific energy demand[a]	kWh/m^3 permeate	0.15–0.5	0.2–0.5
Cleaning	–	Easy	Complex
Packing density		Low	High
Market share[b]	–	99%	1%

[a]Specific energy demand including energy for permeate suction, but excluding biological aeration.
[b]Membrane surface area based, in municipal and industrial (Lesjean, 2008)

membranes and unique processes for improving efficiency and overall performance. Development will continue with the intention of making our world better and safer for generations to come, while reducing the depletion of good water available for the survival of human beings and nature itself.

The development of the Aqua EMBR is one step toward that goal for industrial biological waste, that has been the subject of discussion in industrial wastewater treatment, recycling, and reuse. The MBR module developed is equally useful for sewage waste with differences in their respective operational processes that have also been described.

9.6 AQUATECH MBR SYSTEM

Considering the market demand and application to different kinds of wastewater, Aquatech has developed both side stream (external) and submerged types of MBRs.

9.6.1 Pressured MBR (Aqua EMBR)

Aqua EMBR uses an FS membrane for solid–liquid separation. The membrane's plates are assembled to make a membrane cartridge. These

Table 9.3 Specifications of Aqua EMBR membrane module

Parameters	Figures and Description
Membrane type	Flat polymeric membranes
Module dimensions	$290 \times 290 \times 4500$ mm (L $\times$ W $\times$ H)
Total membrane surface area	42 m^2
Operating mode	Side stream (cross flow)
Material of construction	FRP
Module mounting	Vertical
Type of diffuser for air scouring	Pipe diffuser

membrane cartridges are inserted into a housing to make a membrane module. The membrane module has top and bottom end caps for recirculation of sludge (MLSS). Recirculation occurs on the outside of the membrane system. The product emerges from individual cartridges and is collected in a common product header. The module is placed outside the bioreactor and operated with the requisite velocity of air and MLSS. The velocity of MLSS varies from 0.5 to 1.0 m/s and depends on the characteristics of the sludge formed. Air is provided from the bottom end cap to scrub the membrane surface. Aqua EMBR is most suitable for treatment of critical and various industrial wastewaters because it is more robust and can be operated at high velocity, which helps to sustain flux values, even with difficult-to-treat industrial wastewater. Table 9.3 shows the specifications of the Aqua EMBR membrane module. Overall, Aqua EMBR is an external MBR system with a UF flat plate membrane with airlift arrangements. Thus it combines the advantages of a flat plate member and external side stream MBR.

9.6.1.1 Design of Aqua EMBR System

The Aqua EMBR design has been tested on the laboratory scale for several years to see consistency in membrane performance before testing it with actual wastewater. Flat plate membranes are used in making the MBR module, which is developed with PVDF as a base polymer (PVDF – Polyvinylidene difluoride). PVDF has outstanding oxidative, thermal, and hydrolytic stability, as well as good mechanical and film-forming properties. The membrane is supported by polypropylene fabric for better mechanical strength and has been extensively tested to have uniformity in filterable pores. It is tested separately for consistent pore distribution throughout the membrane. A cross-sectional view of the membrane shows a finger-like structure (Figure 9.6). A uniform and narrow pore size within the membrane surface

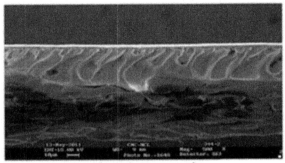
Membrane cross section

Figure 9.6 PVDF flat membrane cross section.

gives stable flux and better solid rejection. The average pore diameter of the membrane is 0.04 μm.

A cartridge assembly consists of membrane elements (MEs) in plate configuration, without using external frames. The external frames result in dead zones and low velocity pockets, which lead to the initiation of deposition of debris and spreading of fouling on the membranes. The MEs have a defined gap according to the type of process to be used. One cartridge could be of 7–10 m² area, and in a system, numbers of cartridges are used.

In a typical system consisting of a 4 m long housing (Figure 9.7), the segment is divided into three parts, and each part has an assembly of two cartridges (Figure 9.8). The housing typically has six cartridges with a typical surface area of 42–60 m². Figure 9.9 shows a 3D view of cartridge placement inside the housing. All six cartridges have independent permeate ports connected in a common header, making a loop from the highest point to utilize the natural suction head. The bottom chamber of the housing has an arrangement for feed flow such that proper distribution takes place between the MEs. There is a separate air bubbling arrangement such that mixing with the feed and distribution takes place uniformly. In a pressurized MBR, the design velocity plays a significant role. The air bubbling along with the feed at specified velocity scrubs the membrane surface while generating the product without choking the surface even under very high MLSS conditions.

The advantage of this design is the ease of replacement of membrane plates at any point if any leakage is found rather than replacing the whole module. The feed water and air are mixed together in the bottom end cap, and then with the sludge it is recirculated through the module and back

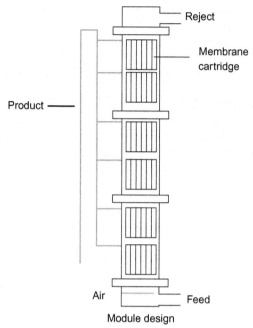

Figure 9.7 Aqua EMBR membrane module.

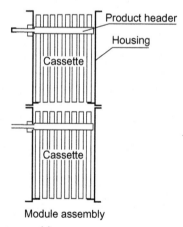

Figure 9.8 Two cartridge assembly.

to the bioreactor tank. The biological water and the air circulate between the gaps of the plates. The gap between the membrane plates and the air/water velocity is standardized based on several years' operating experience. The innovative design of the external FS MBR reduces chemical cleaning frequency, lowers the footprint and power consumption, and offers better

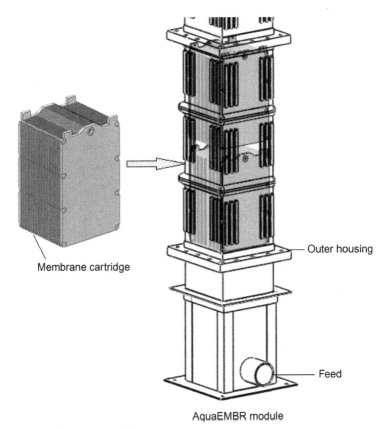

Membrane cartridge

Outer housing

Feed

AquaEMBR module

Figure 9.9 Aqua EMBR module.

outlet product quality. The circulation pump is used to regulate and maintain a relatively constant flow of sludge from the bottom to the top of the membrane.

The module is subjected to periodic chemical cleaning based on the drop of flux using a UF filtrate. The chemicals used for the chemical cleaning are sodium hypochlorite and citric acid/hydrochloric acid.

9.6.1.2 Features of Aqua EMBR System
- Consists of a flat plate membrane system with external (outside) MBR configuration
- Offers high flux
- Offers high product quality
- Offers high packing density
- Vertical module mounting requires smaller footprint

- Provides a separate product line for each cartridge with the ability to monitor product quality for each cartridge as required
- Contains membrane plates that can be replaced if required
- Offers chemical cleaning of the module that can be done *in situ* without disturbing biological activity
- Product is taken by gravity suction so product pump can be avoided
- Offers low power consumption
- Backwashing is not required to maintain the flux because of an innovative operating process of allowing backflow of water in a natural process where the membranes get cleaned *in situ*.

9.6.2 Submerged MBR System

Aquatech has also developed its submerged MBR configuration, considering the increasing demands for treatment of municipal wastewater. The submerged MBR operates at low water velocity and flux compare to the pressurized MBR. This is most suitable for the application of municipal wastewater where the volume of water treated is very high because of its low energy demand and sustainable flux values. Each membrane cartridge contains 10 m^2 areas. Each module consists of membrane cartridges, a steel rack, an air diffuser, a permeate water manifold, and an air inlet header. Each membrane cartridge consists of number of membrane plates that are prepared by welding a FS-type membrane on a designed PP plate. Membrane modules are aerated from the bottom, which continually scrubs the membrane surface. Membranes are backwashed at gravity pressure after certain time intervals, and in addition a cross flow of liquid is provided over the membrane panel. The combination of all these factors prevents membrane fouling, and the module can generate constant product flow for longer time. The compact modular design gives the benefit of a higher membrane surface area and easily removable membrane cartridge during assembly or replacement. A submerged MBR membrane cartridge with module is shown in Figure 9.10. Modules of 80, 160, and 320 m^2 designed membrane area can be used as per the application.

9.7 CASE STUDY

9.7.1 Background

Aquatech took the opportunity to install Aqua EMBR for industrial waste. The significant key features are as follows:

1. The biological waste did not have a typical BOD and COD ratio and was in the range of 0.1–0.2. The wastewater for degradation was only treatable with difficulty.

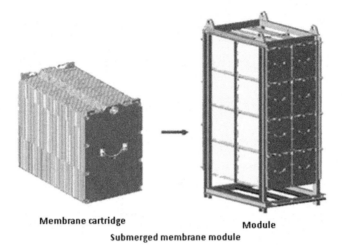

Membrane cartridge   **Module**
Submerged membrane module

Figure 9.10 Submerged MBR module.

2. Fresh water availability was limited.
3. Absence of industrial waste drain making it important to treat and reuse and minimize waste.
4. Treatment was thus necessary to provide useful reuse in the system.
5. A limited footprint is available for installation of a conventional process or sludge handling system.

This was an ideal situation for using Aqua EMBR, which was successfully installed and has been working to satisfactorily.

This chapter deals with one of the challenges that arose.

9.8 TYPICAL CHARACTERISTICS OF POLYMER-BASED CHEMICAL MANUFACTURING INDUSTRIAL WASTEWATER

The wastewater in the present case study was characterized by the presence of high COD and very low BOD. The BOD to COD ratio of effluent was in the range of 0.1–0.2.

Detailed wastewater characterization is given in Table 9.4. The wastewater had a very low BOD/COD ratio, which is in the range of 0.1–0.2, indicating low biodegradability. The wastewater also contained nitrogen in the form of organic nitrogen, which is converted to ammoniacal nitrogen in the process; further conversion of ammoniacal nitrogen to nitrate (nitrification) is an important oxidation step. Otherwise, there is a possibility of free ammonia generation, which might be toxic for the biological culture. The wastewater showed pH and alkalinity in an acceptable range.

Table 9.4 Typical wastewater compositions

Sr. No	Parameter	Average Value	Unit
1	pH	6–8	
2	Temperature	25–35	°C
3	Chemical oxygen demand (COD)	1600–2000	mg/L
4	Biochemical oxygen demand (BOD)	150–200	mg/L
5	Alkalinity	210	mg/L
6	TKN	100–130	mg/L

As the feed COD range is on the high side and also nitrogen is present in the water, the system has to take care of the oxidation of both carbon and nitrogen. The system was operated at high MLSS and high SRT to take the maximum loading in terms of COD and to achieve oxidation of both carbon and nitrogen.

9.9 TECHNOLOGY SELECTION

Treatment technology selection for wastewater takes into account its origin, its constituents, and the formulated treatment objective, which is derived from discharge criteria (Veenstra et al., n.d.).

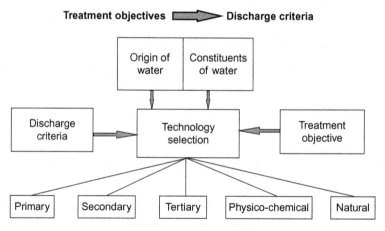

TECHNOLOGY SELECTION

9.9.1 Technologies That Can Be Proposed

Considering the nature of the wastewater and the volume of wastewater to be treated, three processes could be applied.

1. Anaerobic UASB (Upflow anaerobic sludge blanket)
2. Chemical oxidation
3. Activated sludge with membrane filtration (MBR).

9.9.2 Selection Criteria

The treatment objective of the effluent in present the study is to achieve discharge limits in the secondary/tertiary process and to achieve water of a quality lying in the recyclable range with advanced treatment.

Following are the Criteria for Selection of Treatment Technology

➢ *BOD removal efficiency*: Because the objective of treatment is to achieve water quality in the discharge, the selected technology should have the ability to remove biodegradable organic compounds.

➢ *COD removal efficiency*: If the wastewater has a high COD content, the COD has to be reduced to the discharge limit or even lower. Thus, the technology selected should be capable of operating at higher SRT and also reducing chemically oxidizable substances.

➢ *Footprint*: The land area required for the plant should be as small as possible.

➢ *O & M cost*: The cost of operation and maintenance should be low.

➢ *Sludge disposal cost*: The cost of sludge collection and dewatering should be low.

➢ *Ease of installation and commissioning*: The total amount of time required for the installation and commissioning should be minimized.

➢ *Ease of operation*: The technology selected should be easy to operate without the need for constant adjustment once the system becomes stabilized at certain feed conditions.

➢ *Energy efficiency*: The technology provided should be energy efficient.

➢ *Nitrification efficiency*: As wastewater contains excess nitrogen, the process should be capable of nitrifying ammonia.

➢ *Reliability of technology*: The process should preferably be stable against shock loading. For example, it should be able to continue operation and to produce an acceptable effluent under unusual conditions.

Considering the above factors and based on experience in treating such industrial wastewater, Aquatech proposed its own innovative Aqua EMBR technology to treat such complex wastewater.

9.10 SCHEME AND PROCESS DESCRIPTION

A module having 42 m^2 membrane area was used for solid/liquid separation. Wastewater with the characteristics shown in Table 9.4 was fed to the bioreactor after primary screening. During startup the bioreactor with a volume of 20 m^3 was seeded with activated sludge from the local municipal wastewater treatment plant. The sludge had been acclimated with the same feed water for more than a week with continuous aeration to maintain DO 1–2 mg/L. Before startup, the membrane module was tested with clean water, and its permeability was determined. After achieving MLSS of more than 4000 mg/L, sludge filtration was started through the module. The flow scheme for the process is shown in Figure 9.11, where influent wastewater from the feed tank goes to the bioreactor tank containing activated sludge. The biodegradable compounds in feed wastewater get degraded in the bioreactor tank, and then this mixed liquor is fed to the membrane module via sludge feed pump. The recirculated sludge goes back to the bioreactor tank. Permeate from the module is collected in the permeate collection tank. In relation to the scheme, after every 10 minutes service cycle, a rest time of 1 minute was provided as relaxation. During relaxation the permeate from membrane was stopped and only air scouring was done through the membrane. Chemical cleaning of the module was performed once every week with chemicals such as sodium hypochlorite and hydrochloric acid. The performance of the membrane module was assessed in terms of the consistency of the flux and product turbidity while bioreactor stability was analyzed by COD and BOD reductions.

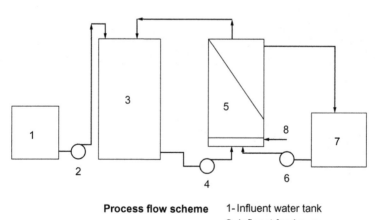

Process flow scheme   1- Influent water tank
2 - Influent feed pump
3 - Bioreactor tank
4 - Sludge feed pump
5 - Membrane module
6 - Cleaning pump
7 - Permeate collection cum cleaning tank
8 - Air supply to membrane module

Figure 9.11 Process flow scheme.

Actual plant

Figure 9.12 Aqua EMBR pilot plant.

Table 9.5 Biological parameters

Sr. No	Parameters	Unit	Value
1	Aeration tank volume	m^3	20
2	HRT	hrs	18
3	MLSS	mg/L	14,000–18,000
4	DO	mg/L	2.0
5	SRT	day	38

Feed and product water quality was monitored by analysis of COD, BOD, pH, alkalinity, and turbidity, while biological parameters such as MLSS, SVI, and DO were measured. All the analytical parameters were computed by standard procedures given in APHA, 21st edition. An illustration of the actual installed plant is shown in Figure 9.12. The biological parameters maintained and operating conditions for the plant are given in Tables 9.5 and 9.6 respectively.

9.11 RESULTS AND DISCUSSION

Initially, a single-stage bioreactor process was proposed for the treatment of wastewater. It was observed that by maintaining biological parameters, more than a 90% COD reduction can be achieved as required for the discharged

Table 9.6 Operating parameters

Sr. No	Parameters	Unit	Value
1	Recirculation flow	m^3/h	35–45
2	Module air flow	m^3/h	12–15
3	Feed pressure	psi	7.0
5	Average membrane flux	LMH	31.0
6	Physical cleaning	One minute rest time after every 10-minute service cycle	
7	Chemical cleaning	Once per week with HCL 0.5% and sodium hypochlorite 1000 ppm	

limit, but poor TKN reduction was achieved. In the first step, BOD gets degraded to form CO_2 and ammonia, and the ammonia formed in the first step and excess ammonia in the feed is supposed to be oxidized to nitrate by a nitrifying culture. But this did not happen, possibly because of the high incoming COD load. Generally, the growth rate of nitrifying autotrophs is considerably lower than that of carbon-reducing heterotrophs. Inhibition or elimination of the autotrophs by these interspecies competition usually leads to a decrease in nitrification efficiency or even to a process failure (Okabe, 1996). Also, the formation and presence of free ammonia in the system is validated by an increase in pH and alkalinity values of the product water.

It was important to address the issue of a lower TKN reduction for a longer sustainable biological process. The issue was overcome by the following modification in the biological process.

9.11.1 Two-Stage Bioreactor for Nitrogen Handling

Biological reactions take place in a bioreactor, and the sludge formation is dependent on the characteristics of feed wastewater and its constituents. In the present case study, the feed water had high TKN values, and therefore the biological reactions do not get completed because of the incomplete oxidation of nitrogen.

In the biological process, the nitrogen conversion takes places in the following steps (Metcalf & Eddy, Inc., 2003).

$$COHNS + O_2 + Nutrient \xrightarrow{Bacteria} CO_2 + NH_3$$
$$+ \underset{\text{New cells}}{C_5H_7NO_2} + Other\ end\ products$$

$$2NH_4^+ + 3O_2 \rightarrow 2NO_2^- + 4H^+ + 2H_2O$$

$$2NO_2^- + O_2 \rightarrow 2NO_3^-$$

To address the nitrogen problem, a two-stage bioreactor method was applied. The first tank was used as a pre-aeration tank in which feed wastewater was added and aerated for 8–10 h. The overflow from this tank was fed into another biological tank connected with the membrane system, which was again aerated maintaining a predetermined HRT. In the first tank 70–75% COD reduction took place, which reduced the COD loading for the second tank. The second tank experienced less feed COD to be degraded and high ammonia, which is a favorable environment for the nitrifying culture to develop and complete the nitrification process. Nitrification was indicated by reduction in TKN, pH, and the alkalinity values of the product. Once sludge was acclimated to the conditions, the bioreactor started showing consistent reduction in pH and alkalinity values.

Table 9.7 shows the process performance with a single stage bioreactor, which clearly shows the increase in pH and alkalinity values of product to 8.94 and 950 mg/L respectively with TKN reduction of only 27%.

After implementation and stabilization of the two-stage bioreactor process (Table 9.8), TKN reduction improved to more than 90%, which resulted in a drop in both pH and alkalinity values. In both the single stage and the two-stage bioreactor process, more than 90% COD reduction was observed.

Table 9.7 Single stage bioreactor performance

Parameters	UOM	Feed	Product
pH		7.04	8.54
COD	mg/L	1970	<150
BOD	mg/L	372	12
Alkalinity	mg/L	160	950
TKN	mg/L	150	109

Table 9.8 Two-stage bioreactor performance

Parameters	UOM	Feed	Pre-aeration Tank	Product
pH		7.02	8.30	7.25
COD	mg/L	2051	620	<150
BOD	mg/L	379	60	8.0
Alkalinity	mg/L	159	500	57
TKN	mg/L	150	–	12

9.11.2 Overall Process Results

The pressurized Aqua EMBR membrane module used has shown consistency in terms of flux and product quality for the rejection of solids by membrane (as shown in Figure 9.13).

Consistent performance can be achieved by maintaining steady-state operating parameters and cleaning the membrane frequently in an overall system. Figure 9.14 shows images of water at different stages marked as (a), (b), and (c), where (a) shows feed water containing average soluble COD 1800 mg/L, (b) shows an image of bioreactor sludge, and (c) shows permeate water from the membrane after separation of sludge.

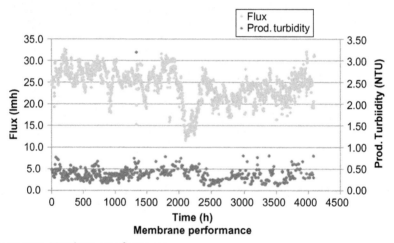

Figure 9.13 Membrane performance.

Figure 9.14 Feed and permeate water quality (see text for details).

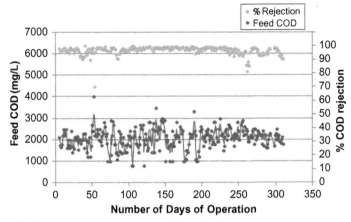

Figure 9.15 Chemical oxygen demand (COD) rejection.

Table 9.9 Overall product quality from the aqua EMBR system

Sr. No	Parameters	Average Value	Unit
1	pH	6.5–7.0	
2	COD	<150	mg/L
3	BOD	<15	mg/L
4	Alkalinity	75	mg/L
5	TKN	<20	mg/L
6	Product turbidity	<0.50	NTU

The COD rejection graph in Figure 9.15 shows a more than 90% COD reduction consistently achieved through the system.

Table 9.9 shows the overall product quality from the Aqua EMBR, COD <150 mg/L, BOD <15 mg/L, and product turbidity <0.3 NTU, and mostly in the range of 0.10–0.15 NTU. Due to proper sludge acclimation and a robust, effective, and innovative MBR membrane system, the permeate produced by the MBR consistently followed the discharge norms and was good enough for non-potable uses like toilet flushing, floor washing, gardening, or cooling water makeup. It can also be reused in the process after treating with reverse osmosis.

9.12 APPLICATION OF SUBMERGED MBR

The submerged MBR system developed by Aquatech operates at a relatively low flux, and the product is collected by application of a low suction pump. Air flowing through the plates creates cross flow velocity and also scrubs the membrane surface. These pilot trials were conducted to ascertain sustained performance for COD reduction, turbidity, power consumption, and other performance parameters. Submerged module testing was done with simulated water in the lab. Simulated wastewater prepared showed 850 mg/L COD and 359 mg/L BOD as given in Table 9.10. Simulated wastewater was fed into the biological system, which had a membrane module inside. The module was operated at a specific operating condition at controlled flux. The quality of MBR product was excellent, as shown in Table 9.11. The turbidity achieved was always <0.1 NTU and COD and BOD reduction was more than 97%.

Figure 9.16 shows the performance of the membrane with respect to flux and indicates the consistent flux under the given operating conditions.

Table 9.10 Influent water (Simulated Wastewater) characteristics for submerged MBR trial

Sr. No	Parameters	Unit	Value
1	pH		7.05
2	COD	mg/lit	850
3	BOD	mg/lit	359
4	Alkalinity	mg/lit	60

Table 9.11 Product Effluent water characteristics

Sr. No	Parameters	Unit	Value
1	pH		6.87
2	COD	mg/lit	<20
3	BOD	mg/lit	<10
4	Alkalinity	mg/lit	50
5	Turbidity	NTU	<0.1

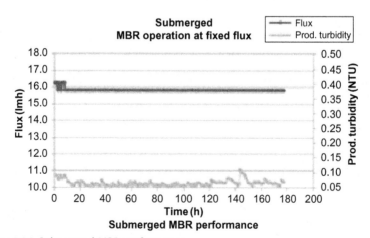

Figure 9.16 Submerged MBR performance.

9.13 SUMMARY

The pilot plant study shows that proper sludge acclimation makes it possible to treat difficult wastewater (BOD/COD <0.4). The study conducted with innovative MBR membrane modules shows that the operating flat membrane in side stream mode is effective and very efficient. It can handle municipal as well as various industrial wastewaters successfully. Pilot scale module operations with wastewater verified the design and also confirmed the flexibility of the biological sludge handling capacity. In the present case study, the objectives of achieving permeate quality were fulfilled by a two-stage sequential bioreactor process, which was used for removal of carbon and reduction of TKN. The biological parameters such as MLSS and SRT were readjusted from the conventional activated sludge, which resulted in achieving COD reduction (>90% COD), along with reduction of BOD and TKN. The Aqua EMBR pressurized membrane module shows consistent solids rejection with turbidity very suitable for reverse osmosis application. The product generated from the wastewater treatment facility is being reused successfully in the process after treatment with reverse osmosis in addition to satisfying other non-potable water needs in the plant.

REFERENCES

Frenkel, V., 2013. MBR: when and why we need it (IDA World Congress on Desalination and Water Reuse desalination: A Promise for the Future). MJCEC Meijiang Convention Center, Tianjin, China.

Judd, S., 2011. The MBR Book – Principles and Application of Membrane Bioreactors in Water and Wastewater. Elsevier.

Metcalf & Eddy, Inc., 2003. Wastewater engineering, Treatment and Reuse, fourth ed. McGraw Hill, New York.

Mittal, A., 2011. Biological wastewater treatment. Water Today, August, Fulltide: 32–44.

Okabe, S., 1996. Spatial microbial distributions of nitrifiers and heterotrophs in mixed-population biofilms. 50 (1), 24–35.

The new Mexico wastewater system operator clarification study chapter 10 "Activated sludge" viewed on 16th February 2014, http://www.nmenv.state.nm.us/swqb/FOT/WastewaterStudyManual/10.pdf.

Sarin, V., 2013. Wastewater Treatment Using Membrane Bio Reactor. PhD Thesis, IIT, Delhi.

Sutton, P.M., 2006. Membrane Bioreactors for Industrial Wastewater Treatment: Applicability and Selection of Optimal System Configuration. Water Environment Federation.

Veenstra, S., Alaerts, G.J., Bijlsma, M., Chapter-3: technology selection. In: Helmer, R., Hespanhol, I. (Eds.), Water Pollution Control - A Guide to the Use of Water Quality Management Principles. Published on behalf of the United Nations Environment Programme, the Water Supply & Sanitation Collaborative Council and the World Health Organization by E. & F. Spon © 1997 WHO/UNEP.

CHAPTER 10

Application of Anaerobic Membrane Bioreactor (AnMBR) for Low-Strength Wastewater Treatment and Energy Generation

Janardhan Bornare, V. Kalyanraman, R.R. Sonde
R.D. Aga Research, Technology and Innovation Centre, Thermax Ltd., Pune, India

10.1 INTRODUCTION

Development of innovative wastewater treatment technologies for India is gaining momentum to meet the changing regulatory requirements, better treatment efficiency, and enhanced sustainability, and for reducing capital as well as operating costs and energy recovery from wastewater. With the increasing population and economic uplift in the country, the water demand for irrigation, domestic use, and power generation is exerting enormous pressure on our water resources because utilization of water has been increasing at a faster pace. The existing demand and supply gap of water has prompted an exclusive path toward recycling and reuse of water. Disposal of domestic sewage from cities and towns is the biggest source of pollution of water bodies in India. Treatment of domestic sewage and subsequent utilization of treated sewage for various applications can prevent pollution of water bodies and reduce the demand for fresh water. The existing sewage treatment facilities are inadequate and operate on obsolete technologies. In India, the activated sludge process (ASP), upflow anaerobic sludge blanket (UASB), and waste stabilization ponds (WSP) are the most commonly employed technologies for sewage treatment. In the current challenging scenario, more emphasis is given to the generation of recyclable quality treated wastewater. Technologies are evolving around the concept of maximum organic reduction and nutrient removal to achieve recyclable quality of treated effluent. Waste to energy generation is one of the thrust areas, and the success of any technology lies in treatment and energy generation from wastewater. Domestic wastewater could be a good source of water if it is treated efficiently to get recyclable quality water. There can be added advantages if energy is generated through this vast resource.

10.1.1 Water Availability and Use

The water resource potential of the country in terms of natural water runoff into rivers is about 1869 BCM (billion cubic meters) as per the estimates of the Central Water Commission of India, considering both surface and groundwater (Source: Central Water Commission, 2010. *Water and Related Statistics. Water Planning and Project Wing, Central Water Commission, India.* December 2010). Due to various constraints of topography and uneven distribution of resources over space and time, the total utilizable water resource in the country has been estimated to be about 1123 BCM (690 BCM from surface and 433 BCM from ground), which is just 28% of the water derived from precipitation. About 85% (688 BCM) of water usage is being diverted for irrigation (Figure 10.1), which may increase to 1072 BCM by 2050. With the present population growth rate (1.9% per year) in India, the population is expected to cross the 1.5 billion mark by 2050. Due to increasing population and all-round development in India, the per capita average annual fresh water availability has been declining since 1951 from 5177 to 1816 m^3 in 2001 and 1588 m^3 in 2010.

Per capita average annual fresh water availability is expected to further decline to 1340 m^3 in 2025 and 1140 m^3 in 2050. Table 10.1 shows the average annual per capita availability of water. While the per capita water availability is decreasing considerably, domestic wastewater is polluting water reservoirs further, and the quality of available water is deteriorating considerably.

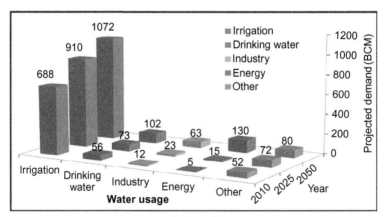

Figure 10.1 Projected water demand by different sectors. *Source: Central Water Commission (2010).*

Table 10.1 Average Annual Water Availability Per Capita

Year	Population (Million)	Per Capita Average Annual Availability (m³/year)
2001	1029 (2001 census)	1816
2011	1210 (2011 census)	1545
2025	1394 (projected)	1340
2050	1640 (projected)	1140

Poor infrastructure and low treatment capacities add further problems to water availability. Because the pressure on water resources is continuously increasing with the rise in population, urbanization, and industrialization, there is an urgent need for efficient water resource management through enhanced water use efficiency and wastewater recycling.

Wastewaters, in general, and domestic wastewater, in particular, could be a good source for helping to meet water demand in certain areas/sectors after efficient treatment, due to the vast quantities available.

10.1.2 Domestic Wastewater (Sewage) Generation and Treatment Scenario in India

There are 35 metropolitan cities in India, each having more than 10 lacs of population and generating 15,644 million liters per day (MLD) of sewage. The treatment capacity exists for 8040 MLD, i.e., 51% of the generated sewage. Figure 10.2 shows the sewage generation and treatment capacity of metropolitan cities in India. Among these cities, Delhi has the maximum treatment capacity, 2330 MLD (30% of the total treatment capacity of metropolitan cities). Next to Delhi, Mumbai has a capacity of 2130 MLD, which is 26% of total capacity in metropolitan cities. Delhi and Mumbai therefore in combination have 55% of the treatment capacity of the metropolitan cities.

In India, there are a projected 498 class-I cities (including metropolitan cities) having a population of more than 1 lac. Nearly 52% of cities (260 out of 498) are located in Andhra Pradesh, Maharashtra, Tamil Nadu, Uttar Pradesh, and West Bengal. The sewage generated in class-I cities is estimated at 35,558.12 MLD, and 93% of total wastewater is generated in class-I cities only. The total sewage treatment capacity of class-I cities is reportedly 11,553.68 MLD, which is 32% of the sewage generation. Figure 10.3 indicates the sewage generation and treatment capacity of class-I cities in India including metropolitan cities.

Out of this 11,553.69 MLD class-I city sewage treatment capacity, 8040 MLD (69%) is treated in 35 metropolitan cities and the remaining (31%) in other class-I cities.

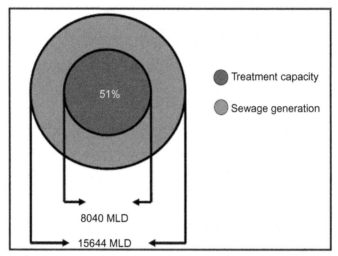

Figure 10.2 Municipal sewage generation and treatment capacity of metropolitan cities.

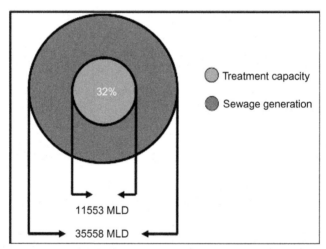

Figure 10.3 Municipal sewage generation and treatment capacity of class-I cities (including metropolitan cities).

The class-II towns are mostly located in Andhra Pradesh, Maharashtra, Tamil Nadu, Uttar Pradesh, and Gujarat. The total sewage generated in class-II towns is 2696.70 MLD. The total sewage treatment capacity in class-II towns is 233.7, which represents 8% of the total sewage generated. Figure 10.4 indicates the sewage generation and treatment capacity of class-II towns in India.

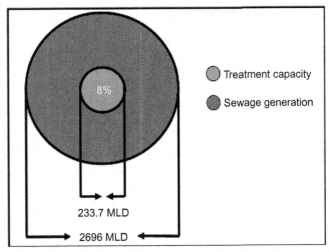

Figure 10.4 Sewage generation and treatment capacity of class-II towns in India.

These data indicate that sewage treatment in India has a huge potential and the treated water can be recycled in a number of applications to meet future water demand.

10.2 EXISTING TECHNOLOGIES FOR THE TREATMENT OF SEWAGE IN INDIA

Since 1985, more than 70 sewage treatment plants have been constructed under the Ganga Action Plan (GAP) and Yamuna Action Plan (YAP). Thirty-two sewage treatment plants (728 MLD treatment capacity) were constructed, and 11 existing sewage treatment plants (151 MLD treatment capacity) were renovated under the GAP phase-I. The ASP was the most preferred technology, accounting for almost 62% of the capacity. Under the YAP phase-I, 26 sewage treatment plants (722 MLD treatment capacity) were set up and the UASB process was the most preferred treatment technology, accounting for 83% of the total installed treatment capacity. Concurrently with YAP, 30 sewage treatment plants having 2325 MLD of sewage treatment capacity were added in Delhi under the river conservation program of the government of NCT Delhi. The river conservation program for Delhi opted for ASP as its treatment technology (Source: *Compendium of Sewage Treatment Technologies*, National River Conservation Directorate Ministry of Environment, & Forests, Government of India, 2009).

Various treatment technologies are employed in sewage treatment plants across India. In class-I cities, ASP is the most commonly employed technology, followed by UASB and WSP technology. ASP and UASB are mostly used as the main treatment system coupled with primary or tertiary treatment units. ASP technology is the most suitable for large cities because it requires less space as compared to UASB technology and WSP technology, and as both of these technologies employ land-intensive ponds in treatment schemes. Air blower/compressor and sludge recycle pump of ASP are major power intensive components which demand for additional operating expenses in comparison to UASB technology. In treatment schemes based on conventional versions of ASP technology, both primary and secondary sludge are commonly treated in anaerobic sludge reactors. This reduces the required area of sludge beds.

Most of the sewage treatment systems using UASB technology include a grit chamber as a preliminary treatment unit and terminal polishing unit. This treatment scheme is simple to operate and has a low-operating cost with an added advantage of biogas generation. Ideally, this makes UASB technology the most favorable treatment option for various cities. However, UASB technology is very sensitive to environmental changes and operational parameters. Shifts in temperature, and variations in volumetric organic loads, toxicity, and inhibitors cause operational difficulties in UASB. These factors affect the retention of a microbial population in the bioreactor and lead to frequent failures in UASB.

A series of WSP technology is being considered for the treatment in class-II towns due to low-operating cost and area availability. The WSP technology is also one of the most economical technologies; however, unlike UASB technology, no resources in the form of biogas are recovered. The advantages of WSP technology over UASB technology include less sensitive operation and greater improvement in bacteriological quality. The most frequently employed configuration of WSP technology uses two parallel streams of at least three stages of ponds, the first stage being the anaerobic pond, the second stage the facultative pond, and the third stage the maturation pond. However, because the class-II towns are getting over-populated, land availability for WSP is becoming major constraint in employing this technology.

10.2.1 Latest Treatment Technologies Adopted in India

A number of newer treatment technologies have come into practice in recent times and are becoming more popular for sewage treatment. These include the moving bed biofilm reactor (MBBR), the fluidized aerobic bioreactor (FAB), the sequencing batch reactor (SBR), and the membrane bioreactor (MBR). In principle, the MBBR uses a combination of suspended and attached growth processes of biological wastewater treatment. Such systems utilize specially designed carrier material for the growth of biofilm, which is held in suspension within the bioreactor. The MBBR system is a great improvement over conventional extended aeration or ASP and provides a robust and compact facility in a fully enclosed design. No sludge recirculation is required to maintain an active biomass in the bioreactor as in the case of conventional ASP. The MBBR system takes higher shock loads without reducing the plant performance because of the large quantity of active biomass available inside the reactor.

The SBR is a fill-and-draw activated sludge system for wastewater treatment. In this system, wastewater is added to a single "batch" reactor, treated to remove undesirable components, and then discharged. Equalization, aeration, and clarification can all be achieved using a single batch reactor. To optimize the performance of the system, two or more batch reactors are used in a predetermined sequence of operations. The unit processes of the SBR and conventional activated sludge systems are the same. The difference between the two technologies is that the SBR performs equalization, biological treatment, and secondary clarification in a single tank using a timed control sequence.

MBBR and SBR technologies are gaining acceptance due to a small footprint area and recyclable quality of effluent. MBR technology, which combines the biological-ASP and membrane filtration, has also become more popular recently. In MBR technology, the suspended solids and microorganisms are separated from the treated water by membrane filtration. Compared with conventional wastewater treatment processes, MBRs offer several advantages including high biodegradation efficiency, excellent effluent quality, smaller sludge production, and compactness. As a result, MBR can offer an attractive option for the treatment and reuse of municipal wastewater. Because of further technical developments and significant cost reductions to some extent, the interest in MBR technology for municipal wastewater treatment has sharply increased; however, deployment of MBR technology in India is facing some challenges because of huge capital requirements and the operating cost of the technology. Other factors, such as membrane fouling, necessity of critical pretreatment, and membrane replacement cost are also impacting on the implementation of MBR technology in India.

10.2.2 Treatment Cost for Various Technologies

All these latest and advanced technologies employed in domestic wastewater treatment are highly energy intensive due to the aeration demand for organic oxidation. The operating cost of aeration systems ranges from Rs. 1 to $9/m^3$, making them unviable options for future use. The treatment cost to produce effluent of a quality better than the Indian standard for discharge into water bodies (biological oxygen demand (BOD_5) < 30 mg/L, SS < 30 mg/L) is different for different technologies. The WSP technology has the lowest treatment cost and treats sewage at a cost of Rs. $1/m^3$. The conventional ASP system has a moderate treatment cost and produces treated water at a cost of Rs. $3.5/m^3$. The treatment cost for an SBR system is Rs. $5/m^3$. Similarly, among treatment options that produce effluents of recyclable quality (BOD_5 < 5 mg/L and SS < 5 mg/L), ASP followed by sand filtration and carbon filtration has the lowest treatment cost, i.e., Rs. $6.5/m^3$ while the MBR process generates treated water at a cost of Rs. $9/m^3$. The SBR system, followed by sand filtration and carbon filtration, produces treated water at a rate of Rs. $7.5/m^3$. The operating cost of a UASB-based domestic sewage treatment system is in the range of Rs. $0.5–1/m^3$. However, it is necessary to evaluate these on a case-by-case basis depending on the specific situation before arriving at a conclusion. The application of UASB technology is a good option for reducing the operating costs and enhancing energy generation from wastewater. However, there are multiple process and operational difficulties associated with the UASB system. There is no single solution for complete wastewater treatment and a downstream aeration system is required for complete treatment. UASB plants installed under various government schemes failed to deliver the desired discharge quality of effluent due to frequent operational failures. These failures occur because of maintenance of biomass in the reactors and frequent washouts. It is very important to maintain proper hydraulic conditions within UASB reactors. Upflow velocity and settling rates need to be balanced for efficient functioning of the bioreactor. Anaerobic microbial species are very sensitive to organic shock loads and temperature variations. In general, UASB technology demands more highly skilled operators compared with other aerobic wastewater treatment technologies.

Considering the above facts, water and energy nexus have become critical factors in developing efficient and sustainable technologies for sewage treatment. A need exists to develop technologies that can provide recyclable quality water with low-operating costs and a smaller footprint. The technology should also afford sustainability and present minimal operational difficulties. The anaerobic MBR (AnMBR) offers such advantages for the treatment of domestic wastewater.

10.3 INTRODUCTION TO THE AnMBR

Relative to aerobic MBRs, little research has been carried out in using membranes to retain the biomass in anaerobic systems. Anaerobic digesters with membrane separation units facilitate retention of microorganisms and allow operation with high biomass concentration. Thus, AnMBRs are expected to provide more efficient digestion, higher methane production, better effluent quality, and smaller size than conventional anaerobic digesters. Various studies have indicated that the main challenge with AnMBR has been fouling of membrane units (Choo et al., 2000; He et al., 2005).

According to how the membrane is integrated with the bioreactor, two AnMBR process configurations can be identified, mainly side stream and submerged. In side stream AnMBRs, membrane modules are placed outside the reactor, and the reactor mixed liquor circulates over a recirculation loop that contains the membrane. The cross flow velocity of the liquid across the surface of the membrane serves as the principal mechanism to disrupt cake formation on the membrane. Due to the higher operating costs involved in side stream MBRs, submerged MBRs, introduced by Yamamoto in 1989, have been the preferred choice in MBR plant installations from the mid-1990s. In submerged AnMBRs, the membranes are placed inside the reactor, submerged in the mixed liquor. In this configuration, a pump or gravity flow due to elevation difference is used to withdraw permeate through the membrane. Because the velocity of the liquid across the membrane cannot be controlled, cake formation can be disrupted by vigorously bubbling gas across the membrane surface. In the case of aerobic MBR, air scouring used also provides aeration in the bioreactor, while for AnMBRs biogas must be used.

Submerged AnMBRs consume much less power than external side stream AnMBRs because of the absence of a high flow recirculation pump. Side stream AnMBRs have a much higher energy requirement because of higher operational transmembrane pressures (TMP) and the elevated volumetric flow required to achieve the desired cross-flow velocity. However, side stream reactors have the advantage that the cleaning operation of membrane modules can be performed more easily in comparison with submerged technology, because membrane decoupling can be done effectively from the bioreactor during membrane cleaning. Submerged AnMBR demands lower energy, due to the fact that it provides a lower level of membrane surface shear and operates at lower permeate flux. This results in higher membrane surface requirements.

10.3.1 Integration Philosophy of Membrane with Anaerobic Bioreactor

The success of anaerobic wastewater treatment can be attributed to an efficient uncoupling of the solids' residence time from the hydraulic residence time through biomass retention, which is usually accomplished through biofilm and/or granule formation. With this strategy, high concentrations of biocatalysts are obtained, leading to high volumetric treatment capacities for a wide variety of wastewaters. At present, close to 80% of the full-scale anaerobic installations are sludge bed reactors in which biomass retention is attained by the formation of methanogenic aggregates. Granule formation is a complex process that involves physicochemical as well as biological interactions. In situations where biofilm or granule formation cannot be guaranteed, such as extreme salinity and high temperatures, or when complete biomass retention must be ensured, membrane-assisted physical separation can be used to achieve the required sludge retention. MBR technology ensures biomass retention by the use of micro-filtration or ultra-filtration membranes. Given that biomass is physically retained inside the reactor, there is no risk of cell washout, and the sludge retention is not dependent on the formation of biofilms or granules. In addition, MBRs offer the possibility of retaining specific microorganisms that, in the generally applied upflow reactors, would wash out. MBRs also appear suitable for the treatment of wastewaters with high organic suspended solids content, because particles are confined inside the reactor, allowing their degradation. As permeate is free of solids or cells, water would require fewer post-treatment steps if reusing or recycling is of interest.

10.3.2 Trends in AnMBR Research

Today the wastewater treatment discharge norms are becoming more and more stringent, and the demand for recyclable-quality treated water is growing. In such circumstances, conventional anaerobic processes are facing limitations, and growing interest in AnMBR is observed. The trend is obviously illustrated by the number of publications on both processes, conventional UASB and AnMBR (Figure 10.5). Starting from the early 2000s the growth

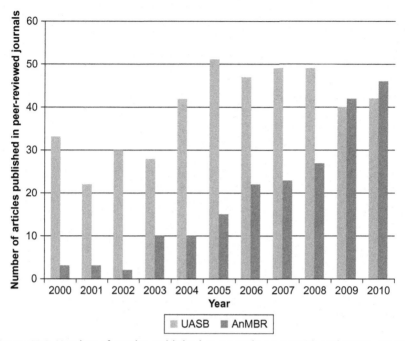

Figure 10.5 Number of articles published in journals on AnMBR and UASB. *Source: Visvanathan and Amila (2012).*

of journal publications on AnMBR has been mounting. On the other hand, studies on UASB have not grown that much.

Several laboratory-scale experiments have been conducted by different scientists treating synthetic and domestic wastewater to investigate the performance of an anaerobic MBR. Table 10.2 presents some literature about the application of AnMBRs for the treatment of different types of wastewater as well as information regarding the biological performance using various parameters such as temperature, mixed liquor suspended solids (MLSS), and organic loading rate (OLR) in the AnMBR. The chemical oxygen demand (COD) and BOD removals reported for various types of wastewaters have generally exceeded those reported from the operation of conventional anaerobic systems. Saddoud and Sayadi (2007) investigated the treatability of slaughterhouse wastewater by an AnMBR at relatively high OLRs between 4.4 and 13.3 kg $COD/m^3/day$. They experienced a process failure at an OLR of 16.3 kg $COD/m^3/day$ due to VFA accumulation.

Li-Mei Yuan et al. (2007) found that the AnMBR exhibited high performance in removing carbon, nitrogen, and phosphorous simultaneously. Under favorable operational conditions, the removal efficiencies for COD,

Table 10.2 Cited References of Biological Performance of AnMBR for Wastewater Treatment

Source	Wastewater	Temperature (°C)	MLSS (g TSS/L)	OLR (g COD/L.d)	Removal (%)
Aquino et al. (2006)	Meat extract and peptone	35	2.6	1.8	96
Fakhru'l-Razi and Noor (1999)	Palm oil mill	35	50–57	14.2–21.7	91.7–94.2
He et al. (2005)	Food industry	37	6–8	4.5	81–94
Wen et al. (1999)	Sewage	24–35	16–22	0.4–11	60–95
Padmasiri et al. (2007)	Swine manure	37	20–40	1–3	—
Kang et al. (2002)	Alcohol fermentation	55	2	3–3.5	90–95
Jeison and van Lier (2007)	Synthetic wastewater (VFA)	55	35	50	98
Akram and Stuckey (2008)	Synthetic wastewater	—	—	16	98
Jeison et al. (2009)	Synthetic wastewater with high salinity	30	30	—	—
Lin et al. (2011)	Thermo–mechanical pulping whitewater	37	6.7–11.3	2.6–4.8	90
Wijekoon et al. (2011)	Molasses-based synthetic wastewater	55	3–8	5–8	70–83

total nitrogen, and total phosphorus were achieved at 93%, 67.4%, and 94.1%, respectively. Van Zyl et al. (2008) studied the treatment of simulated petrochemical industry wastewater using a submerged AnMBR, and in that study, the AnMBR achieved a very high COD removal efficiency at OLRs up to 25 kg $COD/m^3/day$ during long-term operation.

Bailey et al. (1994) studied upflow anaerobic sludge bed reactor performance using cross-flow microfiltration and reported COD removal as high as 99%, with high effluent quality. Many anaerobic MBR systems have been operated at OLRs ranging from less than 5 to over 30 kg $COD/m^3/day$. AnMBRs are expected to provide more efficient digestion with a smaller size than conventional anaerobic digesters. These advantages, together with increased stringency in waste disposal for animal production facilities, have led to pilot scale testing of AnMBRs to treat swine waste (Du Preez et al., 2005; Lee et al., 2001).

The main challenge for AnMBRs has been the fouling of membrane units. High fluid flow velocities resulting in high shear rates at the surface of the membrane can be used in AnMBRs with external membrane units to reduce membrane fouling caused by the adhesion of biomass and colloidal organic matter to the membrane surface (Stephenson et al., 2000). However, the digestion efficiency in AnMBRs may be negatively affected by exposure of the biomass to high shear conditions.

An opportunity in applying AnMBR for low-strength wastewater treatment is identified, and literature indicates that few researchers have attempted studies on AnMBR for low-strength wastewater treatment. Energy recovery and reuse of treated effluents are major considerations of interest in using AnMBR for low-strength wastewater treatment. Even though the energy recovery from low-strength wastewater is fairly low due to low organic content, the reuse option of treated effluent is another eye-opening interest. Current trends in research publications indicate a large number of AnMBR studies in industrial high-strength wastewater treatment. At the same time, the number of studies on municipal wastewater treatment has also increased considerably. Figure 10.6 shows the number of articles published in journals on AnMBR research related to municipal and industrial wastewaters. Given that anaerobic processes have few limitations for complete nutrient removal, the treated effluent consists of higher concentrations of nitrogen and phosphorus. In such cases the treated effluent can be reused for irrigation purposes in agriculture.

Many reviews indicate that the AnMBR has advantages over conventional aerobic-based technologies, and the integration of membrane

Figure 10.6 Number of articles published in journals on AnMBR research related to municipal and industrial wastewaters. *Source: Scopus (2011).*

separation with anaerobic processes through best engineering practices can form commercially viable AnMBR systems.

10.4 DEVELOPMENT OF AnMBR AND EVALUATION STUDIES UNDERTAKEN

Considering all the benefits of the technology, AnMBR was developed with the following objectives in mind:

1. An AnMBR design with the most suitable configuration, using a side stream membrane module for the treatment of low-strength wastewater
2. Optimization studies on the OLR for the treatment of low-strength wastewater, with the goal of achieving effluent COD below 30 mg/L and BOD$_5$ less than 10 mg/L
3. Exploration of the biological performance of the side stream AnMBR for removal of organics during long-term operation
4. Evaluation of the membrane filtration performance and fouling frequency in the long-term operation of AnMBR
5. Identification of the challenges inherent in the continuous operation of the AnMBR
6. Future research scope and directions.

10.4.1 Materials and Methods

A CSTR type of side stream configuration was used in the AnMBR for this study. The pilot scale setup was comprised of a bioreactor with related accessories and a membrane filtration unit. A schematic diagram of the experimental setup used for this study is shown in Figure 10.7. The details of each component are described in the following sections.

10.4.1.1 Bioreactor

A 670-L capacity cylindrical bioreactor was designed and fabricated using stainless steel 304. The bioreactor was fully automatic and equipped with pH, temperature, and agitation indications and controls. A limpet coil jacket was provided around the bioreactor vessel for heating or cooling purposes. A water heating bath and pump (make—Raj Pumps Pvt. Ltd.; type—RBSP SS30) was used for circulating the heated water through the jacket coil and maintaining reactor temperature at the desired level. An electric heater was

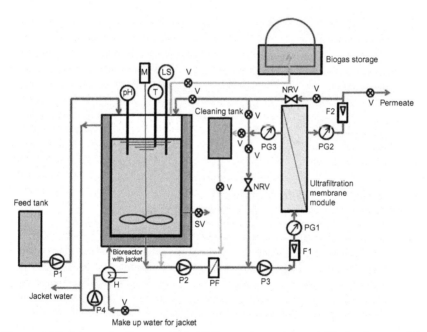

Figure 10.7 Schematic diagram of AnMBR experimental setup. P1: feed pump, P2: membrane feed pump, P3: membrane recirculation pump, P4: jacket water recirculation pump, pH: pH probe, T: temperature sensor, LS: level sensor, M: motor for agitator, H: heater for jacket water, PF: prefilter, F1: flowmeter for membrane recirculation flow, F2: flowmeter for permeate flow, PG1, PG2, PG3: pressure gauges, SV: sampling valve, NRV: non return valve, V: valve for flow control.

used as a heat source in water heating bath. A temperature display along with controller supplied by Electronics Systems and Devices, Pune, was provided in the control panel of the unit. A pH probe supplied by Hanna Instruments was mounted at the bottom of the bioreactor for the measurement of pH. An acid and alkali dosing arrangement was put in place for the control of pH. A Hanna make pH display along with controller was provided in the control panel of the unit. A two-stage blade-type agitator was provided at the top of the bioreactor for agitation purposes. A flange-mounted, 2 HP variable speed motor supplied by Crompton Greaves was used as a drive for the agitator. The agitator speed display (make—Electronics Systems and Devices, Pune; model—ESD 9010) and variable frequency drive (Schneider make) for changing the speed of agitator were used in the control panel of the setup. A 2000L capacity feed tank was fabricated in mild steel and used for feed storage purposes. A drain connection with a ball valve was provided at the bottom of the feed tank for draining and cleaning purposes. A self-priming centrifugal pump (make—Compton Greaves; model—Ministar III) was used for the pumping of feed to the bioreactor. A magnetic-float-operated guided level switch supplied by Pune Techtrol Pvt. Ltd., Pune (model—FGSO-J12O1WWW) was mounted in the bioreactor for maintaining the required water level in the reactor. The electrical signals from the level switch were used for the operation and control of the feed pump. The generated biogas collected in the head space of the bioreactor was taken out of the bioreactor and stored in the variable volume dome arrangement. Stored biogas was burned using a biogas burner on a daily basis.

10.4.1.2 Membrane Filtration Unit

A skid-mounted membrane filtration system consists of membranes, sewage pumps, rotameters, and UPVC piping. Tubular ultra-filtration membrane modules supplied by Berghof Germany were used in this study. Two Berghof membrane modules were used in series configuration in the filtration unit. A half HP sewage pump was used to push the water from the bioreactor into the membrane unit. A 4 HP sewage pump was used for recirculating the treated water through the membrane unit. Rotameters fabricated in acrylic material were used for the measurement of recirculation flow and permeate flow.

The necessary information relating to the membrane modules used for the study is given in Table 10.3 in the form of a membrane data sheet. Figure 10.8 shows the photograph of the experimental setup used for the study.

Table 10.3 Membrane Data Sheet

Module type	MO215G66.03_I8XLV
Membrane area (m^2)	0.49
Module length (mm)	1580 ± 1
Membrane material	PVDF
Membrane MWCO (Dalton or nm)	30 nm
Max. operating pressure (kPa)	-50 to 800
Max. operating temperature (°C)	60 (pH: 5–10)
pH	2–12 (60 °C)
Module diameter	2"
Inlet connections (mm)	50
Outer diameter of module (mm)	50
Permeate connection (mm)	20
Inner diameter of membrane (mm)	8
Module housing material	UPVC
Connection of concentrate	UPVC
Number of permeate connections	2

Figure 10.8 Photograph of the experimental setup.

10.4.1.3 Wastewater Characteristics

Synthetic wastewater was used as a feed for the AnMBR. The concentration of COD for synthetic wastewater was targeted at 500 mg/L by adding glucose, urea, DAP, and other micronutrients. Constituents of synthetic wastewater and their concentrations are given in Table 10.4.

Table 10.4 Constituents of Synthetic Wastewater and Their Concentrations

Constituents	gm/L
Dextrose monohydrate	0.5
Urea	0.021
Di-ammonium phosphate	0.0106
Sodium chloride (NaCl)	0.3
Magnesium chloride (MgCl$_2$)	0.15
Calcium chloride (CaCl$_2$)	0.04
Manganese sulfate (MnSO$_4$)	0.005
Ferrous sulfate (FeSO$_4$)	0.005
Sodium carbonate (Na$_2$CO$_3$)	0.15

Table 10.5 Average Characteristics of the Synthetic Wastewater (Feed)

Parameter	Unit	Reading
COD	mg/L	385
BOD$_5$	mg/L	247
TOC	mg/L	140
TSS	mg/L	23
pH	–	6.9

The average characteristics of the synthetic wastewater are shown in Table 10.5.

10.4.2 Experimental Methodology and Operating Conditions

The initial seeding of the bioreactor was carried out by inoculating 200 L of anaerobic sludge collected from Pimpri Chinchwad Municipal Corporation's sewage treatment plant, Pimpri–Chinchwad Link Road, Pimpri, Pune. The reactor was fed with synthetic wastewater (COD: 500 mg/L) while inoculating. The TSS concentration in the bioreactor immediately after inoculation was 12,500, which reduced later and stabilized in the range of 10,000-11,000 mg/L. A synthetic feed solution having a COD of about 500 mg/L was prepared in the feed tank on daily basis. It was observed that the organic content of the feed was naturally degrading with time and generally more than 25% organic load was getting degraded in one day. To minimize the degradation effects, fresh feed was prepared in the feed tank every day. The complete study was carried out in two phases. During the first phase of 120 days of experimentation, optimization of the OLR to the bioreactor was carried out. In the second phase of 80 days of operation, overall performance evaluation at optimized operating conditions was undertaken.

10.4.2.1 Optimization of OLR

The average COD maintained in the feed tank during the study was about 385 mg/L. Initially, the AnMBR was started with an OLR of 0.62 kg COD/m^3/day. In the first 120 days operation of the bioreactor, the OLR was increased in a stepwise manner, i.e., 0.62, 0.77, 0.92, and 1.32 kg COD/m^3/day with corresponding hydraulic retention time (HRT) of 15, 12, 10, and 7 h, respectively. The bioreactor temperature was constant and maintained at 37 °C for the complete study. Depending upon the OLR, the calculated quantity of sludge from the bioreactor was withdrawn on a daily basis to keep the MLVSS in the range of 7500–8000 mg/L. The treated water from the bioreactor was pumped through two tubular membrane modules arranged in a series for filtration of treated water. The reject from the membrane module was recycled back to the bioreactor. The treated water was recirculated through the membrane unit at a constant flow rate of 10 m^3/h to maintain sufficient cross-flow velocity on the membrane surface. To keep a constant HRT in the bioreactor, the permeate flow rate from the external membrane was always set a little higher than the feed flow rate, recycling the excess permeate because it was difficult to precisely keep a constant flux. Feed, bioreactor effluent, and permeate samples were tested on a daily basis for pH, TSS, VSS, soluble COD, BOD_5, and TOC. The generated biogas was measured on daily basis during the study. All analyses were performed according to Standard Methods for the Examination of Water and Wastewater (APHA, 1992). A standard practice of membrane cleaning was adopted after every two months of operation as per the membrane cleaning protocols.

The experimental operating conditions maintained during the investigation are given in Table 10.6.

10.4.2.2 Membrane-Cleaning Protocols

The membrane filtration unit was disconnected from the bioreactor before starting the chemical cleaning of the membranes. Prior to this cleaning exercise, the membrane modules were rinsed two to three times with tap water for the removal of the sludge layer and solid particles deposited on the membrane surface. One hundred liters of 1.5% citric acid solution was prepared in the cleaning tank and recirculated through the membrane for 30 min at a constant flow of 10 m^3/h. After stopping the recirculation flow, the module was allowed to soak in the acid solution for 60 min. This acid solution was discharged from the tank, and the membranes were flushed with tap water

Table 10.6 Experimental Operating Conditions Maintained During the Study

S. No.	Operating parameter	Unit	Reading
1	Bioreactor volume	Liters	670
2	Working volume of the bioreactor	Liters	550
3	Average feed COD	mg/L	385
4	HRT in the bioreactor	h	15, 12, 10, 7
5	Temperature in the bioreactor	°C	37
6	Organic loading rate to bioreactor	kg COD/m^3/day	0.62, 0.77, 0.92, 1.32
7	MLSS in the bioreactor	mg/L	10,000–11,000
8	MLVSS in the bioreactor	mg/L	7500–8000

two to three times. One hundred liters of alkali solution (0.1% NaOH + 0.2% NaOCl) was prepared in the cleaning tank and recirculated through the membrane modules at a flowrate of 10 m^3/h for 30 min. After stopping the recirculation flow, the membranes were allowed to soak in the alkali solution for 8 h. The recirculation pump was operated for 5 min after ending the soaking cycle and then membrane modules were rinsed with tap water two to three times.

10.4.2.3 Overall Performance Evaluation of AnMBR Under Optimized Operating Conditions

Chemical cleaning of the membranes was carried out before starting the performance evaluation study. During the performance evaluation, organic loading to the bioreactor was set at 1.155 kg COD/m^3/day. Depending upon the MLVSS in the bioreactor, a calculated quantity of the sludge was withdrawn from the bioreactor on a daily basis to keep the MLVSS in the range of 7500–8000 mg/L. The bioreactor temperature was constant and maintained at 37 °C for the complete study. The initial permeate flux was set at 93 L/m^2/h (LMH). Chemical cleaning of the membrane was carried out after 60 days of operation, and the membrane unit was disconnected from the bioreactor for 10 h. Feed flow to the bioreactor was stopped during the membrane cleaning operation; however, an equivalent amount of substrate dose was provided to the bioreactor for stable operation. The partial recycling of permeate flow was continued during this phase of study, and the recycling flow was adjusted throughout the study

Table 10.7 Experimental Operating Conditions Maintained during the Study

S. No.	Operating Parameter	Unit	Reading
1	Bioreactor volume	L	670
2	Working volume of the bioreactor	L	550
3	Average feed COD	mg/L	385
4	HRT in the bioreactor	h	8
5	Temperature in the bioreactor	°C	37
6	Organic loading rate to bioreactor	kg COD/m^3/day	1.155
7	Average MLSS in the bioreactor	mg/L	10,635
8	Average MLVSS in the bioreactor	mg/L	7719
9	Membrane flux (initial)	LMH	93

to keep a constant HRT in the bioreactor. HRT for the feed was 8 h maintained constant throughout the study. Feed, bioreactor effluent, and permeate samples were tested on a regular basis for pH, BOD$_5$, TOC, soluble COD, TSS, and VSS. All analyses were performed according to Standard Methods for the Examination of Water and Wastewater (APHA, 1992). The monitoring of TMP was continuous, which indicated the extent of membrane fouling. The biogas generated in the bioreactor was measured and analyzed on a daily basis. The experimental operating conditions maintained during the investigation are given in Table 10.7.

10.4.3 Results and Discussion

The COD of the synthetic wastewater stored in the feed tank was measured several hours after preparation. From the test results, it was observed that the COD concentration was naturally degrading with time. It was also noted that the average COD concentration of feed solution reduces from 25% to 35% in 24 h. A typical reduction in the COD concentration of feed solution is indicated in Figure 10.9.

10.4.3.1 Optimization of OLR to the Bioreactor

The volumetric OLR was increased in a stepwise manner from 0.62 to 1.32 kg COD/m^3/day. The increase in OLR was implemented by an increase in the feed flow rate to the bioreactor. However, due to the variation of feed COD concentration, the OLR could not be held constant but fluctuated around an average value in different operating periods.

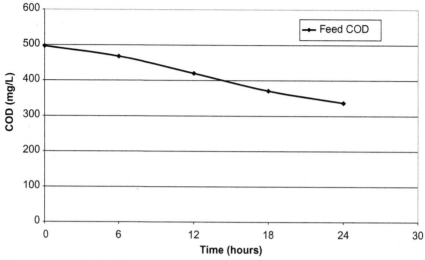

Figure 10.9 Typical reduction in COD concentration of feed solution in feed tank.

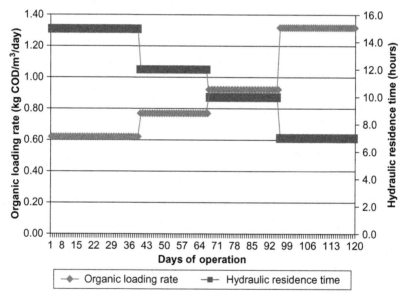

Figure 10.10 OLR and corresponding HRT in the bioreactor.

An average COD concentration of feed solution was maintained at 385 mg/L, and the same was considered for determination of OLR in the bioreactor. Figure 10.10 shows the stepwise increase in OLR and corresponding HRT for the feed.

10.4.3.1.1 The Effect of OLR on COD Removal Efficiency

As the concentration of COD in the feed and permeate varied on daily basis, the average COD removal efficiency was calculated for the AnMBR considering the average COD concentration in feed and permeate during each period of OLR. The average COD removal efficiency as a function of OLR is shown in Figure 10.11. The COD removal efficiency of the AnMBR was 97.67% for an OLR of 0.62 kg COD/m^3/day and corresponding HRT of 15 h. The COD removal efficiency of the AnMBR was reduced to 97.14% and 95.58% for an OLR of 0.77 and 0.92 kg COD/m^3/day, respectively. Within the OLR range from 0.62 to 0.77 kg COD/m^3/day, COD removal efficiency did not vary to a considerable extent and was relatively independent of the OLR. The COD removal efficiency of the AnMBR was 93.77% for an OLR of 1.32 kg COD/m^3/day with a corresponding HRT of 7 h.

The permeate sample was continuously monitored for the concentration of BOD$_5$. Average feed concentration of BOD$_5$ maintained during this phase of study was 247 mg/L. The average concentration of BOD$_5$ in the permeate samples was less than 5 mg/L during the experiments for 15 and 12 h HRT. The average concentrations of BOD$_5$ in permeates were 7 and 11 mg/L

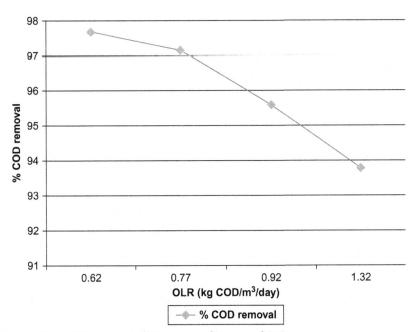

Figure 10.11 COD removal efficiency as a function of OLR.

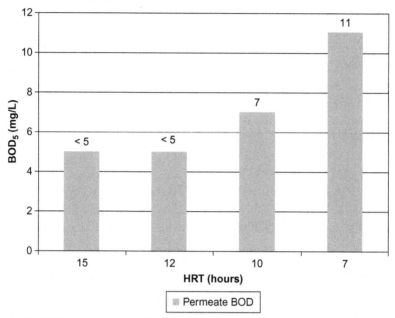

Figure 10.12 Concentrations of BOD_5 in the permeate as a function of HRT.

during the run of 10 and 7 h HRT, respectively. Figure 10.12 shows the evolution of BOD_5 concentrations in permeate as a function of HRT.

10.4.3.1.2 The Effect of OLR on Biogas Generation and Yield

The biogas generation was monitored on a daily basis, and the average biogas generation during each period of specific OLR is presented in Figure 10.13. The average biogas generation was increased from 159 to 193 L/day with the increase in OLR from 0.62 to 0.77 kg $COD/m^3/day$, respectively. The average biogas generations were 219 and 289 L/day with OLRs of 0.92 and 1.32 kg $COD/m^3/day$, respectively. Even though the average biogas generation was increased with an increase in OLR, the biogas yield was continuously decreased from 0.48 m^3 biogas/kg COD removed to 0.42 m^3 biogas/kg COD removed with an increase in OLR from 0.62 to 1.32 kg $COD/m^3/day$. This is due to a better food-to-microbe ratio (0.08 kg COD/kg MLVSS/day) at a lower OLR. At a higher OLR, although the gas generation is increased, the organic degradation efficiency is comparatively lower due to a high food-to-microbe ratio (0.17 kg COD/kg MLVSS/day). Higher concentration of the microbial population can be maintained in the bioreactor for reducing the food-to-microbe ratio at a

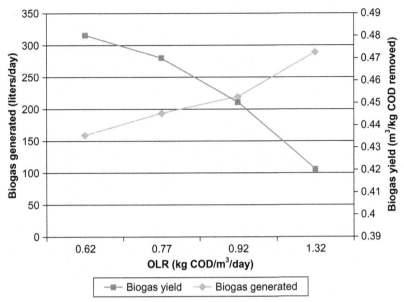

Figure 10.13 Biogas generation and yield as a function of OLR.

higher OLR. An increase in concentration of microbes has an adverse impact on the membrane and may lead to frequent membrane fouling.

10.4.3.2 Overall Performance Evaluation of AnMBR under Optimized Operating Conditions

Even though the average concentration of COD in permeate was well below the target of 30 mg/L during all the experiments of different OLRs, the average concentration of BOD_5 in the permeate was slightly higher than the targeted value of 10 mg/L during the OLR of 1.32 kg $COD/m^3/day$. To achieve the concentration of COD less than 30 mg/L and BOD_5 less than 10 mg/L, it was decided to evaluate the performance of the AnMBR at an HRT of 8 h for feed (1.155 kg $COD/m^3/day$ OLR).

10.4.3.2.1 Biological Performance of the AnMBR

The feed wastewater organic strength was represented by average COD, TOC, and BOD_5 concentrations of 385, 140, and 247 mg/L, respectively. Figure 10.14 shows COD concentration in feed, bioreactor, and permeate from the AnMBR. Even though the targeted COD concentration in the feed was about 500 mg/L, the degradation of organics was observed in the feed tank, and the average COD concentration of the feed was

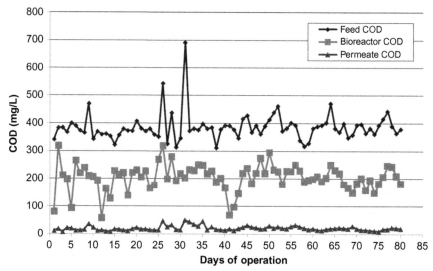

Figure 10.14 Evolution of concentrations of COD in the feed, bioreactor, and permeate during the performance evaluation of the AnMBR.

385 mg/L. The average residual COD, TOC, and BOD$_5$ concentrations in the permeate were 20, 7, and 8, respectively. Because there is a considerable difference in COD concentration for reactor content and permeate, the used ultra-filtration membrane might be rejecting a considerable amount of organics available in the reactor during filtration.

The COD and TOC removal performance of the AnMBR is shown in Figures 10.15 and 10.16, respectively. After stabilization of the AnMBR, the TSS, COD, TOC, and BOD$_5$ average removals were 100%, 95%, 95%, and 97%, respectively.

Figure 10.17 shows the concentrations of BOD$_5$ in the effluent during this phase of performance evaluation of the AnMBR. The average concentration of BOD$_5$ for the feed was 247 mg/L, whereas BOD$_5$ concentration in the permeate was 8 mg/L.

10.4.3.2.2 Biogas Generation and Yield

After stabilization of biological performance, the biogas generation from the AnMBR was studied on a daily basis, and is illustrated in Figure 10.18. A gas leak test was carried out for the bioreactor, gas piping, and gas storage unit before starting the experiment. Biogas generation varied from 229 to 281 L/day with an average generation of 264 L/day. As indicated in Figure 10.18, the observed biogas generation and biogas yield were very

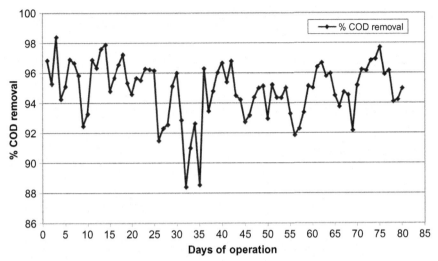

Figure 10.15 COD removal efficiency as a function of time.

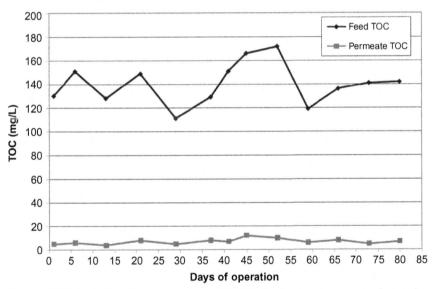

Figure 10.16 Evolution of concentrations of TOC in the feed and permeate during the performance evaluation of the AnMBR.

stable after stabilization of the biological process. The average biogas yield was 0.44 m³/kg COD removed.

The generated biogas was analyzed for methane composition on a periodic basis and found to vary from 69% to 82% with an average

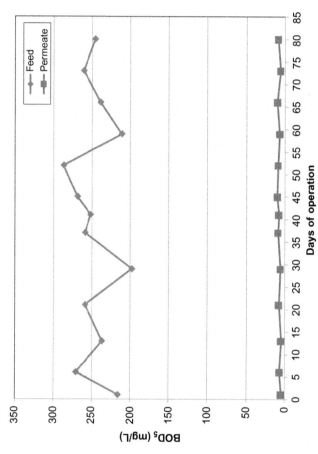

Figure 10.17 Evolution of concentrations of BOD$_5$ during the performance evaluation of the AnMBR.

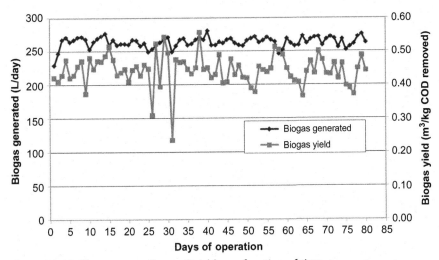

Figure 10.18 Biogas generation and yield as a function of time.

concentration of 75%. The generation of methane–rich biogas can be an indicator of favorable conditions for the methanogenic microbial population in the bioreactor. The methane–rich biogas generation could be due to a good methanogenic and acidogenic population balance in the bioreactor. Figure 10.19 shows the observed methane yield and shows the average methane yield to be 0.33 m^3 methane/kg COD removed, which was close to theoretical value of 0.35 m^3 methane/kg COD removed. The reject flow from the membrane was recycled to the bioreactor and allowed to enter the bioreactor at the top through the gas collected in the headspace of the bioreactor. The continuous contact of the recycled liquid stream with carbon dioxide collected in the headspace may allow absorption of the excess carbon dioxide gas in the liquid and may be one of the reasons for the presence of methane–rich biogas in the headspace.

10.4.3.2.3 Biomass Concentration in the Bioreactor

During the start-up stage of the bioreactor, biomass concentration in the bioreactor was reduced slightly, and then stabilized in the range between 10,000 and 11,000 mg/L. The observed reduction in MLSS could be due to decay of seed sludge during the acclimatization period. After stabilization of the TSS concentration in the bioreactor, solid retention time in the

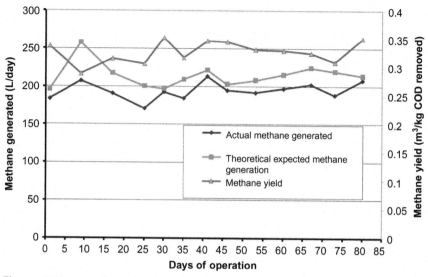

Figure 10.19 Actual methane generation, theoretical methane generation, and observed methane yield as a function of time.

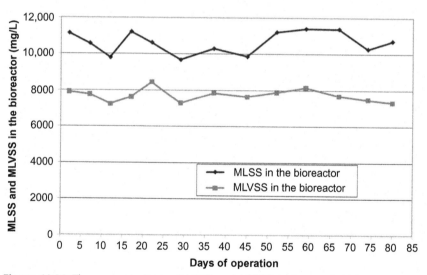

Figure 10.20 The concentrations of MLSS and MLVSS maintained in the bioreactor during performance evaluation.

bioreactor was maintained at 300 days by draining 1.84 L of bioreactor content on a daily basis. As indicated in Figure 10.20, MLSS in the bioreactor was maintained in the range from 9682 to 11,410 mg/L with an average MLSS of 10,635 mg/L. The average MLVSS maintained in the bioreactor during the investigation was 7719 mg/L. The ratio of average VSS to average TSS in the bioreactor was 0.73, and it was almost constant throughout the study.

10.4.3.2.4 Membrane Filtration Performance

Figure 10.21 shows the permeate flux evolution during the performance evaluation phase of the AnMBR. The membrane flux decline was relatively higher in the initial period of 1 week of operation and showed a reduction in flux from 93 to 84 LMH. The membrane flux reduction was very marginal in the next 6-week period of operation. The membrane flux was reduced from 84 to 76 LMH during this period and can be considered as a stable flux operation in membrane filtration. In the last period of 1 week of operation, the flux was very unstable and declining continuously at a faster rate. In this period of operation, the flux declined from 76 to 71 LMH. This flux decline is a clear indication of membrane fouling, and therefore the filtration operation was discontinued for chemical cleaning of the membranes. During first the first 60-day cycle of membrane filtration, the flux declined from 93 to 71 LMH, resulting in an average flux of 80 LMH. The membrane filtration operation was restarted with 99% flux recovery after chemical cleaning of the membranes.

As indicated in Figure 10.22, the TMP was increased from 0.6 to 0.8 bar in the 60-day operation before the chemical cleaning of the membrane. The TMP rise was relatively lower in the initial 50 days of membrane operation; however,

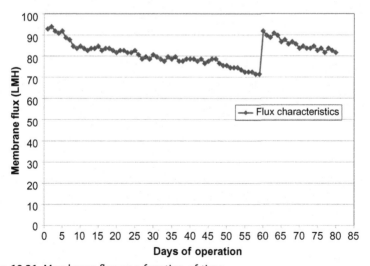

Figure 10.21 Membrane flux as a function of time.

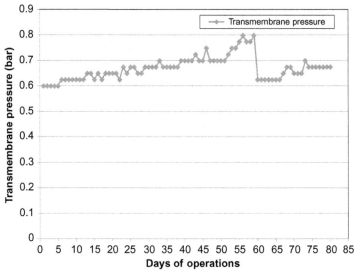

Figure 10.22 The transmembrane pressure as a function of time.

it was very unstable in last 10 days of operation. Probably after this stage if the filtration operation would have been continued, the membrane might have shown some unexpected range of flux and TMP. The rise in TMP across the membrane is the indication of membrane fouling. Figure 10.22 indicates that membrane cleaning is necessary after 60 days of operation to keep the TMP within an acceptable range.

10.5 SUMMARY AND CONCLUSIONS

Considering the population projections for the years 2025 and 2050, the average annual per capita availability of water is estimated to be 1340 and 1140 m^3, respectively. Thus, there is a decreasing trend in the per capita water availability because of increases in population, urbanization, and industrialization, which needs to be addressed. Because there is huge gap between the rate of generation and treatment of sewage in India, regulatory interventions are required for increasing the treatment capacities to bridge the gap. Simultaneously, efforts are necessary to create ground-breaking technologies for the generation of recyclable-quality treated water. Such recyclable water can be used for suitable applications to reduce future fresh water demand.

Many sewage treatment plants in India have been constructed through government funding schemes such as the GAP and the YAP. ASP, UASB, and a series of WSP were the most preferred treatment technologies in earlier days. There are a number of newer treatment technologies that have come into use in recent times that include the MBBR, the FAB, the SBR, and the MBR. Newer technologies are gaining acceptance because of a low footprint and recyclable quality effluent although they are high energy intensive systems. Hence, a need exists for the development of efficient and sustainable technology that can provide recyclable quality treated water with low-operating costs, low capital costs, and a smaller footprint. The AnMBR could be such a technology, which can reduce energy demand for wastewater treatment. The AnMBR facilitates retention of microorganisms in the bioreactor and allows operation with high biomass concentration. Thus, AnMBRs are expected to provide a better quality of treated effluent with energy recovery in terms of biogas. A growing interest in AnMBR research as compared to UASB has been observed. Many researchers have used AnMBR for the treatment of industrial wastewater; however, little research has been carried out for the treatment of low-strength municipal wastewater. Considering the merits of AnMBR, a development plan was made to assess the feasibility of the technology. A study was undertaken on the development of side stream AnMBR for the treatment of low-strength wastewater. The conclusions of the study are as follows:

1. At 37 °C, COD removal efficiency of side stream AnMBR was reduced from 97.67% to 93.77% with an increase in OLR from 0.62 to 1.32 kg $COD/m^3/day$ for a feed COD concentration of 385 mg/L.

2. The average concentration of BOD_5 in permeate was below 5 mg/L for an HRT of 15 and 12 h. The average concentration of BOD_5 in the permeate was increased to 7 and 11 mg/L with an HRT of 10 and 7 h, respectively.

3. The average biogas generation was increased from 159 to 289 L/day with an increase in OLR from 0.62 to 1.32 kg $COD/m^3/day$. Even though the average biogas generation was increased with an increase in OLR, the biogas yield was continuously decreased from 0.48 m^3 biogas/kg COD removed to 0.42 m^3 biogas/kg COD removed.

4. During continuous operation of side stream AnMBR at an OLR of 1.155 kg $COD/m^3/day$, average COD, TOC, and BOD_5 concentrations were 385, 140, and 247 mg/L, for feed, and 20, 7, and 8 mg/L for permeate, respectively. After stabilization of the AnMBR at an

OLR of 1.155 kg COD/m^3/day, the average TSS, COD, TOC, and BOD$_5$ removals were 100%, 95%, 95%, and 97%, respectively.

5. The average biogas and methane yields were 0.44 m^3/kg COD removed and 0.33 m^3/kg COD removed, respectively, for an OLR of 1.155 kg COD/m^3/day.

6. The average flux of 80 LMH was maintained during a 60-day continuous operation of the membrane without chemical cleaning at an average MLSS of 10,635 mg/L in the bioreactor. The TMP was increased from 0.6 to 0.8 bar during the 60-day operation of the AnMBR.

The results of this study demonstrate that use of the AnMBR is feasible for the treatment of dilute municipal wastewater, with more than 90% organic removal efficiency and a biogas yield of more than 0.4 m^3/kg COD removed. The permeate flux of more than 70 LMH is achievable with a TMP up to 0.8 bar through periodic backwashing and cleaning of the membranes.

10.6 FUTURE SCOPE AND RESEARCH NEEDS

Even though the study shows that the treatment of low-strength synthetic wastewater is possible through AnMBR at 37 °C, further studies on actual wastewater treatment at ambient temperature are required to prove the techno-commercial feasibility of the technology. To reduce the operating cost of the technology, further studies on submerged AnMBR need to be conducted. Membrane maintenance and cleaning activities are difficult in submerged AnMBR, and these issues need to be resolved. Studies on the enhancement of membrane flux and control on membrane fouling will help in reducing the operating cost of the treatment. Studies for establishing the membrane flux characteristics for actual wastewater treatment need to be carried out. Because anaerobic treatment is not very effective for removing nutrients, further studies on achieving nutrient removal through AnMBR need to be carried out.

ACKNOWLEDGMENTS

The study was part of the project on "Development of Anaerobic Membrane Bioreactor for Waste to Energy Solutions," partially funded by the Department of Biotechnology, Government of India, under the BIPP scheme. The authors sincerely thank the Department of Biotechnology for its support.

REFERENCES

Akram, A., Stuckey, D.C., 2008. Flux and performance improvement in a submerged anaerobic membrane bioreactor (SAMBR) using powdered activated carbon (PAC). Process Biochem. 43 (1), 93–102.

APHA, 1992. Standard methods for examination of water and wastewater, 18th ed. American Public Health Association, Washington DC, USA.

Aquino, S.F., Hu, A.Y., Akram, A., Stuckey, D.C., 2006. Characterization of dissolved compounds in submerged anaerobic membrane bioreactors (SAMBRs). J. Chem. Technol. Biotechnol. 81 (12), 1894–1904.

Bailey, A.D., Hansford, G.S., Dolod, P.L., 1994. The enhancement of upflow anaerobic sludge bed reactor performance using crossflow microfiltration. Water Res. 28 (2), 291–295.

Central Water Commission, 2010. Water and Related Statistics, Water Planning and Project Wing. Central Water Commission, India, December.

Chettiyappan Visvanathan and Amila Abeynayaka, 2012. Developments and future potentials of anaerobic membrane bioreactors (AnMBRs). Membrane Water Treatment 3 (1), 1–23.

Choo, K.H., Kang, I.J., Yoon, S.H., Park, H., Kim, J.H., Adiya, S., Lee, C.H., 2000. Approaches to membrane fouling control in anaerobic membrane bioreactors. Water Sci. Technol. 41, 363–371.

Du Preez, J., Norddahl, B., Christensen, K., 2005. The BIOREKs concept: a hybrid membrane bioreactor concept for very strong wastewater. Desalination 183 (1–3), 407–415.

Fakhru'l-Razi, A., Noor, M.J.M.M., 1999. Treatment of palm oil mill effluent (POME) with the membrane anaerobic system (MAS). Water Sci. Technol. 39 (10–11), 159–163.

He, Y., Xu, P., Li, C., Zhang, B., 2005. High-concentration food wastewater treatment by an anaerobic membrane bioreactor. Water Res. 39, 4110–4118.

Jeison, D., van Lier, J.B., 2007. Feasibility of thermophilic anaerobic submerged membrane bioreactors for wastewater treatment. Desalination 231 (1–3), 227–235.

Jeison, D., Kremer, B., van Lier, J.B., 2009. Application of membrane enhanced biomass retention to the anaerobic treatment of acidified wastewater under extreme saline condition. Sep. Purif. Technol. 64 (2), 198–205.

Kang, I.-J., Yoon, S.-H., Lee, C.-H., 2002. Comparison of the filtration characteristics of organic and inorganic membranes in a membrane-coupled anaerobic bioreactor. Water Res. 36, 1803–1813.

Lee, S.M., Jung, J.Y., Chung, Y.C., 2001. Novel method for enhancing permeate flux of submerged membrane system in two-phase anaerobic reactor. Water Res. 35 (2), 471–477.

Li-Mei Yuan, Dan-Li Xi, Yan Qiu Zhang, Chuan-Yi Zhang, Yi Ding, Yan-Lin Gao, 2007a. Biological nutrient removal using an alternating of anoxic and anaerobic membrane bioreactor (AAAM) process. In: Proceedings of the 10th International Conference on Environmental Science and Technology, Kos island, Greece, 5-7 September 2007.

Lin, H., Liao, B., Chen, J., Gao, W., Wang, L., Wang, F., et al., 2011. New insights into membrane fouling in a submerged anaerobic membrane bioreactor based on characterization of cake sludge and bulk sludge. Bioresour. Technol. 102 (3), 2373–2379.

Ministry of Urban Development, Government of India, 2012. Recent Trends in Technologies in Sewerage System. March.

National River Conservation Directorate Ministry of Environment & Forests, Government of India, 2009. Compendium of Sewage Treatment Technologies. August.

Padmasiri, S.I., Zhang, J., Fitch, M., Norddahl, B., Morgenroth, E., Raskin, L., 2007. Methanogenic population dynamics and performance of an anaerobic membrane bioreactor (AnMBR) treating swine manure under high shear conditions. Water Res. 41, 134–144.

Saddoud, A., Sayadi, S., 2007. Application of acidogenic fixed-bed reactor prior to anaerobic membrane bioreactor for sustainable slaughterhouse wastewater treatment. J. Hazard. Mater. 149, 700–706.

Stephenson, K.B., Jud, S., Jefferson, B., 2000. Membrane Bioreactors for Wastewater Treatment. IWA Publishing, London.

Van Zyl, P.J., Wentzel, M.C., Ekama, G.A., Riedel, K.J., 2008. Design and start-up of a high rate anaerobic membrane bioreactor for the treatment of a low pH, high strength, dissolved organic wastewater. Water Sci. Technol. 57 (2), 291–295.

Wen, C., Huang, X., Qian, Y., 1999. Domestic wastewater treatment using an anaerobic bioreactor coupled with membrane filtration. Process Biochem. 35 (3–4), 335–340.

Wijekoon, K.C., Visvanathan, C., Abeynayaka, A., 2011. Effect of organic loading rate on VFA production, organic matter removal and microbial activity of a two-stage thermophilic anaerobic membrane bioreactor. Bioresour. Technol. 102 (9), 5353–5360.

Yuan, L.-M., Xi, D.-L., Zhang, Y.Q., Zhang, C.-Y., Yi, D., Gao, Y.-L., 2007. Biological nutrient removal using an alternating of anoxic and anaerobic membrane bioreactor (AAAM) process. In: Proceedings of the 10th International Conference on Environmental Science and Technology, Kos island, Greece, 5–7 September 2007.

FURTHER READING

Bhardwaj, R.M., 2005. Status of Wastewater Generation and Treatment in India. In: IWG-Env, International Work Session on Water Statistics, Vienna, June 20–22, 2005.

Central Ground Water Board, Ministry of Water Resources, Government of India, Faridabad, 2011. Ground Water Year Book–India 2010–11, December.

Central Pollution Control Board, Government of India, 2005. Performance Status of Common Effluent Treatment Plants in India. Central Pollution Control Board, India, October.

Central Pollution Control Board, Ministry of Environment and Forests, Government of India, Delhi, 2009. Status of Water Supply, Wastewater Generation and Treatment in Class-I Cities & Class-II Towns of India. Control of Urban Pollution Series: CUPS/70/2009-10.

Gander, M., Jefferson, B., Judd, S., 2000. Aerobic MBRs for domestic wastewater treatment: a review with cost considerations. Sep. Purif. Technol. 18 (2), 119–130.

Government of India, Ministry of Water Resources, 2011. Restructuring of Central Water Commission (Volume: II), New Delhi, May.

Hilal, N., Al-Zoubi, H., Darwish, N.A., Mohamma, A.W., Arabi, M.A., 2004. A comprehensive review of nanofiltration membranes: treatment, pretreatment, modelling and atomic force microscopy. Desalination 170, 281–308.

Hribljan, M.J., 2007. Water Environment Federation, Webcast at Whittier. Large MBR Design and Residuals Handling, June 12.

Judd, S., 2006. The MBR Book. Elsevier, Oxford, 325 pp.

Ministry of Urban Development, Government of India, 2012. Recent Trends in Technologies in Sewerage System, March.

CHAPTER 11

3D TRASAR™ Technologies for Reliable Wastewater Recycling and Reuse

Manish Singh[1], Ling Liang[2], Atanu Basu[1], Michael A. Belsan[3], G. Anders Hallsby[3], William H. Tripp Morris[3]

[1]Nalco Water India Ltd., Pune, India
[2]Nalco (China) Environment Solutions Co. Ltd, Shanghai, China
[3]Nalco Company, Naperville, Illinois, USA

11.1 INTRODUCTION

With the ever-increasing stress on water resources and increasingly stringent regulatory norms on water consumption and discharge, industries are under pressure to improve their water efficiency through reuse and recycling of process effluents. However, recovery of these effluent streams is a rather daunting task because the quality can have dynamic variations since it is dependent on the production processes upstream of the effluent treatment plants (ETPs). It is also a common practice in several industries to produce batches of different products within the same production plant. Switching from one batch to the other results in variation in the effluent quality and hence adds to the complexity of treating the effluents with the same wastewater treatment plant. Unlike the case of raw water treatment, where the quality of incoming water (such as bore-well water or river water) constantly stays within a certain specification, the case of industrial wastewater treatment is often complicated because there is often little or no control on the quality of the incoming wastewater to be treated. On the other hand, for successful recycling of the treated water within the plant, its quality needs to meet specified criteria. For example, if the treated wastewater is meant for use as boiler makeup water, the treated water would need to meet that quality all the time irrespective of the variations that might affect it. The system designs that allow a proactive response instead of a reactive response in such a dynamic environment would have more operational success.

This chapter focuses on two technologies—3D TRASAR Technology for Sugar and 3D TRASAR Technology for Membranes—that have been developed to automatically handle such dynamic variations and ensure efficient system operation. One of these technologies is specific to the sugar industry, while the other can be applied across several industries wherever a critical reverse osmosis (RO) operation is used. These technologies have demonstrated significant water savings, energy savings, and asset protection and will be described in the following sections.

The sugar industry is under increasing pressure from the environmental regulatory bodies to reduce its fresh water consumption and effluent discharge. Also, several sugar mills are located in regions where the raw water has become scarce and poor in quality over the past several years. It is also interesting to observe that sugar mills can generate power using bagasse, which is a byproduct of the sugarcane milling process. For setting up and operating such co-generation power plants, sugar mills receive subsidies. The surplus power generated is sold back to the central power grid, providing an additional source of revenue for the mills. Most of these co-generation power plants utilize condensing turbines, which require additional water for the cooling water circuits needed for running the condensers. Therefore, in view of the imposed regulations, unavailability of good-quality raw water, and increased water requirements, the mills are pressed to recycle the wastewater.

In the sugar industry, a large amount of water is generated in the form of vapor condensate in the evaporation processes. This vapor condensate is of good quality most of the time and can be reused in low- to medium-pressure boilers and co-generation cooling towers. However, this vapor condensate is largely discarded due to dynamic variation in its quality because of sugar contamination, which cannot be predicted. Wet chemistry tests are still the most widely used methods for detection of juice contamination in the condensates. These are done manually once or a few times per shift and are labor intensive. Another major limitation of relying on these tests is that if a contamination event occurs in between the successive testing, it would most likely be missed. For this reason, many sites have added online conductivity analyzers in conjunction with the manual wet chemistry tests. However, conductivity analyzers are not sensitive toward the detection of trace levels of contamination. The 3D TRASAR Technology for Sugar is a unique online, fluorescence-based monitoring technology that is able to detect these variations with high sensitivity and selectivity, thus providing early detection. This has enabled mill owners to automatically and reliably reuse these condensate streams in their boilers and cooling towers.

Membrane filtration processes are integral to several of the recycling systems. Nanofiltration (NF) and RO filtration systems are being increasingly deployed for industrial wastewater recycling. One of the key attractive features of membrane technologies is that they do not require chemical additions, unlike the other water treatment operations. In this regard, membrane filtration is a relatively greener solution. Additionally, the cost per unit area of membrane has seen a gradual decrease over the past several years, and this has made membrane technology more competitive. Microfiltration (MF) and ultrafiltration (UF) are often used for pretreating the water before it goes to the NF or RO operation. RO and NF are often the most critical parts of the wastewater recycling process because these are the last unit operations in the process and are located just before the point of reuse, such as the boiler or cooling tower. Any failure in their operation would result in unavailability of sufficient water for the points of reuse. Hence, proper operation of these systems is critical. For stable operation of a membrane filtration system, it is essential to minimize fouling because it results in increased cost of operation due to an increase in energy (pumping) demand and additional maintenance costs for cleaning and membrane replacements. This aspect becomes even more important in the case where membranes are used for wastewater treatment. This is because feed water quality variations and operational problems can often lead to unanticipated fouling and scaling. Such system failures can negatively impact plant profitability. This is why it is important to follow a proactive approach where any early indication of fouling and the type of fouling (organic, inorganic, or biological) is captured so that necessary corrective actions can be taken well before it can cause any major operational upset or maintenance need (Huiting et al., 2001). In order to be able to practice such a proactive approach, it is essential to have reliable and accurate operational data that can then be normalized to monitor the fouling trend and the health of the system. To address these challenges, 3D TRASAR Technology for Membranes has been developed for proactive RO membrane performance management. It uses a combination of fluorescence and other probes, together with unique chemistries, to measure the key performance parameters related to system performance, and detect changes and upsets, and communicate them for proactive corrections.

11.2 3D TRASAR TECHNOLOGY FOR SUGAR

11.2.1 Process Description

Cane sugar mills produce sugar (raw, white, or refined) from sugarcane. Sugarcane contains about 70% water. In the milling process, the harvested sugarcane brought to the mill is chopped and passed through crushers to extract juice. This juice is then clarified and sent to evaporators, where the juice is concentrated by evaporating the water that is present in the juice. This water vapor leaves the evaporation process in the form of condensate, which is commonly referred to as the vapor or process condensate. Most of the water coming with the sugarcane is thus removed during the sugar production process, resulting in the generation of a water by-product stream. Therefore, it is rather straightforward to understand that the sugar mills have surplus water. By reusing the vapor condensate, fresh water consumption by the mills can be reduced, thereby improving the water efficiency of the mill.

During the sugar milling process, sugar juice can contaminate the water system. For example, a part of the sugar manufacturing process is the concentration of sugarcane clarified (or thin) juice using multiple effect evaporators (MEEs) (Hugot, 1986). In an MEE, the vapor from the first effect is used to heat the juice in the second effect, the vapor from the second is used to heat the juice in the third effect, and so forth (Figure 11.1). The first effect of the MEE is usually driven by steam, and the condensate from this effect is normally returned to the boiler feed water stream, because it is essentially steam condensate and would normally have no contamination (unless there is a process leak due to tube failure).

The clarified sugar cane juice, with 12–14% sugars, is concentrated to about 55–60% sugars in the MEE. In turn, a lot of water is generated in the form of process condensate. For example, an average-sized sugar plant

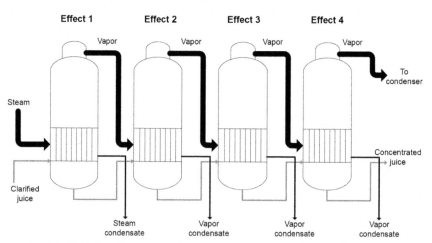

Figure 11.1 Multiple-effect evaporation process used in sugar industry for concentrating juice.

of 5000 tons (cane crushing capacity per day) could generate about 1000 cubic meters per day of process condensate from the second effect alone. This is a large amount of water that could be reused. These higher effect condensates (second effect and higher) could be returned to low-pressure boilers, because there is a threefold benefit in reusing these condensates:

- Process condensate in a sugar mill represents a large quantity of water.
- Most often the quality of process condensate is very good.
- Process condensate contains valuable heat energy.

In regions where water is scarce and power co-generation is practiced by sugar mills, the condensates could also be reused in the co-generation cooling towers. However, many mills send these process condensates to ETPs or are not able to maximize the utilization of these condensates because they do not have online monitoring technologies that can help them reuse these streams reliably.

As this condensate is being generated during an evaporation process, it has heat content and is generally of a quality that is much better than the raw water quality. It is also low in dissolved solids, with conductivity generally in the range 50–250 μS/cm. It has low levels of organics (mostly volatiles), with total organic carbon (TOC) ranging between 30 and 200 ppm. Based on how the mill is set up, the contribution of the process condensate to the total effluent going to the ETP could be as high as 40%. However, the quality of process condensate is much better in comparison with the other wastewater streams, in terms of inorganic as well as organic contaminants. Therefore, if this condensate is sent to the ETP, it results in a high hydraulic but low contaminants load, which makes its treatment in the ETP difficult. Also, if this is directly sent for drainage, it may result in high discharge costs due to the large quantity involved.

11.2.2 Challenges in Reusing Process Condensate

To improve the water efficiency of the sugar mill, it is desirable to reuse the process condensate in boilers or cooling towers. However, the biggest challenge in reusing these condensates is the unpredictable variation in their quality. Occasionally, thin juice will mechanically carry over into the process condensate stream. This event is known as a "sugar shot" and can happen when, for example, the juice levels in the evaporators are high or there are pressure imbalances. The frequency and severity of the sugar shots is unpredictable. The frequency could be as low as 1 or 2 sugars shots per season or as high as 1–2 per day depending upon how well the mill is being operated. When a sugar shot occurs, it is important to discard the contaminated vapor condensate or else the life of assets such as boilers or cooling towers would be compromised.

If a juice-contaminated process condensate is returned to the boiler, the sugar juice can break down in the high temperature environment of the boiler. The juice breakdown results in the formation of organic acids that may rapidly depress boiler water pH (Reid and Dunsmore, 1991). The pH of boiler water is a critical parameter for safe boiler operation. The recommended pH range for boiler water is 9–12 to protect the boiler material of construction (mild steel) from corrosion. At a high temperature under alkaline reducing conditions, mild steel forms a protective thin magnetite layer that is self-limiting and passivates the metal (Flynn, 2009). A depression in boiler water pH, for example due to the presence of organic acids, will damage the passivating layer and increase the mild steel corrosion. Corrosion in the boiler effectively reduces equipment life and creates safety hazards. Severe "sugar shots" may even lead to shutdown of the plant manufacturing process and cause production losses.

In the absence of reliable monitoring and effective chemical treatment programs, reuse of sugar vapor condensates in a cogeneration plant cooling tower causes severe bio-fouling and high corrosion rates. The resulting slime formation in the cooling tower fills and the condenser tube failures may lead to unplanned shutdowns and loss of power production. Therefore, immediate detection and notification of a "sugar shot" is critical. The monitoring methods are discussed in the following section.

11.2.3 Current Technologies

Sugar mills have used various offline and online methods to detect the "sugar shots." These methods have not been very effective in early detection of contamination. A reliable method to detect any contamination will help the mills save energy, protect assets, and reduce water waste, which has become a scarce resource for several sugar mills.

   i. Wet chemistry

   Wet chemistry methods, such as Molisch and phenol (or thymol) sulfuric acid tests, are still widely used due to their simplicity, even though these tests are labor intensive and involve the use of hazardous chemicals. In the α-naphthol or the Molisch test (Molisch, 1886), the presence of sugars is indicated by a violet-colored ring that is formed due to the reaction of α-naphthol with furfural or its derivatives formed by the dehydration of sugars by concentrated sulfuric acid. The test is quick, but only provides qualitative information. Often, the color produced is obscured by charring (Foulger, 1931). At low contamination levels, the method becomes unreliable. For quantitative estimation of sugars, the phenol (or thymol)-sulfuric method is the most frequently used. The sugars give an orange-yellow color when treated with phenol (or thymol) and concentrated sulfuric acid (Dubois et al., 1979). The method is simple, rapid, and sensitive, and gives reproducible results. The color produced is permanent, and it is unnecessary to pay special attention to the control of the conditions. Thus, tests using phenol and sulfuric acid are still the most widely used analytical methods, even though these are labor intensive and not online. There have been attempts to automate these tests with varied success (Reid and Dunsmore, 1991).

ii. Conductivity

Among the online monitoring methods, conductivity is used most widely because it is affordable to implement. However, it is interesting to note that, in the case of sugar process condensate, conductivity is in fact strongly correlated with ammonia concentration (Figure 11.2).

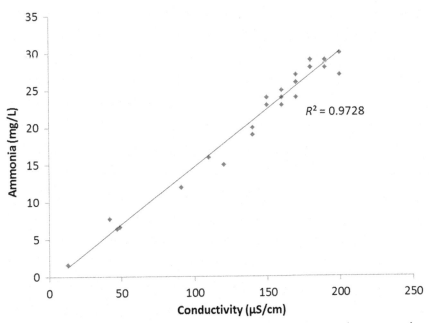

Figure 11.2 Sugar mill process condensate conductivity versus ammonia concentration based on analyses of several samples.

Conductivity has poor correlation with sugar cane juice contamination and performs poorly in providing reliable and early detection.

iii. TOC analyzers

TOC analyzers are usually based on grab sample analyses and are not online in the "true" sense. The samples are automatically collected at regular intervals and analyzed in a separate chamber within the instrument. There are major shortcomings in such a method. For example, even with a 5-minute response time, the TOC analyzer would miss monitoring a lot of condensate. Additionally, the background TOC from volatile organics could be significant and is highly variable. In our studies, it was observed that most of the TOC in condensates is contributed by ethanol, and TOC data showed large variations, even when there were no sugars present in the condensate. This would make TOC analyzers unreliable for monitoring sugarcane juice contamination in condensate. These analyzers require significant amounts of reagents, and hence, similarly to wet chemistry methods, TOC analyzers suffer from environmental health and safety issues.

11.2.4 3D TRASAR Technology for Sugar

Non-sugar materials in the sugarcane juice naturally fluoresce (Carpenter and Wall, 1972) and can be detected with the use of an online fluorometer (McGillivray et al., 1997). The fluorescence phenomenon is shown using the Jablonski diagram in Figure 11.3. In fluorescence, the fluorophore is first excited (through absorption of a photon) from its ground electronic state to one of the various vibrational states in the excited electronic state. A photon is then emitted as the molecule falls to one of the various vibrational levels of the ground electronic state (Lakowicz, 2006).

Taking advantage of the fact that sugarcane juice inherently contains fluorophores, 3D TRASAR[1] Technology for Sugar has been developed for detecting sugarcane juice carryover in condensate. The technology monitors the fluorescence intensity of the condensate. A spike in the fluorescence intensity is indicative of the presence of sugarcane juice in the condensate, and the magnitude of the intensity is relative to the level of contamination. 3D TRASAR Technology for Sugar is fluorescence-based automation that monitors "sugar shots" online, controls the condensate reuse, and helps sugar mills maximize the condensate recovery to boilers (or cooling tower), while maintaining the reliable and efficient operation of the boilers and

[1]3D stands for Detect, Determine, and Deliver.

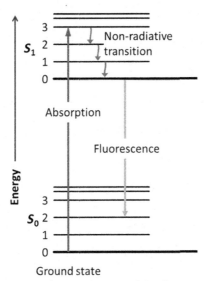

Figure 11.3 Jablonski diagram for visualization of the fluorescence phenomenon.

cooling systems. It detects the sugar contamination level with an online sugar fluorometer, determines the appropriate control action by comparing the sugar reading with the set point, delivers a signal to trigger an alarm, and opens the condensate reject valve during the sugar shot event (Figure 11.4).

For online monitoring of sugarcane juice contamination in condensate, a sugar fluorometer prototype was developed. For this, clarified sugarcane juice and condensate samples from sugar mills in China, India, and Australia were collected and analyzed using fluorescence spectroscopy. Based on the results, the appropriate optical filters and light source for excitation were chosen. The optical and electrical components were assembled in the form of a compact sugar fluorometer unit. The optical schematic is shown in Figure 11.5. The condensate flows through the flow cell housed in the fluorometer. As the sample passes through the flow cell, light of specific wavelength is continuously aimed at the flow cell. The fluorophores get excited and emit photons at a specific wavelength as the fluorophore falls to ground level. These photons are received by the detector and converted into intensity. The fluorometer is a solid-state device that can monitor the juice contamination in condensate in real time. The measurement does not involve the use of any reagents, and, hence, the method is safe for the operators.

The fluorometer and the rest of the components are assembled on a frame. This 3D TRASAR Technology for Sugar equipment is illustrated in Figure 11.6.

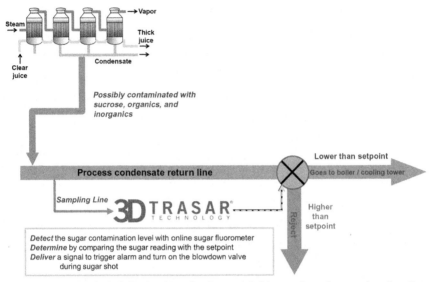

Figure 11.4 3D TRASAR Technology for Sugar reliably monitors "sugar shots" online and controls the condensate reuse.

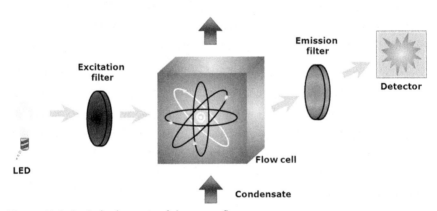

Figure 11.5 Optical schematic of the sugar fluorometer.

The 3D TRASAR® Technology for Sugar unit consists of a 3D TRASAR controller, sugar fluorometer with self-cleaning feature, sampling conditioning system, pH and conductivity monitoring module, and wireless gateway. The controller is the central component in the 3D TRASAR Technology for Sugar that defines and controls system parameters. It is set up and managed by interfacing (locally or remotely) with 3D TRASAR Configurator software. The solid-state fluorometer measures the natural fluorescence in sugar juice.

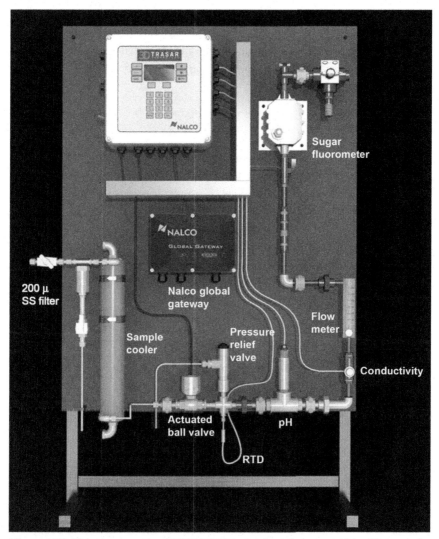

Figure 11.6 The components of 3D TRASAR technology for sugar. *Note: SS stands for Stainless Steel, and RTD stands for Resistance Temperature Detector.*

The assembled self-cleaning feature (optional) increases operational reliability and reduces maintenance of the module. The equipment includes a sample conditioning system, which is designed to safely cool and reduce the pressure of the condensate sample before it enters the fluorometer. The pH and conductivity of the condensate can be continuously measured using this module. Another integral component of the technology is its efficient data management and communication. The technology provides access to comprehensive, real-time data, alarms, and reports that help to speed resolution of issues and promote collaboration on continuous improvement and value creation. This component is instrumental in creating, capturing, and delivering value to the users of the technology.

11.2.5 Case Study

3D TRASAR Technology for Sugar equipment was installed at a large raw sugar mill. The mill faces water scarcity and consequently reuses process condensate in its low-pressure boilers. For monitoring the process condensate quality, an α-naphthol test was done every 15 minutes. In Figure 11.7, the performance of the conductivity sensor is shown. The crosses represent the α-naphthol data points from the operator logs. The results from the test are qualitative and are logged as nil, trace, very faint, and so forth, depending upon the color intensity of the ring that develops for the sample under test. The α-naphthol test data indicates that there is an entrainment event

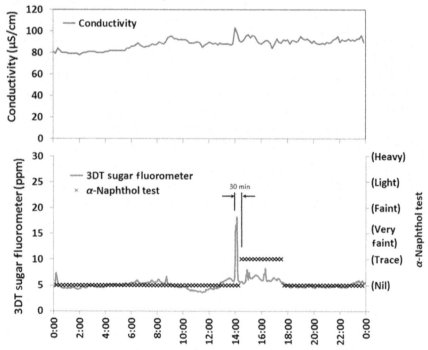

Figure 11.7 Comparison of sugar fluorometer data, conductivity, and the Molisch test. Conductivity shows a poor correlation with sugar concentration.

detected at 14:30 h. As can be observed from the conductivity data, the signal-to-background ratio is poor and it is difficult to conclude if there was any entrainment. It can clearly be seen that there is poor correlation between conductivity and sugars. This is not unexpected, because it is known that low concentrations of sugarcane juice in condensate do not have any significant impact on the conductivity readings. The conductivity sensor becomes sensitive only when a significant amount of contamination occurs.

Figure 11.7 also shows the performance of the sugar fluorometer for the same period. The data clearly shows that sugar fluorometry is significantly more sensitive than conductivity. This is not unexpected, because fluorescence detection is known to be highly sensitive (Lakowicz, 2006). Additionally, the fluorometer was able to detect the entrainment event about 30 minutes prior to its detection by the α-naphthol method. The technology could help in reusing the condensate reliably by preventing such contaminated vapor condensate water from entering the boiler feed water tank, and thereby protecting the boiler from low pH-associated corrosion problems.

3D TRASAR Technology for Sugar continuously monitors process condensate with a fluorescence-based approach, providing early detection of a "sugar shot" and automatically controlling the process condensate reuse and discharge while maintaining reliable and efficient operation of boilers or cooling towers. Based on the extensive data collected during the field applications, the following conclusions can be made:

- 3D TRASAR Technology for Sugar is significantly more sensitive and reliable in detecting "sugar shots" than conductivity and α-naphthol test methods. Conductivity is a strong function of ammonia rather than sugars. At low contamination levels, α-naphthol tests are difficult and unreliable.
- Condensate monitoring based on conductivity or α-naphthol tests could miss sugar entrainment events. Given that the samples are monitored intermittently in the α-naphthol test, short sugar entrainment events are likely to be missed. Conductivity responds to sugar entrainment only when the contamination levels become high. 3D TRASAR Technology for Sugar continuously monitors process condensate online and provides early detection of sugar contamination events.
- 3D TRASAR Technology for Sugar automatically manages process condensate reuse or discharge based on the online detection results. This online monitoring and automation control enhances the effectiveness of condensate reuse and minimizes the negative impact of sugar entrainment on boilers.

11.3 3D TRASAR TECHNOLOGY FOR MEMBRANES

11.3.1 Current Needs of RO Membrane Systems

The most popular membrane technologies practiced currently include MF, UF, NF, and RO. Figure 11.8 shows how these technologies compare in terms of the removal capabilities. MF membranes are suited for suspended particle removal and find applications in the clarification of water and other fluids. The pore size of these membranes ranges from 0.1 to 1 μm. UF membranes have a pore size in the general range of 0.01–0.1 μm. The UF membrane specification is often expressed in terms of molecular weight cut-off (MWCO), which refers to the lowest molecular weight solute (of known molecular weight), which has a 90% rejection by the membrane. The MWCO of UF membranes generally ranges between 1000 and 300,000 Da. These membranes are commonly used for concentrating process streams, removing bacteria and viruses, and removing organics from water. Both MF and UF membranes work on the principle of size exclusion, which means that the species larger than the pore size of the membrane are rejected while the smaller species are allowed to pass through. NF membranes could have a pore size between 1 and 10 nm. In terms of MWCO,

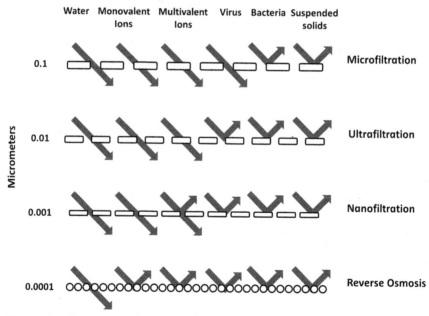

Figure 11.8 Comparison of membrane technologies in terms of removal capability and average pore size.

NF membranes range from 200 to 1000 Da. These can effectively remove divalent and multivalent ions with high efficiency, but have poor to moderate removal rates for monovalent ions. For RO membranes, pore size specification is not used. NF and RO membranes are used for removing dissolved ions and low-molecular-weight dissolved organics. They are also used for concentrating some process streams. These membranes rely on a combination of size exclusion and solution/diffusion permeation. In the solution/diffusion transport model, the solute is absorbed into the membrane, diffuses through the membrane, and finally desorbs out of the membrane. This in turn implies that, other than the size of the solute, the rejection of any given solute by the membrane is also a function of several variables including temperature, pressure, concentration of solute, and the membrane material (ionic charge). This also explains why the operational data of RO membranes needs to be normalized in order to assess whether they are operating as per design. NF and RO filtration require higher operating pressures compared to MF and UF filtration. With respect to the quality of the product, RO membranes produce the cleanest water of all membrane filtration processes.

The focus of this section is RO membranes. RO membrane systems are ubiquitous today. With membrane costs decreasing every year, the systems have become more and more affordable to own and operate (Kucera, 2010). RO membrane systems have been around for several decades now for applications that include desalination and raw water treatment. Increasingly, RO membranes are also playing an important role in wastewater treatment and recycling (Judd and Jefferson, 2003). RO membrane systems, if designed and operated properly, are fairly robust systems. There are, however, several reasons that could limit a RO system from performing in the most optimal manner. These reasons include:

a. Lack of analysis and interpretation of operating data, leading to inappropriate, insufficient, or no corrective action being taken.

It is not uncommon to see installations where operational data is being collected but no interpretation is done. At such installations, the parameters such as flow rates, temperature, pressure, and conductivity are being logged either manually or online, and in some cases even the normalized parameters are being calculated. However, often this monitoring is underutilized because the trends are not closely monitored and inferences are not drawn to arrive at recommended corrective actions.

b. Limited or missing hardware necessary for monitoring of operational variables.

Often necessary sensors, such as the interstage pressure transducers, are not installed. Even if all the necessary sensors are in place, if the regular maintenance and calibration required is neglected, data that is not representative of actual conditions is generated and cannot be used for assessing the health of the system.

c. Insufficient resources to make the proper changes.

The RO system might only be allocated minimal operator time with limited operator expertise.

d. Change in operation or water source since the original installation.

The performance of an RO system can get significantly affected by a variety of changes in the system, its operation, or in water sources. But often due to lack of expertise, these changes get overlooked or ignored, resulting in poor performance.

e. Mechanical limitations of equipment.

Occasionally, the installed equipment has limitations, such as pumps having inadequate capacity, resulting in insufficient flow or pressure, or sensors having improper range, resulting in erroneous data.

f. Bad design.

The key parameters that go into the design of an RO system include the feed water composition, feed temperature, feed pressure, and recovery. Sometimes, in order to reduce costs, system designs are compromised, leading to poor performance. RO membranes need to be protected from contaminants that can cause scaling or fouling. Therefore, a good pretreatment section comprising appropriately selected filters is a must. When designing, it is equally important to consider all details such as the selection of dosing points for the chemicals. The system designers should not compromise on any of these aspects.

Years of learning, gathered from practical experience in running RO systems, have helped in bringing some best practices (Kucera, 2010) to surface, which should be followed when operating any RO system (Table 11.1). The most essential aspect for ensuring that an RO system operates well is to measure proper parameters and use normalization before interpreting the data. Table 11.2 lists the parameters for which the data should be collected, at the minimum (Flynn, 2009). The collected data should then be normalized before it can be used for comparison with previous data or for trending. As mentioned earlier, these parameters are functions of temperature, solute concentrations, and pressure. As these conditions vary continuously, it is impossible to interpret any useful information by directly comparing data collected at different times.

Table 11.1 Best practices for managing an RO system

Practice	Purpose
a. Normalize data	Data normalization is a process where all the operating data is recalculated based on a reference condition. This is necessary, because many of the operating variables are dependent on temperature, operating pressure, and flow rate. For example, one would expect to get more permeate at a higher operating pressure, so if operating pressure is changing, then we would need to normalize this data to an operating reference pressure in order to do a meaningful comparison. The reference condition is a known point where the system was clean, either after a cleaning cycle or when the membranes were replaced.

b. Take action based on normalized trends
   i. Detect fouling
   ii. Detect scaling
   iii. Schedule cleanings
	Cleaning should be performed whenever the normalized data indicates that performance has deteriorated by 10–15%. If this is not followed, it may result in production of permeate water that might be off the specifications, or may result in more frequent cleanings and eventually membrane replacement.

c. Ensure absolute chlorine removal to protect the membranes	Chlorine damages the membranes, and after a finite amount of chlorine exposure, the membranes will fail. It is imperative to monitor the chlorine and ensure that it is appropriately neutralized before the feed water enters the RO train.
d. Optimize recovery without compromising the system	For a given RO system, the optimal recovery is determined by the feed water quality, feed water chemistry, and the required permeate quality. It may be possible to increase recovery, thereby reducing the feed water and wastewater (reject), while still producing water of sufficient quality. However, this "stresses" the RO system, and necessitates that the system is monitored more closely.
e. Monitor feed water changes	Feed water quality needs to be adequate in order to minimize colloidal and organic fouling.

Table 11.2 Minimum data to be collected for monitoring the performance of an RO system

Location	Feed	Interstage	Product	Reject
Pressure	×	×	×	×
Flow rate	×	×	×	
Conductivity	×	×	×	×
Temperature	×			
pH	×			×
Chlorine	×			

In practice, however, these best practices are often not followed due to some of the limitations listed above. If these best practices were automated, it would greatly improve the likelihood that the RO systems would work as designed and fully serve out the designed life.

11.3.2 3D TRASAR Technology for Membranes

3D TRASAR Technology for Membranes is a fully integrated offering (Figure 11.9). Real, tangible value is provided when the information from the membrane system is conveyed continuously to the Nalco enVision web portal and to the Nalco 360™ Expert Center. Nalco 360 Service generates value by making sure the chemistry is being dosed appropriately and by triggering corrective actions that may be supplied by on-site services and supported by chemistry and consumables. And as required, if more water is needed or other improvements are required, one can use the operational data already available to supply equipment uniquely suited for the needs.

The main objective behind the development of 3D TRASAR Technology for Membranes is to drive the performance of the membrane system toward best practices, which would automatically lead to improved efficiency and reduced costs of operation. Necessary instrumentation for collecting the essential operating data is installed if not present already. The technology is designed to

- Provide immediate feedback on problems.
- Predict when cleaning is needed.
- Reduce risk for chlorine exposure.
- Allow safe operation at maximum recovery.
- Improve the chemical control.

These aspects of the technology are explained in more detail in the following list:

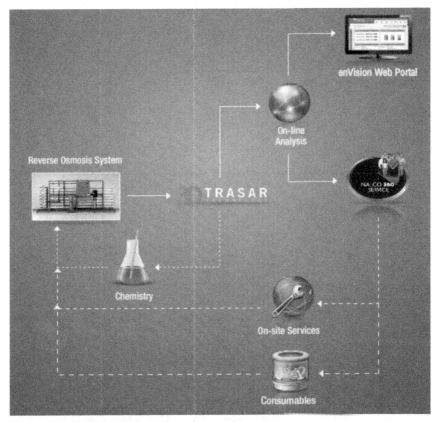

Figure 11.9 3D TRASAR Technology for Membranes systems.

A. Provide immediate feedback on problems.

The main reasons for shortened membrane life are scaling or fouling, deferred cleaning, and halogen attack (or chlorine damage). 3D TRA-SAR Technology for Membranes can detect all of these problems and notify the appropriate people before most damage can occur (Figure 11.10).

B. Predict when cleaning is needed.

Membrane elements must be cleaned whenever there is a 15% loss in normalized permeate flow, a 15% increase in normalized differential pressure, or a 15% increase in salt passage. Failure to follow these rules results in reduced membrane life. It is also important to note that operation of a membrane system at a higher pressure will consume more energy. Cleaning at the right time will increase the membrane life and reduce the operating costs. Figure 11.11 shows the normalized

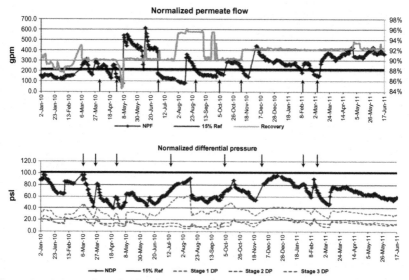

Figure 11.10 Immediate feedback provided by 3D TRASAR Technology for Membranes. The black arrows indicate the moment when alarms were sent out requesting corrective action. *Note: NDP stands for Normalized Differential Pressure; DP stands for Differential Pressure.*

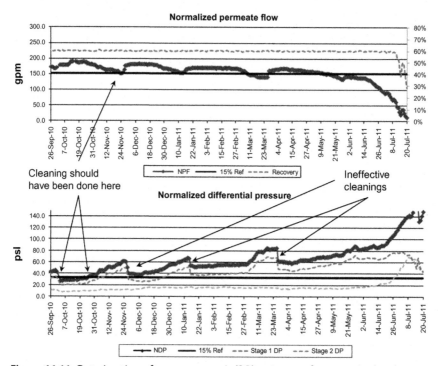

Figure 11.11 Deterioration of reverse osmosis (RO) system performance in the absence of proactive monitoring and missed/improper cleanings. *Note: NDP stands for Normalized Differential Pressure; DP stands for Differential Pressure.*

permeate flow and normalized differential pressure data for an RO system and indicates the times when the system should have been cleaned (based on the rules mentioned earlier). The RO system was not cleaned at the recommended time, while at other instances the cleaning was done but was not enough. This caused steep deterioration in the performance of the RO system, forcing membrane replacement to occur, because the permeate flow diminished to a point where enough water could no longer be produced. 3D TRASAR Membrane Technology monitors these normalized parameters and sends regular reports outlining these changes and upcoming actions.

C. Reduce risk for chlorine exposure.

The bisulfite addition is controlled using the oxidation-reduction potential (ORP) sensor in order to prevent accidental oxidation of the membrane and permanent damage from an unforeseen, sudden increase of chlorine in the feed water. As soon as an increase in the ORP reading is detected, the bisulfite pump is automatically controlled to dose a higher amount of bisulfite to neutralize the chlorine in the feed water (Figure 11.12).

D. Allow safe operation at maximum recovery.

Most RO systems are designed for only 75% recovery, which is traditionally conservative. This figure is also used because many RO

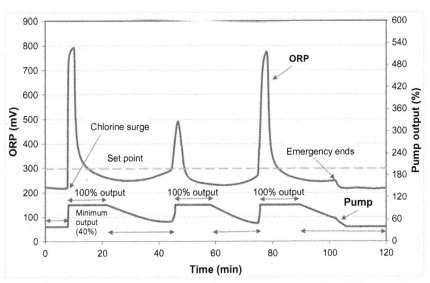

Figure 11.12 Automatic control of bisulfite to prevent membrane damage from sudden surges in chlorine concentration.

systems are not maintained properly since the operators are often not skilled or knowledgeable. Due to these concerns, RO systems are conservatively designed so as to avoid having to replace the membranes more often as opposed to the preservation of water (by running at higher recovery). Lower recovery means more feed water is needed and more water is wasted in the form of reject. On the other hand, a higher recovery means that the concentration factor (in other words, scaling potential) would be higher (Figure 11.13). Therefore, while running at higher recoveries, it is of utmost importance to follow the recommendations for membrane cleaning whenever the system normalized parameters indicate it. Failure to follow this could result in irreversible damage to membranes. 3D TRASAR Technology for Membranes allows one to operate an RO system more efficiently and closer to the membrane limits.

E. Improve chemical control.

The technology uses patented TRASAR™ antiscalant chemistries, which can be monitored accurately using patented fluorescence technology. Figure 11.14 shows an example of how the antiscalant dose rate could be safely decreased from 4 to 1 ppm.

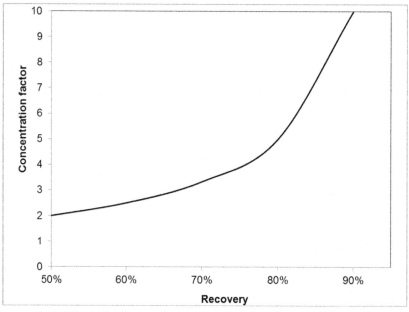

Figure 11.13 Effect of higher recovery on concentration factor in RO systems.

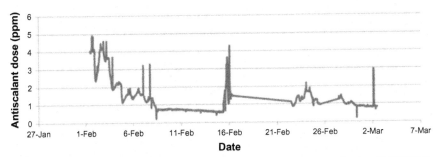

Figure 11.14 Reduction in antiscalant dosage, due to accurate control using 3D TRASAR Technology for Membranes.

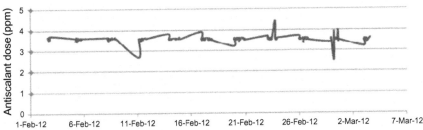

Figure 11.15 Good antiscalant control in an RO system that operates intermittently.

Figure 11.15 shows good control of antiscalants at 3.5 ppm with a system that operates only intermittently. The technology has been designed to allow us to maintain accurate dose rates for antiscalants, even for systems that operate intermittently. Having tight control ensures that the chemical costs are minimized, but with the assurance that there is no risk of scale formation.

Driving these best practices leads to water savings, wastewater savings, chemical savings, consumables savings, and energy savings. Additional benefits include reduced downtime, reduced labor costs, reduced off-specification product, increased capacity, and "green" operation.

For implementation, 3D TRASAR Technology for Membranes equipment is connected to an RO system (Figure 11.16). The signals from the flow sensors at two (usually the feed and permeate) of the three points (as shown in the figure) is transmitted to the equipment. Similarly, the pressure signals from the feed, reject, interstage (if relevant), and permeate pressure transducers are fed through to the equipment. Data from the conductivity sensors for the feed and permeate and the temperature sensor on the feed water are also used. Fluorometers are used for monitoring and controlling the dosing of chemistry and for monitoring the recovery. If needed, pressure

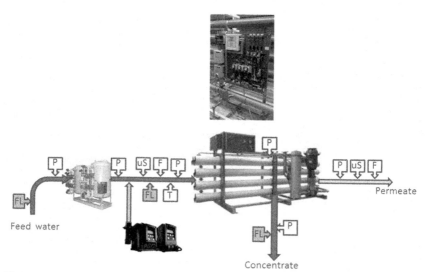

Figure 11.16 Implementation of 3D TRASAR Technology for Membranes. FL=fluorometer, P=pressure transducer, uS=conductivity sensor, F=flow, T=temperature.

sensors before and after the cartridge filter could also be connected. The data from all these sensors is collected by the controller and transmitted to the web portal via a wireless gateway.

11.3.3 Case Study

At a site where the 3D TRASAR Technology for Membranes was implemented, a two-stage RO system was being operated to produce 100 gpm of permeate for process-side application. The system was being run at only 60% recovery and consumed 160 gpm of feed water.

The site was interested in obtaining stable water production, but at reduced water consumption and lower total cost of operation.

11.3.3.1 Halogen Control

Prior to using 3D TRASAR technology, the halogen control was found to be inadequate. A predetermined amount of sodium bisulfate (SBS) was being added to the RO feedwater to control free chlorine. Only spot checks on grab samples were possible due to limited personnel resources. While occasional ORP surges were likely, there was no way to know of their occurrence, because there was no online monitoring in place. ORP is used to detect the existence of residual chlorine in the feed water for RO.

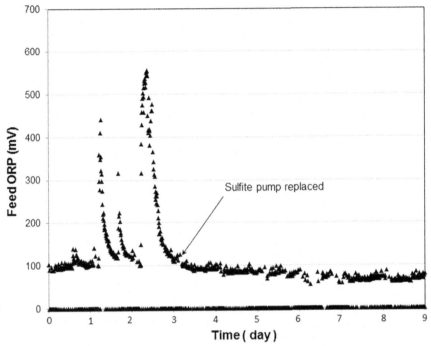

Figure 11.17 ORP control and discovery of problems in sulfite dosing system.

With 3D TRASAR Technology for Membranes in place, the operator was able to get a better handle on the system, which was being monitored 24/7. Nalco 360 Service was able to detect occasional ORP surges and provide further investigative suggestions that led to the discovery that the sulfite dosing pump was malfunctioning (Figure 11.17). This led to protection of $10,600 worth of membranes.

11.3.3.2 Water Recovery Control

Prior to implementing 3D TRASAR technology, recovery rates were sacrificed. The RO system was being operated at only 60% recovery to obtain stable water production. The relatively low recovery was due to the high conductivity in feed water and high scaling potential at the high pH employed by the system (to prevent excess carbon dioxide from leaking into the permeate). There was a desire to increase recovery, but the existing monitoring system was insufficient to track and detect potential system failures.

3D TRASAR Technology for Membranes allows safe operation at higher recoveries. With this technology in place, the recovery was gradually increased to 73% by increasing the partial concentrate recycle. This resulted in economic benefit with mitigated risks (Figure 11.18). Around-the-clock monitoring by Nalco 360 Service allowed more aggressive water recovery without hampering system operation or health. The feed water consumption was reduced to 130 gpm. This resulted in estimated savings of $67,000/year.

The SBS chemical pump failure would have been almost undetectable without the trended data and alarms from 3D TRASAR Technology for Membranes. Additionally, from the unbalanced changes between the flows and pressures in the system, it was possible to identify that the concentrate

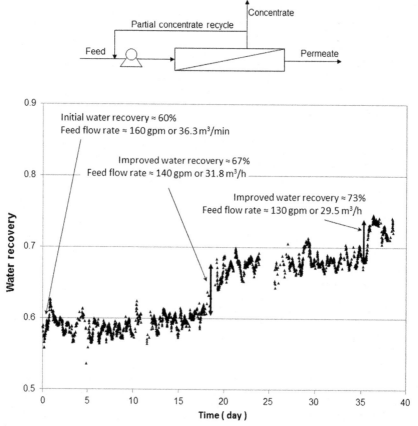

Figure 11.18 Improvement in RO system recovery by using 3D TRASAR Technology for Membranes.

valve had started to fail. This timely detection prevented any RO catastrophic damage, stopping unnecessary water loss, and most importantly, avoiding unplanned RO shut down to fix the valve.

3D TRASAR Technology for Membranes is able to provide immediate tangible benefits, such as increased water and wastewater savings through accurate monitoring of recovery, particularly if the actual recovery of the RO system is less than designed. Through a need-based cleaning approach and this technology, it may be possible to decrease the cleaning frequency and still protect the membranes, thereby resulting in less frequent replacement of membranes.

Because of its ability to proactively monitor the system, the technology helps address issues such as those pertaining to water quality, even before they can impact the process performance or product quality. This is reflected in the minimized costs or consequences associated with poor quality of the final product. The technology allows the RO system to operate at a different (or lower) feed water quality. The downtime can be reduced, thereby lowering the impact of lost production associated with unplanned shutdowns due to membrane fouling. The technology helps to maximize the capital utilization by producing the maximum possible product at just the right quality from the existing RO system.

11.4 SUMMARY

3D TRASAR Technology for Sugar is a fluorescence-based monitoring technology that is able to detect the condensate quality variations with high sensitivity and selectivity and provide early detection. The technology enables reliable reuse of vapor condensate as boiler makeup, cooling tower makeup, or other sections of the plant with the assurance that only good quality condensate is reused.

3D TRASAR Technology for Membranes is a multidimensional solution that integrates multiple components including hardware, chemicals, and remote and on-site services. Hardware collects operational data for performance analysis, web reporting, and data management. The equipment is customized to handle a variety of RO system configurations and can be upgraded as changes are made to operating conditions or process equipment. It controls antiscalant dose and pH. The bisulfite addition is also controlled to prevent accidental oxidation of the membrane. The technology is designed to achieve optimal membrane system performance.

Both technologies aim at providing significant water savings through reduction in wastewater discharge and reduction in raw water consumption as well as energy savings. This provides improvement in overall environmental sustainability. Additionally, the technologies come with a 24/7 remote monitoring and troubleshooting via Nalco 360 Service.

ACKNOWLEDGMENT

The authors are grateful to all those who contributed to the development, evaluation, and deployment of the technologies described in this chapter. Without their hard work and support, this book chapter would not have been possible.

REFERENCES

Carpenter, F.G., Wall, J.H., 1972. Fluorescence in commercial sugars. In: Proceedings of the 1972 Technical Session on Cane Sugar Refining Research. New Orleans, LA, Nov. 13–14, pp. 47–61.

Dubois, M., Gilles, K.A., Hamilton, J.K., Rebers, P.A., Smith, F., 1979. Colorimetric method for the determination of sugars and related substances. Anal. Chem. 28, 350–356.

Flynn, D., 2009. The Nalco Water Handbook. McGraw-Hill, New York.

Foulger, J.H., 1931. The use of the Molisch (α-naphthol) reactions in the study of sugars in biological fluids. J. Biol. Chem. 92 (1931), 345–353.

Hugot, E., 1986. Handbook of Cane Sugar Engineering. Elsevier, New York.

Huiting, H., Kappelhof, J.W.N.M., Bosklopper, Th.G.J., 2001. Operation of NF/RO plants: from reactive to proactive. Desalination 139, 183–189.

Judd, S., Jefferson, B., 2003. Membranes for Industrial Wastewater Recovery and Re-use. Elsevier, New York.

Kucera, J., 2010. Reverse Osmosis: Design, Processes, and Applications for Engineers. Wiley-Scrivener, New York.

Lakowicz, J.R., 2006. Principles of Fluorescence Spectroscopy. Springer, New York.

McGillivray, T., Seaborn, S., Bedford, B.S., Grueneich, D.P., Larson, D.O., Stuart, C.M., 1997. Detection of sugar in multiple effect evaporator condensate systems using fluorescence. In: American Society of Sugar Cane Technologists Meeting. Florida.

Molisch, H., 1886. Zwei neue Zuckerreactionen. Monatshefte für Chemie 7 (1), 198–209.

Reid, M.J., Dunsmore, A., 1991. The protection of boilers fro sugar contamination in feedwater. In: Proceedings of the South African Sugar Technologists' Association, pp. 208–212.

CHAPTER 12

Simulation, Control, and Optimization of Water Systems in Industrial Plants

Jijnasa Panigrahi*, Sharad C. Sharma
Helium Consulting Private Limited, Pune, India
*Corresponding Author: jijnasa.panigrahi@heliumconsulting.com

12.1 INTRODUCTION

Water is a necessary resource to sustain life on this planet. We are blessed to have 70% of our planet covered with this resource. However, at the same time, we are not so fortunate because only 3% of this is available as fresh water for human activity, in the household, agriculture, and industry.

In the process industries, we use water as fresh water, recycled water, as a solvent extraction agent, for washing operations, for steam generation, and for other process uses. This results in large quantities of wastewater, which is also contaminated and therefore difficult to dispose of without causing environmental damage. Thus, it is necessary to look at ways to both reduce consumption of fresh water and minimize discharge—with the ultimate aim of zero discharge—while minimizing the capital and operating costs of doing so. This can be done through a better understanding of the ways in which water is used throughout an industrial complex and finding innovative ways to reduce, reuse, recycle, and treat the water to improve water efficiency.

Hence in light of the above, it is imperative that:

- Water in the process industry is *recycled*.
- Various industries try to achieve *zero discharge* policies.
- *Engineering initiatives* to find water savings potential within existing setups are exploited.
- Advanced *systems and software* techniques available today are deployed to achieve water savings in the areas listed above.

There are different methods for water reuse, recycling, and regeneration in the industrial sector. However, we shall focus on advanced system and software techniques that serve as an important tool in solving the above issues. This chapter will discuss two software techniques, mathematical modeling

Industrial Wastewater Treatment, Recycling, and Reuse

463

and pinch technology. Mathematical modeling is used for design and process change, while pinch technology focuses on water optimization techniques.

The use of computers has provided the opportunity to better understand and assess our water resources through comprehensive numerical model simulations and testing of various schemes or options. In terms of regulation modeling, the modeler can incorporate various user interests as well as historical uses to generate operating scenarios to verify the variations and alternatives of interest to the basin community, as well as the physical environment that has been changed. Assessments of operating policy changes, impacts of floods, and changes in water quality are just a few examples where numerical models are used. The ability to study and make projections with regard to multisector water uses in relation to social, economic, and other considerations and their impacts on the water balance of a basin is an important aspect of water modeling. The importance thereof is further magnified with respect to making projections onto global warming–induced climate change scenarios and/or interjurisdictional basin studies to allow the formulation of interjurisdictional or international water apportionment agreements.

12.1.1 Mathematical Modeling

Among the various methodologies available today for water savings, a class of methodology called mathematical modeling (based on systems and software) has been in use for several decades and is slowly finding maturity in its use and application in the various industries, including in water management. In this section, we talk about the general class of these systems and software and what they do.

A model intended for a simulation study is a mathematical model developed with the help of simulation software. Due to the advancement of computer-based mathematical and process modeling software, modeling plays an important role in design and operation optimization. A remarkable improvement in process design and production systems can be achieved by process modeling. A model may simulate the steady-state, or dynamic (time-varying), behavior of the process, depending on the application. A steady-state model ignores the changes in process variables with time and would typically be used in new process design and retrofits. A dynamic model also has application in design to evaluate control strategy, but it is more likely to be used to simulate start-up/shut-down, for operator training, and for advanced process control. Mathematical modeling helps to simplify a complex system and to understand how the internal constituents interact with each other. Mathematical modeling serves to:

- Reduce tedious and time-consuming experimentation.
- Present the process in a mathematical language suitable for computer solution.
- Apply this representation in several case studies and across software platforms.

A modern chemical plant consists of interconnected units such as heat exchangers, reactors, distillation columns, and mixers with a high degree of integration to achieve energy efficiency. Design and operation of such complex plants is a challenging problem that could be easily carried out by simulation software. In general, whenever there is a need to model and analyze randomness in a system, simulation is the tool of choice. More specifically, situations in which modeling, simulation, and analysis are used include the following:

- When it is impossible or extremely expensive to observe certain processes in the practical world.
- When problems arise in which a mathematical model can be formulated but where analytic solutions are either impossible or too complicated.

In the past, traditional methods have relied heavily on expensive experimentation and the building of scaled models, but now a more flexible and cost-effective approach is available through greater use of mathematical modeling and computer simulation. Computer-aided modeling, simulation, and optimization save time and money by reducing the effort in the configuration of experimental work. In addition, computer simulation and optimization save money in design and operation. The long-term performance and reliability varies from plant to plant and day to day for various reasons such as operating conditions, scaling, and fouling of heat transfer surfaces in boilers and water treatment units. Therefore, better operation of the existing plant depends on the better understanding of the different parameters of the plant. Simulation helps to visualize the ultimate picture and trends of various conditions of existing plant as well as those of a new situation arising out of any process/parameter changes in the plant.

A mathematical model is a collection of (a set of) equations that describe some aspects of the chemical system under investigation. For many complex chemical processes, the model result is a set of nonlinear equations requiring numerical solution. The most common way to deal with this is to program the equations using modeling software such as Aspen Custom Modeler (ACM) or other engineering software.

An industrial process has a lifecycle from conception through R & D, engineering, commissioning, operations, and decommissioning. A good mathematical model (also called a process model) can assist the activities in all these stages, typically providing the ability to develop accurate material and energy balances for the process flowsheet. The model would cover thermodynamics, reaction kinetics, and transport phenomena in the process system, simulating the flow and transfer of material and energy around the process flowsheet, using the tools of mathematics and computing for representation and solution of the system. These tools are made available as a bundle in the form of software packages for process simulation. They commonly comprise a library of unit operations models, component databanks and thermodynamic models, capabilities to define connectivity and other flow sheet features, and a solution engine.

The Aspen Engineering Suite is among the available process simulation software packages and comprises AspenPlus®, AspenDynamics®, and AspenCustom Modeller®.

AspenPlus is a leading process simulation software package in the industrial sector that is widely used to optimize chemical processes. It consists of the world's most extensive property database and handles solids, fluids, and gas phase processes. AspenPlus consists of a wide range of solid, liquid, and gas processing equipment models. If any unit operations are missing in its library, custom models can be built in the ACM or other programming languages. For easier access and data analysis, the model has the facility to be linked to Excel or Visual Basic. In this way the entire flow sheet of the complex can be built and chemical components in the flow stream can be specified from the available component databases. Simple mass and energy balances can be modeled using this steady state software. Minimizing energy-related emissions, determination of the economic impact of design decisions, and identification of energy-saving opportunities can be done. This can be used to model a wide range of industrial processes including those related to power, chemicals, specialty chemicals, polymers, metals, and mining.

Dynamic studies can be done using AspenPlus Dynamics. AspenPlus Dynamics extends Aspen Plus steady-state models into dynamic process models, enabling design and verification of process control schemes, safety studies, relief-valve sizing, failure analysis, and development of startup, shutdown, rate-change, and grade transition policies. Critical or unusual unit operations for which suitable models are not present in Aspen Plus Dynamics can also be modeled using the ACM. Engineers use the ACM to quickly create rigorous models of processing equipment and to apply these equipment models to simulate and optimize continuous, batch, and semi-batch processes. Modeling tools are widely used in the process industry to evaluate the effect of alternate feed stocks, various process and equipment parameters, and operating strategy on key performance indicators such as production, yield, product quality, energy efficiency, and effluent discharge and economics, and thus optimise the process design or operation.

12.1.1.1 Benefits of Modeling, Simulation, and Analysis

According to practitioners, modeling, simulation, and analysis are some of the most frequently used operations research techniques (Maria, 1997). When used judiciously, modeling, simulation, and analysis make it possible to:

- Obtain a better understanding of the system by developing a mathematical model of a system of interest and observing the system's operation in detail over long periods of time.
- Test hypotheses about the system for feasibility.
- Compress time to observe certain phenomena over long periods or expand time to observe a complex phenomenon in detail.
- Study the effects of certain informational, organizational, environmental, and policy changes on the operation of a system by altering the system's model; this can be done without disrupting the real system and significantly reduces the risk of experimenting with the real system.
- Experiment with new or unknown situations about which only weak information is available.
- Identify the "driving" variables—ones that performance measures are most sensitive to—and the interrelationships among them.
- Identify bottlenecks in the flow of entities (e.g., material, energy) or information.
- Use multiple performance metrics for analyzing system configurations.
- Employ a systems approach to problem solving.
- Develop well-designed and robust systems and reduce system development time.

12.1.1.2 What are Some Pitfalls to Guard Against in Simulation?

Simulation can be a time-consuming and complex exercise, from modeling through to output analysis that necessitates the involvement of resident experts and decision makers in the entire process (Maria, 1997). Following is a checklist of pitfalls to guard against.

- Unclear objectives
- Using simulation when an analytic solution is appropriate
- Invalid model
- Simulation model too complex or too simple
- Erroneous assumptions

- Undocumented assumptions. (This is extremely important, and it is strongly suggested that assumptions made at each stage of the modeling, simulation, and analysis exercise be documented thoroughly.)
- Using the wrong input probability distribution
- Replacing a distribution (stochastic) by its mean (deterministic)
- Using the wrong performance measure
- Bugs in the simulation program
- Using standard statistical formulas that assume variable independence
- Initial bias in output data
- Inadequate sensitivity analysis and case studies to cover scenarios of interest
- Poor schedule and budget planning
- Poor communication among the personnel involved in the simulation study.

In a nutshell, a mathematical model can

- Help technical personnel *understand process behavior* in an easy manner.
- Run various scenarios—both steady state and dynamic—in *simulations* without making any actual changes in the plant.
- Effectively and *automatically optimize* the process through the use of sophisticated algorithms.

12.1.2 Water Integration (Pinch Analysis)

In the previous section we discussed the use of process simulation to develop accurate heat and mass balances and derive stream properties for a given flowsheet. The water network is an integral part of the process flowsheet, and thus this information can be used to direct efforts to reduce the consumption of water as a utility.

Figure 12.1 depicts the overview of a water minimization technique in an industry. Opportunities for water minimization can be considered at four levels (Buehner and Rossiter, 1996):

(1) Process changes

Replacing the technology employed in a process can reduce the inherent demand for water. Examples might be replacing a wet cooling system with air coolers or increasing the number of stages in a washing operation. Sometimes it is possible to reduce water demand by changing the way existing equipment is operated, rather than replacing or modifying it.

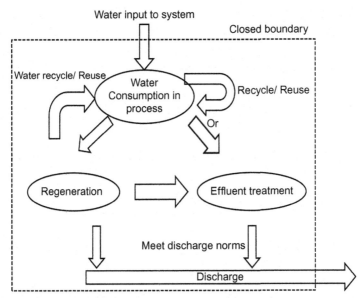

Figure 12.1 Water minimization in plant.

(2)   Water reuses

Wastewater from one operation can be directly used as process water in a following operation, provided the level of contamination from the previous process does not interfere with the subsequent process. This will reduce overall fresh water and wastewater volumes, but not affect contaminant loads in the overall effluent from the system. Generally, reuse excludes returning water, either directly or indirectly, to operations through which it has already passed, in order to avoid buildup of minor contaminants that have not been considered in the analysis.

(3)   Regeneration reuses

Partial treatment of wastewater can remove contaminants that would otherwise prevent reuse. The regeneration process might be filtration, stream-stripping, carbon adsorption, or other such processes including simple pH adjustment and physical removal of unwanted impurities by membrane separation, sour water stripping, or ion exchange.

(4)   Regeneration recycling

Recycling refers to the situation where water is reused in an operation through which it has been already passed. In this case, to avoid buildup in the system, the regeneration step must be capable of removing all contaminants.

Process simulation is used to understand the operating characteristics of a given design. In order to exploit the above opportunities, however, we need to modify the design, analyzing the flowsheet network as a whole, to determine ways to reduce process requirements and reuse, recycle, or regenerate within the water network. The analysis of a network to identify these opportunities should lead to a reduction of fresh water requirements and wastewater discharge, and hence in reduced operating costs. The same basic methods, useful for minimizing cost, can also be used to explore the three-way relationship among capital cost, operating cost, and environmental optimization.

This can be aided by water integration techniques such as pinch analysis—a family of methodologies to reduce consumption of resources (e.g., by increasing recovery) for a whole system such as a plant site. Water integration is one of the aspects of process integration, energy integration being among the others. Water integration considers the quality and quantity of water usage in each unit or area with a view to maximizing water reuse and minimizing wastewater.

A specific technique in process integration is pinch analysis, which can be used to facilitate analysis and design by combining the sources and demands across the several processes that make up the system.

The aim is to minimize the resource requirements in the system, subject to the constraints imposed by thermodynamics, equipment performance, and other physical limitations. There are two main stages in a pinch analysis:

- *Analysis or targeting*: Determining the minimum resource requirement for the process as conceived, without considering the detailed design, and therefore without taking into account the network for recovery
- *Synthesis*: Designing a network, putting the various sub-processes together to achieve, or get close to, the identified target.

Analysis: Pinch provides a visual tool for targeting—the composite curve—that clearly shows, in a single view, the sources and demands for a resource within the system, and consequently the requirement from external sources.

In the case of industrial water as a resource, the process can be considered as a network of mass exchangers. The objective here is to minimize the consumption of fresh water and the disposal of wastewater by maximizing internal water reuse, subject to the limitations of the driving force provided by concentration difference (Wang and Smith, 1994). The curves (Figure 12.2) in this case show water use, for example, as a plot of pollutant load versus flow rate for the source (contaminant-rich) and demand (contaminant-lean) streams, and the non-overlapping sections show freshwater and wastewater targets.

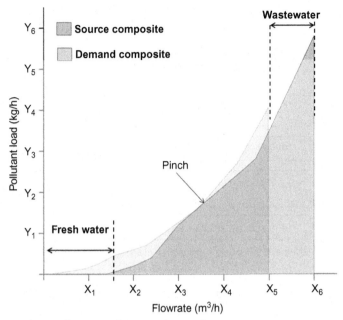

Figure 12.2 Composite curves for water integration.

Synthesis: An appropriate network to get closer to the target can then be synthesized using a hierarchical, evolutionary approach, adopting algorithms such as those that minimize the number of exchange units, for example, by stream bypassing, mixing, splitting, or loop breaking; or simultaneous solution of a mathematical programming problem.

As described above, now we can understand how mathematical modeling helps in a plant not only to study the effect of water reuse and recycling in a process but also as a guide to minimize the water usage cost. Figure 12.3 points out some key benefits of mathematical modeling for water reuse/ recycle in a plant.

Hence through this modern technique of water integration we can:

- Model the entire water network under a single umbrella to *understand the current process*.
- Effectively *analyze current challenges and opportunities* in the network through the use of the model and visual information.
- Synthesize *alternate strategies for optimization* of this water network.

We shall now discuss industrial applications of some of these modeling techniques. We will first discuss generic applications in some common process

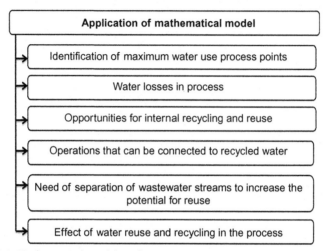

Figure 12.3 Mathematical modeling for water usages.

industries and then take an in-depth look at how such an application worked in real life and delivered clear benefits in the area of energy and water management.

12.2 APPLICABILITY IN VARIOUS INDUSTRIES

The degree of potential for treatment and reuse varies across industries. In the same industry, each processing unit generates waste with varying characteristics and volumes, thus segregation is an important factor in designing reuse technology. A good solution for such industries is recycling waste from one processing unit after some level of treatment and disposal from another processing unit, again after some treatment, to meet required effluent standards. It is essential to decide on the type of waste to be recycled and disposed of after an effective waste inventory. A high volume of wastewater with a low pollutant concentration has the obvious potential for recycling, but it is more difficult to recycle low volumes of water with high concentrations of pollutants. Recycling is governed by factors such as availability of raw water, possibilities of expansion in the processing units, effluent standards, and recovery of products from waste treatment. Based on some of these governing factors, wastewater reuse potential is summarized in Table 12.1.

Mathematical modeling tools can be used in an existing plant, including its water and effluent systems, and for simulations, after validation with plant data.

Table 12.1 Wastewater Reuse Potential for Industries (Visvanathan, C., 2001)

High Potential	Medium Potential	Low Potential
Pulp and paper	Dairy and food processing	Pesticide
Oil refining	Distillery	Rubber
Chemical	Wool textile	Explosive
Fertilizer	Glass and steel	Paint manufacturing
Petroleum	Cotton textile	

- This type of model is useful in looking at the effect of new process units on the water system.
- The model further can be used to optimize overall water systematology within specified physical and regulatory boundaries.
- Water minimization scenarios can be investigated, including unit operations and circuit configuration changes.

Applications of modeling tools vary across industries depending on the type of waste and type of process. A lot of research has been carried out in different parts of the world to minimize the water usage in plants, including using modeling and pinch technology, not only in oil refineries but also in other chemical industries. Some of this is summarized below.

12.2.1 Petroleum Refining, Petrochemical, and Bulk Chemicals

Process industries like refineries, petrochemical plants, and bulk chemical plants can be large consumers of water. There are various sources of process and wastewater in these plants such as desalter effluent, sour water, tank bottom draws, and spent caustic, forming a large network inside these complex unit operations. Many of these streams need to be and are recycled back into the system (IPIEACA, 2010).

Many industries nowadays are using simulation tools for improving stripper performance through changes in operating conditions and process structure. These can be applied not only to reduce steam consumption and lower the contaminate concentrations in the effluents but also to operate the system in a stable fashion. The sour water stripping system in a refinery can be modeled using simulation software to reduce contaminants, thus transforming sour water from a highly concentrated pollutant source to a useful resource.

Figure 12.4 shows a typical flow diagram of sour water stripping system that has been obtained from mathematical modeling. As per the flow diagram, the sour water is converted to useful water by a two-stage stripping and preheating the feed.

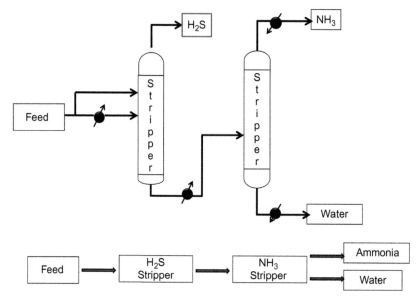

Figure 12.4 Sour water strippers.

Mathematical modeling can be used for solving water allocation prob-lems as well as designing and deploying water purification facilities in these plants. For example, such initiatives using mathematical modeling have helped in the following ways:
• Facilitation of water usage reduction by avoiding using fresh water as wash water in the system.
• Design of better stripper facilities such that the sour water quality can be upgraded.
• Recycling of wastewater that is also upgraded in quality from different parts of the plant to other unit operations within the facility.
Mohammadnejad et al. (2011) did research on two key contaminants, suspended solids and hardness, to analyze the water network using water pinch. They analyzed the network for each contaminant and were able to reduce the amount in fresh water to about 17% and 59.7% in terms of suspended solids and hardness respectively. When the two contaminants were analyzed simultaneously based on their mass transfer, the result shows that the amount of required water was reduced about 42%, indi-cating that the amount of required water is determined by mass transfer of suspended solids.

12.2.2 Power Plants (Thermal and Nuclear), Coal Handling, and Other Utilities

Globally, power plants are one of the largest users of fresh water. The major pollutants from power plants are dissolved solids in cooling water, boiler blowdown, coal drainage, and radioactive material from nuclear power plants. Mathematical models help to identify the water reuse and recycling options in these plants. This practice saves water charges for the plant and effluent discharge costs for treatment.

In addition, certain classes of online advanced control technology can also be used for real-time optimization of the facilities, thus further reducing the usage of precious resources like water as well as creating less effluent due to reduction in specific coal/energy consumption.

12.2.3 Pulp and Paper

The paper-making process is one of the most water-intensive industrial production processes. This is because, without the physical properties of water, it would not be possible for a consistent structure to be achieved when the constituents of paper are processed in sludge. A high level of water consumption is inevitable in the processing of natural raw materials (wood, cellulose vegetable fibers) and also in the process of recycling waste paper. This creates a high level of wastewater for processing. The residues in the wastewater are a problem, particularly in the case of deinking—the process of recycling printed waste paper.

A large volume of process water becomes contaminated from contact with raw materials, by-products, and residues in paper and pulp industries. Basic wastewater sources are evaporator condensate, bleach plant pulp, washer filtrate, condensate from recovery evaporators, and heat exchangers.

The modeling tool uses for this kind of process allow a detailed analysis of the process, including determination of different variables and visualization of the complexity of the process without making direct changes. The application of simulation methodology to study fresh water consumption reduction in the plant contributes to significant saving in water uses.

Closed-loop systems are particularly popular because they enable the pulp and paper industry to recycle and reuse water as well as recover excess pulp fibers in the wastewater. Through the integration tool, water minimization and reuse limits can be established, considering their individual need specificities.

12.2.4 Textiles

The textile industry can be classified into three categories: cotton, woolen, and synthetic fibers, depending upon the raw materials used. The textile dyeing industry consumes large quantities of water and produces large volumes of wastewater from different steps in the dyeing and finishing processes. Wastewater from printing and dyeing units is often rich in color, containing residues of reactive dyes and chemicals, such as complex components, many aerosols, and chemical oxygen demand (COD) and biological oxygen demand (BOD) at high concentrations, as well as much more hard-degradation materials. The toxic effects of dyestuffs and other organic compounds, as well as acidic and alkaline contaminants, from industrial establishments on the general public are widely accepted. Because of the wide variety of process steps, textile wastewater typically contains a complex mixture of organic and inorganic chemicals.

Synthetic fiber mills' use of cooling water is quite high. Final rinse water after mercerizing, bleaching, and dyeing is only slightly contaminated and can be recycled for further rinsing. The modeling tool helps to determine the scope of 15–20% of water recycling without any pre-pretreatment, which represents a relatively small cost for direct reuse of water in the process. The cost-to-benefit ratio is high, and the cost recovery period is low.

12.2.5 Agro-Foods and Starch Industry

The food and beverages processing industry requires a huge amount of water. The waste from agro-food plants contains mostly high organic compounds and bacteriological contamination. Water is used as an ingredient, a cleaning agent, for boiling and cooling purposes, for transportation, and for conditioning of raw materials.

Spent process water in the food industry can be desalinated, and organics can be removed so as to fulfill the requirements for water reuse. The recycling rate is low because organic and bacteriological pollution can affect public health. Given that the quality of water required is high, recycling may be less economical but there is always potential for recycling cooling water.

Mathematical modeling will be a useful tool for finding out the scope of recycling and reuse in industries where water is used exactly. The modeling also helps to improve the process quality by minimizing fresh water use.

In the starch industry, there are multiple contaminants such as total organic content, total dissolved solids, and total suspended solids. It is possible to reduce use of demineralized water and fresh water and consequently the wastewater generation. Mathematical modeling helps to reduce demineralized and freshwater consumptions, which will ultimately affect the cost of water used during processing.

Dakwala et al. (2011) studies the starch industry using water pinch technology to reduce the demineralized (DM) water flow rate and, subsequently, the wastewater flow rate. The wastewater problem is viewed as a single contaminant problem, and all the three modes of water integration, that is, reuse, regeneration-reuse, regeneration-recycle, are demonstrated. The DM water consumption is 50 tph before modification. After modification using water pinch, DM consumption reduces to 31.9 tph (reuse only), 21.6 tph (regeneration-reuse), and 12 tph (regeneration-recycling).

Their procedure for wastewater minimization was based on the concentration interval diagram and the concentration composite curve as introduced by Wang and Smith (1994). The results obtained using that procedure compared well with the results obtained from the well-established software ASPEN WATER, which uses a mathematical programming approach based on MINLP (Mixed-Integer Non-Linear Programming). The cost-benefit analysis illustrates that the profit obtained in the case of reuse is substantial, and the payback periods for the regeneration-reuse and regeneration-recycling are 1.8 and 1.1 months respectively.

12.2.6 Zinc Refinery

The metal and mining industries consume large amounts of energy and water and produce large volumes of waste. Water is probably the most widely used raw material in the process industry.

Bhikha et al. (2011) conducted a study involving reduction of water consumption at a zinc refinery. They used both simulation and water pinch technology for this purpose. The aim of the simulation was to obtain a working mass balance of the operation, from which different water minimization scenarios could be investigated. The simulation process was critical because water quality affects key threshold concentrations that, if exceeded, affect final zinc product purity. Due to the complexity of the process, with numerous recycles and variables, a number of simulations were performed. Each simulation increased in complexity until a final working simulation was obtained that modeled the operating conditions of the plant as closely as possible. The study showed that supplementation of demineralized fresh water with demineralized effluent water results in a 14% reduction in water usage, and supplementation of pump gland seal water with fully treated effluent water results in 19% reduction in water usage.

12.2.7 Alumina Anodizing

Khezri et al. (2010) carried out a study in the alumina anodizing industry for water and wastewater minimization. Water consumption in anodizing is high, and because of high contamination of the anodizing wastewater outflows to the environment, including surface and groundwater, producing an applicable method to reduce water consumption for anodizing and, therefore, reducing wastewater is very important. Waste minimization through pinch technology is effective for reaching this goal. In this research, by applying pinch technology in three rinsing operations, the flow rate reduced up to 1.026 m^3/h (from 7.128 to 6.102 m^3/h). This research is based on the single-contaminant approach as defined by Wang and Smith (1994).

Detailing the application of technology: In the next section we try to use a real-life example to demonstrate how such models have:
- *Reduced CAPEX.*
- *Provided alternate technology routes for waste water management.*
- *Minimized water use by studying unit operations* and routing changes.
- *Balanced* contaminated and fresh water.

12.3 TECHNOLOGY APPLICATION

In this section we will discuss and demonstrate typical applications of the technologies in the alumina industry, as follows:
- Explain the production process.
- Apply simulation in the process use of water in the process.
- Apply simulation for water reuse.
- Provide water optimization.
- Examine further work in this industry.

12.3.1 The Production Process

We will attempt to elaborate these concepts in the context of the Bayer process (Figure 12.5) for obtaining alumina from bauxite. Bauxite is the ore of aluminum. Aluminum oxide is its largest component; the remaining contents are impurities such as silica, iron oxide, and titanium dioxide, and of course free moisture.

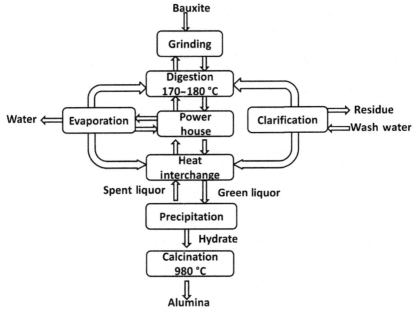

Figure 12.5 Bayer process.

In the Bayer process, bauxite is mixed with a side stream of returning spent caustic liquor and ground down to a size appropriate for extraction. The slurry is then preheated and held in storage tanks to facilitate silica removal (desilication). The main spent liquor stream is also preheated and mixed with the slurry stream. The combined stream is further heated, and the slurry is digested at this elevated temperature under pressure to extract the available alumina from the bauxite. After digestion, the slurry of green liquor (rich in alumina) and inert solids is cooled in a series of flash tanks and reduced to atmospheric pressure. The heating medium for slurry and liquor is a combination of direct and indirect steam, or flash vapor. Thus condensate is produced in this area. The cooled slurry is clarified to remove the unextractable solids from the circuit, which are washed to recover the associated soda.

Further cooling of the green liquor takes place on its way to precipitation, which is favored at lower temperatures, by heat interchange with the spent liquor returning from precipitation. This is again done in a series of flash tanks and heat exchangers, producing more condensate.

Alumina trihydrate (gibbsite) from the green liquor crystallizes in the precipitation area, with the fine crystals acting as seed and the coarse material removed as product hydrate.

This is washed and filtered, and its water of hydration is removed by calcination, resulting in the final primary product, smelter grade alumina.

The liquor leaving the precipitation process, now depleted of alumina and therefore called spent liquor, continues in the liquor circuit and returns to digestion via the heaters in the heat interchange building and evaporation to remove excess liquid due to dilution (added for washing in clarification). Steam is used to heat up the liquor in evaporation, which is then let down in a series of flash tanks and heat exchangers, producing condensate.

12.3.2 Application of Simulation in the Bayer Process

In order to reduce capital and operating costs, increase production, and improve product quality, an organization draws up a portfolio of opportunities to be implemented, depending on business and process conditions at a particular site. Process models assist in prioritizing and refining these opportunities by resolving the tradeoffs between capital, energy, and raw materials using the levers of process topology, operating conditions, and equipment design, accounting for the interactions between various proposed solutions. Here, we mention some of the opportunities in the Bayer process that have benefited from process modeling.

12.3.2.1 Heat Recovery

Reducing boiler energy usage by 2% by designing an appropriate boiler feed water heater network, resulting in hotter boiler feed water and balancing with capital cost.

Absorbing 50–80% of excess energy in digester slurry by installing an additional heater in the last stage of the flash train and balancing with capital cost.

Improving the approach temperature in the first heater in the digester train through more frequent cleaning, 1 °C translating into an energy savings of $100,000 per annum and balancing with maintenance cost.

12.3.2.2 Optimizing Spent Liquor Return Temperature

Optimizing boiler-house/plant steam balance; improving energy utilization by substituting low-grade vapor in place of high-grade vapor where possible, thus increasing availability of high-grade vapor for power generation.

Evaluating impact of flow upgrades/cuts on boiler feed water temperature and availability.

Estimating penalty in delaying cleaning of boiler feed water heaters.

12.3.2.3 Caustic Loss Reduction

Reducing dilution by 30% by transferring some of the heat load in slurry heating from direct to indirect heating; balancing with capital cost required for additional heat exchanger/area.

Optimizing evaporation throughput.

12.3.2.4 Product Quality

Guaranteeing fill temperature for precipitation.

Guaranteeing required temperature after slurry heating for better desilication.

12.3.2.5 Water Recovery

Minimizing cooling water in cooling of green liquor to precipitation by improving heat recovery in heat interchange.

Reducing solids and liquor carryover and piping wear in flash tanks.

12.3.3 Water Use in the Bayer Process

It can be seen that the alumina industry is fairly water intensive and also generates a large amount of waste. Typically, an alumina refinery consumes 2–2.5 tons of raw water per ton of alumina produced (Martin and Howard, 2011). There are opportunities, therefore, for water reuse, depending on the (caustic) contamination level in the water. Zero wastewater discharge can be the ultimate goal. A refinery water balance is presented in Figure 12.6.

About 10% of the total water intake is accounted for by free moisture in the bauxite feed and in the 50% caustic soda solution added as makeup.

12.3.3.1 Water Demand

Water is required for mud and hydrate washing. Cooling water is required, notably in the barometric condensers in evaporation and interstage coolers in precipitation, which is an exothermic reaction. In addition, water is used in pumps as packing and purge water, in other auxiliary equipment, for cleaning, and of course as potable water. There is also considerable water loss around the plant.

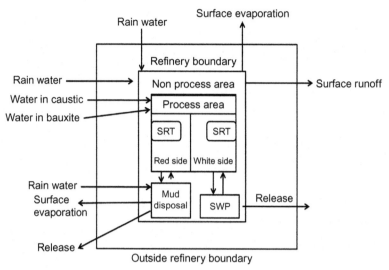

Figure 12.6 Bayer refinery water balance. (Key: SRT = sump relay tank; SWP = storm water pond).

12.3.3.2 Water Supply

Pure condensate generated from steam heat exchangers in digestion and evaporation will be boiler grade and could be directly routed to the power house as boiler feed water.

Contaminated condensate can be routed for further use, depending on the level of contamination. If it contains less contamination, then it could be further used for flocculant preparation, lime slaking, or in the filtration area. If the contamination level is too high, then it can be used in the residue washing area. Highly contaminated water or condensate can be used for hosing, flushing, or for other dilution purposes. Alternatively, it can go to the contaminated condensate tank until the quality is restored by suitably treating the contaminated water to make it reusable.

Evaporation condensate normally contains very low soluble caustic and can therefore be reused in the cooling tower. Other than this, it can be diverted for product hydrate washing or for use in precipitation seed filtration. The blow-down from the cooling tower can be further used in the washing area.

12.3.4 Water Reuse in the Bayer process: Application of Simulation

Chatfield and Sharma (2006) identified a reuse opportunity for digestion condensate through a simulation study. As mentioned above, it is common to route the condensate from steam to the power house as boiler feed water. Flash condensate, however, is of a lower quality. This can be worse with poor flash tank performance, leading to liquor carryover into the condensate. If a flash tank needs to be taken out of service for maintenance, the flashing load needs to be taken up by the next stage, leading to higher vapor velocity, and resulting in more carryover.

A model of the digestion area was applied in various studies to reduce carryover, improve condensate quality, and thus reuse condensate. The model presented options to redistribute vapor loads, reduce flow, or modify flash tank design in order to reduce carryover. It was also used to design a modified condensate system (Figure 12.7), separating good condensate from bad condensate in additional condensate flash pots and routing the good condensate to the boilers, thus reducing energy as well as the environmental costs of condensate discharge.

The AspenPlus® simulation package was used to develop the process model. This package has been used in simulations in various areas of the plant, for energy-saving projects and troubleshooting in existing plants, expansion and new process evaluation, and development and design. Its ability to resolve large flowsheets without limitations in the number of unit operations blocks and process streams enabled the inclusion of detailed

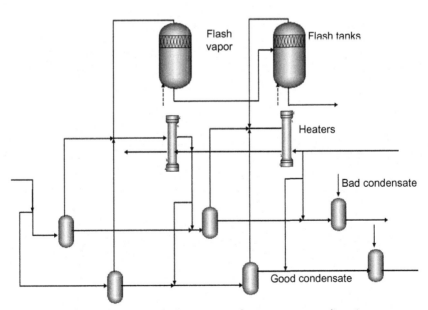

Figure 12.7 Good and bad condensate system for Bayer process digestion.

models of the individual units. The ability to include alternative operating strategies, bypassing entire sequences of operations and mimicking hierarchical control strategies in an Aspen model, enabled development of and simulations with very realistic process models.

12.3.5 Water Optimization in the Bayer Process: Water Integration

Deng and Feng (2009) analyzed the water system of an entire alumina plant using the methods of pinch technology and obtained optimal water regeneration flow rates and contaminant concentrations targets. Based on these, they designed a water–using network that would achieve zero waste discharge, resulting in a freshwater saving rate of 62.7%.

The authors looked at the operations in a water-using network that included losses but no reuse (Figure 12.8). Some of these are fixed-contaminant-load operations (e.g., washing, scrubbing, and extraction) involving mass transfer of contaminant from the source (contaminant-rich) to the demand (contaminant-lean) stream. These operations are designed to pick up a specified amount of contaminant. Other operations are fixed-flow rate (e.g., boiler, cooling tower) that do not involve mass transfer and the main concern is flow rate.

The authors tabulated limiting data for all operations as fixed-contaminant-load operations. This yielded the flow rates required to remove a specified contaminant load, given the inlet and outlet concentration and water loss.

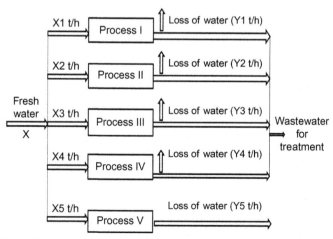

Figure 12.8 Initial water-using network for a process industry.

This was transposed into limiting data for the demands (inlet streams to the water using operations) and sources (outlet streams from the water using operations, i.e., those that could potentially be reused downstream, or recycled).

Using a method called the composite table algorithm, the authors constructed the composite curves for the system and calculated the optimum regeneration flow and concentration.

They then used design rules proposed by Prakash and Shenoy (2005) to design an optimum zero water discharge network for the system.

The zero water discharge network (Figure 12.9) was obtained by superimposing constraints of fixed flow rate for the relevant operations and satisfying these using recycling.

12.3.6 Further Work

From the above discussion and examples, we can see that process modeling helps in better understanding the process and can guide in making improvements and modifications toward optimal water management. Several opportunities may be considered together, with interactions taken into account so that, for example, energy and water savings can be simultaneously targeted. Opportunities in the Bayer process include:

- Reduction of alumina/soda/liquor to residue
- Better condensate management including reuse, rerouting, storage, and dumping
- Wash-water reuse and optimization; concentration targeting
- Managing the dilution–evaporation balance across the plant
- Systematic analysis of contact heating

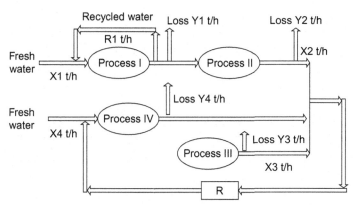

Figure 12.9 Zero water discharge. (Key: R = Regeneration).

- Reduction in cooling water
- Improving integration of co-gen plants with process.

Simultaneous water and heat integration could achieve up to 20% reduction in soda loss and 20% reduction in energy consumption per ton alumina.

12.4 CONCLUSION

In conclusion, we find that there is immense potential in various industries to achieve direct bottom-line benefits through the use of these mathematical models. These will in the coming years become best practice for running water networks inside these process plants. A logical approach to gain the maximum benefits in this area will require:

- *Benefit assessment* through a comprehensive study of the current setup
- Using core knowledge of water networks and purification technology to assess the *specific area of improvement*
- *Application of software and systems* in each of these areas to analyze, design, and operate in the most optimum way
- *Continuous improvement* initiatives through the use of domain knowledge to keep the system performing at the highest levels of optimized operation.

In addition to the above steps, which have been successfully applied by many companies in various industry sectors, some additional modern initiatives that are at an early stage of evaluation include:

- Retrofits of existing plants, where water reuse may be limited by geographical, process, or design constraints
- Uncertainty and flexibility of contaminant loads
- Multiple species/contaminants/properties, because the feasibility of water reuse changes when secondary species, or properties (e.g., conductivity) are brought into the picture and therefore requires a different approach
- Simultaneous heat and water integration, because the issues are related, for example, the tradeoff between wash-water utilization and evaporation
- Simultaneous application of heat integration, exergy analysis, and optimization techniques to look at energy recovery opportunities, that is, the potential for improvement relative to an ideal, taking into account the quality of energy
- Methods for analysis and synthesis.

As we become more globally conscious in our usage of precious resources such as fresh water, it is without a doubt that companies will adopt more modern approaches such as the ones described in this chapter to enhance the quality of the lives of the community as well as be socially responsible corporate citizens.

REFERENCES

Aspen Water User Guide, Aspen Technology Inc.

Bhikha, H., Lewis, A.E., Deglon, D.A., 2011. Reducing water consumption at scorpion zinc. J. S. Afr. Inst. Min. Metall. 111, 437–442.

Buehner, F.W., Rossiter, A.P., 1996. Minimize waste by managing process design. Chemtech by American Chemical Society.

Chatfield, R., Sharma, S.C., 2006. The use of process models to resolve capital and operating cost tradeoffs in Bayer digestion. In: Solymar, K. (Ed.), Proceedings of Seventeenth International Symposium I.C.S.O.B.A., vol. 33(No. 37). OMBKE (Hungarian Mining and Metallurgical Society, Budapest, pp. 114–127, Proceedings of the Bauxite and Alumina related papers, vol. 1.

Dakwala, M., Mohanty, B., Bhargava, R., 2011. Waste water minimization of starch industry using water pinch technology. In: Cleaner Production Initiatives and Challenges for a Sustainable World – 3rd International Workshop, Advances in Cleaner Production, Sao Paulo, Brazil, 18–20 May, 2011.

Deng, C., Feng, X., 2009. Optimal water network with zero wastewater discharge in an alumina plant. WSEAS Transactions on Environment and Development 5 (2), 146–156.

Khezri, S.M., Lotfi, F., Tabibian, S., Erfani, Z., 2010. Application of water pinch technology for water and waste water minimization in alumina anodizing industries. Int. J. Environ. Sci. Tech. 7 (2), 281–290.

Maria, A., 1997. Introduction to modeling and simulation. Proceedings of the 1997 Winter Simulation Conference, pp. 7–13.

Martin, L., Howard, S., 2011. Alumina refinery wastewater management: when zero discharge just isn't feasible. In: Lindsay, S.J. (Ed.), Light Metals 2011. TMS (The Metallurgical Society, Inc.), pp. 191–196.

Mohammadnejad, S., Nabi Bidhendi, G.R., Mehrdadi, N., 2011. Water and wastewater minimization in petroleum refinery through water pinch analysis—single and double contaminants approach. Res. J. Environ. Sci. 5 (1), 88–104.

IPIEACA operation best practice series 2010, Petroleum refining water/wastewater use and management.

Prakash, R., Shenoy, U.V., 2005. Targeting and design of water networks for fixed flow rate and fixed contaminant load operations. Chem. Eng. Sci. 60 (2005), 255–268.

Visvanathan, C., Takashi, Asano. The potential for industrial waste water reuse, Encyclopedia of Life Support Systems, UNESCO Publication, 2001.

Wang, Y.P., Smith, R., 1994. Wastewater minimisation. Chem. Eng. Sci. 49 (7), 981–1006.

CHAPTER 13

Zero Liquid Discharge Solutions

Shrikant Ahirrao
Praj Industries Limited, Pune, India

13.1 INTRODUCTION

India is a large and diverse country. Its land area includes regions with some of the world's highest amounts of rainfall, very dry deserts, coastline, alpine regions, river deltas, and tropical islands. In the past several decades, industrial production has increased in India because of an increasingly open economy and a greater emphasis on industrial development and international trade. Water consumption for industrial use has consequently risen and will continue to rise. Ecological issues are an integral and important part of the environmental issues challenging India. Poor air quality, water pollution, and garbage all affect the quality of food and the environment necessary for ecosystems to thrive. India is recognized as having major issues with water pollution, predominantly because of untreated industrial wastewater and failure to achieve zero liquid discharge (ZLD), defined as no effluent discharged into surface waters from industry (discussed in the following section).

Environmental concerns are steadily increasing, and regulatory authorities are constantly tightening the environmental standards, insisting that industries adopt advanced wastewater treatment technologies, including ZLD solutions.

Today industry is primarily concerned with developing ZLD solutions that address some of the following challenges:

1. Innovative and customized solution offerings
2. Highly corrosive effluent and the selection of metallurgy
3. Recovery of pure process condensate for reuse and recycling
4. Operating temperature and pressures and scaling and fouling tendency
5. Continuous operation of the system with minimization of cleaning-in-place (CIP) effluents
6. Selection of appropriate type of evaporator effects in multi effect evaporation system
7. Optimization of CAPEX and OPEX for ZLD solutions.

This chapter discusses solutions aimed at achieving ZLD. If implemented properly, ZLD has the potential not just to alleviate concerns of effluent

489

discharge but also to lead to water conservation, which is critical for regions experiencing water scarcity. This chapter first outlines the challenges faced in developing and maintaining ZLD and then delves into various aspects of overcoming these challenges and finding practical solutions to avoid liquid effluent. Technological aspects are discussed in this regard, addressing effective recovery of water, along with different case studies pertaining to the greater part of the chemical industry.

13.2 ZERO LIQUID DISCHARGE

Any process or combination of processes, by virtue of which there is no liquid effluent, or discharge from a process plant

It implies that wastewater is treated and effectively recycled and reused such that there is no effluent discharge. ZLD is usually accomplished by concentrating the effluent using various techniques, including membrane-based and multi effect evaporation based systems. ZLD involves:
- The elimination of the liquid waste effluent stream from the plant
- The recycling of recovered water and solids
- The establishment of no liquid pollutant norms.

13.2.1 Challenges of ZLD

There are several challenges facing those trying to develop and implement ZLD. Some of these challenges include:
- Developing a high recovery system to capture more than 95% of wastewater
- Effluent composition
- Effluent characteristics and related corrosion and temperature issues
- Material compatibility
- Operating cost for the treatment.

The challenge is to select the appropriate technologies based on wastewater characteristics and volumes. Thus, techno-economic considerations are important in devising a zero discharge strategy.

13.2.2 Accurate Analysis of Generated Effluent

For zero discharge technology to work efficiently, it is essential to obtain accurate analyses of generated effluents. Some of the key parameters to be determined are as follows:
- *Total dissolved solids* (*TDS*): This parameter determines the evaporation duty and salt handling capacity.

- *Organic matter*: This parameter helps to determine the purity of contained salt, chemical oxygen demand (COD), and biological oxygen demand (BOD) of process condensate, and the organic cut from evaporation.
- *Compositional analysis*: This parameter determines the composition of pure salt and mixed salt, the process schematics, and various other operating parameters.
- *Characteristics at various concentrations*: The characteristics of feed effluent determine the selection of evaporator types and process schematics.
- *Solvents*: In case any low boilers are present in the feed effluent, these substances may be removed by solvent stripper, as low boilers may interfere with evaporation and thus the efficiency of the overall system. If low boilers are present, condensate from multiple effect evaporation will also occur with traces of the low boilers and may not be usable. Praj-patented Rh grid trays are compatible with total suspended solid (TSS) load up to 2000 ppm without affecting the performance.

The analysis procedures for characterization of effluent are well reported and are not discussed here. Previous chapters of this book also discuss the importance of many of the mentioned parameters. As those chapters indicate, the results of effluent characterization form the basis for technology selection and for obtaining or improving operational efficiency.

13.2.3 Design Considerations

The following design factors are to be considered when planning a zero discharge system:
- Temperature
- Salt concentration
- Organic cut
- Crystallization methodology.

The above parameters are first analyzed in the laboratory, and the feasibility of the process is studied, ultimately leading to the selection of processes and the understanding of recovery aspects.

Components of ZLD
- Effluent pre-treatment—filters and clarifiers
- Effluent COD/BOD reduction—biological treatments
- Effluent volume reduction—membranes
- Effluent concentration—evaporation
- Solids separation and discharge—centrifuge and drying
- Treatment of recovered water through distillation and membrane separation.

The previous chapters in this book elaborate on various physico-chemical and biological methodologies for the treatment of industrial wastewaters (See Chapters 1–12) and for effective removal of BOD/COD (Biological/Chemical oxygen demand) or other specific pollutants. This chapter mainly focuses on treatment methodologies that have not been covered earlier, such as the concentration of effluent using an evaporation system followed by crystallization and/or drying by way of a centrifuge and agitated thin film dryer (ATFD).

13.3 EVAPORATION

Evaporation is at the heart of many industrial processes, as well as effluent treatment. It is an extensively energy intensive unit operation. Utility consumption for concentration plays a major role in the overall operating cost of any ZLD plant. Evaporation is a well established chemical engineering operation that has been extensively covered in standard text books and other reference books on the subject. This chapter focuses more on applications in the area of wastewater treatment and complex effluents.

Three principal elements are of concern in evaporator design: heat transfer, vapor–liquid separation, and efficient energy consumption.

13.3.1 Types of Evaporators

The following three types of evaporators are used in industry:
- Falling film evaporators
- Forced circulation evaporators
- Forced falling film evaporators.

Key features of these three types of evaporators are discussed, in brief, in the following section.

13.3.1.1 Falling Film Evaporators

Falling film evaporators are industrially important equipment in which the process fluid to be evaporated flows in the form of falling film. There are a number of advantages and limitations to this design; the important ones are listed below:

Advantages
- Better heat transfer coefficient related to smaller area and lower CAPEX.
- Low power consumption with a lower OPEX.
- Quick start up and transitioning to cleaning mode, leading to low down time

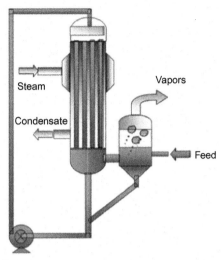

Figure 13.1 Falling film evaporator.

- Useful for heat sensitive chemical solutions, due to lower residence time and lower holdup (see Figure 13.1).

Limitations

- Falling film evaporators are not effective for liquids having higher suspended solid concentrations and higher viscosity.
- Inadequate wetting can lead to tube-fouling in the evaporator's distribution system. Thus, design of distributor system is critical.
- Frequent cleaning in place (CIP) is required and involves actual experience to confirm cleaning frequency.

13.3.1.2 Forced Circulation Evaporators

When the occurrence of boiling at the heating surfaces must be avoided in order to minimize fouling problems or crystallization, force is used to increase the flow velocity, as in the forced circulation evaporators. The advantages and disadvantages of forced circulation evaporators are briefly outlined below:

Advantages

- The forced circulation evaporator is capable of processing suspended slurry, saturated solution, and viscous liquids, and therefore, it can accommodate a wide range of process liquids (see Figure 13.2).

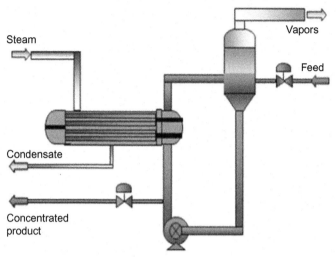

Figure 13.2 Forced circulation evaporator.

- This type of evaporator can operate with minimal temperature differences between the heating media and the boiling liquid, and therefore, it involves a greater number of effects.
- Scaling inside the tubes is less, leading to high operating time.
- Forced circulation evaporators can be installed vertically as well as horizontally, thereby saving space.

Limitations
- Relatively high heat transfer area, leading to higher CAPEX.
- Higher recirculation rates, leading to higher power requirements and thus higher OPEX.

13.3.1.3 Forced Falling Film Evaporator
This design is a variation on the conventional falling film evaporator, and has additional benefits in terms of the following:
- It has all of the advantages of conventional falling film evaporation (see Figure 13.3).
- Proper wetting rate in the tubes is achieved by using spraying nozzles.
- It is less prone to fouling than its conventional counterparts.

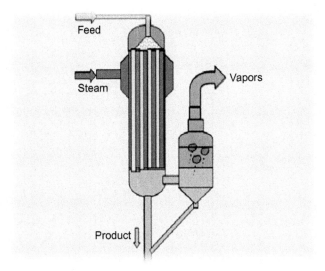

Figure 13.3 Forced falling film evaporator.

Various systems for using these commercially available evaporators have been developed to minimize energy consumption. Three of the most commonly used evaporation systems are:

- The multiple effect evaporation plant (MEE) plant.
- The MEE plant using a thermal vapor recompression (TVR) system
- Evaporation with mechanical vapor recompression (MVR).

Key aspects of these three evaporation systems are outlined in the following sub-sections.

13.3.2 Multiple Effect Evaporation

Multiple effect evaporation, or MEE, involves structured sequencing of the evaporators, with the number of evaporators or effects commonly restricted to four, unless another specific design is required. The salient features are:

- A multiple effect evaporator is an apparatus that efficiently uses the heat from steam to evaporate water.
- In a multiple effect evaporator, water is boiled in a sequence of vessels, each of which is at a lower pressure than the last.
- As the boiling point of the water decreases with pressure, the vapor produced in one vessel can be used to heat the next vessel.
- Only the first vessel requires a source of external heat.

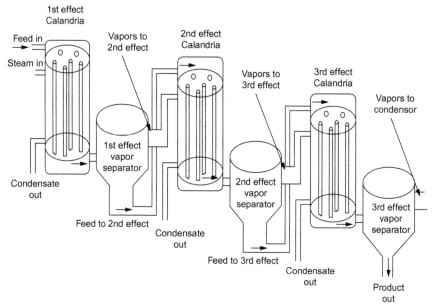

Figure 13.4 Schematic of multiple effect evaporator.

- Evaporators with more than four to five effects are rarely practical (see Figure 13.4 for a schematic of a multiple effect evaporator).

MEE is very simple in operation and the most proven system for evaporation. It is the least energy efficient of all evaporation types, however. The most important parameter for MEE is the feed effluent characteristic, which plays a major role in deciding the number of effects.

13.3.3 Thermal Vapor Recompression

In this evaporation method, compression is performed thermally, as opposed to mechanical vapor compression. The TVR system is shown in Figure 13.5.

- A steam jet ejector is used to raise the temperature and pressure of vapors generated from the first effect of the evaporator.
- The motive steam mixes with a portion of the vapors generated from the first effect of the evaporator.
- The resultant vapors (motive steam plus some vapors from the first effect) are utilized in first effect for condensation.

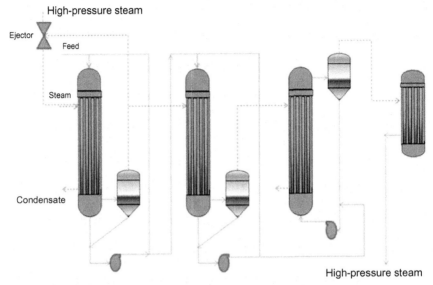

Figure 13.5 Thermal vapor recompression system.

- TVR increases the overall energy efficiency of a multiple effect evaporation plant.
- TVR is a more energy efficient system than multiple effect evaporation.
- TVR requires higher steam pressure.
- In TVR, steam condensate becomes contaminated with process condensate (see Figure 13.5).

13.3.4 Mechanical Vapor Recompression

MVR achieves vapor compression through mechanically driven equipment such as blowers or compressors (refer Figure 13.6).

- The system delivers superheated steam, which is then desuperheated before it is fed back into the evaporator where it was produced.
- The energy input in the system is in the form of energy pumped to the compressor.
- During steady operation, much less steam and cooling water is required, as compared to the requirements of other evaporation systems.
- MVR is the most energy efficient system for evaporation.
- MVR requires almost no additional steam and cooling water during steady operation.

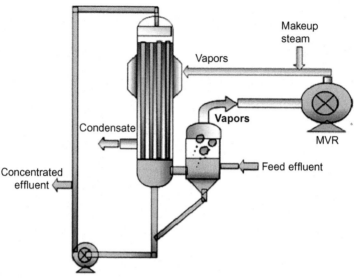

Figure 13.6 Mechanical vapor recompression system.

- MVR can work with any kind of evaporator, including falling film evaporators, forced circulation evaporators, and plastic evaporators.
- MVR requires a marginally higher capital investment than is required by the other evaporation processes.

Advantages
- MVR has low energy consumption.
- MVR has a higher performance coefficient.
- MVR places a reduced load on the cooling tower.
- MVR simplifies the evaporation process and thus plant operation.

Comparison of three evaporation systems is presented in Table 13.1.

Table 13.1 Comparison of typical evaporation systems

Parameter	Multi effect evaporation (MEE)	Multi effect evaporation with TVR	Multi effect evaporation with MVR
Capital investment	Medium	Medium	Relatively high
Operating cost	High	Medium	Low
Electrical energy	Low	Low	High
Thermal energy	High	Low	Low

13.4 SOLIDS SEPARATION EQUIPMENT

Although good in terms of water recovery and the elimination of effluent discharge, the ZLD process does generate solids. This section discusses aspects of solids separation inherent in all ZLD systems.

13.4.1 Centrifuge

Here the solid gets separated through the application of mechanical energy. The important types of centrifuge to be considered for this process are pusher centrifuges.

13.4.1.1 Pusher Centrifuge

This type of centrifuge provides continuous filtering used for solid-liquid separation in the chemical and mineral industries. The capacity sizing of a pusher centrifuge completely depends upon the type of solids or salts being handled. There are mainly two types of inorganic salts we look at for pusher centrifuge operation. These salts are:

- NaCl (sodium chloride)
- Na_2SO_4 (sodium sulfate).

The operation of the pusher centrifuge is a function of many parameters such as particle size, viscosity, solids concentration, cake quality, and particle attrition.

13.4.2 Dryer

In dryers, the separation of solids is accomplished through the application of thermal energy. The major types of dryers are:

- Rotary drum dryer
- Spray dryers
- Agitated thin film dryer.

13.4.2.1 Agitated Thin Film Dryers

ATFDs are characterized by a mechanically agitated thin product layer inside the dryer. The dryer consists of a vertical, cylindrical body with a heating jacket and a rotor inside of a shell which is equipped with rows of blades all over the length of the dryer. The hinged blades spread the wet feed product in a thin film over the surface of the heated wall. The turbulence increases as the product passes through the clearance before entering a calming zone situated behind the blades. The volatile component evaporates continuously. The product layer is typically less than a millimeter in thickness.

The following case studies discuss attempts to implement zero discharge solutions.

13.5 CASE STUDIES

In this section we present three case studies related to important sections of the chemical industry responsible for major environmental pollution: the chemical, fertilizer, textile, and bulk drug industries.

13.5.1 Case Study 1: Pharmaceutical API (Active Pharmaceutical Ingredient) Industry

The pharmaceutical API (active pharmaceutical ingredient) industry, which is also known as the bulk drug industry, manufactures drug intermediates and end-products. In the majority of cases, API industries are involved in contract manufacturing, meaning that their products are manufactured according to client requirements.

API industries manufacture multiple products, and the characteristics of these products differ in terms of quality and quantity. Thus, effluents from the industries vary in composition, and hence, the parameters associated with the effluents also vary.

13.5.1.1 Effluent from the API Industry:

The types of effluents produced by API industries are listed below, along with the manufacturing activities or other sources that produce those effluents.
1. Lean stream (low COD–low TDS effluents)
    a. Process
        i. Reactor washings from the second wash onwards
    b. Utility
        i. DM plant regeneration waste
        ii. RO (reverse osmosis) reject stream
        iii. Cooling tower blow down
        iv. Boiler blow down
    c. Other
        i. Floor washings
        ii. Various drains.
2. ML stream (high COD–high TDS effluent)
    a. Product separation
    b. First washings of reactor
    c. Traces of solvents from plant.

13.5.1.2 Lean Stream

The lean stream results from the majority of washings and utility waste, and as a result, it has moderate concentrations of COD and TDS, which are treated using physico-chemical and biological methods.

Table 13.2 Parameters of lean stream

Parameters	Unit	Concentration
pH		7–8
TDS	ppm	3000–4000
COD	ppm	3500–5000
BOD	ppm	1500–2000
Ammoniacal nitrogen	ppm	50–100

The effluent in the lean stream is collected in the equalization tank (EQT). The parameters of the lean streams are listed in Table 13.2.

The utility waste that does not have high COD levels is mixed in the equalization tank or it is treated for TSS & Turbidity. Mixing after biological treatment saves the hydraulic flow rate of overall ETP (Effluent Treatment Plant).

13.5.1.3 Primary Treatment

In the plant, primary treatment consists of the removal of TSS and turbidity. This process is accomplished through the use of the following units.

1. Oil and grease trap

   Wastewater from floor washing may contain oil and grease, and hence, this system includes a trap for removing oil and grease from the raw effluent. The oil and gas traps comprise top and bottom flow cascades, through which effluent moves, while oil floats to the top of the medium to be skimmed off. After oil removal, the effluent flows to the EQT.

2. Equalization tank

   As suggested by its name, the EQT retains the effluent until its flow and characteristics have equalized. A coarse bubble air grid in the tank gently mixes the contents to encourage the process, and once equalized, the effluent is pumped to the tube settler for TSS removal.

3. Flash mixer

   In the flash mixer, coagulant is introduced into the effluent. The effluent and the coagulant must be mixed quickly, and an agitator is employed to assist with this process.

4. Flocculator

   The flocculator is the device in which the coagulant-dosed effluent is mixed with high density polyelectrolyte and mechanically agitated.

5. Tube settler

   The suspended solids load of the effluent is reduced in the tube settler. The solids are agglomerated in the settler, and the clarified overflow is sent to a clarified water tank. The sludge from the hopper bottom is then sent to the sludge handling system.Compared with all other clarifiers available today, the tube settler provides maximum surface area, and as a result it requires a smaller footprint than competing clarifiers do, while ensuring maximized TSS removal from the effluent.

13.5.1.4 Secondary Treatment

The objective of secondary treatment is to reduce the organic load in the form of COD and BOD. It also reduces ammoniacal nitrogen in effluent. The secondary treatment process relies on the following units:

1. Biological system

    Until recently, the conventional activated sludge process was used for the treatment of BOD, as it was a proven method. As space has become an issue, however, the size of the aeration tank has been reduced by using advanced systems such as moving bed bioreactors (MBBRs). A two stage MBBR is often used to ensure the maximum efficiency of the system. MBBR biofilm technology is based on specially designed plastic biofilm carriers or biocarriers that are suspended and continuously moving within a tank or reactor. An aeration grid located at the bottom of the reactor supplies oxygen to the biofilm, along with the mixing energy required to keep the biocarriers suspended and completely mixed within the reactor.

2. Secondary settling tank

    Biomass carried from the MBBR is taken, along with the overflow effluent, to a secondary settling tank. Unlike the primary settler, the secondary settling tank does not employ any dosing aid, as effluent from the secondary tank will eventually be taken to the second stage MBBR. Thus, the solids are agglomerated in the secondary settler, and the clarified overflow is sent to the filter feed tank. The sludge from the hopper bottom is sent to the sludge handling system.

3. Filter feed tank

    Clarified effluent from tube settler III is collected in a filter feed tank, which is an reinforced cement concrete (RCC) tank.

13.5.1.5 Membrane Recycling System

The biologically treated effluent from ETP passes through the following units:

1. Filter feed pump

    The effluent from the clarified effluent tank is first pumped through a pressure sand filter (PSF), a multigrade filter (MGF), and an ultrafiltration (UF) system for removal of suspended solids.

2. **Multigrade filter**

The MGF removes the traces of suspended impurities retained after clarification in the tube settler. It separates out TSS larger than 100 μm, and the outflow from the MGF is sent to the UF system.

The MGF is backwashed periodically.

3. **UF system**

The UF system is a skid-mounted membrane system that includes hollow fiber UF membrane modules, re-circulation valves, and pneumatically actuated valves. All of these components are neatly interconnected into a compact, modular train equipped with a self-control system.

Each UF membrane module consists of thousands of hollow fiber membranes that are capable of clearing the feed stream of virtually all suspended solids, colloids, bacteria, and viruses larger than 0.01–0.015 μm.

4. **UF backwash pump**

In the filtration process, the rejected suspended solids form a layer on the surface of the membrane, reducing the membrane's permeation rate. This layer must be dislodged to maintain the membrane permeation rate. Dislodgement can be achieved through a method called backwash or reverse flow. The service interval between backwashing is set at intervals from 30 to 60 minutes, and the backwash interval lasts for 120 seconds before service resumes. The backwash stream is then returned to the EQT.

5. **UF CIP unit**

With every filtration cycle, the membrane permeability decreases slightly due to membrane fouling. When the allowable transmembrane pressure reaches the maximum limit or set-point, the membrane must be chemically cleaned. The purpose of the membrane cleaning process is to remove foulants and to restore productivity.

Cleaning effectiveness is a function of agents used, chemical concentration, and factors such as cleaning time, frequency, temperature, circulation rate, and pressure.

Typically, the chemical cleaning solution in a CIP consists of acid wash (pH 3), usually using hydrochloric acid or citric acid, followed by an alkaline wash, usually using NaOCl (sodium hypochlorite) at 500 ppm (pH 10.5 to 11).

6. **UF permeate tank**

Permeate from the UF system is collected in this tank.

7. Priming pump

The UF permeate is pumped via a priming pump to the cartridge filter.

8. Antiscalent dosing unit

During the RO process, dissolved salts are concentrated on the reject side as permeate water is being drawn at specific recovery. As a result of this arrangement, the dissolved salts concentration increases on the reject end, and salts may precipitate if the saturation limit is exceeded. As a result, salt scaling (deposition) occurs on the membrane surface, leading to inferior performance of the RO system. To prevent scaling, an appropriate dose of special antiscalant formulation is required. Normally, a 5% concentrated solution of the antiscalant is prepared, and dosing pumps in the line introduce about 3–5 mg/L of the antiscalant solution into the effluent, as needed.

9. Cartridge filter

The filtered stream then passes through the cartridge filter, which arrests any suspended solids that remain.

10. High pressure pump

Exiting the cartridge filter, the clear stream is transported via high pressure pumps to the membrane modules for treatment.

11. Membrane skid

The membrane skid consists of membranes mounted in membrane housings. The logistics of feeding the effluent through the various membrane housings depend on the system configuration. Part of the reject from the membrane ETP is recycled back due to high pressure pump suction, whereas the balance is suitably discharged for further treatment.

12. RO CIP system

A CIP system is provided to clean both RO systems at regular intervals. The CIP system consists of a tank and pumps.

13. Permeate tank

The permeate coming out of the RO system is collected in this tank.

14. Decanter centrifuge

Excess sludge from the anaerobic and aerobic digestion processes, as well as from tube settler I, enters the sludge sump, which then pumps the sludge to drying beds.

13.5.1.6 ML Stream
13.5.1.6.1 Solvent Stripping System
As stated above, the ML stream contains mixed solvent traces which need to be removed prior to the feed entering the system.

The system primarily consists of the following units:
1. Stripper
2. Stripper reboiler
3. Stripper condensers.

Feed effluent is fed to the preheater where it is preheated before being introduced to the distillation column. The distillation column operates under vacuum. Heat required for distillation is provided through indirect steam heating in the stripper reboiler. Vapors from the top of the column are condensed in stripper condensers. Part of the condensate is sent back as reflux, while the remaining condensate is taken out as distillate.

Distinct advantages of the distillation:

- Operation-specific design
- Energy efficient distillation system
- Optimal heat usage to conserve energy
- Vacuum operation reduces chances of scaling
- Well-engineered plants with high efficiency trays to ensure thorough separation and removal of impurities with almost 100% separation efficiency
- Proper care in the design of the calming zone, focusing on trays, the large area of the down-comers, gas-liquid separators with tangential entry, and vapor bottles (such focus ensures proper gas-liquid disengagement and eliminates chances of liquid carryover)
- Column with hyper-stat Rh grid trays ensure high on-tray turbulence and thus minimize chances of scaling and choking. (This special construction of trays and tray access points also aids in the cleaning of column internals. Comparisons between the packed column design and the tray column design are given in Table 13.3)
- Condensers designed with multiple passes to ensure high velocity and to minimize scaling inside tubes
- Preheated feed promotes proper stripping of impurities from water

13.5.1.7 Evaporation System: Multiple Effect Evaporator

"ECOVAP" Evaporation System. In this system, feed effluent is preheated by process condensate, vapor from first, second and third effect followed by vapors which are coming out from TVR through series of preheaters respectively. The preheated effluent is then fed into the first effect. The effluent flows in the forward feed manner in evaporators, all four effects being of the forced circulation type.

A live steam of 8 kg/cm^2 (g) is fed to the inlet of the TVR system as a motive fluid. The TVR system then sucks up a portion of the vapors from

Table 13.3 Comparison between packed column and tray column designs

Sr. No.	Packed column	Tray column
1.	As only one feed inlet is possible, the column needs to operate at design conditions, giving no flexibility in operation.	The column can have multiple feed inlets, providing operational flexibility, depending on variation in the feed condition.
2.	Being a packed column, the pressure drop is greater and results in a higher steam requirement.	The pressure drop is comparatively lower than the drop in the packed column. Hence, less steam is consumed.
3.	Packed columns have a narrower operating range.	Tray columns have a wider operating range.
4.	If the liquid flow rate is too low, the wetting of the packing material may not be adequate.	If the liquid flow rate is too high, treatment in tray columns is more economical.
5.	If solid particles are present in the liquid, cleaning a packing section is much too complicated and expensive, compared with plate discharge.	Tray columns can handle suspended solids without performance being affected.
6.	When stressed by temperature variations and pressure, the packing elements are easily breakable.	Trays are made up of SS316, which has no chance of breaking due to temperature and pressure variation.
7.	Packed columns possess no special design elements for addressing TSS handling.	 The tray columns are specially designed to take care of all the above points in a more effective manner. To this end, the columns use an Rh grid type tray.

effect 1, mixing these vapors with the live steam, which is given to the shell side of the first effect. Vapors generated in the first effect are subsequently condensed on the shell side of second effect. The vapors generated in second effect are then condensed on the shell side of the third effect, and so on.

Vent vapors and vapors from the last effect are condensed in the surface condenser. Condensate from the first effect is flashed on the shell side of the second effect, and collective condensate from the first and second effects is flashed on the shell side of the third effect, and so on. This sequence of shell-side flashing helps to increase steam economy. Finally, process condensate is collected in a tank and transferred to the battery limit by way of a process condensate pump through the feed preheater. Concentrated product from the last effect is transferred to the ATFD for further removal of moisture. The entire evaporation system operates under a vacuum.

13.5.1.8 Dryer System

The concentrated product from the evaporator is fed into the feed balance tank of the ATFD. This feed flows into the top of the ATFD rotor by means of a feed pump. The feed is then uniformly distributed with the help of the rotor blades, which are rotating at high speed. Due to this rotation, the feed is sprayed onto the surface of the tank wall.

Dry saturated steam is used as a heating medium for each jacket. Steam condensate is removed from each jacket through the bottom nozzles of a steam trap. The dry concentrated product is then recovered from the bottom of the equipment.

The evaporated vapors are conveyed to the surface condenser where they are condensed using cooling water. Process condensate from the ATFD is transferred to the battery limit by a process condensate pump.

The dryer system operates under vacuum. To further illustrate the described processes, lean stream parameters are given in Table 13.4. ML streams parameters are given in Table 13.5, and utilities required for the system are given in Table 13.6.

Block diagram (see Figure 13.7).
Primary treatment (see Figure 13.8).
Secondary treatment (see Figure 13.9).
Recycling system (see Figure 13.10).
Mixed solvent stripping (see Figure 13.11).
Multiple effect evaporator (see Figure 13.12).
Overall ETP schematic (see Figure 13.13).

Table 13.4 Lean stream parameters

Sr. No.	Parameters	Characteristics	Units
1.	Flow	145	KLD
2.	COD	<3600	mg/L
3.	BOD	<1500	mg/L
4.	COD:BOD Ratio	<3	
5.	TSS	<400	mg/L
6.	TDS	1000–2000	mg/L
7.	pH	6.5–8.5	–
8.	Chloride	<255	mg/L
9.	Sulfate	<45	mg/L
10.	Calcium	<34	mg/L
11.	Magnesium	<76	mg/L
12.	Silica	<31	mg/L
13.	Sodium	<98	mg/L
14.	Total alkalinity	<200	mg/L

Table 13.5 ML stream parameters

1.	Flow rate	m^3/day	18
2.	Feed rate	kg/h	927
3.	Solvents present in feed	% w/w	5 (maximum)
4.	Moisture content in the stripper distillate	% w/w	50 (maximum)
5.	Steam consumption	kg/h	240
6.	Hours of operation	h	20
7.	Solids content % w/w TS	% w/w	4.5%
8.	Viscosity	cP	<1 @ 25 °C
9.	Specific gravity		1.03
10.	Specific heat	(kcal/kg °C)	0.95
11.	Suspended solids	ppm	<100
12.	Oil and grease	ppm	<5
13.	pH		7.0
14.	Temperature	°C	Ambient
15.	TDS	ppm	50,000
16.	Total hardness	mg/L	<50
17.	Total silica	mg/L	<50
18.	Ammonical nitrogen	mg/L	<50
19.	Dissolved gases	mg/L	Nil

Table 13.5 ML stream parameters—cont'd

	Solvents		
1.	IPA	% w/w	<1.25%
2.	Acetone	% w/w	<0.50%
3.	Methanol	% w/w	<0.60%
4.	Ethanol	% w/w	<0.05%
5.	TBME	% w/w	<0.05%
6.	Ethyl acetate	% w/w	<1.00%
7.	Toluene	% w/w	<0.25%
8.	DMF	% w/w	<0.35%
9.	DMSO	% w/w	<0.25%
10.	Heptane	% w/w	<0.10%
11.	Hexane	% w/w	<0.10%
12.	THF	% w/w	<0.50%
13.	Water	% w/w	95%

Table 13.6 Utilities required

Description	Unit	ETP & recycling system	ZLD system
Operating electrical load	kW	80	100
Chemical cost	Rs/m³	25–27	For CIP
Cooling water (in circulation)	m³/h	–	90
Steam	kg/h	–	860

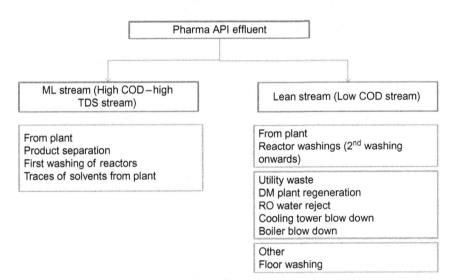

Figure 13.7 Typical pharmaceutical API effluent.

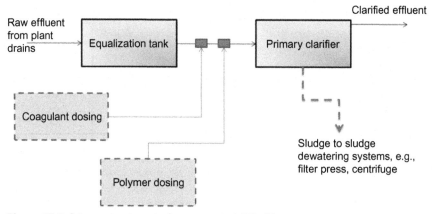

Figure 13.8 Primary treatment pharmaceutical API effluent.

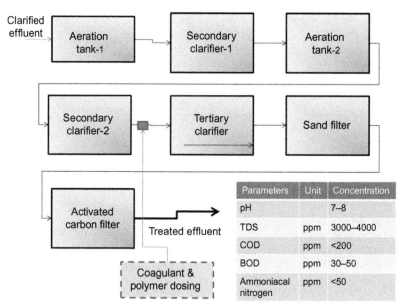

Figure 13.9 Secondary treatment pharmaceutical API effluent.

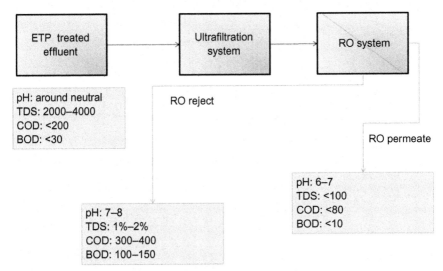

Figure 13.10 Recycling system for pharmaceutical API effluent.

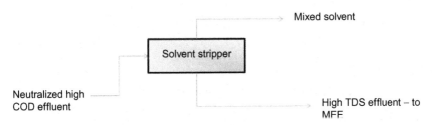

Figure 13.11 Mixed solvent stripping.

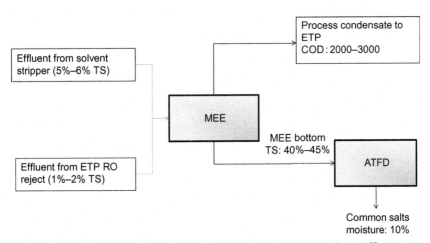

Figure 13.12 Multiple effect evaporator scheme for pharmaceutical API effluent.

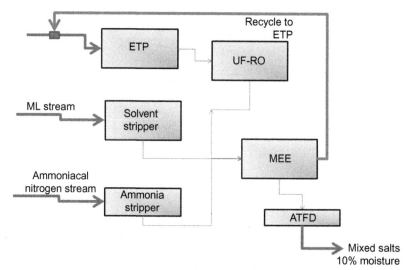

Figure 13.13 Zero liquid discharge (ZLD) scheme for pharmaceutical API effluent.

13.5.1.9 Concluding Remarks

In the pharmaceutical API Industry, failures of recycling and ZLD systems are primarily due to the mismanagement of toxic solvents at the source, which increases the overall load on the ETP, recycling, and ZLD systems. A ZLD system should be implemented at the source of the effluent. Presently, inefficient stripping and multiple effect evaporators result in failures of driers and thereby ZLD systems. The current technique represents the unique concept of the system audit, tray design for strippers, and optimum heat transfer areas in evaporators, thus ensuring optimum integration and longer system operation for a given feed.

13.5.2 Case Study 2: ZLD Solutions for Textile Common Effluent Treatment Plant (CETP)

This plant in this case study uses a ZLD system based on MEE to treat effluent generated by textile CETP. The final end products are process condensate and salts of reusable quality. The reject of the existing RO system is first preheated and then fed to the evaporation system in a forward feed manner.

In this plant, MEE is achieved via the ECOVAP system (Praj's proprietary). ECOVAP is suitable for operations with effluents having higher solids concentration and viscosity. Suspended solids slurry and precipitating salts can be handled by this system with no choking issues and with fewer cleaning cycles.

A block diagram of ZLD is given in Figure 13.14.

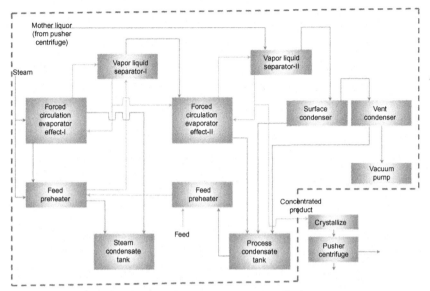

Figure 13.14 Block diagram for ZLD in treating textile effluent.

13.5.2.1 Process Description

Feed effluent is preheated by process condensate and steam, before being fed into the first effect. The effluent flows in a forward feed manner in the evaporators, and both evaporators are of the forced circulation type.

Low-pressure steam is introduced to the shell side of the first effect. The vapors generated in the first effect are then condensed on the shell side of the second effect. Vent vapors and vapors forming in the second effect are condensed in the surface condenser and vent condenser, and the condensate is collected in a tank and transferred to the battery limit by the process condensate transfer pump. Concentrated slurry from the final effect is transferred to the battery limit by the product transfer pump.

The major advantage of this innovative technology is that it allows for the recovery of sodium sulfate by selective crystallization. The sodium sulfate recovered from this plant is more than 90% pure. Hence, the client has obtained good payback. The entire evaporation system operates under a vacuum.

13.5.2.2 Advantages of ECOVAP

- *Long cleaning frequency cycle of 30–45 days*
- *Process condensate purity* (process condensate is pure enough to be reused directly for utility or any other purpose).

- *Constantly getting 25% w/w solids after 2nd effect* (these salts are recovered and reused in the process, resulting in the reduced consumption of salts in the plant and thus benefiting the environment. Prior to the installation of ECOVAP, the salt quality in terms of w/w solids was not consistent, resulting in the choking of the equipment, so that complete recovery of the salt was not possible).

For photographs see Figure 13.15.

Figure 13.15 Photographs of installations in the textile industry.

13.5.2.3 Concluding Remarks

The textile industry in India faces environmental issues due to color and nonusable high TDS brine solutions present in its effluent. The major issue for plant operation is the mixed salts coming out of the evaporator systems, as those salts cannot be reused in the dyeing process. The present technique represents a unique solution for the selective crystallization and recovery of Glauber's salt followed by mixed salts, which can then be reused in the dyeing process. The recovery and reuse of salts results in handsome paybacks for the evaporation system.

13.5.3 Case Study 3: Innovative Wastewater Treatment solutions for Chemicals and Fertilizers Industry

In this particular case, effluent is from Fertilizer industry. The effluent is rich in Ammonical Nitrogen content hence it was not feasible to treat it biologically. Thus, the scheme recommended was steam stripping of Ammonia followed by Sequential Batch Reactor for COD & BOD reduction.

Client: chemicals and fertilizers industry

Scheme offered:

Ammonia stripper + 2 stage sequential batch reactor (SBR) + tertiary treatment

(PSF + ACF)

Capacity: 600 m^3/day.

13.5.3.1 Process Description
13.5.3.1.1 Equalization

The effluents from all industrial streams is collected in an EQT and provided adequate retention time. Jet mixing encourages flow and characteristics equalization. The pH of the effluent is also adjusted to 11.5 in this tank. The EQT is housed in a shed with a vent for the release of excess gases.

13.5.3.1.2 Online NaOH Dosing System Before Stripper

An online alkali dosing system with auto control is provided to increase the pH of effluent before it passes into the ammonia stripper.

13.5.3.1.3 ECOFINE Ammonia Stripper

The "ECOFINE" ammonia stripper is designed to remove excess ammonia from the effluent before it is sent to aerobic biological treatment.

Raw water effluent, which contains around 2000 ppm of ammonia and VOCs (Volatile Organic Compounds), enters the ammonia stripper plant at 11.5 pH. The feed is preheated in a heat exchanger, using column bottom stream. This preheated stream is then fed to the stripper.

Low pressure steam is supplied to the ammonia stripper reboiler for use as a heating medium. Vapors coming out of the stripper column contain mostly water with ammonia and other VOCs, which are then partially condensed in the stripper condenser. During the process, around *40–45 m³/day of steam condensate will be generated via stripping.*

The condensed stream is recycled back to the column, and the remaining vapors (emissions) are sent to the existing incinerator system or flare unit. The treated water from the stripper column is subsequently cooled in the shell and tube heat exchanger before being sent for further treatment.

The ammonia stripping plant is designed as a skid-mounted unit, so that the plant can be transported in state with all parts assembled, painted, and erected. The total plant package involves small skid-mounted units to be assembled on site. The structural material requirements are based on Indian industry standards, however. Thus, the design for the structure complies with such standards, and all static equipment, pressure vessels, pumps, and heat exchangers are erected on the skid.

Pipes and fittings are also mounted on the skid and meet ANSI standards.
- Steam is supplied through the reboiler for higher condensate recycling.
- Raw effluent at 30 °C is first heated with the help of hot treated water taken from the column bottom using the heat exchanger.
- Hot raw effluent from the heat exchanger is fed to the top of the column for ammonia stripping.
- Cold treated water is available from the heat exchanger outlet for further treatment.

13.5.3.1.3.1 Benefits of ECOFINE—Ammonia Stripper Plant
- Optimal heat usage to conserve energy
- Well-engineered plants with high efficiency trays to ensure elaborate separation
- Proper care in the design of calming zone on trays, the large area of down-comers, gas-liquid separators with tangential entry, and vapor bottles. (This design ensures proper gas-liquid disengagement and eliminates risk of liquid carryover.)

- Condensers designed with multiple passes to ensure high velocity and to minimize scaling inside tubes
- Preheating of feed aid the process of stripping ammonia from the water.

13.5.3.1.4 Online Acid Dosing System After Stripper

An online acid dosing system is provided after the stripper to decrease the effluent pH before it proceeds to the aerobic biological treatment system.

13.5.3.2 Sequential Batch Reactor (I and II)
13.5.3.2.1 SBR-I Feed Tank

Effluent from the pH correction tank is collected in the SBR-I feed tank, the where it is retained for a sufficient period of time. The effluent is then pumped to SBR-I from the feed tank.

13.5.3.2.2 Sequential Batch Reactor (ECOCAT-SBR)

The SBR uses an activated sludge process operated in batch mode with precise sequencing and control. The complete process takes place in a single reactor. No additional settling unit or secondary clarifier is required. The complete biological operation is divided into cycles, during which all treatment steps take place.

A basic cycle comprises the following four steps:
- Fill and aeration
- Aeration
- Sludge settlement
- Clear liquid decanting.

Occuring in sequence, these phases constitute a cycle, which is then repeated. The time required for each phase depends on the characteristics of the effluent.

During various phases, the operation of the equipment, such as the feed pump, sludge recycle pump, air blower, and decanter, is controlled automatically using on-off control valves, a variable frequency drive, and a DO meter, with the help of a DCS/SCADA system (DO Meter – Dissolved Oxygen Meter; DCS – Distributed Control System; SCADA – Supervisory Control & Data Acquisition).

13.5.3.2.3 SBR-II Feed Tank

Treated effluent from SBR-I is decanted into the SBR-II feed tank, where it is provided with adequate retention time. The effluent is then pumped to SBR-II for further treatment.

13.5.3.2.4 Sludge Thickener

Excess sludge from SBR-I and SBR-II is periodically pumped to the sludge sump, which delivers it to the sludge thickener. The thickener is a hopper bottom circular tank with a centrally driven mechanism where suspended solids become settled at the hopper bottom and the clear liquid comes out as overflow which is then recycled back to the SBR-I feed tank.

13.5.3.2.5 Thickened Sludge Sump

The thickened sludge from the thickener bottom is sent to the thickened sludge sump.

13.5.3.2.6 Decanter

The thickened sludge from the thickened sludge sump is now pumped to the sludge decanter for dewatering. The centrate from the decanter centrifuge is again taken back to the SBR-I feed tank for further treatment. The dewatered sludge is sent for further disposal.

13.5.3.2.7 Filter Feed Tank

Overflow from SBR-II is collected in this tank and provided with adequate retention time. It is then pumped to the PSF and activated carbon filter (ACF) via filter feed pumps.

13.5.3.3 Pressure Sand Filter

The filter feed pump transports effluent from the filter feed tank to the sand filter. The suspended solids present in the wastewater get trapped in the sand media, and clear water goes to ACF for further treatment. Online NaOCl dosing is achieved via the inlet feed line to the activated carbon column. The filter backwash water is then fed back into the SBR-I feed tank for further treatment.

13.5.3.3.1 Activated Carbon Filter

The filtered water from the sand filter is fed into the ACF. Activated carbon uses the physical adsorption process whereby attractive van der Waals forces pull the solute out of solution and onto its surface. This filter helps to reduce dissolved solids, color, and odor from the wastewater. The filter backwash water is fed back into the SBR-I feed tank for further treatment. Outlet parameters for this process are given in Table 13.7. Installation photographs of the plant are shown in Figure 13.16.

Table 13.7 Outlet parameters achieved

Sr. No.	Parameters	Value
1	pH	6.5–8.5
2	COD, mg/L	<250
3	BOD, mg/L	<30
4	TSS, mg/L	<100
5	NH_3, mg/L	<50

Figure 13.16 Photographs of installations in the chemical and fertilizer Industry.

13.5.3.4 Concluding Remarks

In the chemical and fertilizer industry, understanding effluent characteristics is very important because of the complexity of the effluent. Praj has developed a strong research and development backup for the analysis of various complex effluents. In this particular case, biological treatment was possible after effectively removing ammonia from the effluent. The present technique provides steam stripping for effective removal of ammonia. The SBR is one of the proven technologies for further ammonia reduction through nitrification and denitrification.

13.6 SUMMARY

In this chapter, some steps toward achieving ZLD have been discussed. Primarily current ZLD developments are based on evaporation and discharge of solid waste. This chapter also discusses recent advances in efficient evaporation systems, as well as the use of better solids handling equipment in ZLD processes. The three case studies included in this chapter clearly demonstrate the possibility of implementing such solutions in practice. There are several opportunities to use effluent treatment technologies discussed in this book in a synergistic fashion to enhance the effectiveness and efficacy of remedies proposed for ZLD.

CHAPTER 14

Industrial Wastewater Treatment, Recycling, and Reuse—Past, Present and Future

Vivek V. Ranade[1], Vinay M. Bhandari[1]

Chemical Engineering and Process Development Division, CSIR-National Chemical Laboratory, Pune, India
[1]Corresponding author: vv.ranade@ncl.res.in; vm.bhandari@ncl.res.in

14.1 INTRODUCTION

The world went through a major transformation in the twentieth century, and it has witnessed unprecedented advances in physics, chemistry, biology, and engineering sciences. There have been spectacular changes in materials, industrial operations, and computational processes, which themselves are just broad classes for thousands of advances over the last century. Industrialization has resulted in significant life style improvements for human beings in all spheres of living. Unfortunately, there is another side to this rosy picture in the form of increased risks to humans and the environment caused by the industrialization and development that has taken place. The effects of industrialization can be seen in the form of air, water, and soil pollution that may threaten the very existence of living species on earth if remedial measures are not taken. This is a Himalayan task and demands contributions from individuals, groups, communities, industries, and governments.

We believe that in the coming years the newer forms of existing methodologies and emerging technologies will dominate continuing developments worldwide and will contribute in a major way to alleviating the risks posed by environmental pollution. Water is a critical element in this scheme for safeguarding the environment, and, as a result, industrial wastewater treatment is an issue that will remain in the forefront in the future. Another facet to this problem is water scarcity around the world, especially in developing countries. Water management is therefore one of the most important problems facing those trying to ensure environmental protection and sustainability, and this problem can best be resolved through effective industrial wastewater treatment, recycling, and reuse.

A big question is: Are we prepared to take on the challenge? From a technological viewpoint, we respond with a positive and hopeful "yes." However, significant efforts are required to change the mindsets of people! This book has made an attempt to compile some of the available technologies and to present a coherent picture of the current state of industrial water treatment, recycling, and reuse. This last chapter represents an epilogue in which we discuss some aspects of the past, present, and future of this subject.

14.2 THE PAST

Wastewater treatment philosophy dates back thousands of years and was part of many ancient civilizations, such as those of Rome and the Indus Valley (Vigneswaran et al., 2011; http://www.lenntech.com/history-water-treatment.htm). Located near the Indus River in the ancient Indian region, Mohenjo-Daro represents one of the oldest known systems of wastewater management, as it is believed to date to around 1500 BC (Wiesmann, 2007). At that time, the local population settled in the vicinity of rivers that ensured easier supply and discharge of waters. These ancient settlements were believed to have water supplies and sewerage facilities. The development of water transport systems of various forms was crucial and was supplemented by treatment methods. The primary focus in the past was sewage wastewater treatment, and major treatment processes included segregation, dilution, and filtration in different forms. Though India is facing severe wastewater treatment problems, and water scarcity today, it is believed to have history and prior art of water management, disposal of waste, and health issues that find mention in various forms of literature, beginning with the Vedas, one the most ancient pieces of literature on Earth. However, modern world wastewater treatment arrived only a few centuries back, in about the sixteenth century. The development of physical, chemical, and biological treatments started gradually. The twentieth century saw the main thrust in this area, and the understanding of wastewater evolved through the 1900s to the present date.

In the early 1900s, wastewater treatment mainly involved filtration, settling, and septic tank use. The design of wastewater treatment plants was also considered, along with the application of disinfection methods, largely in the form of chlorination. The biological treatment method also emerged during this period, and the first use of the activated sludge process (ASP) was reported in 1916 in the USA (http://civil.colorado.edu/~silverst/cven5534/History%20of%20Wastewater%20Treatment%

20in%20the%20US.pdf;   http://www.activatedsludgeconference.com/downloads/OneHundredYearsofActivatedSludge-Abriefhistory.pdf). In the following years, increased understanding of wastewaters led to the evolution of various treatment methods, especially for the removal of odor, color, and solids. The main methods included adsorption, ion exchange, coagulation, and chemical treatment. The application of the ASP created a secondary treatment problem in the form of sludge generation, and methods were researched to minimize sludge production, compacting, and other undesired results. The importance of dissolved oxygen was also realized during this period.

After 1950, waste water treatment scenario changed dramatically. Global industrial development also increased in speed and scale, largely due to the emergence of technologies in the form of novel materials, processes, and equipment configurations and devices. During this period, communities also developed new regulations for the release and treatment of wastewaters, fuelling further advances in water management. An increased understanding of biological oxygen demand and chemical oxygen demand propelled the development of methodologies for reducing the organic load in wastewaters, as well as the classification of pollutants in different forms. The membrane separation process also started in this period, with simple symmetric membranes acting merely as filtration medium. Asymmetric membranes followed, allowing for increased flux, as did membranes with diverse materials, better characteristics, and improved mechanical and chemical properties. Eventually, more complex hybrid membranes systems emerged, providing facilitated transport and reactive membranes for highly specific applications. This development, needless to say, widened the scope of application tremendously for industrial wastewater treatment (Microfiltration, Ultrafiltration, Nanofiltration, Reverse Osmosis, emulsion liquid membranes, membrane distillation) and also altered biological treatments in the form of membrane bioreactors. Aside from the developments in technology and instrumentation, overall awareness about environmental pollution and more particularly water pollution increased tremendously in the late twentieth century.

14.3 THE PRESENT

Present day industrial wastewater treatment involves primary, secondary, and tertiary treatment stages. It also tends to employ a combination of chemical and biological treatment methods in order to meet the discharge norms for treated water. In general, industrial wastewater treatment requires a large amount of chemicals, multiple operations and designs, a fairly high degree of process control, and regular maintenance. In fact, in many plants maintenance has been such a serious issue that effluent treatment plants fail to operate as per the desired standards or have to close down. Stringent pollution control norms require maintaining complex systems that are difficult to oversee and thus require trained operators, especially for the maintenance and operation of biological treatments (anaerobic systems, in particular). This complexity also leads to an escalating cost of treatment.

As a rule of thumb, complex effluent treatment plants designed for industrial wastewater treatment, recycling, and reuse are difficult to operate and hard to maintain, and, as a result, these plants are cost intensive in terms of plant space, equipment, and labor. For smaller chemical process industries, an effluent treatment plant also tends to require more space than a manufacturing plant does. Therefore, in developing countries, common effluent treatment plant (CETP) facilities have been provided for facilitating the treatment of combined effluent from various industries. However, an intelligent combination of centralized and decentralized wastewater treatment facilities (decentralized: collection, treatment, and recycling or reuse of wastewater at or near the source of generation) can significantly aid not only in cost reduction and increased effectiveness, but also in apt use of water recycling and reuse.

Contemporary economic development has also led to the growth of many water intensive industries that tend to represent the most water-polluting sectors of the economy. The most notable examples of these industries include the iron and steel industry (mainly cooling water contaminated with pollutants such as ammonia; cyanide; complex polycyclic aromatic hydrocarbons such as benzene, phenols, cresols; acids; salts; and oil); the mining industry (high volumes of water containing fine particulates, metals, surfactants, and oils); the chemical, petroleum, and fertilizer industries (medium to large volumes of water containing a variety of organic compounds, oil, pesticides, dyes, and solvents); the dye and textile industry (medium volumes of water containing reactive and nonreactive dyes and many refractory pollutants that are difficult to degrade); the pulp and paper industry (large volumes of water containing organics, chlorinated compounds, color, and Biological Oxygen Demand [BOD]); the distillery industry (medium–high volumes of water containing high BOD, Chemical Oxygen Demand [COD], color, and organics); and the food and dairy industry (low to medium volumes of water, generally biodegradable and non-toxic). The diverse nature of these industries demands effluent treatment in similarly diverse forms, employing a variety of technologies and for small to very large scale operations. Typically, present day wastewater treatment activities involve initial pH adjustment; particulate removal; oil and grease removal; removal of metals; removal of biodegradable pollutants using biological processes (if the ratio of BOD to COD is favorable); chemical treatment of wastewaters for removal of organics, acids, non-biodegradable pollutants, and toxic pollutants, among other substances; and finally, polishing of treated water to make it reusable. This book considers a variety of conventional, advanced, and specialized forms of industrial wastewater treatment methods in Chapters 1–13, with a specific focus on industrial case studies and applications.

Present day industrial wastewater treatment emphasizes the drastic reduction of pollutant levels, and each pollutant class often has its own discharge limits specified by the respective government or pollution control agencies, with an eye towards minimizing the cost of treatment. Interestingly, most of the present day operations do not specifically intend to recycle or reuse wastewater unless faced with severe constraints. The likelihood of water scarcity in many regions in the near future, combined with the escalating cost of water, is driving many industries to adopt industrial wastewater treatment, recycling, and reuse.

The discussion of various treatment technologies in the previous chapters can be used to identify present day wastewater treatment needs, as well as treatment guidelines for the future. For a quick analysis, these needs and guidelines can be summarized as follows.

14.3.1 Overall Effluent Treatment Plant

- Proper sequencing of different technologies and distribution of the pollutant load
- Optimization of the cost of overall effluent treatment plant (ETP) operations with recovery or recycling options
- Increased flexibility to handle different effluents, loads, and process variations
- Less space requirement
- Reducing the cost of water treatment technologies and plant operation through recycling, reuse, and novel energy generation methods.

14.3.2 Adsorption and Ion Exchange

- Newer materials as adsorbents or resins, natural or synthetic, hydrophilic and/or hydrophobic properties that have high capacity for removal of pollutants with selectivity, where required; as well as tailor-made structures for improved performance
- New processes and equipment designs
- Better theoretical models for *a priori* prediction of process performance and evaluation
- New processes, methods, and equipment for regeneration that employ less chemicals and energy and avoid producing secondary waste streams
- Hybrid or composite materials and processes or process intensification resulting in increased efficiency and efficacy of the operation
- Low cost of materials, processes, and regeneration
- Performance guarantees for commercial operations involving wastewaters from different industries and for a variety of pollutants.

14.3.3 Coagulation

- New coagulant materials: natural and synthetic substances, inorganic and organic substances, such as polymers of different molecular weight and functionality
- Hybrid or composite materials and coagulant formulations for better performance in terms of COD or pollutant removal, settling characteristics, and sludge volume
- New designs and theoretical models for predicting performance
- New processes, methods, and equipment that require less plant space and energy and minimize the secondary waste stream.
- Low cost of materials, equipment, and operation
- Performance guarantees for commercial operations involving wastewaters from different industries and for a variety of pollutants.

14.3.4 Extractions

- New solvents, designer solvents, and biodegradable solvents suitable for selective removal of pollutants
- New equipment and process design for extraction and recovery
- Development of extractant-impregnated substances, including resins, adsorbents, and membranes
- Low cost and new applications.

14.3.5 Membranes

- Newer membranes: hydrophilic and hydrophobic membranes and natural and synthetic membranes that have high capacity for removal of pollutants and also exhibit selectivity with tailor made structures for improved performance; better mechanical, chemical; and thermal properties; and less fouling
- New processes, equipment designs, and membrane modules
- Better theoretical models for *a priori* prediction of process performance and evaluation
- New processes, methods, and equipment for pre-treatment and regeneration, employing less chemicals and energy and avoiding generation of a secondary waste stream
- Hybrid and composite materials and processes and process intensification that result in the increased efficiency and efficacy of the operation
- Low cost of materials, processes, and regeneration
- Performance guarantees for commercial operations involving wastewaters from different industries and for a variety of pollutants.

14.3.6 Advanced Oxidations

- New catalysts and reagents that have high capacity for removal and destruction of pollutants by employing mild process conditions of temperature and pressure
- New processes and equipment designs
- Better theoretical models for *a priori* prediction of process performance and evaluation
- Hybrid processes and process intensification that result in the increased efficiency and efficacy of the operation
- Low cost of materials and processes
- Performance guarantees for commercial operations involving wastewaters from different industries and for a variety of pollutants.

14.3.7 Cavitation

- Newer applications with high capacities for removal of pollutants and also exhibiting selectivity for refractory pollutants
- New processes and equipment designs, such as cavitating reactors
- Better theoretical models for a priori prediction of process performance and evaluation
- Computational fluid dynamics (CFD) for better understanding of flows in various devices, cavities formation, and collapse leading to disintegration of pollutants

- Hybrid processes and process intensification that result in the increased efficiency and efficacy of the operation
- Low cost of operation and energy
- Performance guarantees for commercial operations involving wastewaters from different industries and for a variety of pollutants.

14.3.8 Biological Processes

- New microorganisms that can degrade pollutants more effectively
- New processes and equipment designs, especially for increased effectiveness and less space
- Hybrid processes and process intensification that result in the increased efficiency and efficacy of the operation
- Low cost of operation and energy
- Performance guarantees for commercial operations involving wastewaters from different industries and for a variety of pollutants.

14.3.9 Advanced Modelling, Instrumentation, and Process Control Tools

- Water networks and pinch analysis: minimize water requirements
- Computational fluid dynamics: enhance the performance of individual process equipment
- On-line analysis and monitoring instruments
- Wireless sensing and monitoring
- System optimization and process control tools and artificial intelligence tools.

Other than demanding the general and specific recommendations pertaining to each process, present day operations also require several specific solutions for smooth operations and some of these could be:

- *Pumps*: Several pumps are employed for the primary, secondary, and tertiary treatment stages, and these pumps require protection, especially from grit, to safeguard and extend life of the equipment. Conventional pumps are being replaced by suitable submersible pumps that are easy to install, operate, maintain, and repair.
- *Piping and new materials*: A variety of new materials, such as a specialized form of stainless steel composites, can be employed for better corrosion resistance, avoidance of depositions, and weight reduction.
- *Solar energy*: More efficient photo-catalytic processes for wastewater treatment must be developed in order to make efficient use of solar

energy using equipment that is easily maintained and cost effective compared to existing treatment methods.

• *Trained manpower.* Trained manpower is essential for the smooth assimilation, installation, operation, and maintenance of any technology. Wastewater treatment, recycling, and reuse are interdisciplinary endeavors that require input from chemistry and engineering. In view of the complex ETP of today, training is essential so that plants operate efficiently without cost intensive maintenance and breakdowns. Often, a gap is observed between the plant personnel and the people responsible for instruments and computers. Different groups, often lack good understanding of each other's role in the larger operation and in issues such as technological knowledge and software application. A suitable training can help to bridge this gap for smooth plant operation.

At this juncture, it might be useful to re-examine the knowledge and developments resulting from our experience using industrial wastewater treatment technologies. Despite several advances in various technologies and mathematical modelling, present day water treatment still involves a set of sequential treatment operations applied to a single wastewater stream consisting of the wastewater from all process operations. In the absence of the concept and technologies of wastewater reuse, these processes are fed by freshwater (see Figure 14.1).

The described treatment process can be improved in several ways. Decentralizing water treatment and exploring opportunities to treat the effluent as close to the point of generation as possible is one of the most promising ones. It is also possible to reuse the wastewater from one process in another part of the process, without providing additional prior treatment or after treating it immediately after the process step in which it is generated. In such a scheme, some pollutants are removed from streams closer to the point of generation, and the partially treated water can then be reused in later stages. This treatment arrangement is schematically illustrated in Figure 14.2.

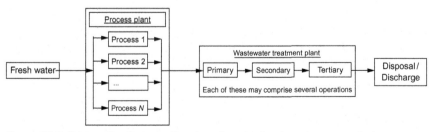

Figure 14.1 Schematic of conventional wastewater treatment arrangement.

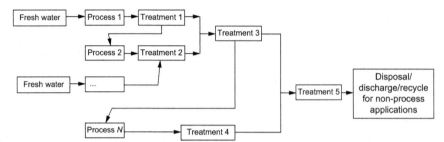

Figure 14.2 Decentralized wastewater treatment with possible recycling and reuse.

All the technological ingredients for conceptualizing, designing, and implementing such a decentralized wastewater treatment scheme are available today. The technologies discussed in this book can be very well implemented using such a scheme. Advanced modeling tools discussed in this book can also be harnessed to create optimal wastewater treatment arrangements that will minimize fresh water requirements by maximizing recycle and reuse of treated water. We hope that the mindset of combining wastewater streams and treating them sequentially will change in coming years. Some comments on future trends and needs may be appropriate at this juncture.

14.4 THE FUTURE

The future is difficult to predict with certainty. Some developments can be on predictable lines, while others can hardly be imagined. One thing is clear that we have a long way to go before we implement the most efficient, economic, reliable, and environmentally sustainable processes for wastewater treatment, recycling, and reuse. Thus, the future will likely have better technologies and designs for efficient effluent processing, and water recycle and reuse.

Every advance in the wastewater treatment technologies has led to a corresponding increase in the society's expectations. The future of wastewater treatment lies not just in enhancing quality of water being discharged to the environment, but also in recycling, effective reuse, and the recovery of valuable chemicals, bio-energy, and biomaterials. New ideas can emerge for converting pollutants to useful materials rather than just removing or destroying them. This is illustrated in Figure 14.3.

Wastewater treatment is conventionally considered as a non–productive operation that only eats into profit and adds to the stress of meeting statutory

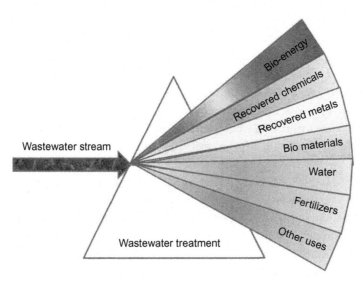

Figure 14.3 Wastewater as a resource.

regulations. Yet, generating energy from wastewaters, which is not new to present day operations, provides an arena for developing profit-making schemes related to industrial wastewater treatment. New generation technologies such as microbial fuel cells can be developed in the future to extract electricity from wastewaters. Apart from effectively treating the effluent, these developments can transform "energy consuming wastewater treatment processes" to "energy producing wastewater treatment processes." This option has the potential to overtake conventional anaerobic processes that produce methane for energy. In the future such options will receive increased attention worldwide, as communities attempt to modify or even replace existing biological treatment methodologies and to develop a more "energy-environment sustainable" platform for industrial wastewater treatment, recycling, and reuse.

Wastewater treatment can also generate added value through the recovery of important materials and chemicals, or even the water itself in water stressed regions. However, significant efforts are still required to improve the process efficiencies and to develop newer separation options for recovery and reuse. For example, newer separation technologies can be developed for recovering phosphorus, phosphates, acids, metals, and other substances from wastewaters. "Smart" or "intelligent" materials will one day have the capacity to identify and target valuable or toxic pollutants, recovering, removing, or destroying these chemicals in an effective and efficient manner. In addition, the wastewater treatment field will likely dominate coupled operations, using enzymes; immobilized resins and adsorbents; extractant impregnated adsorbents, resins, and membranes; or facilitated transport mechanisms for targeted species.

In a broader sense, wastewater treatment involves dilute separations, and, as a consequence, it requires specific inputs and adaptations based on the conventional separation process approach. Naturally, the chemistry of the materials removed and the materials used for the removal is crucial, and our understanding of this chemistry must grow and improve as newer material are being processed or explored. Such developments demand updates to the physical and chemical properties databases, increased understanding of chemical interactions for effective removal, mathematical models for new processes or better prediction of process performance, and better mathematical tools and process optimization software that are easy to adapt and understood for treatment operations. The analysis of dilute solutions with respect to priority pollutants is especially important, and treatment plants will increasingly demand on-line sensors that can provide data in real time.

A number of new technologies are emerging with the aim of refurbishing present day wastewater treatment, and these technologies should become the main focus of the industry in the near future. The newer variations of the treatment processes include electrodeposition, electroflotation, electro-coagulation, cavigulation (coagulation plus cavitation), electro-oxidation, hybrid membranes, newer bioprocesses involving microorganisms that can "eat" specific pollutants, and algae-based wastewater treatments. While advanced oxidation processes often concentrate on Fenton oxidation, catalytic oxidation using different catalysts, or photo-oxidation, developments will also occur in new areas such as supercritical water oxidation and the application of nano-catalysts. Extractive membrane bioreactors are also likely to see further development. The current established treatment methodologies, such as adsorption and ion exchange, will also change and improve through reactive adsorption and similar operations.

Apart from newer forms of processes, the industry will also experience the future development of novel reactors, reactor configurations, and devices that operate with improved efficiency and reduced cost (capital and operating). This development is indicated by new terminologies such as anaerobic membrane bioreactor (AnMBR), enhanced membrane bioreactor (EMBR), anaerobic migrating blanket reactor (AMBR), membrane biofilm reactor, and sequencing batch biofilter granular reactor (SBBGR).

Nanotechnology is expected to play crucial role in wastewater treatment. There has been a lot of research on using nano-deposition or nano-coating in photocatalytic degradation. Nano-materials are also expected to play an important role in the modification of existing adsorbents and catalysts used in wastewater treatment processes.

In the future, there will be an increased demand for newer materials, methods, and sensors that can be used online for accurate and reliable effluent characterization, along with process control, to eliminate fluctuations in effluent quality and character. Such sensors can track quality and variations through the measurement of physical, chemical, or biological parameters. This kind of measurement can be quite complicated, requiring various devices for variety of parameters and pollutants. Also, the measurement system needs to be developed at an affordable cost. Apart from conventional measurements, such as pH, dissolved oxygen, and COD, real time measurement and process control measures are required for solids, ammonia, nitrates, phosphates, microorganisms, pathogens, and toxins in wastewaters. Wireless sensors, which are mainly battery driven, can provide new opportunities for cheap and easy installations without wiring, and several such sensors can improve plant performance. A number of constraints, such as number and nature of sensors, energy, memory, computational speed, communications bandwidth, and software development, need to be resolved for the effective use of wireless sensor networks in wastewater treatment, recycling, and reuse. Novel network protocols are likely to be invented in the future to address key requirements in wireless communication and to overcome difficulties posed by actual implementation of wastewater treatment processes. Satellite based monitoring systems are also likely to gain prominence in this regard.

Industrial wastewater reduction can be effectively explored using a systematic water networks design based on water pinch analysis, which can minimize fresh water use and wastewater generation for any plant (see methods discussed by Shenoy, 2012). A suitable parameter or parameters, such as COD or total solids, can be chosen for analysis, with the objective of minimizing water consumption and wastewater generation. The water pinch analysis can significantly reduce the operating cost by engineering appropriate modifications. Over time, mathematical and computational tools improve on the existing knowledgebase, and these tools can substantially aid the water pinch analysis, which becomes complicated when multiple parameters are considered. Advanced tools, such as CFD, can help to intensify desired processes and reduce the spatial footprint of the water treatment equipment. Newer treatment technologies and the decentralized implementation of these technologies through network analysis and optimization tools will allow plants to use closed loop cycles for water (minimizing fresh water requirements and towards zero liquid discharge).

Looking at the current pace of technological developments, it is futile to speculate on what the future holds. We do, however, anticipate the following trends:

- Developments based on utilization of living species: microorganisms, algae, plants, or aquatic species such as fish. The objective here would be to find biological natural resources that can consume, degrade, or otherwise remove pollutants with or without specificity.
- Satellite monitoring, wireless sensing, and BIG data: Satellite monitoring of water discharges and quality, as well as a large number of distributed wireless sensors, is expected to increase dramatically in the future. The BIG data generated through these efforts and corresponding data analysis tools have a potential to dramatically impact industrial wastewater treatment technologies and implementation.
- Public–private partnerships: Increased numbers of public–private partnerships across the world are expected. Major cities are expected to establish renewable energy and waste management facilities that can effectively treat domestic wastewaters and generate electricity for the populace.
- Start-up companies: New, small, and dynamic start-up companies are expected to play a much larger role in developing new wastewater treatment technologies. Traditionally, wastewater treatment technologies have been developed by companies which were in the business of providing turn-key plants. In the coming years, start-up companies dedicated to developing new IP and technologies are expected to play a significant role.
- Collaboration and joint research programs: There are a number of governmental and non-governmental agencies working actively in the area of wastewater treatment. There has also been an increasing trend for joint collaborations between institutes and governments. The future will certainly see a rise in these activities focusing on various research needs and implementation involving cooperation for exchange of ideas, results, and technologies.

The future is expected to see socio-economic and environmentally sustainable solutions to both domestic and industrial wastewater treatment problems, with increased partnership in the two areas. Future wastewater treatment plants will not only treat wastewaters to meet discharge standards, but they will also generate power to meet the cost of their own operations, while recovering materials and energy so as to make them profitable. We hope that key issues mentioned in this book and the creative use of technologies discussed here will make useful contributions towards enhancing the effectiveness of industrial wastewater treatment, recycling, and reuse. New advances may be assimilated using the framework discussed in this text. We hope that this book will stimulate further developments and provide an impetus for the wider implementation of industrial wastewater, recycling, and reuse technologies in practice.

REFERENCES

http://www.activatedsludgeconference.com/downloads/
  OneHundredYearsofActivatedSludge-Abriefhistory.pdf (viewed on 23 February, 2014).
http://civil.colorado.edu/ silverst/cven5534/History%20of%20Wastewater%20Treatment
  %20in%20the%20US.pdf (viewed on 23 February, 2014).
http://www.lenntech.com/history-water-treatment.htm (viewed on 20 February, 2014).
Shenoy, U.V., 2012. Enhanced nearest neighbours algorithm for design of water networks.
  Chem. Eng. Sci. 84, 197–206.
Vigneswaran, S., Davis, C., Kandasamy, J., Chanan, A., 2011. Water and wastewater treatment technologies, Vol. 1. Urban Wastewater Treatment: Past, Present and Future. Encyclopedia of Life Support Systems (EOLSS) (viewed on 20 February, 2014, www.eolss.net/Sample-Chapters/C07/E6-144.pdf).
Wiesmann, U., 2007. Historical Development of Wastewater Collection and Treatment. In: Choi, Su, Dombrowski, Eva-Maria (Eds.), Fundamentals of Biological Wastewater Treatment. Wiley-vch Verlag GmbH & Co. KGaA, Weinheim.

NOTATIONS

SYMBOLS

a_L constant in Langmuir equation (mg g^{-1})

a_R constant in Redlich-Peterson equation

B_D constant in Dubunin-Radushkevich equation (mol^2 J^{-1})

b_L constant in Langmuir equation (L g^{-1})

C concentration (kmol m^{-3})

C_e equilibrium concentration in solution phase (kmol m^{-3})

C_V cavitation number

D_{HA} diffusion coefficient (m^2 s^{-1})

d diameter (m)

d grain size of the crystallite (Chapter 8)

d_T, d_{TO} tangential nozzle diameter at diode inlet and pipe inlet of vortex diode

d_c diameter of diode chamber (m)

E energy of adsorption (kJ mol^{-1})

f_T frequency of turbulence

HA acid species

$[HA]$ concentration of acid (kmol m^{-3})

$[HA]_i$ initial concentration of acid (kmol m^{-3})

$[HA]_o$ outlet concentration of acid (kmol m^{-3})

$[HA]_o^b$ breakthrough concentration of acid (kmol m^{-3})

h height of diode chamber (m)

K constant in Freundlich equation

K_R constant in Redlich-Peterson equation

k reaction rate constant (s^{-1})

$k_s a$ film mass transfer coefficient (s^{-1})

N_p number of passes

$1/n$ constant in Freundlich equation

P pressure (kg cm^{-2})

P_v vapor pressure of liquid

P_s, P_i power (J s^{-1}) (Chapter 2)

Q flow rate (m^3 s^{-1})

Q_e/q_e equilibrium concentration in adsorbent phase (mg g^{-1})

q_D theoretical saturation capacity in Dubunin-Radushkevich equation (mol g^{-1})

R molar gas constant

R_b radius of bead (m)

$Rs.$ rupees (Indian)

r_c radius of curvature (m)

T temperature

t time (s)

t_B breakthrough time (s)

t_D diffusion time (s)

t_E film transfer time (s)

t_F	flow time (s)
t_R	reaction time (s)
u	velocity (m s^{-1})
v_o	velocity at the throat of the cavitating constriction
V	volume (m^3)
α	ratio of throat perimeter to the throat area (Chapter 3)
β	constant in Redlich-Peterson equation (Chapter 2)
β	ratio of throat area to the cross-sectional area of pipe (Chapter 3)
β	full width at half maximum (Chapter 8)
η	pump efficiency
θ	angle of reflection
λ	wavelength of the X-ray used
ρ	density
ΔP	pressure drop

ABBREVIATIONS

4-CP	4-chlorophenol
4-NP	4-nitrophenol
AAJSC	acid activated jute stick chars
AB	acidogenic bacteria
ACF	activated carbon filter
AES	activated egg shells
AF	anaerobic filter
AMBR	anaerobic migrating blanket reactor
ANL	activated neem leaf
AnMBR	anaerobic membrane bio reactor
AOP	advanced oxidation process
API	American Petroleum Institute
API	active pharmaceutical ingredients (Chapter 13)
APPC	activated potato peel chars
ASP	activated sludge process
ASTM	American standard testing method
ATFD	agitated thin film dryer
ATW	activated tea waste
BCM	billion cubic meters
BET	Brunauer-Emmett-Teller (Chapter 8)
BET	bio-electrochemical treatment (Chapter 6)
B-DWW	biomethanated distillery wastewater
BDD	boron doped diamond
BI	biodegradability index
BIOFOR	biological filter oxygenated reactor
BJH	Barret-Joyner-Halenda
BMO	binary mixed oxide of Al-Fe
Bn	billion
BNR	biological nutrient removal

BOD	biological oxygen demand
BOD$_5$	5-day BOD
BP	banana peel
CAD	command area development
CADWM	command area development and water management
CAPEX	capital expenditure
CETP	common effluent treatment plant
CFD	computational fluid dynamics
CIP	cleaning in place
COD	chemical oxygen demand
CPCB	Central Pollution Control Board (India)
CPI	corrugated plate interceptor
CSTR	continuous stirred tank reactor
CWAO	catalytic wet air oxidation
CWC	central water commission
DAF	dissolved air flotation
DM	demineralized
DMF	dual media filter
DO	dissolved oxygen
EGSB	expanded granular sludge blanket
EMBR	enhanced membrane bioreactor
EMP	Embden Meyerhof pathway
ETC	electron transport chain
ETP	effluent treatment plant
FAB	fluidized aerobic bed
FABR	fluidized aerated bed reactor
FAME	fatty acid methyl esters
FHL	formate hydrogen lyase
FTIR	Fourier transform infrared
GP	growth phase
HB	hydroxybutyrate
HC	hydrodynamic cavitation
HRT	hydraulic retention time
HV	hydroxyvalerate
ICID	International Commission on Irrigation and Drainage
IGF	induced gas flotation
INCID	Indian National Committee of Irrigation and Drainage
IPCC	Intergovernmental Panel on Climate Change
JSC	jute stick chars
LOPROX	low pressure wet oxidaton
MBBR	moving bed bio reactor
MBR	membrane bio reactor
ME	membrane element
MEC	microbial electrolysis cell
MEE	multiple effect evaporators/evaporation
MF	microfiltration
MFC	microbial fuel cell
MFL	Madras Fertilizer Ltd.

MGF	multigrade filter
MINLP	mixed-integer non-linear programming
MLD	million liters per day
MLSS	mixed liquor suspended solids
MLVSS	mixed liquor volatile suspended solids
MMF	multimedia filter
MoEF	Ministry of Environment and Forests
MRL	Madras Refineries Ltd.
MVR	mechanical vapor recompression
MWCO	molecular weight cut off
MWR	Ministry of Water Resources
NF	nanofiltration
NGRMP	National Groundwater Recharge Master Plan
NL	neem leaf
OECD	Organisation for Economic Co-operation and Development
OLR	organic loading rate
OPEX	operating expenditure
ORP	oxidation-reduction potential
PEM	proton exchange membrane
PFL	pyruvate formate lyase
PFOR	pyruvate ferredoxin oxidoreductase
PHA	polyhydroxyalkanoates
PHB	polyhydroxybutyrate
PHV	poly-B-hydroxy valerate
PP	polishing pond
PPC	potato peel char
PPP	pentose phosphate pathway
PSF	pressure sand filter
RAS	return activated sludge
RCF	Rashtriya Chemicals and Fertilizers
RO	reverse osmosis
SAFF	submerged aeration fixed film reactor
SBBGR	sequencing batch biofilter granular reactor
SBR	sequencing batch reactor
SDI	silt density index
SEM	scanning electron micrograph
SRT	sludge retention time
STP	sewage treatment plant
TAG	triacylglycerides
TCA	tricarboxylic acid
TCD	thermal conductivity detector
TDS	total dissolved solids
TEA	terminal electron acceptor
TKN	total Kjeldahl nitrogen
TMP	trans-membrane pressure
TN	total nitrogen
TOC	total organic carbon
TSS	total suspended solids

TTP	tertiary treatment plant
TVR	thermal vapor recompression
TW	tea waste
UASB	upflow anaerobic sludge blanket
UF	ultrafiltration
UN	United Nations
UNDP	United National Development Program
UNEP	United Nations Environment Program
UNESCO	United Nations Educational, Scientific and Cultural Organization
UN-HABITAT	United Nations Human Settlements Program
US	ultrasonication (Chapter 3)
VFA	volatile fatty acids
VOC	volatile organic carbon
WAO	wet air oxidation
WHO	World Health Organization
WPO	wet peroxide oxidation
WSP	waste stabilization pond
WUAs	Water Users' Associations
WWAP	World Water Assessment Program
WWTP	wastewater treatment plant
XRD	X-ray diffraction
ZLD	zero liquid discharge
ZPC	zero point charge

INDEX

Note: Page numbers followed by *f* indicate figures and *t* indicate tables.

Printed in the United States
By Bookmasters